Data-Centric Systems and Applications

Bing Liu

Web Data Mining

Exploring Hyperlinks,
Contents, and Usage Data

With 177 Figures

 Springer

Bing Liu

Department of Computer Science
University of Illinois at Chicago
851 S. Morgan Street
Chicago, IL 60607-7053
USA
liub@cs.uic.edu

Library of Congress Control Number: 2006937132

ACM Computing Classification (1998): H.2, H.3, I.2, I.5, E.5

Corrected 2nd printing 2008

ISBN-10 3-540-37881-2 Springer Berlin Heidelberg New York
ISBN-13 978-3-540-37881-5 Springer Berlin Heidelberg New York

Springer is a part of Springer Science+Business Media
springer.com

Cover Design: KünkelLopka, Heidelberg
Typesetting: by the Author
Production: le-tex publishing services oHG, Leipzig

Printed on acid-free paper   45/3180/YL   5 4 3   SPIN 12637102

To my parents, my wife Yue and children Shelley and Kate

Preface

The rapid growth of the Web in the last decade makes it the largest publicly accessible data source in the world. Web mining aims to discover useful information or knowledge from Web hyperlinks, page contents, and usage logs. Based on the primary kinds of data used in the mining process, Web mining tasks can be categorized into three main types: Web structure mining, Web content mining and Web usage mining. Web structure mining discovers knowledge from hyperlinks, which represent the structure of the Web. Web content mining extracts useful information/knowledge from Web page contents. Web usage mining mines user access patterns from usage logs, which record clicks made by every user.

The goal of this book is to present these tasks, and their core mining algorithms. The book is intended to be a text with a comprehensive coverage, and yet, for each topic, sufficient details are given so that readers can gain a reasonably complete knowledge of its algorithms or techniques without referring to any external materials. Four of the chapters, structured data extraction, information integration, opinion mining, and Web usage mining, make this book unique. These topics are not covered by existing books, but yet they are essential to Web data mining. Traditional Web mining topics such as search, crawling and resource discovery, and link analysis are also covered in detail in this book.

Although the book is entitled *Web Data Mining*, it also includes the main topics of data mining and information retrieval since Web mining uses their algorithms and techniques extensively. The data mining part mainly consists of chapters on association rules and sequential patterns, supervised learning (or classification), and unsupervised learning (or clustering), which are the three most important data mining tasks. The advanced topic of partially (semi-) supervised learning is included as well. For information retrieval, its core topics that are crucial to Web mining are described. This book is thus naturally divided into two parts. The first part, which consists of Chaps. 2–5, covers data mining foundations. The second part, which contains Chaps. 6–12, covers Web specific mining.

Two main principles have guided the writing of this book. First, the basic content of the book should be accessible to undergraduate students, and yet there are sufficient in-depth materials for graduate students who plan to pursue Ph.D. degrees in Web data mining or related areas. Few assumptions are made in the book regarding the prerequisite knowledge of readers. One with a basic understanding of algorithms and probability concepts should have no problem with this book. Second, the book should examine the Web mining technology from a practical point of view. This is important because most Web mining tasks have immediate real-world applications. In the past few years, I was fortunate to have worked directly or indirectly with many researchers and engineers in several search engine and e-commerce companies, and also traditional companies that are interested in exploiting the information on the Web in their businesses. During the process, I gained practical experiences and first-hand knowledge of real-world problems. I try to pass those non-confidential pieces of information and knowledge along in the book. The book, thus, should have a good balance of theory and practice. I hope that it will not only be a learning text for students, but also a valuable source of information/knowledge and even ideas for Web mining researchers and practitioners.

Acknowledgements

Many researchers have assisted me technically in writing this book. Without their help, this book might never have become reality. My deepest thanks goes to Filippo Menczer and Bamshad Mobasher, who were so kind to have helped write two essential chapters of the book. They are both experts in their respective fields. Filippo wrote the chapter on Web crawling and Bamshad wrote the chapter on Web usage mining. I am also very grateful to Wee Sun Lee, who helped a great deal in the writing of Chap. 5 on partially supervised learning.

Jian Pei helped with the writing of the PrefixSpan algorithm in Chap. 2, and checked the MS-PS algorithm. Eduard Dragut assisted with the writing of the last section of Chap. 10 and also read the chapter many times. Yuanlin Zhang gave many great suggestions on Chap. 9. I am indebted to all of them.

Many other researchers also assisted in various ways. Yang Dai and Rudy Setiono helped with Support Vector Machines (SVM). Chris Ding helped with link analysis. Clement Yu and ChengXiang Zhai read Chap. 6, and Amy Langville read Chap. 7. Kevin C.-C. Chang, Ji-Rong Wen and Clement Yu helped with many aspects of Chap 10. Justin Zobel helped clarify some issues related to index compression, and Ion Muslea helped clarify some issues on wrapper induction. Divy Agrawal, Yunbo Cao, Edward Fox, Hang Li, Xiaoli Li, Zhaohui Tan, Dell Zhang and Zijian Zheng helped check various chapters or sections. I am very grateful.

Discussions with many researchers helped shape the book as well: Amir Ashkenazi, Imran Aziz, Roberto Bayardo, Wendell Baker, Ling Bao, Jeffrey Benkler, AnHai Doan, Byron Dom, Michael Gamon, Robert Grossman, Jiawei Han, Wynne Hsu, Ronny Kohavi, David D. Lewis, Ian McAllister, Wei-Ying Ma, Marco Maggini, Llew Mason, Kamel Nigan, Julian Qian, Yan Qu, Thomas M. Tirpak, Andrew Tomkins, Alexander Tuzhilin, Weimin Xiao, Gu Xu, Philip S. Yu, and Mohammed Zaki.

My former and current students, Gao Cong, Minqing Hu, Nitin Jindal, Xin Li, Yiming Ma, Yanhong Zhai and Kaidi Zhao checked many algorithms and made numerous corrections. Some chapters of the book have been used in my graduate classes at the University of Illinois at Chicago. I thank the students in these classes for implementing several algorithms. Their questions helped me improve and, in some cases, correct the algorithms. It is not possible to list all their names. Here, I would particularly like to thank John Castano, Xiaowen Ding, Murthy Ganapathibhotla, Cynthia Kersey, Hari Prasad Divyakotti, Ravikanth Turlapati, Srikanth Tadikonda, Makio Tamura, Haisheng Wang, and Chad Williams for pointing out errors in texts, examples or algorithms. Michael Bombyk from DePaul University also found several typing errors.

It was a pleasure working with the helpful staff at Springer. I thank my editor Ralf Gerstner who asked me in early 2005 whether I was interested in writing a book on Web mining. It has been a wonderful experience working with him since. I also thank my copyeditor Mike Nugent for helping me improve the presentation, and my production editor Michael Reinfarth for guiding me through the final production process. Two anonymous reviewers also gave me many insightful comments.

The Department of Computer Science at the University of Illinois at Chicago provided computing resources and a supportive environment for this project.

Finally, I thank my parents, brother and sister for their constant supports and encouragements. My greatest gratitude goes to my own family: Yue, Shelley and Kate. They have helped me in so many ways. Despite their young ages, Shelley and Kate actually read many parts of the book and caught numerous typing errors. My wife has taken care of almost everything at home and put up with me and the long hours that I have spent on this book. I dedicate this book to them.

Bing Liu

Table of Contents

Part II: Web Mining

1 Introduction

When you read this book, you, without doubt, already know what the **World Wide Web** is and have used it extensively. The World Wide Web (or the **Web** for short) has impacted on almost every aspect of our lives. It is the biggest and most widely known information source that is easily accessible and searchable. It consists of billions of interconnected documents (called **Web pages**) which are authored by millions of people. Since its inception, the Web has dramatically changed our information seeking behavior. Before the Web, finding information means asking a friend or an expert, or buying/borrowing a book to read. However, with the Web, everything is only a few clicks away from the comfort of our homes or offices. Not only can we find needed information on the Web, but we can also easily share our information and knowledge with others.

The Web has also become an important channel for conducting businesses. We can buy almost anything from online stores without needing to go to a physical shop. The Web also provides convenient means for us to communicate with each other, to express our views and opinions on anything, and to discuss with people from anywhere in the world. The Web is truly a **virtual society**. In this first chapter, we introduce the Web, its history, and the topics that we will study in the book.

1.1 What is the World Wide Web?

The World Wide Web is officially defined as a "wide-area hypermedia information retrieval initiative aiming to give universal access to a large universe of documents." In simpler terms, the Web is an Internet-based computer network that allows users of one computer to access information stored on another through the world-wide network called the **Internet**.

The Web's implementation follows a standard **client-server** model. In this model, a user relies on a program (called the **client**) to connect to a remote machine (called the **server**) where the data is stored. Navigating through the Web is done by means of a client program called the **browser**, e.g., Netscape, Internet Explorer, Firefox, etc. Web browsers work by sending requests to remote servers for information and then interpreting the returned documents written in HTML and laying out the text and graphics on the user's computer screen on the client side.

The operation of the Web relies on the structure of its **hypertext** documents. Hypertext allows Web page authors to link their documents to other related documents residing on computers anywhere in the world. To view these documents, one simply follows the links (called **hyperlinks**).

The idea of hypertext was invented by Ted Nelson in 1965 [403], who also created the well known hypertext system Xanadu (http://xanadu. com/). Hypertext that also allows other media (e.g., image, audio and video files) is called **hypermedia**.

1.2   A Brief History of the Web and the Internet

Creation of the Web: The Web was invented in 1989 by **Tim Berners-Lee**, who, at that time, worked at CERN (Centre European pour la Recherche Nucleaire, or European Laboratory for Particle Physics) in Switzerland. He coined the term "World Wide Web," wrote the first World Wide Web server, httpd, and the first client program (a browser and editor), "**WorldWideWeb**".

It began in March 1989 when Tim Berners-Lee submitted a proposal titled "Information Management: A Proposal" to his superiors at CERN. In the proposal, he discussed the disadvantages of hierarchical information organization and outlined the advantages of a hypertext-based system. The proposal called for a simple protocol that could request information stored in remote systems through networks, and for a scheme by which information could be exchanged in a common format and documents of individuals could be linked by hyperlinks to other documents. It also proposed methods for reading text and graphics using the display technology at CERN at that time. The proposal essentially outlined a **distributed hypertext system**, which is the basic architecture of the Web.

Initially, the proposal did not receive the needed support. However, in 1990, Berners-Lee re-circulated the proposal and received the support to begin the work. With this project, Berners-Lee and his team at CERN laid the foundation for the future development of the Web as a distributed hypertext system. They introduced their server and browser, the protocol used for communication between clients and the server, the HyperText Transfer Protocol (**HTTP**), the HyperText Markup Language (**HTML**) used for authoring Web documents, and the Universal Resource Locator (**URL**). And so it began.

Mosaic and Netscape Browsers: The next significant event in the development of the Web was the arrival of **Mosaic**. In February of 1993, Marc Andreesen from the University of Illinois' NCSA (National Center for Supercomputing Applications) and his team released the first "Mosaic for X" graphical Web browser for UNIX. A few months later, different versions of Mosaic were released for Macintosh and Windows operating systems. This was an important event. For the first time, a Web client, with a consistent and simple point-and-click graphical user interface, was implemented for the three most popular operating systems available at the time. It soon made big splashes outside the academic circle where it had begun. In mid-1994, Silicon Graphics founder Jim Clark collaborated with Marc Andreessen, and they founded the company **Mosaic Communications** (later renamed as **Netscape Communications**). Within a few months, the **Netscape** browser was released to the public, which started the explosive growth of the Web. The **Internet Explorer** from Microsoft entered the market in August, 1995 and began to challenge Netscape.

The creation of the World Wide Web by Tim Berners-Lee followed by the release of the Mosaic browser are often regarded as the two most significant contributing factors to the success and popularity of the Web.

Internet: The Web would not be possible without the Internet, which provides the communication network for the Web to function. The **Internet** started with the computer network **ARPANET** in the Cold War era. It was produced as the result of a project in the United States aiming at maintaining control over its missiles and bombers after a nuclear attack. It was supported by Advanced Research Projects Agency (ARPA), which was part of the Department of Defense in the United States. The first ARPANET connections were made in 1969, and in 1972, it was demonstrated at the First International Conference on Computers and Communication, held in Washington D.C. At the conference, ARPA scientists linked computers together from 40 different locations.

In 1973, Vinton Cerf and Bob Kahn started to develop the protocol later to be called **TCP/IP (Transmission Control Protocol/Internet Protocol)**. In the next year, they published the paper "Transmission Control Protocol", which marked the beginning of TCP/IP. This new protocol allowed diverse computer networks to interconnect and communicate with each other. In subsequent years, many networks were built, and many competing techniques and protocols were proposed and developed. However, ARPANET was still the backbone to the entire system. During the period, the network scene was chaotic. In 1982, the TCP/IP was finally adopted, and the **Internet**, which is a connected set of networks using the TCP/IP protocol, was born.

Search Engines: With information being shared worldwide, there was a need for individuals to find information in an orderly and efficient manner. Thus began the development of search engines. The search system **Excite** was introduced in 1993 by six Stanford University students. **EINet Galaxy** was established in 1994 as part of the MCC Research Consortium at the University of Texas. Jerry Yang and David Filo created **Yahoo!** in 1994, which started out as a listing of their favorite Web sites, and offered directory search. In subsequent years, many search systems emerged, e.g., **Lycos, Inforseek, AltaVista, Inktomi, Ask Jeeves, Northernlight**, etc.

Google was launched in 1998 by Sergey Brin and Larry Page based on their research project at Stanford University. Microsoft started to commit to search in 2003, and launched the **MSN** search engine in spring 2005. It used search engines from others before. **Yahoo!** provided a general search capability in 2004 after it purchased Inktomi in 2003.

W3C (The World Wide Web Consortium): W3C was formed in the December of 1994 by MIT and CERN as an international organization to lead the development of the Web. W3C's main objective was "to promote standards for the evolution of the Web and interoperability between WWW products by producing specifications and reference software." The first **International Conference on World Wide Web (WWW)** was also held in 1994, which has been a yearly event ever since.

From 1995 to 2001, the growth of the Web boomed. Investors saw commercial opportunities and became involved. Numerous businesses started on the Web, which led to irrational developments. Finally, the bubble burst in 2001. However, the development of the Web was not stopped, but has only become more rational since.

1.3 Web Data Mining

The rapid growth of the Web in the last decade makes it the largest publicly accessible data source in the world. The Web has many unique characteristics, which make mining useful information and knowledge a fascinating and challenging task. Let us review some of these characteristics.

1. The amount of data/information on the Web is huge and still growing. The coverage of the information is also very wide and diverse. One can find information on almost anything on the Web.
2. Data of all types exist on the Web, e.g., structured tables, semi-structured Web pages, unstructured texts, and multimedia files (images, audios, and videos).

3. Information on the Web is **heterogeneous**. Due to the diverse author-ship of Web pages, multiple pages may present the same or similar in-formation using completely different words and/or formats. This makes integration of information from multiple pages a challenging problem.

4. A significant amount of information on the Web is linked. Hyperlinks exist among Web pages within a site and across different sites. Within a site, hyperlinks serve as information organization mechanisms. Across different sites, hyperlinks represent implicit conveyance of authority to the target pages. That is, those pages that are linked (or pointed) to by many other pages are usually high quality pages or **authoritative pages** simply because many people trust them.

5. The information on the Web is noisy. The **noise** comes from two main sources. First, a typical Web page contains many pieces of information, e.g., the **main content** of the page, navigation links, advertisements, copyright notices, privacy policies, etc. For a particular application, only part of the information is useful. The rest is considered noise. To per-form fine-grain Web information analysis and data mining, the noise should be removed. Second, due to the fact that the Web does not have quality control of information, i.e., one can write almost anything that one likes, a large amount of information on the Web is of low quality, erroneous, or even misleading.

6. The Web is also about services. Most commercial Web sites allow people to perform useful operations at their sites, e.g., to purchase products, to pay bills, and to fill in forms.

7. The Web is dynamic. Information on the Web changes constantly. Keeping up with the change and monitoring the change are important is-sues for many applications.

8. The Web is a virtual society. The Web is not only about data, informa-tion and services, but also about interactions among people, organiza-tions and automated systems. One can communicate with people any-where in the world easily and instantly, and also express one's views on anything in Internet forums, blogs and review sites.

All these characteristics present both challenges and opportunities for min-ing and discovery of information and knowledge from the Web. In this book, we only focus on mining textual data. For mining of images, videos and audios, please refer to [143, 441].

To explore information mining on the Web, it is necessary to know data mining, which has been applied in many Web mining tasks. However, Web mining is not entirely an application of data mining. Due to the rich-ness and diversity of information and other Web specific characteristics discussed above, Web mining has developed many of its own algorithms.

1.3.1   What is Data Mining?

Data mining is also called **knowledge discovery in databases (KDD)**. It is commonly defined as the process of discovering useful **patterns** or knowledge from data sources, e.g., databases, texts, images, the Web, etc. The patterns must be valid, potentially useful, and understandable. Data mining is a multi-disciplinary field involving machine learning, statistics, databases, artificial intelligence, information retrieval, and visualization.

There are many data mining tasks. Some of the common ones are **supervised learning** (or **classification**), **unsupervised learning** (or **clustering**), **association rule mining**, and **sequential pattern mining**. We will study all of them in this book.

A data mining application usually starts with an understanding of the application domain by **data analysts (data miners)**, who then identify suitable data sources and the target data. With the data, data mining can be performed, which is usually carried out in three main steps:

- **Pre-processing**: The raw data is usually not suitable for mining due to various reasons. It may need to be cleaned in order to remove noises or abnormalities. The data may also be too large and/or involve many irrelevant attributes, which call for data reduction through sampling and attribute selection. Details about data pre-processing can be found in any standard data mining textbook.
- **Data mining**: The processed data is then fed to a data mining algorithm which will produce patterns or knowledge.
- **Post-processing**: In many applications, not all discovered patterns are useful. This step identifies those useful ones for applications. Various evaluation and visualization techniques are used to make the decision.

The whole process (also called the **data mining process**) is almost always iterative. It usually takes many rounds to achieve final satisfactory results, which are then incorporated into real-world operational tasks.

Traditional data mining uses structured data stored in relational tables, spread sheets, or flat files in the tabular form. With the growth of the Web and text documents, **Web mining** and **text mining** are becoming increasingly important and popular. Web mining is the focus of this book.

1.3.2   What is Web Mining?

Web mining aims to discover useful information or knowledge from the **Web hyperlink structure**, **page content**, and **usage data**. Although Web mining uses many data mining techniques, as mentioned above it is not purely an application of traditional data mining due to the heterogeneity and semi-structured or unstructured nature of the Web data. Many new mining tasks and algorithms were invented in the past decade. Based on the primary kinds of data used in the mining process, Web mining tasks can be categorized into three types: Web structure mining, Web content mining and Web usage mining.

- **Web structure mining**: Web structure mining discovers useful knowledge from hyperlinks (or links for short), which represent the structure of the Web. For example, from the links, we can discover important Web pages, which, incidentally, is a key technology used in search engines. We can also discover communities of users who share common interests. Traditional data mining does not perform such tasks because there is usually no link structure in a relational table.
- **Web content mining**: Web content mining extracts or mines useful information or knowledge from Web page contents. For example, we can automatically classify and cluster Web pages according to their topics. These tasks are similar to those in traditional data mining. However, we can also discover patterns in Web pages to extract useful data such as descriptions of products, postings of forums, etc, for many purposes. Furthermore, we can mine customer reviews and forum postings to discover consumer sentiments. These are not traditional data mining tasks.
- **Web usage mining**: Web usage mining refers to the discovery of user access patterns from Web usage logs, which record every click made by each user. Web usage mining applies many data mining algorithms. One of the key issues in Web usage mining is the pre-processing of click-stream data in usage logs in order to produce the right data for mining.

In this book, we will study all these three types of mining. However, due to the richness and diversity of information on the Web, there are a large number of Web mining tasks. We will not be able to cover them all. We will only focus on some important tasks and their algorithms.

The **Web mining process** is similar to the data mining process. The difference is usually in the data collection. In traditional data mining, the data is often already collected and stored in a data warehouse. For Web mining, data collection can be a substantial task, especially for Web structure and content mining, which involves crawling a large number of target Web pages. We will devote a whole chapter on crawling.

Once the data is collected, we go through the same three-step process: data pre-processing, Web data mining and post-processing. However, the techniques used for each step can be quite different from those used in traditional data mining.

1.4 Summary of Chapters

This book consists of two main parts. The first part, which includes Chaps. 2–5, covers the major topics of data mining. The second part, which comprises the rest of the chapters, covers Web mining (including a chapter on Web search). In the Web mining part, Chaps. 7 and 8 are on Web structure mining, which are closely related to Web search (Chap. 6). Since it is difficult to draw a boundary between Web search and Web mining, Web search and mining are put together. Chaps 9–11 are on Web content mining, and Chap. 12 is on Web usage mining. Below we give a brief introduction to each chapter.

Chapter 2 – Association Rules and Sequential Patterns: This chapter studies two important data mining models that have been used in many Web mining tasks, especially in Web usage and content mining. Association rule mining finds sets of data items that occur together frequently. Sequential pattern mining finds sets of data items that occur together frequently in some sequences. Clearly, they can be used to find regularities in the Web data. For example, in Web usage mining, association rule mining can be used to find users' visit and purchase patterns, and sequential pattern mining can be used to find users' navigation patterns.

Chapter 3 – Supervised Learning: Supervised learning is perhaps the most frequently used mining/learning technique in both practical data mining and Web mining. It is also called **classification**, which aims to learn a classification function (called a **classifier**) from data that are labeled with pre-defined classes or categories. The resulting classifier is then applied to classify future data instances into these classes. Due to the fact that the data instances used for learning (called the **training data**) are labeled with pre-defined classes, the method is called supervised learning.

Chapter 4 – Unsupervised Learning: In unsupervised learning, the data used for learning has no pre-defined classes. The learning algorithm has to find the hidden structures or regularities in the data. One of the key unsupervised learning techniques is **clustering**, which organizes data instances into **groups** or **clusters** according to their similarities (or differences). Clustering is widely used in Web mining. For example, we can cluster Web pages into groups, where each group may represent a particular topic. We can also cluster documents into a hierarchy of clusters, which may represent a topic hierarchy.

Chapter 5 – Partially Supervised Learning: Supervised learning requires a large number of labeled data instances to learn an accurate classifier. Labeling, which is often done manually, is labor intensive and time consuming. To reduce the manual labeling effort, **learning from labeled and unlabeled examples** (or **LU learning**) was proposed to use a small set of labeled examples (data instances) and a large set of unlabeled examples for learning. This model is also called **semi-supervised learning**.

Another learning model that we will study is called **learning from positive and unlabeled examples** (or **PU learning**), which is for two-class classification. However, there are no labeled negative examples for learning. This model is useful in many situations. For example, we have a set of Web mining papers and we want to identify other Web mining papers in a research paper repository which contains all kinds of papers. The set of Web mining papers can be treated as the positive data, and the papers in the research repository can be treated as the unlabeled data.

Chapter 6 – Information Retrieval and Web Search: Search is probably the largest application on the Web. It has its root in **information retrieval** (or IR for short), which is a field of study that helps the user find needed information from a large collection of text documents. Given a query (e.g., a set of **keywords**), which expresses the user's information need, an IR system finds a set of documents that is relevant to the query from its underlying collection. This is also how a Web search engine works.

Web search brings IR to a new height. It applies some IR techniques, but also presents a host of interesting problems due to special characteristics of the Web data. First of all, Web pages are not the same as plain text documents because they are semi-structured and contain hyperlinks. Thus, new methods have been designed to produce better Web IR (or search) systems. Another major issue is efficiency. Document collections used in traditional IR systems are not large, but the number of pages on the Web is huge. For example, Google claimed that it indexed more than 8 billion pages when this book was written. Web users demand very fast responses. No matter how effective a retrieval algorithm is, if the retrieval cannot be done extremely efficiently, few people will use it. In the chapter, several other search related issues will also be discussed.

Chapter 7 – Link Analysis: Hyperlinks are a special feature of the Web, which have been exploited for many purposes, especially for Web search. Google's success is largely attributed to its hyperlink-based ranking algorithm called **PageRank**, which is originated from **social network analysis**. In this chapter, we will first introduce some main concepts of social network analysis and then describe two most well known Web link analysis algorithms, PageRank and HITS. In addition, we will also study several community finding algorithms. When Web pages link to one another, they form Web communities, which are groups of content creators that share some common interests. Communities not only manifest in hyperlinks, but also in other contexts such as emails and Web page contents.

Chapter 8 – Web Crawling: A Web **crawler** is a program that automatically traverses the Web's hyperlink structure and downloads each linked page to a local storage. Crawling is often the first step of Web mining or in building a Web search engine. Although conceptually easy, building a practical crawler is by no means simple. Due to efficiency and many other concerns, it involves a great deal of engineering. There are two types of crawlers: **universal crawlers** and **topic crawlers**. A universal crawler downloads all pages irrespective of their contents, while a topic crawler downloads only pages of certain topics. The difficulty in topic crawling is how to recognize such pages. We will study several techniques for this purpose.

Chapter 9 – Structured Data Extraction: Wrapper Generation: A large number of pages on the Web contain structured data, which are usually data records retrieved from underlying databases and displayed in Web pages following some fixed templates. Structured data often represent their host pages' essential information, e.g., lists of products and services. Extracting such data allows one to provide value added services, e.g., comparative shopping, and meta-search. There are two main approaches to extraction. One is the supervised approach, which uses supervised learning to learn data extraction rules. The other is the unsupervised pattern discovery approach, which finds repeated patterns (hidden templates) in Web pages for data extraction.

Chapter 10 – Information Integration: Due to diverse authorships of the Web, different Web sites typically use different words or terms to express the same or similar information. In order to make use of the data or information extracted from multiple sites to provide value added services, we need to semantically integrate the data/information from these sites in order to produce consistent and coherent databases. Intuitively, integration means to match columns in different data tables that contain the same type of information (e.g., product names) and to match data values that are semantically the same but expressed differently in different sites.

Chapter 11 – Opinion Mining: Apart from structured data, the Web also contains a huge amount of unstructured text. Analyzing such text is also of great importance. It is perhaps even more important than extracting structured data because of the sheer volume of valuable information of almost any imaginable types contained in it. This chapter will only focus on mining people's **opinions** or **sentiments** expressed in **product reviews, forum discussions** and **blogs**. The task is not only technically challenging, but also very useful in practice because businesses and organizations always want to know consumer opinions on their products and services.

Chapter 12 – Web Usage Mining: Web usage mining aims to study user clicks and their applications to e-commerce and business intelligence. The objective is to capture and model **behavioral patterns** and **profiles** of users who interact with a Web site. Such patterns can be used to better understand the behaviors of different user segments, to improve the organization and structure of the site, and to create **personalized experiences** for users by providing dynamic **recommendations** of products and services.

1.5 How to Read this Book

This book is a textbook although two chapters are contributed by two other researchers. The contents of the two chapters have been carefully edited and integrated into the common framework of the whole book. The book is suitable for both graduate students and senior undergraduate students in the fields of computer science, information science, engineering, statistics, and social science. It can also be used as a reference by researchers and practitioners who are interested in or are working in the field of Web mining, data mining or text mining.

As mentioned earlier, the book is divided into two parts. Part I (Chaps. 2–5) covers the major topics of data mining. Text classification and clustering are included in this part as well. Part II, which includes the rest of the chapters, covers Web mining (and search). In general, all chapters in Part II require some techniques in Part I. Within each part, the dependency is minimal except Chap. 5, which needs several techniques from Chap. 4.

To Instructors: This book can be used as a class text for a one-semester course on Web data mining. In this case, there are two possibilities. If the students already have data mining or machine learning background, the chapters in Part I can be skipped. If the students do not have any data mining background, I recommend covering some selected sections from each chapter of Part I before going to Part II. The chapters in Part II can be covered in any sequence. You can also select a subset of the chapters according to your needs.

The book may also be used as a class text for an introductory data mining course where Web mining concepts and techniques are introduced. In this case, I recommend first covering all the chapters in Part I and then selectively covering some chapters or sections from each chapter in Part II depending on needs. It is usually a good idea to cover some sections of

Chaps. 6 and 7 as search engines fascinate most students. I also recommend including one or two lectures on data pre-processing for data mining since the topic is important for practical data mining applications but is not covered in this book. You can find teaching materials on data pre-processing from most introductory data mining books.

Supporting Materials: Updates to chapters and teaching materials, including lecture slides, data sets, implemented algorithms, and other resources, are available at http://www.springer.com/3-540-37881-2.

Bibliographic Notes

The W3C Web site (http://www.w3.org/) is the most authoritative resource site for information on Web developments, standards and guidelines. The history of the Web and hypertext, and Tim Berners-Lee's original proposal can all be found there. Many other sites also contain information about the history of the Web, the Internet and search engines, e.g., http://www.elsop. com/wrc/h_web.htm, http://www.zeltser.com/web-history/, http://www.isoc. org/internet/history/, http://www.livinginternet.com, http://www.w3c.rl.ac.uk/ primers/history/origins.htm and http://searchenginewatch.com/.

There are some earlier introductory texts on Web mining, e.g., those by Baldi et al. [33] and Chakrabarti [85]. There are also several application oriented books, e.g., those by Linoff and Berry [338], and Thuraisingham [515], and edited volumes by Djeraba et al. [143], Scime [480], and Zhong et al. [617].

On data mining, there are many textbooks, e.g., those by Duda et al. [155], Dunham [156], Han and Kamber [218], Hand et al. [221], Larose [305], Langley [302], Mitchell [385], Roiger and Geatz [467], Tan et al. [512], and Witten and Frank [549]. Application oriented books include those by Berry and Linoff [49], Pyle [450], Parr Rud [468], and Tang and MacLennan [514]. Several edited volumes exist as well, e.g., those by Fayyad et al. [174], Grossman et al. [208], and Wang et al. [533].

Latest research results on Web mining can be found in a large number of conferences and journals (too many to list) due to the interdisciplinary nature of the field. All the journals and conferences related to the Web technology, information retrieval, data mining, databases, artificial intelligence, and machine learning may contain Web mining related papers.

2 Association Rules and Sequential Patterns

Association rules are an important class of regularities in data. Mining of association rules is a fundamental data mining task. It is perhaps the most important model invented and extensively studied by the database and data mining community. Its objective is to find all **co-occurrence** relationships, called **associations**, among data items. Since it was first introduced in 1993 by Agrawal et al. [9], it has attracted a great deal of attention. Many efficient algorithms, extensions and applications have been reported.

The classic application of association rule mining is the **market basket** data analysis, which aims to discover how items purchased by customers in a supermarket (or a store) are associated. An example association rule is

Cheese → Beer   [support = 10%, confidence = 80%].

The rule says that 10% customers buy Cheese and Beer together, and those who buy Cheese also buy Beer 80% of the time. Support and confidence are two measures of rule strength, which we will define later.

This mining model is in fact very general and can be used in many applications. For example, in the context of the Web and text documents, it can be used to find word co-occurrence relationships and Web usage patterns as we will see in later chapters.

Association rule mining, however, does not consider the sequence in which the items are purchased. Sequential pattern mining takes care of that. An example of a sequential pattern is "5% of customers buy bed first, then mattress and then pillows". The items are not purchased at the same time, but one after another. Such patterns are useful in Web usage mining for analyzing **clickstreams** in server logs. They are also useful for finding **language** or **linguistic patterns** from natural language texts.

2.1 Basic Concepts of Association Rules

The problem of mining association rules can be stated as follows: Let $I = \{i_1, i_2, ..., i_m\}$ be a set of **items**. Let $T = (t_1, t_2, ..., t_n)$ be a set of **transactions** (the database), where each transaction t_i is a set of items such that $t_i \subseteq I$. An **association rule** is an implication of the form,

$X \rightarrow Y$, where $X \subset I$, $Y \subset I$, and $X \cap Y = \varnothing$.

X (or Y) is a set of items, called an **itemset**.

Example 1: We want to analyze how the items sold in a supermarket are related to one another. I is the set of all items sold in the supermarket. A transaction is simply a set of items purchased in a basket by a customer. For example, a transaction may be:

{Beef, Chicken, Cheese},

which means that a customer purchased three items in a basket, Beef, Chicken, and Cheese. An association rule may be:

Beef, Chicken $\rightarrow$ Cheese,

where {Beef, Chicken} is X and {Cheese} is Y. For simplicity, brackets "{" and "}" are usually omitted in transactions and rules.   ∎

A transaction $t_i \in T$ is said to **contain** an itemset X if X is a subset of t_i (we also say that the itemset X **covers** t_i). The **support count** of X in T (denoted by $X.count$) is the number of transactions in T that contain X. The strength of a rule is measured by its **support** and **confidence**.

Support: The support of a rule, $X \rightarrow Y$, is the percentage of transactions in T that contains $X \cup Y$, and can be seen as an estimate of the probability, $Pr(X \cup Y)$. The rule support thus determines how frequent the rule is applicable in the transaction set T. Let n be the number of transactions in T. The support of the rule $X \rightarrow Y$ is computed as follows:

$$support = \frac{(X \cup Y).count}{n}. \qquad (1)$$

Support is a useful measure because if it is too low, the rule may just occur due to chance. Furthermore, in a business environment, a rule **covering** too few cases (or transactions) may not be useful because it does not make business sense to act on such a rule (not profitable).

Confidence: The confidence of a rule, $X \rightarrow Y$, is the percentage of transactions in T that contain X also contain Y. It can be seen as an estimate of the conditional probability, $Pr(Y \mid X)$. It is computed as follows:

$$confidence = \frac{(X \cup Y).count}{X.count}. \qquad (2)$$

Confidence thus determines the **predictability** of the rule. If the confidence of a rule is too low, one cannot reliably infer or predict Y from X. A rule with low predictability is of limited use.

Objective: Given a transaction set T, the problem of mining association rules is to discover all association rules in T that have support and confidence greater than or equal to the user-specified **minimum support** (denoted by **minsup**) and **minimum confidence** (denoted by **minconf**).

The keyword here is "all", i.e., association rule mining is complete. Previous methods for rule mining typically generate only a subset of rules based on various heuristics (see Chap. 3).

Example 2: Figure 2.1 shows a set of seven transactions. Each transaction t_i is a set of items purchased in a basket in a store by a customer. The set I is the set of all items sold in the store.

t_1:   Beef, Chicken, Milk
t_2:   Beef, Cheese
t_3:   Cheese, Boots
t_4:   Beef, Chicken, Cheese
t_5:   Beef, Chicken, Clothes, Cheese, Milk
t_6:   Chicken, Clothes, Milk
t_7:   Chicken, Milk, Clothes

Fig. 2.1. An example of a transaction set

Given the user-specified minsup = 30% and minconf = 80%, the following association rule (**sup** is the support, and **conf** is the confidence)

Chicken, Clothes → Milk   [sup = 3/7, conf = 3/3]

is valid as its support is 42.86% (> 30%) and its confidence is 100% (> 80%). The rule below is also valid, whose consequent has two items:

Clothes → Milk, Chicken   [sup = 3/7, conf = 3/3].

Clearly, more association rules can be discovered, as we will see later. ■

We note that the data representation in the transaction form of Fig. 2.1 is a simplistic view of shopping baskets. For example, the quantity and price of each item are not considered in the model.

We also note that a text document or even a sentence in a single document can be treated as a transaction without considering word sequence and the number of occurrences of each word. Hence, given a set of documents or a set of sentences, we can find word co-occurrence relations.

A large number of association rule mining algorithms have been reported in the literature, which have different mining efficiencies. Their resulting sets of rules are, however, all the same based on the definition of association rules. That is, given a transaction data set T, a minimum support and a minimum confidence, the set of association rules existing in T is uniquely determined. Any algorithm should find the same set of rules although their computational efficiencies and memory requirements may be different. The best known mining algorithm is the **Apriori** algorithm proposed in [11], which we study next.

2.2 Apriori Algorithm

The Apriori algorithm works in two steps:

1. **Generate all frequent itemsets:** A frequent itemset is an itemset that has transaction support above minsup.
2. **Generate all confident association rules from the frequent itemsets:** A confident association rule is a rule with confidence above minconf.

We call the number of items in an itemset its **size**, and an itemset of size k a k-itemset. Following Example 2 above, {Chicken, Clothes, Milk} is a frequent 3-itemset as its support is 3/7 (minsup = 30%). From the itemset, we can generate the following three association rules (minconf = 80%):

Rule 1:	Chicken, Clothes → Milk	[sup = 3/7, conf = 3/3]
Rule 2:	Clothes, Milk → Chicken	[sup = 3/7, conf = 3/3]
Rule 3:	Clothes → Milk, Chicken	[sup = 3/7, conf = 3/3].

Below, we discuss the two steps in turn.

2.2.1 Frequent Itemset Generation

The Apriori algorithm relies on the *apriori* or **downward closure** property to efficiently generate all frequent itemsets.

Downward Closure Property: If an itemset has minimum support, then every non-empty subset of this itemset also has minimum support.

The idea is simple because if a transaction contains a set of items X, then it must contain any non-empty subset of X. This property and the minsup threshold prune a large number of itemsets that cannot be frequent.

To ensure efficient itemset generation, the algorithm assumes that the items in I are sorted in **lexicographic order** (a total order). The order is used throughout the algorithm in each itemset. We use the notation {$w[1]$, $w[2]$, ..., $w[k]$} to represent a k-itemset w consisting of items $w[1]$, $w[2]$, ..., $w[k]$, where $w[1] < w[2] < ... < w[k]$ according to the total order. The Apriori algorithm for frequent itemset generation, which is given in Fig. 2.2, is based on **level-wise search**. It generates all frequent itemsets by

Algorithm Apriori(*T*)
1    $C_1 \leftarrow$ init-pass(*T*);   // the first pass over *T*
2    $F_1 \leftarrow \{f \mid f \in C_1, f.count/n \geq minsup\}$;   // *n* is the no. of transactions in *T*
3   **for** ($k = 2; F_{k-1} \neq \varnothing; k++$) **do**   // subsequent passes over *T*
4        $C_k \leftarrow$ candidate-gen(F_{k-1});
5   **for** each transaction $t \in T$ **do**   // scan the data once
6   **for** each candidate $c \in C_k$ **do**
7   **if** *c* is contained in *t* **then**
8   *c*.count++;
9   **endfor**
10   **endfor**
11       $F_k \leftarrow \{c \in C_k \mid c.count/n \geq minsup\}$
12   **endfor**
13   **return** $F \leftarrow \bigcup_k F_k$;

Fig. 2.2. The Apriori algorithm for generating frequent itemsets

Function candidate-gen(F_{k-1})
1    $C_k \leftarrow \varnothing$;   // initialize the set of candidates
2   **forall** $f_1, f_2 \in F_{k-1}$ // find all pairs of frequent itemsets
3   with $f_1 = \{i_1, \dots, i_{k-2}, i_{k-1}\}$ // that differ only in the last item
4   and $f_2 = \{i_1, \dots, i_{k-2}, i'_{k-1}\}$
5   and $i_{k-1} < i'_{k-1}$ **do**   // according to the lexicographic order
6            $c \leftarrow \{i_1, \dots, i_{k-1}, i'_{k-1}\}$;   // join the two itemsets f_1 and f_2
7            $C_k \leftarrow C_k \cup \{c\}$;   // add the new itemset *c* to the candidates
8   **for** each ($k-1$)-subset *s* of *c* **do**
9   **if** ($s \notin F_{k-1}$) **then**
10   delete *c* from C_k;   // delete *c* from the candidates
11   **endfor**
12   **endfor**
13   **return** C_k;   // return the generated candidates

Fig. 2.3. The candidate-gen function

making multiple passes over the data. In the first pass, it counts the supports of individual items (line 1) and determines whether each of them is frequent (line 2). F_1 is the set of frequent 1-itemsets. In each subsequent pass *k*, there are three steps:

1. It starts with the seed set of itemsets F_{k-1} found to be frequent in the ($k-1$)-th pass. It uses this seed set to generate **candidate itemsets** C_k (line 4), which are possible frequent itemsets. This is done using the candidate-gen() function.
2. The transaction database is then scanned and the actual support of each candidate itemset *c* in C_k is counted (lines 5–10). Note that we do not need to load the whole data into memory before processing. Instead, at

any time, only one transaction resides in memory. This is a very important feature of the algorithm. It makes the algorithm scalable to huge data sets, which cannot be loaded into memory.
3. At the end of the pass or scan, it determines which of the candidate itemsets are actually frequent (line 11).

The final output of the algorithm is the set F of all frequent itemsets (line 13). The candidate-gen() function is discussed below.

Candidate-gen function: The candidate generation function is given in Fig. 2.3. It consists of two steps, the **join step** and the **pruning step**.

Join step (lines 2–6 in Fig. 2.3): This step joins two frequent $(k–1)$-itemsets to produce a possible candidate c (line 6). The two frequent itemsets f_1 and f_2 have exactly the same items except the last one (lines 3–5). c is added to the set of candidates C_k (line 7).
Pruning step (lines 8–11 in Fig. 2.3): A candidate c from the join step may not be a final candidate. This step determines whether all the $k–1$ subsets (there are k of them) of c are in F_{k-1}. If anyone of them is not in F_{k-1}, c cannot be frequent according to the downward closure property, and is thus deleted from C_k.

The correctness of the candidate-gen() function is easy to show (see [11]). Here, we use an example to illustrate the working of the function.

Example 3: Let the set of frequent itemsets at level 3 be

$$F_3 = \{\{1, 2, 3\}, \{1, 2, 4\}, \{1, 3, 4\}, \{1, 3, 5\}, \{2, 3, 4\}\}.$$

For simplicity, we use numbers to represent items. The join step (which generates candidates for level 4) will produce two candidate itemsets, {1, 2, 3, 4} and {1, 3, 4, 5}. {1, 2, 3, 4} is generated by joining the first and the second itemsets in F_3 as their first and second items are the same respectively. {1, 3, 4, 5} is generated by joining {1, 3, 4} and {1, 3, 5}.

After the pruning step, we have only:

$$C_4 = \{\{1, 2, 3, 4\}\}$$

because {1, 4, 5} is not in F_3 and thus {1, 3, 4, 5} cannot be frequent.

Example 4: Let us see a complete running example of the Apriori algorithm based on the transactions in Fig. 2.1. We use minsup = 30%.

F_1:   {{Beef}:4, {Cheese}:4, {Chicken}:5, {Clothes}:3, {Milk}:4}

Note: the number after each frequent itemset is the support count of the itemset, i.e., the number of transactions containing the itemset. A minimum support count of 3 is sufficient because the support of 3/7 is greater than 30%, where 7 is the total number of transactions.

C_2:   {{Beef, Cheese}, {Beef, Chicken}, {Beef, Clothes}, {Beef, Milk},
          {Cheese, Chicken}, {Cheese, Clothes}, {Cheese, Milk},
          {Chicken, Clothes}, {Chicken, Milk}, {Clothes, Milk}}

F_2:   {{Beef, Chicken}:3, {Beef, Cheese}:3, {Chicken, Clothes}:3,
          {Chicken, Milk}:4, {Clothes, Milk}:3}

C_3:   {{Chicken, Clothes, Milk}}

          Note: {Beef, Cheese, Chicken} is also produced in line 6 of Fig. 2.3.
          However, {Cheese, Chicken} is not in F_2, and thus the itemset {Beef,
          Cheese, Chicken} is not included in C_3.

F_3:   {{Chicken, Clothes, Milk}:3}.   ∎

Finally, some remarks about the Apriori algorithm are in order:

- Theoretically, this is an exponential algorithm. Let the number of items in I be m. The space of all itemsets is $O(2^m)$ because each item may or may not be in an itemset. However, the mining algorithm exploits the sparseness of the data and the high minimum support value to make the mining possible and efficient. The **sparseness** of the data in the context of market basket analysis means that the store sells a lot of items, but each shopper only purchases a few of them.
- The algorithm can scale up to large data sets as it does not load the entire data into the memory. It only scans the data K times, where K is the size of the largest itemset. In practice, K is often small (e.g., < 10). This scale-up property is very important in practice because many real-world data sets are so large that they cannot be loaded into the main memory.
- The algorithm is based on level-wise search. It has the flexibility to stop at any level. This is useful in practice because in many applications, long frequent itemsets or rules are not needed as they are hard to use.
- As mentioned earlier, once a transaction set T, a minsup and a minconf are given, the set of frequent itemsets that can be found in T is uniquely determined. Any algorithm should find the same set of frequent itemsets. This property about association rule mining does not hold for many other data mining tasks, e.g., classification or clustering, for which different algorithms may produce very different results.
- The main problem with association rule mining is that it often produces a huge number of itemsets (and rules), tens of thousands, or more, which makes it hard for the user to analyze them to find those useful ones. This is called the **interestingness** problem. Researchers have proposed several methods to tackle this problem (see Bibliographic Notes).

An efficient implementation of the Apriori algorithm involves sophisticated data structures and programming techniques, which are beyond the scope of this book. Apart from the Apriori algorithm, there is a large number of other algorithms, e.g., FP-growth [220] and many others.

2.2.2 Association Rule Generation

In many applications, frequent itemsets are already useful and sufficient. Then, we do not need to generate association rules. In applications where rules are desired, we use frequent itemsets to generate all association rules.

Compared with frequent itemset generation, rule generation is relatively simple. To generate rules for every frequent itemset f, we use all non-empty subsets of f. For each such subset α, we output a rule of the form

$$(f - \alpha) \to \alpha, \text{ if}$$

$$confidence = \frac{f.count}{(f - \alpha).count} \geq minconf, \tag{3}$$

where $f.count$ (or $(f-\alpha).count$) is the support count of f (or $(f - \alpha)$). The support of the rule is $f.count/n$, where n is the number of transactions in the transaction set T. All the support counts needed for confidence computation are available because if f is frequent, then any of its non-empty subsets is also frequent and its support count has been recorded in the mining process. Thus, no data scan is needed in rule generation.

This exhaustive rule generation strategy is, however, inefficient. To design an efficient algorithm, we observe that the support count of f in the above confidence computation does not change as α changes. It follows that for a rule $(f - \alpha) \to \alpha$ to hold, all rules of the form $(f - \alpha_{sub}) \to \alpha_{sub}$ must also hold, where α_{sub} is a non-empty subset of α, because the support count of $(f - \alpha_{sub})$ must be less than or equal to the support count of $(f - \alpha)$. For example, given an itemset {A, B, C, D}, if the rule (A, B → C, D) holds, then the rules (A, B, C → D) and (A, B, D → C) must also hold.

Thus, for a given frequent itemset f, if a rule with consequent α holds, then so do rules with consequents that are subsets of α. This is similar to the downward closure property that, if an itemset is frequent, then so are all its subsets. Therefore, from the frequent itemset f, we first generate all rules with one item in the consequent. We then use the consequents of these rules and the function candidate-gen() (Fig. 2.3) to generate all possible consequents with two items that can appear in a rule, and so on. An algorithm using this idea is given in Fig. 2.4. Note that all 1-item consequent rules (rules with one item in the consequent) are first generated in line 2 of the function genRules(). The confidence is computed using (3).

Algorithm genRules(F)   // F is the set of all frequent itemsets
1   **for** each frequent k-itemset f_k in F, $k \geq 2$ **do**
2   output every 1-item consequent rule of f_k with confidence $\geq$ *minconf* and
            support $\leftarrow f_k.count / n$ // n is the total number of transactions in T
3        $H_1 \leftarrow$ {consequents of all 1-item consequent rules derived from f_k above};
4   ap-genRules(f_k, H_1);
5   **endfor**

Procedure ap-genRules(f_k, H_m)   // H_m is the set of m-item consequents
1   **if** $(k > m + 1)$ AND $(H_m \neq \varnothing)$ **then**
2        $H_{m+1} \leftarrow$ candidate-gen(H_m);
3   **for** each h_{m+1} in H_{m+1} **do**
4            $conf \leftarrow f_k.count / (f_k - h_{m+1}).count$;
5   **if** $(conf \geq minconf)$ **then**
6   output the rule $(f_k - h_{m+1}) \rightarrow h_{m+1}$ with confidence = *conf* and
                support $= f_k.count / n$;   // n is the total number of transactions in T
7   **else**
8   delete h_{m+1} from H_{m+1};
9   **endfor**
10   ap-genRules(f_k, H_{m+1});
11   **endif**

Fig. 2.4. The association rule generation algorithm

Example 5: We again use transactions in Fig. 2.1, minsup = 30% and min-conf = 80%. The frequent itemsets are as follows (see Example 4):

F_1:   {{Beef}:4, {Cheese}:4, {Chicken}:5, {Clothes}:3, {Milk}:4}
F_2:   {{Beef, Cheese}:3, {Beef, Chicken}:3, {Chicken, Clothes}:3,
        {Chicken, Milk}:4, {Clothes, Milk}:3}
F_3:   {{Chicken, Clothes, Milk}:3}.

We use only the itemset in F_3 to generate rules (generating rules from each itemset in F_2 can be done in the same way). The itemset in F_3 generates the following possible 1-item consequent rules:

Rule 1:   Chicken, Clothes $\rightarrow$ Milk   [sup = 3/7, conf = 3/3]
Rule 2:   Chicken, Milk $\rightarrow$ Clothes   [sup = 3/7, conf = 3/4]
Rule 3:   Clothes, Milk $\rightarrow$ Chicken   [sup = 3/7, conf = 3/3].

Due to the minconf requirement, only Rule 1 and Rule 3 are output in line 2 of the algorithm genRules(). Thus, H_1 = {{Chicken}, {Milk}}. The function ap-genRules() is then called. Line 2 of ap-genRules() produces H_2 = {{Chicken, Milk}}. The following rule is then generated:

Rule 4:   Clothes $\rightarrow$ Milk, Chicken   [sup = 3/7, conf = 3/3].

Thus, three association rules are generated from the frequent itemset {Chicken, Clothes, Milk} in F_3, namely Rule 1, Rule 3 and Rule 4. ■

2.3 Data Formats for Association Rule Mining

So far, we have used only transaction data for mining association rules. Market basket data sets are naturally of this format. Text documents can be seen as transaction data as well. Each document is a transaction, and each distinctive word is an item. Duplicate words are removed.

However, mining can also be performed on relational tables. We just need to convert a table data set to a transaction data set, which is fairly straightforward if each attribute in the table takes **categorical** values. We simply change each value to an **attribute–value** pair.

Example 6: The table data in Fig. 2.5(A) can be converted to the transaction data in Fig. 2.5(B). Each attribute–value pair is considered an **item**. Using only values is not sufficient in the transaction form because different attributes may have the same values. For example, without including attribute names, value a's for Attribute1 and Attribute2 are not distinguishable. After the conversion, Fig. 2.5(B) can be used in mining. ■

If an attribute takes numerical values, it becomes complex. We need to first discretize its value range into intervals, and treat each interval as a categorical value. For example, an attribute's value range is from 1–100. We may want to divide it into 5 equal-sized intervals, 1–20, 21–40, 41–60, 61–80, and 81–100. Each interval is then treated as a categorical value. Discretization can be done manually based on expert knowledge or automatically. There are several existing algorithms [151, 501].

A point to note is that for a table data set, the join step of the candidate generation function (Fig. 2.3) needs to be slightly modified in order to ensure that it does not join two itemsets to produce a candidate itemset containing two items from the same attribute.

Clearly, we can also convert a transaction data set to a table data set using a binary representation and treating each item in I as an attribute. If a transaction contains an item, its attribute value is 1, and 0 otherwise.

2.4 Mining with Multiple Minimum Supports

The key element that makes association rule mining practical is the minsup threshold. It is used to prune the search space and to limit the number of frequent itemsets and rules generated. However, using only a single min-

Attribute1	Attribute2	Atribute3
a	a	x
b	n	y

(A) Table data

t_1: (Attribute1, a), (Attribute2, a), (Attribute3, x)

t_2: (Attribute1, b), (Attribute2, n), (Attribute3, y)

(B) Transaction data

Fig. 2.5. From a table data set to a transaction data set

sup implicitly assumes that all items in the data are of the same nature and/or have similar frequencies in the database. This is often not the case in real-life applications. In many applications, some items appear very frequently in the data, while some other items rarely appear. If the frequencies of items vary a great deal, we will encounter two problems [344]:

1. If the minsup is set too high, we will not find rules that involve infrequent items or **rare items** in the data.
2. In order to find rules that involve both frequent and rare items, we have to set the minsup very low. However, this may cause combinatorial explosion and make mining impossible because those frequent items will be associated with one another in all possible ways.

Let us use an example to illustrate the above problem with a very low minsup, which will actually introduce another problem.

Example 7: In a supermarket transaction data set, in order to find rules involving those infrequently purchased items such as FoodProcessor and CookingPan (they generate more profits per item), we need to set the minsup very low. Let us use only frequent itemsets in this example as they are generated first and rules are produced from them. They are also the source of all the problems. Now assume we set a very low minsup of 0.005%. We find the following meaningful frequent itemset:

{FoodProcessor, CookingPan}   [sup = 0.006%].

However, this low minsup may also cause the following two meaningless itemsets being discovered:

f_1: {Bread, Cheese, Egg, Bagel, Milk, Sugar, Butter}   [sup = 0.007%],

f_2: {Bread, Egg, Milk, CookingPan}   [sup = 0.006%].

Knowing that 0.007% of the customers buy the seven items in f_1 together is useless because all these items are so frequently purchased in a supermarket. Worst still, they will almost certainly cause combinatorial explosion! For itemsets involving such items to be useful, their supports have to be much higher. Similarly, knowing that 0.006% of the customers buy the four items in f_2 together is also meaningless because Bread, Egg and Milk are purchased on almost every grocery shopping trip.   ∎

This dilemma is called the **rare item problem**. Using a single minsup for the whole data set is inadequate because it cannot capture the inherent natures and/or frequency differences of items in the database. By the natures of items we mean that some items, by nature, appear more frequently than others. For example, in a supermarket, people buy FoodProcessor and CookingPan much less frequently than Bread and Milk. The situation is the same for online stores. In general, those durable and/or expensive goods are bought less often, but each of them generates more profit. It is thus important to capture rules involving less frequent items. However, we must do so without allowing frequent items to produce too many meaningless rules with very low supports and cause combinatorial explosion [344].

One common solution to this problem is to partition the data into several smaller blocks (subsets), each of which contains only items of similar frequencies. Mining is then done separately for each block using a different minsup. This approach is, however, not satisfactory because itemsets or rules that involve items across different blocks will not be found.

A better solution is to allow the user to specify multiple minimum supports, i.e., to specify a different **minimum item support** (**MIS**) to each item. Thus, different itemsets need to satisfy different minimum supports depending on what items are in the itemsets. This model thus enables us to achieve our objective of finding itemsets involving rare items without causing frequent items to generate too many meaningless itemsets. This method helps solve the problem of f_1. To deal with the problem of f_2, we prevent itemsets that contain both very frequent items and very rare items from being generated. A **constraint** will be introduced to realize this.

An **interesting by-product** of this extended model is that it enables the user to easily instruct the algorithm to generate only itemsets that contain certain items but not itemsets that contain only the other items. This can be done by setting the MIS values to more than 100% (e.g., 101%) for these other items. This capability is very useful in practice because in many applications the user is only interested in certain types of itemsets or rules.

2.4.1   Extended Model

To allow multiple minimum supports, the original model in Sect. 2.1 needs to be extended. In the extended model, the minimum support of a rule is expressed in terms of **minimum item supports (MIS)** of the items that appear in the rule. That is, each item in the data can have a MIS value specified by the user. By providing different MIS values for different items, the user effectively expresses different support requirements for different rules. It seems that specifying a MIS value for each item is a difficult task. This is not so as we will see at the end of Sect. 2.4.2.

Let MIS(i) be the MIS value of item i. The **minimum support** of a rule R is the lowest MIS value among the items in the rule. That is, a rule R,

$$i_1, i_2, \ldots, i_k \rightarrow i_{k+1}, \ldots, i_r,$$

satisfies its minimum support if the rule's actual support in the data is greater than or equal to:

$$min(\text{MIS}(i_1), \text{MIS}(i_2), \ldots, \text{MIS}(i_r)).$$

Minimum item supports thus enable us to achieve the goal of having higher minimum supports for rules that involve only frequent items, and having lower minimum supports for rules that involve less frequent items.

Example 8: Consider the set of items in a data set, {Bread, Shoes, Clothes}. The user-specified MIS values are as follows:

MIS(Bread) = 2%   MIS(Clothes) = 0.2%   MIS(Shoes) = 0.1%.

The following rule doesn't satisfy its minimum support:

Clothes → Bread   [sup = 0.15%, conf = 70%].

This is so because min(MIS(Bread), MIS(Clothes)) = 0.2%. The following rule satisfies its minimum support:

Clothes → Shoes   [sup = 0.15%, conf = 70%].

because min(MIS(Clothes), MIS(Shoes)) = 0.1%.   ■

As we explained earlier, the **downward closure property** holds the key to pruning in the Apriori algorithm. However, in the new model, if we use the Apriori algorithm to find all frequent itemsets, the downward closure property no longer holds.

Example 9: Consider the four items 1, 2, 3 and 4 in a data set. Their minimum item supports are:

MIS(1) = 10%   MIS(2) = 20%   MIS(3) = 5%   MIS(4) = 6%.

If we find that itemset {1, 2} has a support of 9% at level 2, then it does not satisfy either MIS(1) or MIS(2). Using the Apriori algorithm, this itemset is discarded since it is not frequent. Then, the potentially frequent itemsets {1, 2, 3} and {1, 2, 4} will not be generated for level 3. Clearly, itemsets {1,

2, 3} and {1, 2, 4} may be frequent because MIS(3) is only 5% and MIS(4) is 6%. It is thus wrong to discard {1, 2}. However, if we do not discard {1, 2}, the downward closure property is lost.   ■

Below, we present an algorithm to solve this problem. The essential idea is to sort the items according to their MIS values in ascending order to avoid the problem.

Note that MIS values prevent low support itemsets involving only frequent items from being generated because their individual MIS values are all high. To prevent very frequent items and very rare items from appearing in the same itemset, we introduce the **support difference constraint**.

Let $sup(i)$ be the actual support of item i in the data. For each itemset s, the support difference constraint is as follows:

$$max_{i \in s}\{sup(i)\} - min_{i \in s}\{sup(i)\} \leq \varphi,$$

where $0 \leq \varphi \leq 1$ is the user-specified **maximum support difference**, and it is the same for all itemsets. The constraint basically limits the difference between the largest and the smallest actual supports of items in itemset s to φ. This constraint can reduce the number of itemsets generated dramatically, and it does not affect the downward closure property.

2.4.2   Mining Algorithm

The new algorithm generalizes the Apriori algorithm for finding frequent itemsets. We call the algorithm, **MS-Apriori**. When there is only one MIS value (for all items), it reduces to the Apriori algorithm.

Like Apriori, MS-Apriori is also based on level-wise search. It generates all frequent itemsets by making multiple passes over the data. However, there is an exception in the second pass as we will see later.

The key operation in the new algorithm is the sorting of the items in I in ascending order of their MIS values. This order is fixed and used in all subsequent operations of the algorithm. The items in each itemset follow this order. For example, in Example 9 of the four items 1, 2, 3 and 4 and their given MIS values, the items are sorted as follows: 3, 4, 1, 2. This order helps solve the problem identified above.

Let F_k denote the set of frequent k-itemsets. Each itemset w is of the following form, $\{w[1], w[2], ..., w[k]\}$, which consists of items, $w[1]$, $w[2]$, ..., $w[k]$, where $MIS(w[1]) \leq MIS(w[2]) \leq ... \leq MIS(w[k])$. The algorithm MS-Apriori is given in Fig. 2.6. Line 1 performs the sorting on I according to the MIS value of each item (stored in MS). Line 2 makes the first pass over the data using the function init-pass(), which takes two arguments, the

Algorithm MS-Apriori(T, MS, φ)　　　// MS stores all MIS values
1　　$M \leftarrow$ sort(I, MS);　　　　　　　　// according to MIS(i)'s stored in MS
2　　$L \leftarrow$ init-pass(M, T);　　　　　　// make the first pass over T
3　　$F_1 \leftarrow \{\{l\} \mid l \in L, l.count/n \geq \text{MIS}(l)\}$;　// n is the size of T
4　　**for** ($k = 2$; $F_{k-1} \neq \varnothing$; k++) **do**
5　　　　**if** $k = 2$ **then**
6　　　　　　$C_k \leftarrow$ level2-candidate-gen(L, φ)　// $k = 2$
7　　　　**else** $C_k \leftarrow$ MScandidate-gen(F_{k-1}, φ)
8　　　　**endif**;
9　　　　**for** each transaction $t \in T$ **do**
10　　　　　　**for** each candidate $c \in C_k$ **do**
11　　　　　　　　**if** c is contained in t **then**　　　// c is a subset of t
12　　　　　　　　　　$c.count$++
13　　　　　　　　**if** $c - \{c[1]\}$ is contained in t **then**　// c without the first item
14　　　　　　　　　　($c - \{c[1]\}$).$count$++
15　　　　　　**endfor**
16　　　　**endfor**
17　　　　$F_k \leftarrow \{c \in C_k \mid c.count/n \geq \text{MIS}(c[1])\}$
18　　**endfor**
19　　**return** $F \leftarrow \bigcup_k F_k$;

Fig. 2.6. The MS-Apriori algorithm

data set T and the sorted items M, to produce the seeds L for generating candidate itemsets of length 2, i.e., C_2. init-pass() has two steps:

1. It first scans the data once to record the support count of each item.
2. It then follows the sorted order to find the first item i in M that meets MIS(i). i is inserted into L. For each subsequent item j in M after i, if $j.count/n \geq \text{MIS}(i)$, then j is also inserted into L, where $j.count$ is the support count of j, and n is the total number of transactions in T.

Frequent 1-itemsets (F_1) are obtained from L (line 3). It is easy to show that all frequent 1-itemsets are in F_1.

Example 10: Let us follow Example 9 and the given MIS values for the four items. Assume our data set has 100 transactions (not limited to the four items). The first pass over the data gives us the following support counts: $\{3\}.count = 6$, $\{4\}.count = 3$, $\{1\}.count = 9$ and $\{2\}.count = 25$. Then,

$$L = \{3, 1, 2\}, \text{ and } F_1 = \{\{3\}, \{2\}\}.$$

Item 4 is not in L because $4.count/n < \text{MIS}(3)$ ($= 5\%$), and $\{1\}$ is not in F_1 because $1.count / n < \text{MIS}(1)$ ($= 10\%$).　■

For each subsequent pass (or data scan), say pass k, the algorithm performs three operations.

1. The frequent itemsets in F_{k-1} found in the $(k-1)th$ pass are used to generate the candidates C_k using the MScandidate-gen() function (line 7). However, there is a special case, i.e., when $k = 2$ (line 6), for which the candidate generation function is different, i.e., level2-candidate-gen().
2. It then scans the data and updates various support counts of the candidates in C_k (line 9–16). For each candidate c, we need to update its support count (lines 11–12) and also the support count of c without the first item (lines 13–14), i.e., $c - \{c[1]\}$, which is used in rule generation and will be discussed in Sect. 2.4.3. If rule generation is not required, lines 13 and 14 can be deleted.
3. The frequent itemsets (F_k) for the pass are identified in line 17.

We present candidate generation functions level2-candidate-gen() and MScandidate-gen() below.

Level2-candidate-gen function: It takes an argument L, and returns a superset of the set of all frequent 2-itemsets. The algorithm is given in Fig. 2.7. Note that in line 5, we use $|sup(h) - sup(l)| \leq \varphi$ because $sup(l)$ may not be lower than $sup(h)$, although MIS(l) ≤ MIS(h).

Example 11: Let us continue with Example 10. We set $\varphi = 10\%$. Recall the MIS values of the four items are (in Example 9):

MIS(1) = 10%	MIS(2) = 20%
MIS(3) = 5%	MIS(4) = 6%.

The level2-candidate-gen() function in Fig. 2.7 produces

$$C_2 = \{\{3, 1\}\}.$$

{1, 2} is not a candidate because the support count of item 1 is only 9 (or 9%), less than MIS(1) (= 10%). Hence, {1, 2} cannot be frequent. {3, 2} is not a candidate because $sup(3) = 6\%$ and $sup(2) = 25\%$ and their difference is greater than $\varphi = 10\%$ ∎

Note that we must use L rather than F_1 because F_1 does not contain those items that may satisfy the MIS of an earlier item (in the sorted order) but not the MIS of itself, e.g., item 1 in the above example. Using L, the problem discussed in Sect. 2.4.1 is solved for C_2.

MScandidate-gen function: The algorithm is given in Fig. 2.8, which is similar to the candidate-gen function in the Apriori algorithm. It also has two steps, the **join step** and the **pruning step**. The join step (lines 2–6) is the same as that in the candidate-gen() function. The pruning step (lines 8–12) is, however, different.

For each $(k-1)$-subset s of c, if s is not in F_{k-1}, c can be deleted from C_k. However, there is an exception, which is when s does not include $c[1]$

Function level2-candidate-gen(L, φ)

1	$C_2 \leftarrow \varnothing$; // initialize the set of candidates
2	**for** each item l in L in the same order **do**
3	**if** $l.count/n \geq \text{MIS}(l)$ **then**
4	**for** each item h in L that is after l **do**
5	**if** $h.count/n \geq \text{MIS}(l)$ and $|sup(h) - sup(l)| \leq \varphi$ **then**
6	$C_2 \leftarrow C_2 \cup \{\{l, h\}\}$; // insert the candidate $\{l, h\}$ into C_2

Fig. 2.7. The level2-candidate-gen function

Function MScandidate-gen(F_{k-1}, φ)

1	$C_k \leftarrow \varnothing$; // initialize the set of candidates
2	**forall** $f_1, f_2 \in F_k$ // find all pairs of frequent itemsets
3	with $f_1 = \{i_1, \dots, i_{k-2}, i_{k-1}\}$ // that differ only in the last item
4	and $f_2 = \{i_1, \dots, i_{k-2}, i'_{k-1}\}$
5	and $i_{k-1} < i'_{k-1}$ and $|sup(i_{k-1}) - sup(i'_{k-1})| \leq \varphi$ **do**
6	$c \leftarrow \{i_1, \dots, i_{k-1}, i'_{k-1}\}$; // join the two itemsets f_1 and f_2
7	$C_k \leftarrow C_k \cup \{c\}$; // insert the candidate itemset c into C_k
8	**for** each $(k{-}1)$-subset s of c **do**
9	**if** $(c[1] \in s)$ or $(\text{MIS}(c[2]) = \text{MIS}(c[1]))$ **then**
10	**if** $(s \notin F_{k-1})$ **then**
11	delete c from C_k; // delete c from the set of candidates
12	**endfor**
13	**endfor**
14	**return** C_k; // return the generated candidates

Fig. 2.8. The MScandidate-gen function

(there is only one such s). That is, the first item of c, which has the lowest MIS value, is not in s. Even if s is not in F_{k-1}, we cannot delete c because we cannot be sure that s does not satisfy MIS($c[1]$), although we know that it does not satisfy MIS($c[2]$), unless MIS($c[2]$) = MIS($c[1]$) (line 9).

Example 12: Let $F_3 = \{\{1, 2, 3\}, \{1, 2, 5\}, \{1, 3, 4\}, \{1, 3, 5\}, \{1, 4, 5\}, \{1, 4, 6\}, \{2, 3, 5\}\}$. Items in each itemset are in the sorted order. The join step produces (we ignore the support difference constraint here)

$$\{1, 2, 3, 5\}, \{1, 3, 4, 5\} \text{ and } \{1, 4, 5, 6\}.$$

The pruning step deletes $\{1, 4, 5, 6\}$ because $\{1, 5, 6\}$ is not in F_3. We are then left with $C_4 = \{\{1, 2, 3, 5\}, \{1, 3, 4, 5\}\}$. $\{1, 3, 4, 5\}$ is not deleted although $\{3, 4, 5\}$ is not in F_3 because the minimum support of $\{3, 4, 5\}$ is MIS(3), which may be higher than MIS(1). Although $\{3, 4, 5\}$ does not satisfy MIS(3), we cannot be sure that it does not satisfy MIS(1). However, if MIS(3) = MIS(1), then $\{1, 3, 4, 5\}$ can also be deleted. ■

The problem discussed in Sect. 2.4.1 is solved for C_k $(k > 2)$ because, due to the sorting, we do not need to extend a frequent $(k-1)$-itemset with any item that has a lower MIS value. Let us see a complete example.

Example 13: Given the following seven transactions,

> Beef, Bread
> Bread, Clothes
> Bread, Clothes, Milk
> Cheese, Boots
> Beef, Bread, Cheese, Shoes
> Beef, Bread, Cheese, Milk
> Bread, Milk, Clothes

and MIS(Milk) = 50%, MIS(Bread) = 70%, and 25% for all other items. Again, the support difference constraint is not used. The following frequent itemsets are produced:

F_1 = {{Beef}, {Cheese}, {Clothes}, {Bread}}
F_2 = {{Beef, Cheese}, {Beef, Bread}, {Cheese, Bread}
        {Clothes, Bread}, {Clothes, Milk}}
F_3 = {{Beef, Cheese, Bread}, {Clothes, Milk, Bread}}.   ∎

To conclude this sub-section, let us further discuss two important issues:

1. Specify MIS values for items: This is usually done in two ways:
   - Assign a MIS value to each item according to its actual support/frequency in the data set T. For example, if the actual support of item i in T is $sup(i)$, then the MIS value for i may be computed with $\lambda \times sup(i)$, where λ is a parameter $(0 \leq \lambda \leq 1)$ and is the same for all items in T.
   - Group items into clusters (or blocks). Items in each cluster have similar frequencies. All items in the same cluster are given the same MIS value. We should note that in the extended model frequent itemsets involving items from different clusters will be found.
2. Generate itemsets that must contain certain items: As mentioned earlier, the extended model enables the user to instruct the algorithm to generate itemsets that must contain certain items, or not to generate any itemsets consisting of only the other items. Let us see an example.

Example 14: Given the data set in Example 13, if we want to generate frequent itemsets that must contain at least one item in {Boots, Bread, Cheese, Milk, Shoes}, or not to generate itemsets involving only Beef and/or Clothes, we can simply set

> MIS(Beef) = 101%, and MIS(Clothes) = 101%

Then the algorithm will not generate the itemsets, {Beef}, {Clothes} and {Beef, Clothes}. However, it will still generate such frequent itemsets as {Cheese, Beef} and {Cheese, Bread, Beef}.   ■

In many applications, this feature comes quite handy because the user is often only interested in certain types of itemsets or rules.

2.4.3 Rule Generation

Association rules are generated using frequent itemsets. In the case of a single minsup, if f is a frequent itemset and f_{sub} is a subset of f, then f_{sub} must also be a frequent itemset. All their support counts are computed and recorded by the Apriori algorithm. Then, the confidence of each possible rule can be easily calculated without seeing the data again.

However, in the case of MS-Apriori, if we only record the support count of each frequent itemset, it is not sufficient. Let us see why.

Example 15: Recall in Example 8, we have

MIS(Bread) = 2%   MIS(Clothes) = 0.2%   MIS(Shoes) = 0.1%.

If the actual support for the itemset {Clothes, Bread} is 0.15%, and for the itemset {Shoes, Clothes, Bread} is 0.12%, according to MS-Apriori, {Clothes, Bread} is not a frequent itemset since its support is less than MIS(Clothes). However, {Shoes, Clothes, Bread} is a frequent itemset as its actual support is greater than

min(MIS(Shoes), MIS(Clothes), MIS(Bread)) = MIS(Shoes)).

We now have a problem in computing the confidence of the rule,

Clothes, Bread → Shoes

because the itemset {Clothes, Bread} is not a frequent itemset and thus its support count is not recorded. In fact, we may not be able to compute the confidences of the following rules either:

Clothes → Shoes, Bread
Bread → Shoes, Clothes

because {Clothes} and {Bread} may not be frequent.   ■

Lemma: The above problem may occur only when the item that has the lowest MIS value in the itemset is in the consequent of the rule (which may have multiple items). We call this problem the **head-item problem**.

Proof by contradiction: Let f be a frequent itemset, and $a \in f$ be the item with the lowest MIS value in f (a is called the **head item**). Thus, f uses

MIS(a) as its minsup. We want to form a rule, $X \rightarrow Y$, where $X, Y \subset f, X \cup Y = f$ and $X \cap Y = \emptyset$. Our examples above already show that the head-item problem may occur when $a \in Y$. Now assume that the problem can also occur when $a \in X$. Since $a \in X$ and $X \subset f$, a must have the lowest MIS value in X and X must be a frequent itemset, which is ensured by the MS-Apriori algorithm. Hence, the support count of X is recorded. Since f is a frequent itemset and its support count is also recorded, then we can compute the confidence of $X \rightarrow Y$. This contradicts our assumption. ∎

The lemma indicates that we need to record the support count of $f - \{a\}$. This is achieved by lines 13–14 in MS-Apriori (Fig. 2.6). All problems in Example 15 are solved. A similar rule generation function as genRules() in Apriori can be designed to generate rules with multiple minimum supports.

2.5 Mining Class Association Rules

The mining models studied so far do not use any targets. That is, any item can appear as a consequent or condition of a rule. However, in some applications, the user is interested in only rules with some fixed **target items** on the right-hand side. For example, the user has a collection of text documents from some topics (target items), and he/she wants to know what words are correlated with each topic. In [352], a data mining system based entirely on such rules (called **class association rules)** is reported, which is in production use in Motorola for many different applications. In the Web environment, class association rules are also useful because many types of Web data are in the form of transactions, e.g., search queries issued by users, and pages clicked by visitors. There are often target items as well, e.g., advertisements. Web sites want to know how user activities are associated with advertisements that they may like to view. This touches the issue of classification or prediction, which we will study in the next chapter.

2.5.1 Problem Definition

Let T be a transaction data set consisting of n transactions. Each transaction is labeled with a class y. Let I be the set of all items in T, Y be the set of all **class labels** (or target items) and $I \cap Y = \emptyset$. A **class association rule (CAR)** is an implication of the form

$X \rightarrow y$, where $X \subseteq I$, and $y \in Y$.

The definitions of **support** and **confidence** are the same as those for normal association rules. In general, a class association rule is different from a normal association rule in two ways:

1. The consequent of a CAR has only a single item, while the consequent of a normal association rule can have any number of items.
2. The consequent y of a CAR can only be from the class label set Y, i.e., $y \in Y$. No item from I can appear as the consequent, and no class label can appear as a rule condition. In contrast, a normal association rule can have any item as a condition or a consequent.

Objective: The problem of mining CARs is to generate the complete set of CARs that satisfies the user-specified minimum support (minsup) and minimum confidence (minconf) constraints.

Example 16: Figure 2.9 shows a data set which has seven text documents. Each document is a transaction and consists of a set of keywords. Each transaction is also labeled with a topic class (education or sport).

I = {Student, Teach, School, City, Game, Baseball, Basketball, Team, Coach, Player, Spectator}
Y = {Education, Sport}.

	Transactions	**Class**
doc 1:	Student, Teach, School	: Education
doc 2:	Student, School	: Education
doc 3:	Teach, School, City, Game	: Education
doc 4:	Baseball, Basketball	: Sport
doc 5:	Basketball, Player, Spectator	: Sport
doc 6:	Baseball, Coach, Game, Team	: Sport
doc 7:	Basketball, Team, City, Game	: Sport

Fig. 2.9. An example of a data set for mining class association rules

Let minsup = 20% and minconf = 60%. The following are two examples of class association rules:

Student, School $\rightarrow$ Education   [sup= 2/7, conf = 2/2]
Game $\rightarrow$ Sport   [sup= 2/7, conf = 2/3].   ∎

A question that one may ask is: can we mine the data by simply using the Apriori algorithm and then perform a post-processing of the resulting rules to select only those class association rules? In principle, the answer is yes because CARs are a special type of association rules. However, in practice this is often difficult or even impossible because of combinatorial explosion, i.e., the number of rules generated in this way can be huge.

2.5.2 Mining Algorithm

Unlike normal association rules, CARs can be mined directly in a single step. The key operation is to find all **ruleitems** that have support above minsup. A **ruleitem** is of the form:

(*condset*, *y*),

where **condset** $\subseteq I$ is a set of items, and $y \in Y$ is a class label. The support count of a condset (called **condsupCount**) is the number of transactions in T that contain the condset. The support count of a ruleitem (called **rulesupCount**) is the number of transactions in T that contain the condset and are labeled with class y. Each ruleitem basically represents a rule:

condset $\rightarrow y$,

whose **support** is (*rulesupCount / n*), where n is the total number of transactions in T, and whose **confidence** is (*rulesupCount / condsupCount*).

Ruleitems that satisfy the minsup are called **frequent** ruleitems, while the rest are called infrequent ruleitems. For example, ({Student, School}, Education) is a ruleitem in T of Fig. 2.9. The support count of the condset {Student, School} is 2, and the support count of the ruleitem is also 2. Then the support of the ruleitem is 2/7 (= 28.6%), and the confidence of the ruleitem is 100%. If minsup = 10%, then the ruleitem satisfies the minsup threshold. We say that it is frequent. If minconf = 80%, then the ruleitem satisfies the minconf threshold. We say that the ruleitem is **confident**. We thus have the class association rule:

Student, School $\rightarrow$ Education   [sup= 2/7, conf = 2/2].

The rule generation algorithm, called **CAR-Apriori**, is given in Fig. 2.10, which is based on the Apriori algorithm. Like the Apriori algorithm, CAR-Apriori generates all the frequent ruleitems by making multiple passes over the data. In the first pass, it computes the support count of each 1-ruleitem (containing only one item in its condset) (line 1). The set of all 1-candidate ruleitems considered is:

$C_1 = \{(\{i\}, y) \mid i \in I, \text{and } y \in Y\}$,

which basically associates each item in I (or in the transaction data set T) with every class label. Line 2 determines whether the candidate 1-ruleitems are frequent. From frequent 1-ruleitems, we generate 1-condition CARs (rules with only one condition) (line 3). In a subsequent pass, say k, it starts with the seed set of $(k–1)$-ruleitems found to be frequent in the $(k–1)$-th pass, and uses this seed set to generate new possibly frequent k-ruleitems, called **candidate k-ruleitems** (C_k in line 5). The actual support

Algorithm CAR-Apriori(T)

```
1    C₁ ← init-pass(T);                          // the first pass over T
2    F₁ ← {f | f ∈ C₁, f. rulesupCount / n ≥ minsup};
3    CAR₁ ← {f | f ∈ F₁, f.rulesupCount / f.condsupCount ≥ minconf};
4    for (k = 2; Fₖ₋₁ ≠ ∅; k++) do
5        Cₖ ← CARcandidate-gen(Fₖ₋₁);
6        for each transaction t ∈ T do
7            for each candidate c ∈ Cₖ do
8                if c.condset is contained in t then   // c is a subset of t
9                    c.condsupCount++;
10                   if t.class = c.class then
11                       c.rulesupCount++
12           endfor
13       end-for
14       Fₖ ← {c ∈ Cₖ | c.rulesupCount / n ≥ minsup};
15       CARₖ ← {f | f ∈ Fₖ, f.rulesupCount / f.condsupCount ≥ minconf};
16   endfor
17   return CAR ← ∪ₖ CARₖ;
```

Fig. 2.10. The CAR-Apriori algorithm

counts, both *condsupCount* and *rulesupCount*, are updated during the scan of the data (lines 6–13) for each candidate k-ruleitem. At the end of the data scan, it determines which of the candidate k-ruleitems in C_k are actually frequent (line 14). From the frequent k-ruleitems, line 15 generates k-condition CARs (class association rules with k conditions).

One interesting note about ruleitem generation is that if a ruleitem/rule has a confidence of 100%, then extending the ruleitem with more conditions (adding items to its condset) will also result in rules with 100% confidence although their supports may drop with additional items. In some applications, we may consider these subsequent rules **redundant** because additional conditions do not provide any more information. Then, we should not extend such ruleitems in candidate generation for the next level, which can reduce the number of generated rules substantially. If desired, redundancy handling can be added in the CAR-Apriori algorithm easily.

The CARcandidate-gen() function is very similar to the candidate-gen() function in the Apriori algorithm, and it is thus omitted. The only difference is that in CARcandidate-gen() ruleitems with the same class are joined by joining their condsets.

Example 17: Let us work on a complete example using our data in Fig. 2.9. We set minsup = 15%, and minconf = 70%

F_1:   {({School}, Education):(3, 3),   ({Student}, Education):(2, 2),
         ({Teach}, Education):(2, 2),   ({Baseball}, Sport):(2, 2),

({Basketball}, Sport):(3, 3),   ({Game}, Sport):(3, 2),
({Team}, Sport):(2, 2)}

Note: The two numbers within the parentheses after each ruleitem are its *condSupCount* and *ruleSupCount* respectively.

CAR_1: School → Education   [sup = 3/7, conf = 3/3]
        Student → Education   [sup = 2/7, conf = 2/2]
        Teach → Education   [sup = 2/7, conf = 2/2]
        Baseball → Sport   [sup = 2/7, conf = 2/2]
        Basketball → Sport   [sup = 3/7, conf = 3/3]
        Game → Sport   [sup = 2/7, conf = 2/3]
        Team → Sport   [sup = 2/7, conf = 2/2]

Note: We do not deal with rule redundancy in this example.

C_2: {({School, Student}, Education),   ({School, Teach}, Education),
        ({Student, Teach}, Education),   ({Baseball, Basketball}, Sport),
        ({Baseball, Game}, Sport),   ({Baseball, Team}, Sport),
        ({Basketball, Game}, Sport),   ({Basketball, Team}, Sport),
        ({Game, Team}, Sport)}

F_2: {({School, Student}, Education):(2, 2),
        ({School, Teach}, Education):(2, 2), ({Game, Team}, Sport):(2, 2)}

CAR_2: School, Student → Education   [sup = 2/7, conf = 2/2]
        School, Teach → Education   [sup = 2/7, conf = 2/2]
        Game, Team → Sport   [sup = 2/7, conf = 2/2]   ■

We note that for many applications involving target items, the data sets used are relational tables. They need to be converted to transaction forms before mining. We can use the method in Sect. 2.3 for the purpose.

Example 18: In Fig. 2.11(A), the data set has three data attributes and a class attribute with two possible values, positive and negative. It is converted to the transaction data in Fig. 2.11(B). Notice that for each class, we only use its original value. There is no need to attach the attribute "Class"

Attribute1	Attribute2	Atribute3	Class
a	a	x	positive
b	n	y	negative

(A) Table data

t_1: (Attribute1, a), (Attribute2, a), (Attribute3, x)   : Positive
t_2: (Attribute1, b), (Attribute2, n), (Attribute3, y)   : negative

(B) Transaction data

Fig. 2.11. Converting a table data set (A) to a transaction data set (B)

because there is no ambiguity. As discussed in Sect. 2.3, for each numeric attribute, its value range needs to be discretized into intervals either manually or automatically before conversion and rule mining. There are many discretization algorithms. Interested readers are referred to [151].   ■

2.5.3 Mining with Multiple Minimum Supports

The concept of mining with multiple minimum supports discussed in Sect. 2.4 can be incorporated in class association rule mining in two ways:

1. **Multiple minimum class supports:** The user can specify different minimum supports for different classes. For example, the user has a data set with two classes, Yes and No. Based on the application requirement, he/she may want all rules of class Yes to have the minimum support of 5% and all rules of class No to have the minimum support of 20%.
2. **Multiple minimum item supports:** The user can specify a minimum item support for every item (either a class item/label or a non-class item). This is more general and is similar to normal association rule mining discussed in Sect. 2.4.

For both approaches, similar mining algorithms to that given in Sect. 2.4 can be devised. The *support difference constraint* in Sect. 2.4.1 can be incorporated as well. Like normal association rule mining with multiple minimum supports, by setting minimum class and/or item supports to more than 100% for some items, the user effectively instructs the algorithm not to generate rules involving only these items.

Finally, although we have discussed only multiple minimum supports so far, we can easily use different minimum confidences for different classes as well, which provides an additional flexibility in applications.

2.6 Basic Concepts of Sequential Patterns

Association rule mining does not consider the order of transactions. However, in many applications such orderings are significant. For example, in market basket analysis, it is interesting to know whether people buy some items in sequence, e.g., buying bed first and then buying bed sheets some time later. In Web usage mining, it is useful to find navigational patterns in a Web site from sequences of page visits of users (see Chap. 12). In text mining, considering the ordering of words in a sentence is vital for finding linguistic or language patterns (see Chap. 11). For these applications, association rules will not be appropriate. Sequential patterns are needed. Below, we define the problem of mining sequential patterns and introduce the main concepts involved.

Let $I = \{i_1, i_2, ..., i_m\}$ be a set of items. A **sequence** is an ordered list of itemsets. Recall an **itemset** X is a non-empty set of items $X \subseteq I$. We denote a sequence s by $\langle a_1 a_2 ... a_r \rangle$, where a_i is an itemset, which is also called an **element** of s. We denote an element (or an itemset) of a sequence by $\{x_1, x_2, ..., x_k\}$, where $x_j \in I$ is an item. We assume without loss of generality that items in an element of a sequence are in **lexicographic order**. An item can occur only once in an element of a sequence, but can occur multiple times in different elements. The **size** of a sequence is the number of elements (or itemsets) in the sequence. The **length** of a sequence is the number of items in the sequence. A sequence of length k is called a **k-sequence**. If an item occurs multiple times in different elements of a sequence, each occurrence contributes to the value of k. A sequence $s_1 = \langle a_1 a_2 ... a_r \rangle$ is a **subsequence** of another sequence $s_2 = \langle b_1 b_2 ... b_v \rangle$, or s_2 is a **supersequence** of s_1, if there exist integers $1 \leq j_1 < j_2 < ... < j_{r-1} < j_r \leq v$ such that $a_1 \subseteq b_{j_1}$, $a_2 \subseteq b_{j_2}$, ..., $a_r \subseteq b_{j_r}$. We also say that s_2 **contains** s_1.

Example 19: Let $I = \{1, 2, 3, 4, 5, 6, 7, 8, 9\}$. The sequence $\langle \{3\}\{4, 5\}\{8\} \rangle$ is contained in (or is a subsequence of) $\langle \{6\} \{3, 7\}\{9\}\{4, 5, 8\}\{3, 8\} \rangle$ because $\{3\} \subseteq \{3, 7\}$, $\{4, 5\} \subseteq \{4, 5, 8\}$, and $\{8\} \subseteq \{3, 8\}$. However, $\langle \{3\}\{8\} \rangle$ is not contained in $\langle \{3, 8\} \rangle$ or vice versa. The size of the sequence $\langle \{3\}\{4, 5\}\{8\} \rangle$ is 3, and the length of the sequence is 4.   ∎

Objective: Given a set S of input **data sequences** (or **sequence database**), the problem of mining sequential patterns is to find all sequences that have a user-specified **minimum support**. Each such sequence is called a **frequent sequence**, or a **sequential pattern**. The **support** for a sequence is the fraction of total data sequences in S that contains this sequence.

Example 20: We use the market basket analysis as an example. Each sequence in this context represents an ordered list of transactions of a particular customer. A transaction is a set of items that the customer purchased at a time (called the transaction time). Then transactions in the sequence are ordered by increasing transaction time. Table 2.1 shows a transaction database which is already sorted according to customer ID (the major key) and transaction time (the minor key). Table 2.2 gives the data sequences (also called **customer sequences**). Table 2.3 gives the output sequential patterns with the minimum support of 25%, i.e., two customers.   ∎

Table 2.1. A set of transactions sorted by customer ID and transaction time

Customer ID	Transaction Time	Transaction (items bought)
1	July 20, 2005	30
1	July 25, 2005	90
2	July 9, 2005	10, 20
2	July 14, 2005	30
2	July 20, 2005	10, 40, 60, 70
3	July 25, 2005	30, 50, 70, 80
4	July 25, 2005	30
4	July 29, 2005	30, 40, 70, 80
4	August 2, 2005	90
5	July 12, 2005	90

Table 2.2. The sequence database produced from the transactions in Table 2.1.

Customer ID	Data Sequence
1	⟨{30} {90}⟩
2	⟨{10, 20} {30} {10, 40, 60, 70}⟩
3	⟨{30, 50, 70, 80}⟩
4	⟨{30} {30, 40, 70, 80} {90}⟩
5	⟨{90}⟩

Table 2.3. The final output sequential patterns

	Sequential Patterns with Support ≥ 25%
1-sequences	⟨{30}⟩, ⟨{40}⟩, ⟨{70}⟩, ⟨{80}⟩, ⟨{90}⟩
2-sequences	⟨{30} {40}⟩, ⟨{30} {70}⟩, ⟨{30}, {90}⟩, ⟨{30, 70}⟩, ⟨{30, 80}⟩, ⟨{40, 70}⟩, ⟨{70, 80}⟩
3-sequences	⟨{30} {40, 70}⟩, ⟨{30, 70, 80}⟩

2.7 Mining Sequential Patterns Based on GSP

This section describes two algorithms for mining sequential patterns based on the GSP algorithm in [500]: the original GSP, which uses a single minimum support, and MS-GSP, which uses multiple minimum supports.

2.7.1 GSP Algorithm

GSP works in almost the same way as the Apriori algorithm. We still use F_k to store the set of all frequent k-sequences, and C_k to store the set of all

Algorithm GSP(S)

1    $C_1 \leftarrow$ init-pass(S);   // the first pass over S

2    $F_1 \leftarrow \{\langle\{f\}\rangle | f \in C_1, f.count/n \geq minsup\};$ // n is the number of sequences in S

3   **for** ($k = 2; F_{k-1} \neq \varnothing; k{+}{+}$) **do**   // subsequent passes over S

4        $C_k \leftarrow$ candidate-gen-SPM(F_{k-1});

5   **for** each data sequence $s \in S$ **do**   // scan the data once

6   **for** each candidate $c \in C_k$ **do**

7   **if** c is contained in s **then**

8                    $c.count{+}{+}$;   // increment the support count

9   **endfor**

10   **endfor**

11     $F_k \leftarrow \{c \in C_k \mid c.count/n \geq minsup\}$

12   **endfor**

13   **return** $\bigcup_k F_k$;

Fig. 2.12. The GSP Algorithm for generating sequential patterns

Function candidate-gen-SPM(F_{k-1})   // SPM: Sequential Pattern Mining

1. **Join step.** Candidate sequences are generated by joining F_{k-1} with F_{k-1}. A sequence s_1 joins with s_2 if the subsequence obtained by dropping the first item of s_1 is the same as the subsequence obtained by dropping the last item of s_2. The candidate sequence generated by joining s_1 with s_2 is the sequence s_1 extended with the last item in s_2. There are two cases:
   - the added item forms a separate element if it was a separate element in s_2, and is appended at the end of s_1 in the merged sequence, and
   - the added item is part of the last element of s_1 in the merged sequence otherwise.

   When joining F_1 with F_1, we need to add the item in s_2 both as part of an itemset and as a separate element. That is, joining $\langle\{x\}\rangle$ with $\langle\{y\}\rangle$ gives us both $\langle\{x, y\}\rangle$ and $\langle\{x\}\{y\}\rangle$. Note that x and y in $\{x, y\}$ are ordered.
2. **Prune step.** A candidate sequence is pruned if any one of its ($k{-}1$)-subsequences is infrequent (without minimum support).

Fig. 2.13. The candidate-gen-SPM function

candidate k-sequences. The algorithm is given in Fig. 2.12. The main difference is in the candidate generation, candidate-gen-SPM(), which is given in Fig. 2.13. We use an example to illustrate the function.

Example 21: Table 2.4 shows F_3, and C_4 after the join and prune steps. In the join step, the sequence $\langle\{1, 2\}\{4\}\rangle$ joins with $\langle\{2\}\{4, 5\}\rangle$ to produce $\langle\{1, 2\}\{4, 5\}\rangle$, and joins with $\langle\{2\}\{4\}\{6\}\rangle$ to produce $\langle\{1, 2\}\{4\}\{6\}\rangle$. The other sequences cannot be joined. For instance, $\langle\{1\}\{4, 5\}\rangle$ does not join with any sequence since there is no sequence of the form $\langle\{4, 5\}\{x\}\rangle$ or $\langle\{4, 5, x\}\rangle$. In the prune step, $\langle\{1, 2\}\{4\}\{6\}\rangle$ is removed since $\langle\{1\}\{4\}\{6\}\rangle$ is not in F_3. ■

Table 2.4. Candidate generation: an example

Frequent 3-sequences	Candidate 4-sequences	
	after joining	after pruning
$\langle\{1, 2\} \{4\}\rangle$	$\langle\{1, 2\} \{4, 5\}\rangle$	$\langle\{1, 2\} \{4, 5\}\rangle$
$\langle\{1, 2\} \{5\}\rangle$	$\langle\{1, 2\} \{4\} \{6\}\rangle$	
$\langle\{1\} \{4, 5\}\rangle$		
$\langle\{1, 4\} \{6\}\rangle$		
$\langle\{2\} \{4, 5\}\rangle$		
$\langle\{2\} \{4\} \{6\}\rangle$		

2.7.2   Mining with Multiple Minimum Supports

As in association rule mining, using a single minimum support in sequential pattern mining is also a limitation for many applications because some items appear very frequently in the data, while some others appear rarely.

Example 22: One of the Web mining tasks is the mining of comparative sentences such as *"the picture quality of camera X is better than that of camera Y."* from product reviews, forum postings and blogs (see Chap. 11). Such a sentence usually contains a comparative indicator word such as better in the example. We want to discover linguistic patterns involving a set of given comparative indicators, e.g., better, more, less, ahead, win, superior, etc. Some of these indicators (e.g., more and better) appear very frequently in natural language sentences, while some others (e.g., win and ahead) appear rarely. In order to find patterns that contain such rare indicators, we have to use a very low minsup. However, this causes patterns involving frequent indicators to generate a huge number of spurious patterns. Moreover, we need a way to tell the algorithm that we only want patterns that contain at least one comparative indicator. Using GSP with a single minsup is no longer appropriate. The multiple minimum supports model solves both problems nicely.   ■

We again use the concept of **minimum item supports** (**MIS**). The user is allowed to assign each item a MIS value. By providing different MIS values for different items, the user essentially expresses different support requirements for different sequential patterns. To ease the task of specifying many MIS values by the user, the same strategies as those for mining association rules can also be applied here (see Sect. 2.4.2).

Let MIS(i) be the MIS value of item i. The **minimum support** of a sequential pattern P is the lowest MIS value among the items in the pattern. Let the set of items in P be: $i_1, i_2, \ldots, i_r$. The minimum support for P is:

Algorithm MS-GSP(S, MS)   // MS stores all MIS values
1    $M \leftarrow$ sort(I, MS);   // according to MIS(i)'s stored in MS
2    $L \leftarrow$ init-pass(M, S);   // make the first pass over S
3    $F_1 \leftarrow \{\langle\{l\}\rangle \mid l \in L, l.count/n \geq \text{MIS}(l)\}$;   // n is the size of S
4   **for** ($k = 2$; $F_{k-1} \neq \varnothing$; k++) **do**
5   **if** $k = 2$ **then**
6            $C_k \leftarrow$ level2-candidate-gen-SPM(L)
7   **else** $C_k \leftarrow$ MScandidate-gen-SPM(F_{k-1})
8   **endif**
9   **for** each data sequence $s \in S$ **do**
10   **for** each candidate $c \in C_k$ **do**
11   **if** c is contained in s **then**
12                   $c.count$++
13   **if** c' is contained in s, where c' is c after an occurrence of
                        $c.minMISItem$ is removed from c **then**
14                   $c.rest.count$++   // $c.rest$: c without $c.minMISItem$
15   **endfor**
16   **endfor**
17       $F_k \leftarrow \{c \in C_k \mid c.count/n \geq \text{MIS}(c.minMISItem)\}$
18   **endfor**
19   **return** $F \leftarrow \bigcup_k F_k$;

Fig. 2.14. The MS-GSP algorithm

$$\text{minsup}(P) = min(\text{MIS}(i_1), \text{MIS}(i_2), \ldots, \text{MIS}(i_r)).$$

The new algorithm, called **MS-GSP**, is given in Fig. 2.14. It generalizes the GSP algorithm in Fig. 2.12. Like GSP, MS-GSP is also based on level-wise search. Line 1 sorts the items in ascending order according to their MIS values stored in MS. Line 2 makes the first pass over the sequence data using the function init-pass(), which performs the same function as that in MS-Apriori to produce the seeds set L for generating the set of candidate sequences of length 2, i.e., C_2. Frequent 1-sequences (F_1) are obtained from L (line 3).

For each subsequent pass, the algorithm works similarly to MS-Apriori. The function level2-candidate-gen-SPM() can be designed based on level2-candidate-gen in MS-Apriori and the join step in Fig. 2.13. MScandidate-gen-SPM() is, however, complex, which we will discuss shortly.

In line 13, $c.minMISItem$ gives the item that has the lowest MIS value in the candidate sequence c. Unlike that in MS-Apriori, where the first item in each itemset has the lowest MIS value, in sequential pattern mining the item with the lowest MIS value may appear anywhere in a sequence. Similar to those in MS-Apriori, lines 13 and 14 are used to ensure that all sequential rules can be generated after MS-GSP without scanning the original data. Note that in traditional sequential pattern mining, sequential rules are not defined. We will define several types in Sect. 2.9.

Let us now discuss MScandidate-gen-SPM(). In MS-Apriori, the ordering of items is not important and thus we put the item with the lowest MIS value in each itemset as the first item of the itemset, which simplifies the join step. However, for sequential pattern mining, we cannot artificially put the item with the lowest MIS value as the first item in a sequence because the ordering of items is significant. This causes problems for joining.

Example 23: Assume we have a sequence $s_1 = \langle\{1, 2\}\{4\}\rangle$ in F_3, from which we want to generate candidate sequences for the next level. Suppose that item 1 has the lowest MIS value in s_1. We use the candidate generation function in Fig. 2.13. Assume also that the sequence $s_2 = \langle\{2\}\{4, 5\}\rangle$ is not in F_3 because its minimum support is not satisfied. Then we will not generate the candidate $\langle\{1, 2\}\{4, 5\}\rangle$. However, $\langle\{1, 2\}\{4, 5\}\rangle$ can be frequent because items 2, 4, and 5 may have higher MIS values than item 1.   ∎

To deal with this problem, let us make an observation. The problem only occurs when the first item in the sequence s_1 or the last item in the sequence s_2 is the only item with the lowest MIS value, i.e., no other item in s_1 (or s_2) has the same lowest MIS value. If the item (say x) with the lowest MIS value is not the first item in s_1, then s_2 must contain x, and the candidate generation function in Fig. 2.13 will still be applicable. The same reasoning goes for the last item of s_2. Thus, we only need special treatment for these two cases.

Let us see how to deal with the first case, i.e., the first item is the only item with the lowest MIS value. We use an example to develop the idea. Assume we have the frequent 3-sequence of $s_1 = \langle\{1, 2\}\{4\}\rangle$. Based on the algorithm in Fig. 2.13, s_1 may be extended to generate two possible candidates using $\langle\{2\}\{4\}\{x\}\rangle$ and $\langle\{2\}\{4, x\}\rangle$

$$c_1 = \langle\{1, 2\}\{4\}\{x\}\rangle \quad \text{and} \quad c_2 = \langle\{1, 2\}\{4, x\}\rangle,$$

where x is an item. However, $\langle\{2\}\{4\}\{x\}\rangle$ and $\langle\{2\}\{4, x\}\rangle$ may not be frequent because items 2, 4, and x may have higher MIS values than item 1, but we still need to generate c_1 and c_2 because they can be frequent. A different join strategy is thus needed.

We observe that for c_1 to be frequent, the subsequence $s_2 = \langle\{1\}\{4\}\{x\}\rangle$ must be frequent. Then, we can use s_1 and s_2 to generate c_1. c_2 can be generated in a similar manner with $s_2 = \langle\{1\}\{4, x\}\rangle$. s_2 is basically the subsequence of c_1 (or c_2) without the second item. Here we assume that the MIS value of x is higher than item 1. Otherwise, it falls into the second case.

Let us see the same problem for the case where the last item has the only lowest MIS value. Again, we use an example to illustrate. Assume we have the frequent 3-sequence $s_2 = \langle\{3, 5\}\{1\}\rangle$. It can be extended to produce two possible candidates based on the algorithm in Fig. 2.13,

Function MScandidate-gen-SPM(F_{k-1})

1   **Join Step**. Candidate sequences are generated by joining F_{k-1} with F_{k-1}.

2   **if** the MIS value of the first item in a sequence (denoted by s_1) is less than ($<$) the MIS value of every other item in s_1 **then**   // s_1 and s_2 can be equal

Sequence s_1 joins with s_2 if (1) the subsequences obtained by dropping the second item of s_1 and the last item of s_2 are the same, and (2) the MIS value of the last item of s_2 is greater than that of the first item of s_1. Candidate sequences are generated by extending s_1 with the last item of s_2:

- **if** the last item l in s_2 is a separate element **then**
    $\{l\}$ is appended at the end of s_1 as a separate element to form a candidate sequence c_1.
    if (the length and the size of s_1 are both 2) AND (the last item of s_2 is greater than the last item of s_1) **then** // maintain lexicographic order
    l is added at the end of the last element of s_1 to form another candidate sequence c_2.
- **else if** ((the length of s_1 is 2 and the size of s_1 is 1) AND (the last item of s_2 is greater than the last item of s_1)) OR (the length of s_1 is greater than 2) **then**
    the last item in s_2 is added at the end of the last element of s_1 to form the candidate sequence c_2.

3   **elseif** the MIS value of the last item in a sequence (denoted by s_2) is less than ($<$) the MIS value of every other item in s_2 **then**
    A similar method to the one above can be used in the reverse order.

4   **else**   use the **Join Step** in Fig. 2.13

5   **Prune step**: A candidate sequence is pruned if any one of its ($k-1$)-subsequences is infrequent (without minimum support) except the subsequence that does not contain the item with strictly the lowest MIS value.

Fig. 2.15. The MScandidate-gen-SPM function

$c_1 = \langle\{x\}\{3, 5\}\{1\}\rangle$, and $c_2 = \langle\{x, 3, 5\}\{1\}\rangle$.

For c_1 to be frequent, the subsequence $s_1 = \langle\{x\}\{3\}\{1\}\rangle$ has to be frequent (we assume that the MIS value of x is higher than that of item 1). Thus, we can use s_1 and s_2 to generate c_1. c_2 can be generated with $s_1 = \langle\{x, 3\}\{1\}\rangle$. s_1 is basically the subsequence of c_1 (or c_2) without the second last item.

The MScandidate-gen-SPM() function is given in Fig. 2.15, which is self-explanatory. Some special treatments are needed for 2-sequences because the same s_1 (or s_2) may generate two candidate sequences. We use two examples to show the working of the function.

Example 24: Consider the items 1, 2, 3, 4, 5, and 6 with their MIS values,

MIS(1) = 0.03	MIS(2) = 0.05	MIS(3) = 0.03
MIS(4) = 0.07	MIS(5) = 0.08	MIS(6) = 0.09.

The data set has 100 sequences. The following frequent 3-sequences are in F_3 with their actual support counts attached after ":":

(a). $\langle\{1\}\{4\}\{5\}\rangle$:4   (b). $\langle\{1\}\{4\}\{6\}\rangle$:5   (c). $\langle\{1\}\{5\}\{6\}\rangle$:6

(d). $\langle\{1\}\{5, 6\}\rangle$:5   (e). $\langle\{1\}\{6\}\{3\}\rangle$:4   (f). $\langle\{6\}\{3\}\{6\}\rangle$:9

(g). $\langle\{5, 6\}\{3\}\rangle$:5   (h). $\langle\{5\}\{4\}\{3\}\rangle$:4   (i). $\langle\{4\}\{5\}\{3\}\rangle$:7.

For sequence (a) (= s_1), item 1 has the lowest MIS value. It cannot join with sequence (b) because condition (1) in Fig. 2.15 is not satisfied. However, (a) can join with (c) to produce the candidate sequence, $\langle\{1\}\{4\}\{5\}\{6\}\rangle$. (a) can also join with (d) to produce $\langle\{1\}\{4\}\{5, 6\}\rangle$. (b) can join with (e) to produce $\langle\{1\}\{4\}\{6\}\{3\}\rangle$, which is pruned subsequently because $\langle\{1\}\{4\}\{3\}\rangle$ is infrequent. (d) and (e) can be joined to give $\langle\{1\}\{5, 6\}\{3\}\rangle$, but it is pruned because $\langle\{1\}\{5\}\{3\}\rangle$ does not exist. (e) can join with (f) to produce $\langle\{1\}\{6\}\{3\}\{6\}\rangle$ which is done in line 4 because both item 1 and item 3 in (e) have the same MIS value. However, it is pruned because $\langle\{1\}\{3\}\{6\}\rangle$ is infrequent. We do not join (d) and (g), although they can be joined based on the algorithm in Fig. 2.13, because the first item of (d) has the lowest MIS value and we use a different join method for such sequences.

Now we look at 3-sequences whose last item has strictly the lowest MIS value. (i) (= s_1) can join with (h) (= s_2) to produce $\langle\{4\}\{5\}\{4\}\{3\}\rangle$. However, it is pruned because $\langle\{4\}\{4\}\{3\}\rangle$ is not in F_3. ∎

Example 25: Now we consider generating candidates from frequent 2-sequences, which is special as we noted earlier. We use the same items and MIS values in Example 24. The following frequent 2-sequences are in F_2 with their actual support counts attached after ":":

(a). $\langle\{1\}\{5\}\rangle$:6   (b). $\langle\{1\}\{6\}\rangle$:7   (c) $\langle\{5\}\{4\}\rangle$:8

(d). $\langle\{1, 5\}\rangle$:6   (e). $\langle\{1, 6\}\rangle$:6.

(a) can join with (b) to produce both $\langle\{1\}\{5\}\{6\}\rangle$ and $\langle\{1\}\{5, 6\}\rangle$. (b) can join with (d) to produce $\langle\{1, 5\}\{6\}\rangle$. (e) can join with (a) to produce $\langle\{1, 6\}\{5\}\rangle$. Clearly, there are other joins. Again, (a) will not join with (c). ∎

Note that the **support difference constraint** in Sect. 2.4.1 can also be included. We omitted it to simplify the algorithm as it is already complex. Also, the user can instruct the algorithm to generate only certain sequential patterns or not to generate others by setting the MIS values suitably.

2.8 Mining Sequential Patterns Based on PrefixSpan

We now introduce another sequential pattern mining algorithm, called PrefixSpan [439], which does not generate candidates. Different from the GSP algorithm [500], which can be regarded as performing breadth-first search to find all sequential patterns, PrefixSpan performs depth-first search.

2.8.1 PrefixSpan Algorithm

It is easy to introduce the original PrefixSpan algorithm using an example.

Example 26: Consider again mining sequential patterns from Table 2.2 with minsup = 25%. PrefixSpan first sorts all items in each element (or itemset) as shown in the table. Then, by one scan of the sequence database, it finds all frequent items, i.e., 30, 40, 70, 80 and 90. The corresponding length one sequential patterns are ⟨{30}⟩, ⟨{40}⟩, ⟨{70}⟩, ⟨{80}⟩ and ⟨{90}⟩.

We notice that the complete set of sequential patterns can actually be divided into five mutually exclusive subsets: the subset with prefix ⟨{30}⟩, the subset with prefix ⟨{40}⟩, the subset with prefix ⟨{70}⟩, the subset with prefix ⟨{80}⟩, and the subset with prefix ⟨{90}⟩. We only need to find the five subsets one by one.

To find sequential patterns having prefix ⟨{30}⟩, the algorithm extends the prefix by adding items to it one at a time. To add the next item x, there are two possibilities, i.e., x joining the last itemset of the prefix (i.e., ⟨{30, x}⟩) and x forming a separate itemset (i.e., ⟨{30}{x}⟩). PrefixSpan performs the task by first forming the ⟨{30}⟩-projected database and then finding all the cases of the two types in the projected database. The projected database is produced as follows: If a sequence contains item 30, then the suffix following the first 30 is extracted as a sequence in the projected database. Furthermore, since infrequent items cannot appear in a sequential pattern, all infrequent items are removed from the projection. The first sequence in our example, ⟨{30}{90}⟩, is projected to ⟨{90}⟩. The second sequence, ⟨{10, 20}{30}{10, 40, 60, 70}⟩, is projected to ⟨{40, 70}⟩, where the infrequent items 10 and 60 are removed. The third sequence ⟨{30, 50, 70, 80}⟩ is projected to ⟨{_, 70, 80}⟩, where the infrequent item 50 is removed. Note that the underline symbol "_" in this projection denotes that the items (only 30 in this case) in the last itemset of the prefix are in the same itemset as items 50, 70 and 80 in the sequence. The fourth sequence is projected to ⟨{30, 40, 70, 80}{90}⟩. The projection of the last sequence is empty since it does not contain item 30. The final projected database for prefix ⟨{30}⟩ contains the following sequences:

⟨{90}⟩, ⟨{40, 70}⟩, ⟨{_, 70, 80}⟩, and ⟨{30, 40, 70, 80}{90}⟩

By scanning the projected database once, PrefixSpan finds all possible one item extensions to the prefix, i.e., all x's for ⟨{30, x}⟩ and all x's for ⟨{30}{x}⟩. Let us discuss the details.

Find All Frequent Patterns of the Form ⟨{30, x}⟩: Two templates {_, x} and {30, x} are used to match each projected sequence to accumulate the support count for each possible x (here x matches any item). If in the same sequence multiple matches are found with the same x, they are only counted once. Note that in general, the second template should use the last itemset in the prefix rather than only its last item. In our example, they are the same because there is only one item in the last itemset of the prefix.

Find All Frequent Patterns of the Form ⟨{30}{x}⟩: In this case, x's are frequent items in the projected database that are not in the same itemset as the last item of the prefix.

Let us continue with our example. It is easy to check that both items 70 and 80 are in the same itemset as 30. That is, we have two frequent sequences ⟨{30, 70}⟩ and ⟨{30, 80}⟩. The support count of ⟨{30, 70}⟩ is 2 based on the projected database; one from the projected sequence ⟨{_, **70**, 80}⟩ (a {_, x} match) and one from the projected sequence ⟨{**30**, 40, **70**, 80}{90}⟩ (a {30, x} match). In both cases, the x's are the same, i.e., 70. Similarly, the support count of ⟨{30, 80}⟩ is 2 as well and thus frequent.

It is also easy to check that items 40, 70, and 90 are also frequent but not in the same itemset as 30. Thus, ⟨{30}{40}⟩, ⟨{30}{70}⟩, and ⟨{30}{90}⟩ are three sequential patterns. The set of sequential patterns having prefix ⟨{30}⟩ can be further divided into five mutually exclusive subsets: the ones with prefixes ⟨{30, 70}⟩, ⟨{30, 80}⟩, ⟨{30}{40}⟩, ⟨{30}{70}⟩, and ⟨{30}{90}⟩.

We can recursively find the five subsets by forming their corresponding projected databases. For example, to find sequential patterns having prefix ⟨{30}{40}⟩, we can form the ⟨{30}{40}⟩-projected database containing projections ⟨{_, 70}⟩ and ⟨{_, 70, 80}{90}⟩. Template ⟨{_, x}⟩ has two matches and in both cases x is 70. Thus, ⟨{30}{40, 70}⟩ is output as a sequential pattern. Since there is no other frequent item in this projected database, the prefix cannot grow longer. The depth-first search returns from this branch.

After completing the mining of the ⟨{30}⟩-projected database, we find all sequential patterns with prefix ⟨{30}⟩, i.e., ⟨{30}⟩, ⟨{30}{40}⟩, ⟨{30}{40, 70}⟩, ⟨{30}{70}⟩, ⟨{30}{90}⟩, ⟨{30, 70}⟩, ⟨{30, 80}⟩ and ⟨{30, 70, 80}⟩

By forming and mining the ⟨{40}⟩-, ⟨{70}⟩-, ⟨{80}⟩- and ⟨{90}⟩-projected databases, the remaining sequential patterns can be found. ∎

The pseudo code of PrefixSpan can be found in [439]. Comparing to the breadth-first search of GSP, the key advantage of PrefixSpan is that it does not generate any candidates. It only counts the frequency of local items. With a low minimum support, a huge number of candidates can be generated by GSP, which can cause memory and computational problems.

2.8.2 Mining with Multiple Minimum Supports

The PrefixSpan algorithm can be adapted to mine with multiple minimum supports. Again, let MIS(i) be the user-specified **minimum item support** of item i. Let φ be the user-specified support difference threshold in the **support difference constraint** (Sect. 2.4.1), i.e., $|sup(i) - sup(j)| \leq \varphi$, where i and j are items in the same sequential pattern, and $sup(x)$ is the actual support of item x in the sequence database S. PrefixSpan can be modified as follows. We call the modified algorithm **MS-PS**.

1. Find every item i whose actual support in the sequence database S is at least MIS(i). i is called a frequent item.
2. Sort all the discovered frequent items in ascending order according to their MIS values. Let $i_1, ..., i_u$ be the frequent items in the sorted order.
3. For each item i_k in the above sorted order,
   (i) identify all the data sequences in S that contain i_k and at the same time remove every item j in each sequence that does not satisfy $|sup(j) - sup(i_k)| \leq \varphi$. The resulting set of sequences is denoted by S_k. Note that we are not using i_k as the prefix to project the database S.
   (ii) call the function r-PrefixSpan(i_k, S_k, count(MIS(i_k))) (restricted *PrefixSpan*), which finds all sequential patterns that contain i_k, i.e., no pattern that does not contain i_k should be generated. r-PrefixSpan() uses count(MIS(i_k)) (the minimum support count in terms of the number of sequences) as the only minimum support for mining in S_k. The sequence count is easier to use than the MIS value in percentage, but they are equivalent. Once the complete set of such patterns is found from S_k, All occurrences of i_k are removed from S.

r-PrefixSpan() is almost the same as PrefixSpan with one important difference. During each recursive call, either the prefix or every sequence in the projected database must contain i_k because, as we stated above, this function finds only those frequent sequences that contain i_k. Another minor difference is that the support difference constraint needs to be checked during each projection as $sup(i_k)$ may not be the lowest in the pattern.

Example 27: Consider mining sequential patterns from Table 2.5. Let MIS(20) = 30% (3 sequences in minimum support count), MIS(30) = 20% (2 sequences), MIS(40) = 30% (3 sequences), and the MIS values for the rest of the items be 15% (2 sequences). We ignore the support difference constraint as it is simple. In step 1, we find three frequent items, 20, 30 and 40. After sorting in step 2, we have (30, 20, 40). We then go to step 3.

In the first iteration of step 3, we work on $i_1 = 30$. Step 3(i) gives us the second, fourth and sixth sequences in Table 2.5, i.e.,

Table 2.5. An example of a sequence database

Sequence ID	Data Sequence
1	⟨{20, 50}⟩
2	⟨{40}{30}{40, 60}⟩
3	⟨{40, 90, 120}⟩
4	⟨{30}{20, 40}{40, 100}⟩
5	⟨{20, 40}{10}⟩
6	⟨{40}{30}{110}⟩
7	⟨{20}{80}{70}⟩

$S_1 = \{\langle\{40\}\{30\}\{40, 60\}\rangle, \langle\{30\}\{20, 40\}\{40, 100\}\rangle, \langle\{40\}\{30\}\{110\}\rangle\}$.

We then run r-PrefixSpan(30, S_1, 2) in step 3(ii). The frequent items in S_1 are 30, and 40. They both have the support of 3 sequences. The length one frequent sequence is only ⟨{30}⟩. ⟨{40}⟩ is not included because we require that every frequent sequence must contain 30. We next find frequent sequences having prefix ⟨{30}⟩. The database S_1 is projected to give ⟨{40}⟩ and ⟨{40}{40}⟩. 20, 60 and 100 have been removed because their supports in S_1 are less than the required support for item 30 (i.e., 2 sequences). For the same reason, the projection of ⟨{40}{30}{110}⟩ is empty. Thus, we find a length two frequent sequence ⟨{30}{40}⟩. In this case, there is no item in the same itemset as 30 to form a frequent sequence of the form ⟨{30, x}⟩.

Next, we find frequent sequences with prefix ⟨{40}⟩. We again project S_1, which gives us only ⟨{30}{40}⟩ and ⟨{30}⟩. ⟨{40, 100}⟩ is not included because it does not contain 30. This projection gives us another length two frequent sequence ⟨{40}{30}⟩. The first iteration of step 3 ends.

In the second iteration of step 3, we work on $i_2 = 20$. Step 3(i) gives us the first, fourth, fifth and seventh sequences in Table 2.5 with item 30 removed, $S_2 = \{\langle\{20, 50\}\rangle, \langle\{20, 40\}\{40, 100\}\rangle, \langle\{20, 40\}\{10\}\rangle, \langle\{20\}\{80\}\{70\}\rangle\}$. It is easy to see that only item 20 is frequent, and thus only a length one frequent sequence is generated, ⟨{20}⟩.

In the third iteration of step 3, we work on $i_3 = 40$. We can verify that again only one frequent sequence, i.e., ⟨{40}⟩, is found.

The final set of sequential patterns generated from the sequence database in Table 2.5 is $\{\langle\{30\}\rangle, \langle\{20\}\rangle, \langle\{40\}\rangle, \langle\{40\}\{30\}\rangle, \langle\{30\}\{40\}\rangle\}$.   ∎

2.9   Generating Rules from Sequential Patterns

In classic sequential pattern mining, no rules are generated. It is, however, possible to define and generate many types of rules. This section introduces only three types, **sequential rules**, **label sequential rules** and **class sequential rules**, which have been used in Web usage mining and Web content mining (see Chaps. 11 and 12).

2.9.1   Sequential Rules

A **sequential rule (SR)** is an implication of the form, $X \to Y$, where Y is a sequence and X is a **proper subsequence** of Y, i.e., X is a subsequence of Y and the length Y is greater than the length of X. The **support** of a sequential rule, $X \to Y$, in a sequence database S is the fraction of sequences in S that contain Y. The **confidence** of a sequential rule, $X \to Y$, in S is the proportion of sequences in S that contain X also contain Y.

Given a minimum support and a minimum confidence, according to the downward closure property, all the rules can be generated from frequent sequences without going to the original sequence data. Let us see an example of a sequential rule found from the data sequences in Table 2.6.

Table 2.6. An example of a sequence database for mining sequential rules

	Data Sequence
1	$\langle\{1\}\{3\}\{5\}\{7, 8, 9\}\rangle$
2	$\langle\{1\}\{3\}\{6\}\{7, 8\}\rangle$
3	$\langle\{1, 6\}\{7\}\rangle$
4	$\langle\{1\}\{3\}\{5, 6\}\rangle$
5	$\langle\{1\}\{3\}\{4\}\rangle$

Example 28: Given the sequence database in Table 2.6, the minimum support of 30% and the minimum confidence of 60%, one of the sequential rules found is the following,

$\langle\{1\}\{7\}\rangle \to \langle\{1\}\{3\}\{7, 8\}\rangle$ [sup = 2/5, conf = 2/3]

Data sequences 1, 2 and 3 contain $\langle\{1\}\{7\}\rangle$, and data sequences 1 and 2 contain $\langle\{1\}\{3\}\{7, 8\}\rangle$.   ■

If multiple minimum supports are used, we can employ the results of multiple minimum support pattern mining to generate all the rules.

2.9.2   Label Sequential Rules

Sequential rules may not be restrictive enough in some applications. We introduce a special kind of sequential rules called **label sequential rules**. A label sequential rule (LSR) is of the form, $X \to Y$, where Y is a sequence and X is a sequence produced from Y by replacing some of its items with wildcards. A wildcard is denoted by an "*" which matches any item. These replaced items are usually very important and are called **labels**. The labels are a small subset of all the items in the data.

Example 29: Given the sequence database in Table 2.6, the minimum support of 30% and the minimum confidence of 60%, one of the label sequential rules found is the following,

$$\langle\{1\}\{*\}\{7, *\}\rangle \to \langle\{1\}\{3\}\{7, 8\}\rangle \quad [\text{sup} = 2/5, \text{conf} = 2/2].$$

Notice the confidence change compared to the rule in Example 28. The supports of the two rules are the same. In this case, data sequences 1 and 2 contain $\langle\{1\}\{*\}\{7, *\}\rangle$, and they also contain $\langle\{1\}\{3\}\{7, 8\}\rangle$. Items 3 and 8 are labels. ∎

LSRs are useful because in some applications we need to predict the labels in an input sequence, e.g., items 3 and 8 above. The confidence of the rule simply gives us the estimated probability that the two "*"s are 3 and 8 given that an input sequence contains $\langle\{1\}\{*\}\{7, *\}\rangle$. We will see an application of LSRs in Chap. 11, where we want to predict whether a word in a comparative sentence is an entity (e.g., a product name), which is a label.

Note that due to the use of wildcards, frequent sequences alone are not sufficient for computing rule confidences. Scanning the data is needed. Notice also that the same pattern may appear in a data sequence multiple times. Rule confidences thus can be defined in different ways according to application needs. The wildcards may also be restricted to match only certain types of items to make the label prediction meaningful and unambiguous (see some examples in Chap. 11).

2.9.3   Class Sequential Rules

Class sequential rules (CSR) are analogous to class association rules (CAR). Let S be a set of data sequences. Each sequence is also labeled with a class y. Let I be the set of all items in S, and Y be the set of all class labels, $I \cap Y = \varnothing$. Thus, the input data D for mining is represented with $\{(s_1, y_1), (s_2, y_2), \ldots, (s_n, y_n)\}$, where s_i is a sequence in S and $y_i \in Y$ is its class. A **class sequential rule (CSR)** is of the form

$$X \to y, \text{ where } X \text{ is a sequence, and } y \in Y.$$

A data instance (s_i, y_i) is said to **cover** a CSR, $X \to y$, if X is a subsequence of s_i. A data instance (s_i, y_i) is said to **satisfy** a CSR if X is a subsequence of s_i and $y_i = y$.

Example 30: Table 2.7 gives an example of a sequence database with five data sequences and two classes, c_1 and c_2. Using the minimum support of 30% and the minimum confidence of 60%, one of the discovered CSRs is:

$$\langle\{1\}\{3\}\{7, 8\}\rangle \rightarrow c_1 \quad [\text{sup} = 2/5, \text{conf} = 2/3].$$

Data sequences 1 and 2 satisfy the rule, and data sequences 1, 2 and 5 cover the rule.   ∎

Table 2.7. An example of a sequence database for mining CSRs

	Data Sequence	Class
1	$\langle\{1\}\{3\}\{5\}\{7, 8, 9\}\rangle$	c_1
2	$\langle\{1\}\{3\}\{6\}\{7, 8\}\rangle$	c_1
3	$\langle\{1, 6\}\{9\}\rangle$	c_2
4	$\langle\{3\}\{5, 6\}\rangle$	c_2
5	$\langle\{1\}\{3\}\{4\}\{7, 8\}\rangle$	c_2

As in class association rule mining, we can modify the GSP and Prefix-Span algorithms to produce algorithms for mining all CSRs. Similarly, we can also use multiple minimum class supports and/or multiple minimum item supports as in class association rule mining.

Bibliographic Notes

Association rule mining was introduced in 1993 by Agrawal et al. [9]. Since then, numerous research papers have been published on the topic. This short chapter only introduces some basics, and it, by no means, does justice to the huge body of work in the area. The bibliographic notes here should help you explore further.

Since given a data set, a minimum support and a minimum confidence, the solution (the set of frequent itemsets or the set of rules) is unique, most papers improve the mining efficiency. The most well-known algorithm is the Apriori algorithm proposed by Agrawal and Srikant [11], which has been studied in this chapter. Another important algorithm is the **FP-growth** algorithm proposed by Han et al. [220]. The algorithm compresses the data and stores it in memory using a frequent pattern tree. It then mines all frequent itemsets without candidate generation. Other notable algorithms include those by Agarwal et al. [2], Mannila et al. [361], Park et al. [435], Zaki et al. [589], etc. An efficiency comparison of various algorithms was reported by Zheng et al. [616].

Apart from performance improvements, several variations of the original model were also proposed. Srikant and Agrawal [499], and Han and Fu

[217] proposed two algorithms for mining **generalized association rules** or **multi-level association rules**. Liu et al. [344] extended the original model to take **multiple minimum supports**, which was also studied by Wang et al. [534], Seno and Karypis [482], Xiong et al. [562], etc. Srikant et al. [502] proposed to mine association rules with **item constraints**. The model restricts the rules that should be generated. Ng et al. [408] generalized the idea, which was followed by many subsequent papers on the topic of **constrained rule mining**.

It is well known that association rule mining often generates a huge number of frequent itemsets and rules. Bayardo [42], and Lin and Kedem [334] introduced the problem of mining **maximal frequent itemsets**, which are itemsets with no frequent supersets. Improved algorithms are reported in many papers [e.g., 2, 73]. Since maximal pattern mining only finds longest patterns, the support information of their subsets, which are obviously also frequent, is not found. As a result, association rules cannot be generated. The next significant development was the mining of **closed frequent itemsets** studied by Pasquier et al. [436], Zaki and Hsiao [588], and Wang et al. [529]. Closed itemsets are better than maximal frequent itemsets because closed frequent itemsets provide a lossless concise representation of all frequent itemsets.

Other developments on association rules include **cyclic association rules** proposed by Ozden et al. [420], **periodic patterns** by Yang et al. [571], **negative association rules** by Savasere [476] and Wu et al. [560], **weighted association rules** by Wang et al. [539], **association rules with numerical variables** by Webb [541], **class association rules** by Liu et al. [343], **high-performance rule mining** by Buehrer et al. [72] and many others. Recently, Cong et al. [112, 113] introduced association rule mining from bioinformatics data, which typically have a very large number of attributes (more than ten thousands) but only a very small number of records or transactions (less than 100).

Another major research area on association rules is the **interestingness** of discovered rules. Since an association rule miner often generates a huge number of rules, it is very difficult, if not impossible, for human users to inspect them in order to find those truly interesting rules. Researchers have proposed many techniques to help users identify such rules easily [e.g., 43, 283, 342, 345, 346, 352, 421, 492, 511, 522, 535, 565]. A deployed data mining system that uses some of the techniques is reported in [352].

Regarding sequential pattern mining, the first algorithm was proposed by Agrawal and Srikant [12], which was a direct application of the Apriori algorithm. Improvements were made subsequently by several researchers, e.g., Ayres et al. [29], Pei et al. [439], Srikant and Agrawal [500], Zaki [586], etc. The MS-GSP and MS-PS algorithms for mining sequential patterns with multiple minimum supports and the support difference constraint are introduced in this book. Label and class sequential rules have been used in [255, 256] for mining comparative sentences from text documents.

There are several publicly available implementations of algorithms for mining frequent itemsets, maximal frequent itemsets, closed frequent itemsets, and sequential patterns from various research groups, most notably from those of Jiawei Han, Johnanne Gehrke, and Mohammed Zaki. There were also two workshops dedicated to frequent itemset mining organized by Roberto Bayardo, Bart Goethals, and Mohammed J. Zaki, which reported many efficient implementations. The workshop Web sites are http://fimi.cs.helsinki.fi/fimi03/ and http://fimi.cs.helsinki.fi/fimi04/.

3 Supervised Learning

Supervised learning has been a great success in real-world applications. It is used in almost every domain, including text and Web domains. Supervised learning is also called **classification** or **inductive learning** in machine learning. This type of learning is analogous to human learning from past experiences to gain new knowledge in order to improve our ability to perform real-world tasks. However, since computers do not have "experiences", machine learning learns from data, which are collected in the past and represent past experiences in some real-world applications.

There are several types of supervised learning tasks. In this chapter, we focus on one particular type, namely, learning a target function that can be used to predict the values of a discrete class attribute. This type of learning has been the focus of the machine learning research and is perhaps also the most widely used learning paradigm in practice. This chapter introduces a number of such supervised learning techniques. They are used in almost every Web mining application. We will see their uses from Chaps. 6–12.

3.1 Basic Concepts

A data set used in the learning task consists of a set of data records, which are described by a set of attributes $A = \{A_1, A_2, ..., A_{|A|}\}$, where $|A|$ denotes the number of attributes or the size of the set A. The data set also has a special target attribute C, which is called the **class** attribute. In our subsequent discussions, we consider C separately from attributes in A due to its special status, i.e., we assume that C is not in A. The class attribute C has a set of discrete values, i.e., $C = \{c_1, c_2, ..., c_{|C|}\}$, where $|C|$ is the number of classes and $|C| \geq 2$. A class value is also called a **class label**. A data set for learning is simply a relational table. Each data record describes a piece of "past experience". In the machine learning and data mining literature, a data record is also called an **example**, an **instance**, a **case** or a **vector**. A data set basically consists of a set of examples or instances.

Given a data set D, the objective of learning is to produce a **classification/prediction function** to relate values of attributes in A and classes in C. The function can be used to predict the class values/labels of the future data. The function is also called a **classification model**, a **predictive model** or simply a **classifier**. We will use these terms interchangeably in this book. It should be noted that the function/model can be in any form, e.g., a decision tree, a set of rules, a Bayesian model or a hyperplane.

Example 1: Table 3.1 shows a small loan application data set. It has four attributes. The first attribute is Age, which has three possible values, young, middle and old. The second attribute is Has_Job, which indicates whether an applicant has a job. Its possible values are true (has a job) and false (does not have a job). The third attribute is Own_house, which shows whether an applicant owns a house. The fourth attribute is Credit_rating, which has three possible values, fair, good and excellent. The last column is the Class attribute, which shows whether each loan application was approved (denoted by Yes) or not (denoted by No) in the past.

Table 3.1. A loan application data set

ID	Age	Has_job	Own_house	Credit_rating	Class
1	young	false	false	fair	No
2	young	false	false	good	No
3	young	true	false	good	Yes
4	young	true	true	fair	Yes
5	young	false	false	fair	No
6	middle	false	false	fair	No
7	middle	false	false	good	No
8	middle	true	true	good	Yes
9	middle	false	true	excellent	Yes
10	middle	false	true	excellent	Yes
11	old	false	true	excellent	Yes
12	old	false	true	good	Yes
13	old	true	false	good	Yes
14	old	true	false	excellent	Yes
15	old	false	false	fair	No

We want to learn a classification model from this data set that can be used to classify future loan applications. That is, when a new customer comes into the bank to apply for a loan, after inputting his/her age, whether he/she has a job, whether he/she owns a house, and his/her credit rating, the classification model should predict whether his/her loan application should be approved.   ∎

Our learning task is called **supervised learning** because the class labels (e.g., Yes and No values of the class attribute in Table 3.1) are provided in the data. It is as if some teacher tells us the classes. This is in contrast to the **unsupervised learning**, where the classes are not known and the learning algorithm needs to automatically generate classes. Unsupervised learning is the topic of the next chapter.

The data set used for learning is called the **training data** (or **the training set**). After a **model** is learned or built from the training data by a **learning algorithm**, it is evaluated using a set of **test data** (or **unseen data)** to assess the model accuracy.

It is important to note that the test data is not used in learning the classification model. The examples in the test data usually also have class labels. That is why the test data can be used to assess the accuracy of the learned model because we can check whether the class predicted for each test case by the model is the same as the actual class of the test case. In order to learn and also to test, the available data (which has classes) for learning is usually split into two disjoint subsets, the training set (for learning) and the test set (for testing). We will discuss this further in Sect. 3.3.

The accuracy of a classification model on a test set is defined as:

$$Accuracy = \frac{\text{Number of correct classifications}}{\text{Total number of test cases}}, \tag{1}$$

where a correct classification means that the learned model predicts the same class as the original class of the test case. There are also other measures that can be used. We will discuss them in Sect. 3.3.

We pause here to raises two important questions:

1. What do we mean by learning by a computer system?
2. What is the relationship between the training and the test data?

We answer the first question first. Given a data set D representing past "experiences", a task T and a performance measure M, a computer system is said to **learn** from the data to perform the task T if after learning the system's performance on the task T improves as measured by M. In other words, the learned model or knowledge helps the system to perform the task better as compared to no learning. Learning is the process of building the model or extracting the knowledge.

We use the data set in Example 1 to explain the idea. The task is to predict whether a loan application should be approved. The performance measure M is the accuracy in Equation (1). With the data set in Table 3.1, if there is no learning, all we can do is to guess randomly or to simply take the majority class (which is the Yes class). Suppose we use the majority class and announce that every future instance or case belongs to the class Yes. If the future data are drawn from the same distribution as the existing training data in Table 3.1, the estimated classification/prediction accuracy on the future data is 9/15 = 0.6 as there are 9 Yes class examples out of the total of 15 examples in Table 3.1. The question is: can we do better with learning? If the learned model can indeed improve the accuracy, then the learning is said to be effective.

The second question in fact touches the **fundamental assumption of machine learning**, especially the theoretical study of machine learning. The assumption is that the distribution of training examples is identical to the distribution of test examples (including future unseen examples). In practical applications, this assumption is often violated to a certain degree. Strong violations will clearly result in poor classification accuracy, which is quite intuitive because if the test data behave very differently from the training data then the learned model will not perform well on the test data. To achieve good accuracy on the test data, training examples must be sufficiently representative of the test data.

We now illustrate the steps of learning in Fig. 3.1 based on the preceding discussions. In step 1, a learning algorithm uses the training data to generate a classification model. This step is also called the **training step** or **training phase**. In step 2, the learned model is tested using the test set to obtain the classification accuracy. This step is called the **testing step** or **testing phase**. If the accuracy of the learned model on the test data is satisfactory, the model can be used in real-world tasks to predict classes of new cases (which do not have classes). If the accuracy is not satisfactory, we need to go back and choose a different learning algorithm and/or do some further processing of the data (this step is called **data pre-processing**, not shown in the figure). A practical learning task typically involves many iterations of these steps before a satisfactory model is built. It is also possible that we are unable to build a satisfactory model due to a high degree of randomness in the data or limitations of current learning algorithms.

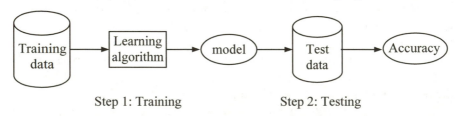

Step 1: Training   Step 2: Testing

Fig. 3.1. The basic learning process: training and testing

From the next section onward, we study several supervised learning algorithms, except Sect. 3.3, which focuses on model/classifier evaluation.

We note that throughout the chapter we assume that the training and test data are available for learning. However, in many text and Web page related learning tasks, this is not true. Usually, we need to collect raw data, design attributes and compute attribute values from the raw data. The reason is that the raw data in text and Web applications are often not suitable for learning either because their formats are not right or because there are no obvious attributes in the raw text documents or Web pages.

3.2   Decision Tree Induction

Decision tree learning is one of the most widely used techniques for classification. Its classification accuracy is competitive with other learning methods, and it is very efficient. The learned classification model is represented as a tree, called a **decision tree**. The techniques presented in this section are based on the C4.5 system from Quinlan [453].

Example 2: Figure 3.2 shows a possible decision tree learnt from the data in Table 3.1. The tree has two types of nodes, **decision nodes** (which are internal nodes) and **leaf nodes**. A decision node specifies some test (i.e., asks a question) on a single attribute. A leaf node indicates a class.

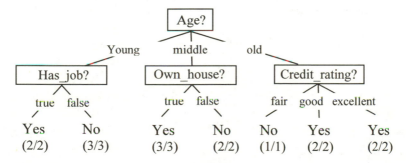

Fig. 3.2. A decision tree for the data in Table 3.1

The root node of the decision tree in Fig. 3.2 is Age, which basically asks the question: what is the age of the applicant? It has three possible answers or **outcomes**, which are the three possible values of Age. These three values form three tree branches/edges. The other internal nodes have the same meaning. Each leaf node gives a class value (Yes or No). (x/y) below each class means that x out of y training examples that reach this leaf node have the class of the leaf. For instance, the class of the left most leaf node is Yes. Two training examples (examples 3 and 4 in Table 3.1) reach here and both of them are of class Yes. ■

To use the decision tree in **testing**, we traverse the tree top-down according to the attribute values of the given test instance until we reach a leaf node. The class of the leaf is the predicted class of the test instance.

Example 3: We use the tree to predict the class of the following new instance, which describes a new loan applicant.

Age	Has_job	Own_house	Credit-rating	Class
young	false	false	good	?

Going through the decision tree, we find that the predicted class is No as we reach the second leaf node from the left. ∎

A decision tree is constructed by partitioning the training data so that the resulting subsets are as pure as possible. A **pure subset** is one that contains only training examples of a single class. If we apply all the training data in Table 3.1 on the tree in Fig. 3.2, we will see that the training examples reaching each leaf node form a subset of examples that have the same class as the class of the leaf. In fact, we can see that from the x and y values in (x/y). We will discuss the decision tree building algorithm in Sect. 3.2.1.

An interesting question is: Is the tree in Fig. 3.2 unique for the data in Table 3.1? The answer is no. In fact, there are many possible trees that can be learned from the data. For example, Fig. 3.3 gives another decision tree, which is much smaller and is also able to partition the training data perfectly according to their classes.

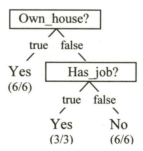

Fig. 3.3. A smaller tree for the data set in Table 3.1

In practice, one wants to have a small and accurate tree for many reasons. A smaller tree is more general and also tends to be more accurate (we will discuss this later). It is also easier to understand by human users. In many applications, the user understanding of the classifier is important. For example, in some medical applications, doctors want to understand the model that classifies whether a person has a particular disease. It is not satisfactory to simply produce a classification because without understanding why the decision is made the doctor may not trust the system and/or does not gain useful knowledge.

It is useful to note that in both Fig. 3.2 and Fig. 3.3, the training examples that reach each leaf node all have the same class (see the values of

(x/y at each leaf node). However, for most real-life data sets, this is usually not the case. That is, the examples that reach a particular leaf node are not of the same class, i.e., $x \leq y$. The value of x/y is, in fact, the **confidence** (conf) value used in association rule mining, and x is the **support count**. This suggests that a decision tree can be converted to a set of if-then rules.

Yes, indeed. The conversion is done as follows: Each path from the root to a leaf forms a rule. All the decision nodes along the path form the conditions of the rule and the leaf node or the class forms the consequent. For each rule, a support and confidence can be attached. Note that in most classification systems, these two values are not provided. We add them here to see the connection of association rules and decision trees.

Example 4: The tree in Fig. 3.3 generates three rules. "," means "and".

Own_house = true → Class =Yes   [sup=6/15, conf=6/6]
Own_house = false, Has_job = true → Class = Yes [sup=3/15, conf=3/3]
Own_house = false, Has_job = false → Class = No [sup=6/15, conf=6/6].

We can see that these rules are of the same format as association rules. However, the rules above are only a small subset of the rules that can be found in the data of Table 3.1. For instance, the decision tree in Fig. 3.3 does not find the following rule:

Age = young, Has_job = false → Class = No   [sup=3/15, conf=3/3].

Thus, we say that a decision tree only finds a subset of rules that exist in data, which is sufficient for classification. The objective of association rule mining is to find all rules subject to some minimum support and minimum confidence constraints. Thus, the two methods have different objectives. We will discuss these issues again in Sect. 3.5 when we show that association rules can be used for classification as well, which is obvious.

An interesting and important property of a decision tree and its resulting set of rules is that the tree paths or the rules are **mutually exclusive** and **exhaustive**. This means that every data instance is **covered** by a single rule (a tree path) and a single rule only. By **covering** a data instance, we mean that the instance satisfies the conditions of the rule.

We also say that a decision tree **generalizes** the data as a tree is a smaller (more compact) description of the data, i.e., it captures the key regularities in the data. Then, the problem becomes building the best tree that is small and accurate. It turns out that finding the best tree that models the data is a NP-complete problem [248]. All existing algorithms use heuristic methods for tree building. Below, we study one of the most successful techniques.

. **Algorithm** decisionTree(D, A, T)
1   **if** D contains only training examples of the same class $c_j \in C$ **then**
2   make T a leaf node labeled with class c_j;
3   **elseif** $A = \varnothing$ **then**
4   make T a leaf node labeled with c_j, which is the most frequent class in D
5   **else**   // D contains examples belonging to a mixture of classes. We select a single
6   // attribute to partition D into subsets so that each subset is purer
7   p_0 = impurityEval-1(D);
8   **for** each attribute $A_i \in A$ (=$\{A_1, A_2, ..., A_k\}$) **do**
9   p_i = impurityEval-2(A_i, D)
10   **endfor**
11   Select $A_g \in \{A_1, A_2, ..., A_k\}$ that gives the biggest impurity reduction,
                computed using $p_0 - p_i$;
12   **if** $p_0 - p_g < threshold$ **then**   // A_g does not significantly reduce impurity p_0
13   make T a leaf node labeled with c_j, the most frequent class in D.
14   **else**   // A_g is able to reduce impurity p_0
15   Make T a decision node on A_g;
16   Let the possible values of A_g be v_1, v_2, ..., v_m. Partition D into m
                    disjoint subsets $D_1, D_2, ..., D_m$ based on the m values of A_g.
17   **for** each D_j in $\{D_1, D_2, ..., D_m\}$ **do**
18   **if** $D_j \neq \varnothing$ **then**
19   create a branch (edge) node T_j for v_j as a child node of T;
20   decisionTree(D_j, $A-\{A_g\}$, T_j)   // A_g is removed
21   **endif**
22   **endfor**
23   **endif**
24   **endif**

Fig. 3.4. A decision tree learning algorithm

3.2.1   Learning Algorithm

As indicated earlier, a decision tree T simply partitions the training data set D into disjoint subsets so that each subset is as pure as possible (of the same class). The learning of a tree is typically done using the **divide-and-conquer** strategy that recursively partitions the data to produce the tree. At the beginning, all the examples are at the root. As the tree grows, the examples are sub-divided recursively. A decision tree learning algorithm is given in Fig. 3.4. For now, we assume that every attribute in D takes discrete values. This assumption is not necessary as we will see later.

The **stopping criteria** of the recursion are in lines 1–4 in Fig. 3.4. The algorithm stops when all the training examples in the current data are of the same class, or when every attribute has been used along the current tree path. In tree learning, each successive recursion chooses the **best attribute** to partition the data at the current node according to the values of the attribute. The best attribute is selected based on a function that aims to minimize the impurity after the partitioning (lines 7–11). In other words, it maximizes the purity. The key in decision tree learning is thus the choice of the **impurity function**, which is used in lines 7, 9 and 11 in Fig. 3.4. The recursive recall of the algorithm is in line 20, which takes the subset of training examples at the node for further partitioning to extend the tree.

This is a greedy algorithm with no backtracking. Once a node is created, it will not be revised or revisited no matter what happens subsequently.

3.2.2 Impurity Function

Before presenting the impurity function, we use an example to show what the impurity function aims to do intuitively.

Example 5: Figure 3.5 shows two possible root nodes for the data in Table 3.1.

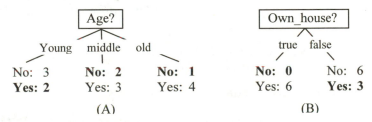

Fig. 3.5. Two possible root nodes or two possible attributes for the root node

Fig. 3.5(A) uses Age as the root node, and Fig. 3.5(B) uses Own_house as the root node. Their possible values (or outcomes) are the branches. At each branch, we listed the number of training examples of each class (No or Yes) that land or reach there. Fig. 3.5(B) is obviously a better choice for the root. From a prediction or classification point of view, Fig. 3.5(B) makes fewer mistakes than Fig. 3.5(A). In Fig. 3.5(B), when Own_house = true every example has the class Yes. When Own_house = false, if we take majority class (the most frequent class), which is No, we make three mistakes/errors. If we look at Fig. 3.5(A), the situation is worse. If we take the majority class for each branch, we make five mistakes (marked in bold). Thus, we say that the impurity of the tree in Fig. 3.5(A) is higher than the tree in Fig. 3.5(B). To learn a decision tree, we prefer Own_house to Age to be the root node. Instead of counting the number of mistakes or errors, C4.5 uses a more principled approach to perform this evaluation on every attribute in order to choose the best attribute to build the tree. ∎

The most popular impurity functions used for decision tree learning are **information gain** and **information gain ratio**, which are used in C4.5 as two options. Let us first discuss information gain, which can be extended slightly to produce information gain ratio.

The information gain measure is based on the **entropy** function from **information theory** [484]:

$$entropy(D) = -\sum_{j=1}^{|C|} \Pr(c_j) \log_2 \Pr(c_j) \tag{2}$$

$$\sum_{j=1}^{|C|} \Pr(c_j) = 1,$$

where $\Pr(c_j)$ is the probability of class c_j in data set D, which is the number of examples of class c_j in D divided by the total number of examples in D. In the entropy computation, we define $0\log0 = 0$. The unit of entropy is **bit**. Let us use an example to get a feeling of what this function does.

Example 6: Assume we have a data set D with only two classes, positive and negative. Let us see the entropy values for three different compositions of positive and negative examples:

1. The data set D has 50% positive examples ($\Pr(positive) = 0.5$) and 50% negative examples ($\Pr(negative) = 0.5$).

$$entropy(D) = -0.5 \times \log_2 0.5 - 0.5 \times \log_2 0.5 = 1.$$

2. The data set D has 20% positive examples ($\Pr(positive) = 0.2$) and 80% negative examples ($\Pr(negative) = 0.8$).

$$entropy(D) = -0.2 \times \log_2 0.2 - 0.8 \times \log_2 0.8 = 0.722.$$

3. The data set D has 100% positive examples ($\Pr(positive) = 1$) and no negative examples, ($\Pr(negative) = 0$).

$$entropy(D) = -1 \times \log_2 1 - 0 \times \log_2 0 = 0.$$

We can see a trend: When the data becomes purer and purer, the entropy value becomes smaller and smaller. In fact, it can be shown that for this binary case (two classes), when $\Pr(positive) = 0.5$ and $\Pr(negative) = 0.5$ the entropy has the maximum value, i.e., 1 bit. When all the data in D belong to one class the entropy has the minimum value, 0 bit. ∎

It is clear that the entropy measures the amount of impurity or disorder in the data. That is exactly what we need in decision tree learning. We now describe the information gain measure, which uses the entropy function.

Information Gain

The idea is the following:

1. Given a data set D, we first use the entropy function (Equation 2) to compute the impurity value of D, which is $entropy(D)$. The **impurityEval-1** function in line 7 of Fig. 3.4 performs this task.
2. Then, we want to know which attribute can reduce the impurity most if it is used to partition D. To find out, every attribute is evaluated (lines 8–10 in Fig. 3.4). Let the number of possible values of the attribute A_i be v. If we are going to use A_i to partition the data D, we will divide D into v disjoint subsets $D_1, D_2, ..., D_v$. The entropy after the partition is

$$entropy_{A_i}(D) = \sum_{j=1}^{v} \frac{|D_j|}{|D|} \times entropy(D_j). \qquad (3)$$

The **impurityEval-2** function in line 9 of Fig. 3.4 performs this task.
3. The information gain of attribute A_i is computed with:

$$gain(D, A_i) = entropy(D) - entropy_{A_i}(D). \qquad (4)$$

Clearly, the gain criterion measures the reduction in impurity or disorder. The *gain* measure is used in line 11 of Fig. 3.4, which chooses attribute A_g resulting in the largest reduction in impurity. If the gain of A_g is too small, the algorithm stops for the branch (line 12). Normally a threshold is used here. If choosing A_g is able to reduce impurity significantly, A_g is employed to partition the data to extend the tree further, and so on (lines 15–21 in Fig. 3.4). The process goes on recursively by building sub-trees using $D_1, D_2, ..., D_m$ (line 20). For subsequent tree extensions, we do not need A_g any more, as all training examples in each branch has the same A_g value.

Example 7: Let us compute the gain values for attributes Age, Own_house and Credit_Rating using the whole data set D in Table 3.1, i.e., we evaluate for the root node of a decision tree.

First, we compute the entropy of D. Since D has 6 **No** class training examples, and 9 **Yes** class training examples, we have

$$entropy(D) = -\frac{6}{15} \times \log_2 \frac{6}{15} - \frac{9}{15} \times \log_2 \frac{9}{15} = 0.971.$$

We then try Age, which partitions the data into 3 subsets (as Age has three possible values) D_1 (with Age=young), D_2 (with Age=middle), and D_3 (with Age=old). Each subset has five training examples. In Fig. 3.5, we also see the number of **No** class examples and the number of **Yes** examples in each subset (or in each branch).

$$entropy_{Age}(D) = -\frac{5}{15} \times entropy(D_1) - \frac{5}{15} \times entropy(D_2) - \frac{5}{15} \times entropy(D_3)$$

$$= \frac{5}{15} \times 0.971 + \frac{5}{15} \times 0.971 + \frac{5}{15} \times 0.722 = 0.888.$$

Likewise, we compute for Own_house, which partitions D into two subsets, D_1 (with Own_house=true) and D_2 (with Own_house=false).

$$entropy_{Own\_house}(D) = -\frac{6}{15} \times entropy(D_1) - \frac{9}{15} \times entropy(D_2)$$

$$= \frac{6}{15} \times 0 + \frac{9}{15} \times 0.918 = 0.551.$$

Similarly, we obtain $entropy_{Has\_job}(D) = 0.647$, and $entropy_{Credit\_rating}(D) = 0.608$. The gains for the attributes are:

$gain(D, \text{Age}) = 0.971 - 0.888 = 0.083$
$gain(D, \text{Own\_house}) = 0.971 - 0.551 = 0.420$
$gain(D, \text{Has\_job}) = 0.971 - 0.647 = 0.324$
$gain(D, \text{Credit\_rating}) = 0.971 - 0.608 = 0.363.$

Own_house is the best attribute for the root node. Figure 3.5(B) shows the root node using Own_house. Since the left branch has only one class (Yes) of data, it results in a leaf node (line 1 in Fig. 3.4). For Own_house = false, further extension is needed. The process is the same as above, but we only use the subset of the data with Own_house = false, i.e., D_2. ∎

Information Gain Ratio

The gain criterion tends to favor attributes with many possible values. An extreme situation is that the data contain an *ID* attribute that is an identification of each example. If we consider using this *ID* attribute to partition the data, each training example will form a subset and has only one class, which results in $entropy_{ID}(D) = 0$. So the gain by using this attribute is maximal. From a prediction point of review, such a partition is useless.

Gain ratio (Equation 5) remedies this bias by normalizing the gain using the entropy of the data with respect to the values of the attribute. Our previous entropy computations are done with respect to the class attribute:

$$gainRatio(D, A_i) = \frac{gain(D, A_i)}{-\sum_{j=1}^{s} \left(\frac{|D_j|}{|D|} \times \log_2 \frac{|D_j|}{|D|} \right)} \tag{5}$$

where s is the number of possible values of A_i, and D_j is the subset of data that has the jth value of A_i. $|D_j|/|D|$ corresponds to the probability of Equation (2). Using Equation (5), we simply choose the attribute with the highest gainRatio value to extend the tree.

This method works because if A_i has too many values the denominator will be large. For instance, in our above example of the *ID* attribute, the denominator will be $\log_2|D|$. The denominator is called the **split info** in C4.5. One note is that the split info can be 0 or very small. Some heuristic solutions can be devised to deal with it (see [453]).

3.2.3 Handling of Continuous Attributes

It seems that the decision tree algorithm can only handle discrete attributes. In fact, continuous attributes can be dealt with easily as well. In a real life data set, there are often both discrete attributes and continuous attributes. Handling both types in an algorithm is an important advantage.

To apply the decision tree building method, we can divide the value range of attribute A_i into intervals at a particular tree node. Each interval can then be considered a discrete value. Based on the intervals, gain or gainRatio is evaluated in the same way as in the discrete case. Clearly, we can divide A_i into any number of intervals at a tree node. However, two intervals are usually sufficient. This **binary split** is used in C4.5. We need to find a **threshold** value for the division.

Clearly, we should choose the threshold that maximizes the gain (or gainRatio). We need to examine all possible thresholds. This is not a problem because although for a continuous attribute A_i the number of possible values that it can take is infinite, the number of actual values that appear in the data is always finite. Let the set of distinctive values of attribute A_i that occur in the data be $\{v_1, v_2, ..., v_r\}$, which are sorted in ascending order. Clearly, any threshold value lying between v_i and v_{i+1} will have the same effect of dividing the training examples into those whose value of attribute A_i lies in $\{v_1, v_2, ..., v_i\}$ and those whose value lies in $\{v_{i+1}, v_{i+2}, ..., v_r\}$. There are thus only $r-1$ possible splits on A_i, which can all be evaluated.

The threshold value can be the middle point between v_i and v_{i+1}, or just on the "right side" of value v_i, which results in two intervals $A_i \le v_i$ and $A_i > v_i$. This latter approach is used in C4.5. The advantage of this approach is that the values appearing in the tree actually occur in the data. The threshold value that maximizes the gain (gainRatio) value is selected. We can modify the algorithm in Fig. 3.4 (lines 8–11) easily to accommodate this computation so that both discrete and continuous attributes are considered.

A change to line 20 of the algorithm in Fig. 3.4 is also needed. For a continuous attribute, we do not remove attribute A_g because an interval can be further split recursively in subsequent tree extensions. Thus, the same continuous attribute may appear multiple times in a tree path (see Example 9), which does not happen for a discrete attribute.

From a geometric point of view, a decision tree built with only continuous attributes represents a partitioning of the data space. A series of splits from the root node to a leaf node represents a hyper-rectangle. Each side of the hyper-rectangle is an axis-parallel hyperplane.

Example 8: The hyper-rectangular regions in Fig. 3.6(A), which partitions the space, are produced by the decision tree in Fig. 3.6(B). There are two classes in the data, represented by empty circles and filled rectangles.   ■

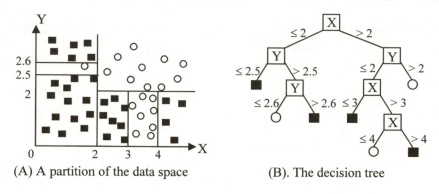

(A) A partition of the data space   (B). The decision tree

Fig. 3.6. A partitioning of the data space and its corresponding decision tree

Handling of continuous (numeric) attributes has an impact on the efficiency of the decision tree algorithm. With only discrete attributes the algorithm grows linearly with the size of the data set D. However, sorting of a continuous attribute takes $|D|\log|D|$ time, which can dominate the tree learning process. Sorting is important as it ensures that gain or gainRatio can be computed in one pass of the data.

3.2.4   Some Other Issues

We now discuss several other issues in decision tree learning.

Tree Pruning and Overfitting: A decision tree algorithm recursively partitions the data until there is no impurity or there is no attribute left. This process may result in trees that are very deep and many tree leaves may cover very few training examples. If we use such a tree to predict the training set, the accuracy will be very high. However, when it is used to classify unseen test set, the accuracy may be very low. The learning is thus not effective, i.e., the decision tree does not **generalize** the data well. This phenomenon is called **overfitting**. More specifically, we say that a classifier f_1 **overfits** the data if there is another classifier f_2 such that f_1 achieves a higher accuracy on the training data than f_2, but a lower accuracy on the unseen test data than f_2 [385].

Overfitting is usually caused by noise in the data, i.e., wrong class values/labels and/or wrong values of attributes, but it may also be due to the complexity and randomness of the application domain. These problems cause the decision tree algorithm to refine the tree by extending it to very deep using many attributes.

To reduce overfitting in the context of decision tree learning, we perform pruning of the tree, i.e., to delete some branches or sub-trees and replace them with leaves of majority classes. There are two main methods to do this, **stopping early** in tree building (which is also called **pre-pruning**) and **pruning** the tree after it is built (which is called **post-pruning**). Post-pruning has been shown more effective. Early-stopping can be dangerous because it is not clear what will happen if the tree is extended further (without stopping). Post-pruning is more effective because after we have extended the tree to the fullest, it becomes clearer which branches/sub-trees may not be useful (overfit the data). The general idea of post-pruning is to estimate the error of each tree node. If the estimated error for a node is less than the estimated error of its extended sub-tree, then the sub-tree is pruned. Most existing tree learning algorithms take this approach. See [453] for a technique called the pessimistic error based pruning.

Example 9: In Fig. 3.6(B), the sub-tree representing the rectangular region

$$X \le 2,\ Y > 2.5,\ Y \le 2.6$$

in Fig. 3.6(A) is very likely to be overfitting. The region is very small and contains only a single data point, which may be an error (or noise) in the data collection. If it is pruned, we obtain Fig. 3.7(A) and (B). ∎

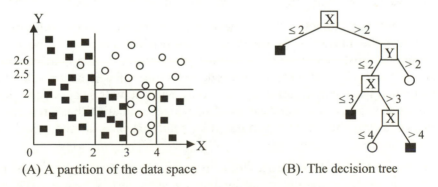

(A) A partition of the data space   (B). The decision tree

Fig. 3.7. The data space partition and the decision tree after pruning

Another common approach to pruning is to use a separate set of data called the **validation set**, which is not used in training and neither in testing. After a tree is built, it is used to classify the validation set. Then, we can find the errors at each node on the validation set. This enables us to know what to prune based on the errors at each node.

Rule Pruning: We noted earlier that a decision tree can be converted to a set of rules. In fact, C4.5 also prunes the rules to simplify them and to reduce overfitting. First, the tree (C4.5 uses the unpruned tree) is converted to a set of rules in the way discussed in Example 4. Rule pruning is then performed by removing some conditions to make the rules shorter and fewer (after pruning some rules may become redundant). In most cases, pruning results in a more accurate rule set as shorter rules are less likely to overfit the training data. Pruning is also called **generalization** as it makes rules more **general** (with fewer conditions). A rule with more conditions is more **specific** than a rule with fewer conditions.

Example 10: The sub-tree below $X \leq 2$ in Fig. 3.6(B) produces these rules:

    Rule 1: $X \leq 2, Y > 2.5, Y > 2.6 \rightarrow \blacksquare$
    Rule 2: $X \leq 2, Y > 2.5, Y \leq 2.6 \rightarrow O$
    Rule 3: $X \leq 2, Y \leq 2.5 \rightarrow \blacksquare$

Note that $Y > 2.5$ in Rule 1 is not useful because of $Y > 2.6$, and thus Rule 1 should be

    Rule 1: $X \leq 2, Y > 2.6 \rightarrow \blacksquare$

In pruning, we may be able to delete the conditions $Y > 2.6$ from Rule 1 to produce:

    $X \leq 2 \rightarrow \blacksquare$

Then Rule 2 and Rule 3 become redundant and can be removed.   ■

A useful point to note is that after pruning the resulting set of rules may no longer be **mutually exclusive** and **exhaustive**. There may be data points that satisfy the conditions of more than one rule, and if inaccurate rules are discarded, of no rules. An ordering of the rules is thus needed to ensure that when classifying a test case only one rule will be applied to determine the class of the test case. To deal with the situation that a test case does not satisfy the conditions of any rule, a **default class** is used, which is usually the majority class.

Handling Missing Attribute Values: In many practical data sets, some attribute values are missing or not available due to various reasons. There are many ways to deal with the problem. For example, we can fill each missing value with the special value "unknown" or the most frequent value of the attribute if the attribute is discrete. If the attribute is continuous, use the mean of the attribute for each missing value.

The decision tree algorithm in C4.5 takes another approach. At a tree node, distribute the training example with missing value for the attribute to each branch of the tree proportionally according to the distribution of the training examples that have values for the attribute.

Handling Skewed Class Distribution: In many applications, the proportions of data for different classes can be very different. For instance, in a data set of intrusion detection in computer networks, the proportion of intrusion cases is extremely small ($< 1\%$) compared with normal cases. Directly applying the decision tree algorithm for classification or prediction of intrusions is usually not effective. The resulting decision tree often consists of a single leaf node "normal", which is useless for intrusion detection. One way to deal with the problem is to over sample the intrusion examples to increase its proportion. Another solution is to rank the new cases according to how likely they may be intrusions. The human users can then investigate the top ranked cases.

3.3   Classifier Evaluation

After a classifier is constructed, it needs to be evaluated for accuracy. Effective evaluation is crucial because without knowing the approximate accuracy of a classifier, it cannot be used in real-world tasks.

There are many ways to evaluate a classifier, and there are also many measures. The main measure is the classification **accuracy** (Equation 1), which is the number of correctly classified instances in the test set divided by the total number of instances in the test set. Some researchers also use the **error rate**, which is $1 - accuracy$. Clearly, if we have several classifiers, the one with the highest accuracy is preferred. Statistical significance tests may be used to check whether one classifier's accuracy is significantly better than that of another given the same training and test data sets. Below, we first present several common methods for classifier evaluation, and then introduce some other evaluation measures.

3.3.1   Evaluation Methods

Holdout Set: The available data D is divided into two disjoint subsets, the **training set** D_{train} and the **test set** D_{test}, $D = D_{train} \cup D_{test}$ and $D_{train} \cap D_{test} =$

∅. The test set is also called the holdout set. This method is mainly used when the data set D is large. Note that the examples in the original data set D are all labeled with classes.

As we discussed earlier, the training set is used for learning a classifier while the test set is used for evaluating the resulting classifier. The training set should not be used to evaluate the classifier as the classifier is biased toward the training set. That is, the classifier may overfit the training set, which results in very high accuracy on the training set but low accuracy on the test set. Using the unseen test set gives an unbiased estimate of the classification accuracy. As for what percentage of the data should be used for training and what percentage for testing, it depends on the data set size. 50–50 and two thirds for training and one third for testing are commonly used.

To partition D into training and test sets, we can use a few approaches:

1. We randomly sample a set of training examples from D for learning and use the rest for testing.
2. If the data is collected over time, then we can use the earlier part of the data for training/learning and the later part of the data for testing. In many applications, this is a more suitable approach because when the classifier is used in the real-world the data are from the future. This approach thus better reflects the dynamic aspects of applications.

Multiple Random Sampling: When the available data set is small, using the above methods can be unreliable because the test set would be too small to be representative. One approach to deal with the problem is to perform the above random sampling n times. Each time a different training set and a different test set are produced. This produces n accuracies. The final estimated accuracy on the data is the average of the n accuracies.

Cross-Validation: When the data set is small, the **n-fold cross-validation** method is very commonly used. In this method, the available data is partitioned into n equal-size disjoint subsets. Each subset is then used as the test set and the remaining $n-1$ subsets are combined as the training set to learn a classifier. This procedure is then run n times, which gives n accuracies. The final estimated accuracy of learning from this data set is the average of the n accuracies. 10-fold and 5-fold cross-validations are often used.

A special case of cross-validation is the **leave-one-out cross-validation**. In this method, each fold of the cross validation has only a single test example and all the rest of the data is used in training. That is, if the original data has m examples, then this is m-fold cross-validation. This method is normally used when the available data is very small. It is not efficient for a large data set as m classifiers need to be built.

In Sect. 3.2.4, we mentioned that a validation set can be used to prune a decision tree or a set of rules. If **a validation set** is employed for that purpose, it should not be used in testing. In that case, the available data is divided into three subsets, a training set, a validation set and a test set. Apart from using a validation set to help tree or rule pruning, a validation set is also used frequently to estimate parameters in learning algorithms. In such cases, the values that give the best accuracy on the validation set are used as the final values of the parameters. Cross-validation can be used for parameter estimating as well. Then a separate validation set is not needed. Instead, the whole training set is used in cross-validation.

3.3.2   Precision, Recall, F-score and Breakeven Point

In some applications, we are only interested in one class. This is particularly true for text and Web applications. For example, we may be interested in only the documents or web pages of a particular topic. Also, in classification involving skewed or highly imbalanced data, e.g., network intrusion and financial fraud detection, we are typically interested in only the minority class. The class that the user is interested in is commonly called the **positive class**, and the rest **negative classes** (the negative classes may be combined into one negative class). Accuracy is not a suitable measure in such cases because we may achieve a very high accuracy, but may not identify a single intrusion. For instance, 99% of the cases are normal in an intrusion detection data set. Then a classifier can achieve 99% accuracy without doing anything by simply classifying every test case as "not intrusion". This is, however, useless.

Precision and **recall** are more suitable in such applications because they measure how precise and how complete the classification is on the positive class. It is convenient to introduce these measures using a **confusion matrix** (Table 3.2). A confusion matrix contains information about actual and predicted results given by a classifier.

Table 3.2. Confusion matrix of a classifier

	Classified positive	Classified negative
Actual positive	TP	FN
Actual negative	FP	TN

where

TP: the number of correct classifications of the positive examples (**true positive**)
FN: the number of incorrect classifications of positive examples (**false negative**)
FP: the number of incorrect classifications of negative examples (**false positive**)
TN: the number of correct classifications of negative examples (**true negative**)

Based on the confusion matrix, the precision (p) and recall (r) of the positive class are defined as follows:

$$p = \frac{TP}{TP + FP}, \qquad r = \frac{TP}{TP + FN}. \tag{6}$$

In words, precision p is the number of correctly classified positive examples divided by the total number of examples that are classified as positive. Recall r is the number of correctly classified positive examples divided by the total number of actual positive examples in the test set. The intuitive meanings of these two measures are quite obvious.

However, it is hard to compare classifiers based on two measures, which are not functionally related. For a test set, the precision may be very high but the recall can be very low, and vice versa.

Example 11: A test data set has 100 positive examples and 1000 negative examples. After classification using a classifier, we have the following confusion matrix (Table 3.3),

Table 3.3. Confusion matrix of a classifier

	Classified positive	Classified negative
Actual positive	1	99
Actual negative	0	1000

This confusion matrix gives the precision $p = 100\%$ and the recall $r = 1\%$ because we only classified one positive example correctly and classified no negative examples wrongly. ∎

Although in theory precision and recall are not related, in practice high precision is achieved almost always at the expense of recall and high recall is achieved at the expense of precision. In an application, which measure is more important depends on the nature of the application. If we need a single measure to compare different classifiers, the **F-score** is often used:

$$F = \frac{2pr}{p + r} \tag{7}$$

The F-score (also called the **F₁-score**) is the harmonic mean of precision and recall.

$$F = \frac{2}{\dfrac{1}{p} + \dfrac{1}{r}} \tag{8}$$

The harmonic mean of two numbers tends to be closer to the smaller of the two. Thus, for the F-score to be high, both p and r must be high.

There is also another measure, called **precision and recall breakeven point,** which is used in the information retrieval community. The breakeven point is when the precision and the recall are equal. This measure assumes that the test cases can be ranked by the classifier based on their likelihoods of being positive. For instance, in decision tree classification, we can use the confidence of each leaf node as the value to rank test cases.

Example 12: We have the following ranking of 20 test documents. 1 represents the highest rank and 20 represents the lowest rank. "+" ("–") represents an actual positive (negative) documents.

1	2	3	4	5	6	7	8	9	10	11	12	13	14	15	16	17	18	19	20
+	+	+	–	+	–	+	–	+	+	–	–	+	–	–	–	+	–	–	+

Assume that the test set has 10 positive examples.

At rank 1:	$p = 1/1 = 100\%$	$r = 1/10 = 10\%$
At rank 2:	$p = 2/2 = 100\%$	$r = 2/10 = 20\%$
...	...	...
At rank 9:	$p = 6/9 = 66.7\%$	$r = 6/10 = 60\%$
At rank 10:	$p = 7/10 = 70\%$	$r = 7/10 = 70\%$

The breakeven point is $p = r = 70\%$. Note that interpolation is needed if such a point cannot be found. ∎

3.4 Rule Induction

In Sect. 3.2, we showed that a decision tree can be converted to a set of rules. Clearly, the set of rules can be used for classification as the tree. A natural question is whether it is possible to learn classification rules directly. The answer is yes. The process of learning such rules is called **rule induction** or **rule learning**. We study two approaches in the section.

3.4.1 Sequential Covering

Most rule induction systems use an algorithm called **sequential covering**. A classifier built with this algorithm consists of a list of rules, which is also called a **decision list** [463]. In the list, the ordering of the rules is significant.

The basic idea of sequential covering is to learn a list of rules sequentially, one at a time, to cover the training data. After each rule is learned, the training examples covered by the rule are removed. Only the remaining data are used to find subsequent rules. Recall that a rule covers an example if the example satisfies the conditions of the rule. We study two specific algorithms based on this general strategy. The first algorithm is based on the CN2 system [104], and the second algorithm is based on the ideas in FOIL [452], I-REP [189], REP [70], and RIPPER [106] systems. Many ideas are also taken from [385].

Algorithm 1 (*Ordered Rules*)

This algorithm learns each rule without pre-fixing a class. That is, in each iteration, a rule of any class may be found. Thus rules of different classes may intermix in the final rule list. The ordering of rules is important.

This algorithm is given in Fig. 3.8. D is the training data. *RuleList* is the list of rules, which is initialized to empty set (line 1). *Rule* is the best rule found in each iteration. The function learn-one-rule-1() learns the *Rule* (lines 2 and 6). The stopping criteria for the while-loop can be of various kinds. Here we use $D = \varnothing$ or *Rule* is NULL (a rule is not learned). Once a rule is learned from the data, it is inserted into *RuleList* at the end (line 4). All the training examples that are covered by the rule are removed from the data (line 5). The remaining data is used to find the next rule and so on. After rule learning ends, a **default class** is inserted at the end of *RuleList*. This is because there may still be some training examples that are not covered by any rule as no good rule can be found from them, or because some test cases may not be covered by any rule and thus cannot be classified. The final list of rules is as follows:

$$<r_1, r_2, ..., r_k, default\text{-}class> \qquad (9)$$

where r_i is a rule.

Algorithm 2 (*Ordered Classes*)

This algorithm learns all rules for each class together. After rule learning for one class is completed, it moves to the next class. Thus all rules for each class appear together in the rule list. The sequence of rules for each class is unimportant, but the rule subsets for different classes are ordered. Typically, the algorithm finds rules for the least frequent class first, then the second least frequent class and so on. This ensures that some rules are learned for rare classes. Otherwise, they may be dominated by frequent classes and end up with no rules if considered after frequent classes.

The algorithm is given in Fig. 3.9. The data set D is split into two subsets, *Pos* and *Neg*, where *Pos* contains all the examples of class c from D,

Algorithm sequential-covering-1(*D*)
1   *RuleList* ← ∅;
2   *Rule* ← learn-one-rule-1(*D*);
3   **while** *Rule* is not NULL AND *D* ≠ ∅ **do**
4   *RuleList* ← insert *Rule* at the end of *RuleList*;
5   Remove from *D* the examples covered by *Rule*;
6   *Rule* ← learn-one-rule-1(*D*)
7   **endwhile**
8   insert a default class *c* at the end of *RuleList*, where *c* is the majority class
         in *D*;
9   **return** *RuleList*

Fig. 3.8. The first rule learning algorithm based on sequential covering

Algorithm sequential-covering-2(*D*, *C*)
1   *RuleList* ← ∅;   // empty rule set at the beginning
2   **for** each class *c* ∈ *C* **do**
3   prepare data (*Pos, Neg*), where *Pos* contains all the examples of class
              c from *D*, and *Neg* contains the rest of the examples in *D*;
4   **while** Pos ≠ ∅ **do**
5   *Rule* ← learn-one-rule-2(*Pos, Neg, c*);
6   **if** *Rule* is NULL **then**
7   **exit-while-loop**
8   **else** *RuleList* ← insert *Rule* at the end of *RuleList*;
9   Remove examples covered by *Rule* from (*Pos, Neg*)
10   **endif**
11   **endwhile**
12   **endfor**
13   **return** *RuleList*

Fig. 3.9. The second rule learning algorithm based on sequential covering

and *Neg* the rest of the examples in *D* (line 3). *c* is the class that the algorithm is working on now. Two stopping conditions for rule learning of
each class are in line 4 and line 6. The other parts of the algorithm are
quite similar to those of the first algorithm in Fig. 3.8. Both learn-one-rule-
1() and learn-one-rule-2() functions are described in Sect. 3.4.2.

Use of Rules for Classification

To use a list of rules for classification is straightforward. For a test case,
we simply try each rule in the list sequentially. The class of the first rule
that covers this test case is assigned as the class of the test case. Clearly, if
no rule applies to the test case, the default class is used.

3.4.2   Rule Learning: Learn-One-Rule Function

We now present the function learn-one-rule(), which works as follows: It starts with an empty set of conditions. In the first iteration, one condition is added. In order to find the best condition to add, all possible conditions are tried, which form **candidate rules**. A **condition** is of the form A_i *op* v, where A_i is an attribute and v is a value of A_i. We also called it an **attribute-value** pair. For a discrete attribute, *op* is "=". For a continuous attribute, $op \in \{>, \leq\}$. The algorithm evaluates all the candidates to find the best one (the rest are discarded). After the first best condition is added, it tries to add the second condition and so on in the same fashion until some stopping condition is satisfied. Note that we omit the rule class here because it is implied, i.e., the majority class of the data covered by the conditions.

This is a heuristic and greedy algorithm in that after a condition is added, it will not be changed or removed through backtracking. Ideally, we would want to try all possible combinations of attributes and values. However, this is not practical as the number of possibilities grows exponentially. Hence, in practice, the above greedy algorithm is used. However, instead of keeping only the best set of conditions, we can improve the function a little by keeping k best sets of conditions ($k > 1$) in each iteration. This is called the **beam search** (k beams), which ensures that a larger space is explored. Below, we present two specific implementations of the algorithm, namely learn-one-rule-1() and learn-one-rule-2(). learn-one-rule-1() is used in the sequential-covering-1 algorithm, and learn-one-rule-2() is used in the sequential-covering-2 algorithm.

Learn-One-Rule-1

This function uses beam search (Fig. 3.10). The number of beams is k. *BestCond* stores the conditions of the rule to be returned. The class is omitted as it is the majority class of the data covered by *BestCond*. *candidate-CondSet* stores the current best condition sets (which are the frontier beams) and its size is less than or equal to k. Each condition set contains a set of conditions connected by "and" (conjunction). *newCandidateCondSet* stores all the new candidate condition sets after adding each attribute-value pair (a possible condition) to every candidate in *candidateCondSet* (lines 5–11). Lines 13–17 update the *BestCond*. Specifically, an evaluation function is used to assess whether each new candidate condition set is better than the existing best condition set *BestCond* (line 14). If so, it replaces the current *BestCond* (line 15). Line 18 updates *candidateCondSet*, which selects k new best condition sets (new beams).

Once the final *BestCond* is found, it is evaluated to see if it is significantly better than without any condition ($\varnothing$) using a *threshold* (line 20). If

Function learn-one-rule-1(D)
1 $BestCond \leftarrow \varnothing$;   // rule with no condition.
2 $candidateCondSet \leftarrow \{BestCond\}$;
3 $attributeValuePairs \leftarrow$ the set of all attribute-value pairs in D of the form
     (A_i op v), where A_i is an attribute and v is a value or an interval;
4 **while** $candidateCondSet \neq \varnothing$ **do**
5      $newCandidateCondSet \leftarrow \varnothing$;
6   **for** each candidate $cond$ in $candidateCondSet$ **do**
7   **for** each attribute-value pair a in $attributeValuePairs$ **do**
8              $newCond \leftarrow cond \cup \{a\}$;
9              $newCandidateCondSet \leftarrow newCandidateCondSet \cup \{newCond\}$
10   **endfor**
11   **endfor**
12   remove duplicates and inconsistencies, e.g., $\{A_i = v_1, A_i = v_2\}$;
13   **for** each candidate $newCond$ in $newCandidateCondSet$ **do**
14   **if** evaluation($newCond, D$) > evaluation($BestCond, D$) **then**
15              $BestCond \leftarrow newCond$
16   **endif**
17   **endfor**
18      $candidateCondSet \leftarrow$ the k best members of $newCandidateCondSet$
              according to the results of the evaluation function;
19 **endwhile**
20 **if** evaluation($BestCond, D$) – evaluation($\varnothing, D$) > $threshold$ **then**
21   **return** the rule: "$BestCond \rightarrow c$" where is c the majority class of the data
              covered by $BestCond$
22 **else return** NULL
23 **endif**

Fig. 3.10. The learn-one-rule-1 function

Function evaluation($BestCond, D$)
1 $D' \leftarrow$ the subset of training examples in D covered by $BestCond$;
2 $entropy\,(D') = -\sum_{j=1}^{|C|} \Pr(c_j)\log_2 \Pr(c_j)$；
3 **return** $- entropy(D')$ // since entropy measures impurity.

Fig. 3.11. The entropy based evaluation function

yes, a rule will be formed using $BestCond$ and the most frequent (or the majority) class of the data covered by $BestCond$ (line 21). If not, NULL is returned to indicate that no significant rule is found.

The evaluation() function (Fig. 3.11) uses the entropy function as in the decision tree learning. Other evaluation functions are possible too. Note that when $BestCond = \varnothing$, it covers every example in D, i.e., $D = D'$.

Function learn-one-rule-2(*Pos, Neg, class*)
1   split (*Pos, Neg*) into (*GrowPos, GrowNeg*) and (*PrunePos, PruneNeg*)
2   *BestRule* ← GrowRule(*GrowPos, GrowNeg, class*)   // grow a new rule
3   *BestRule* ← PruneRule(*BestRule, PrunePos, PruneNeg*) // prune the rule
4   **if** the error rate of *BestRule* on (*PrunePos, PruneNeg*) exceeds 50% **then**
5   **return** NULL
6   **endif**
7   **return** *BestRule*

Fig. 3.12. The learn-one-rule-2() function

Learn-One-Rule-2

In the learn-one-rule-2() function (Fig. 3.12), a rule is first generated and then it is pruned. This method starts by splitting the positive and negative training data *Pos* and *Neg*, into growing and pruning sets. The growing sets, *GrowPos* and *GrowNeg*, are used to generate a rule, called *BestRule*. The pruning sets, *PrunePos* and *PruneNeg* are used to prune the rule because *BestRule* may overfit the data. Note that *PrunePos* and *PruneNeg* are actually validation sets discussed in Sects. 3.2.4 and 3.3.1.

growRule() function: growRule() generates a rule (called *BestRule*) by repeatedly adding a condition to its condition set that maximizes an evaluation function until the rule covers only some positive examples in *GrowPos* but no negative examples in *GrowNeg*. This is basically the same as lines 4–17 in Fig. 3.10, but without beam search (i.e., only the best rule is kept in each iteration). Let the current partially developed rule be *R*:

$$R: \quad av_1, .., av_k \rightarrow class$$

where each av_j is a condition (an attribute-value pair). By adding a new condition av_{k+1}, we obtain the rule R^+: $av_1, .., av_k, av_{k+1} \rightarrow class$. The evaluation function for R^+ is the following **information gain** criterion (which is different from the gain function used in decision tree learning):

$$gain(R, R^+) = p_1 \times \left(\log_2 \frac{p_1}{p_1 + n_1} - \log_2 \frac{p_0}{p_0 + n_0} \right) \tag{10}$$

where p_0 (respectively, n_0) is the number of positive (negative) examples covered by R in *Pos* (*Neg*), and p_1 (n_1) is the number of positive (negative) examples covered by R^+ in *Pos* (*Neg*). The GrowRule() function simply returns the rule R^+ that maximizes the gain.

PruneRule() function: To prune a rule, we consider deleting every subset of conditions from the *BestRule*, and choose the deletion that maximizes:

$$v(BestRule, PrunePos, PruneNeg) = \frac{p-n}{p+n},$$ (11)

where p (respectively n) is the number of examples in *PrunePos* (*PruneNeg*) covered by the current rule (after a deletion).

3.4.3 Discussion

Separate-and-Conquer vs. Divide-and-Conquer: Decision tree learning is said to use the *divide-and-conquer* strategy. At each step, all attributes are evaluated and one is selected to partition/divide the data into m disjoint subsets, where m is the number of values of the attribute. Rule induction discussed in this section is said to use the *separate-and-conquer* strategy, which evaluates all attribute-value pairs (conditions) (which are much larger in number than the number of attributes) and selects only one. Thus, each step of divide-and-conquer expands m rules, while each step of separate-and-conquer expands only one rule. Due to both effects, the separate-and-conquer strategy is much slower than the divide-and-conquer strategy.
Rule Understandability: If-then rules are easy to understand by human users. However, a word of caution about rules generated by sequential covering is in order. Such rules can be misleading because the covered data are removed after each rule is generated. Thus the rules in the rule list are not independent of each other. A rule r may be of high quality in the context of the data D' from which r was generated. However, it may be a weak rule with a very low accuracy (confidence) in the context of the whole data set D ($D' \subseteq D$) because many training examples that can be covered by r have already been removed by rules generated before r. If you want to understand the rules and possibly use them in some real-world tasks, you should be aware of this fact.

3.5 Classification Based on Associations

In Sect. 3.2, we showed that a decision tree can be converted to a set of rules, and in Sect. 3.4, we saw that a set of rules may also be found directly for classification. It is thus only natural to expect that association rules, in particular **class association rules** (**CAR**), may be used for classification too. Yes, indeed! In fact, normal association rules can be employed for classification as well as we will see in Sect. 3.5.3. CBA, which stands for *Classification Based on Associations*, is the first reported system that uses association rules for classification [343]. In this section, we describe three approaches to employing association rules for classification:

1. Using class association rules for classification directly.
2. Using class association rules as features or attributes.
3. Using normal (or classic) association rules for classification.

The first two approaches can be applied to tabular data or transactional data. The last approach is usually employed for transactional data only. All these methods are useful in the Web environment as many types of Web data are in the form of transactions, e.g., search queries issued by users, and Web pages clicked by visitors. Transactional data sets are difficult to handle by traditional classification techniques, but are very natural for association rules. Below, we describe the three approaches in turn. We should note that various sequential rules can be used for classification in similar ways as well if sequential data sets are involved.

3.5.1 Classification Using Class Association Rules

Recall that a class association rule (CAR) is an association rule with only a class label on the right-hand side of the rule as its consequent (Sect. 2.5). For instance, from the data in Table 3.1, the following rule can be found:

Own_house = false, Has_job = true → Class = Yes [sup=3/15, conf=3/3],

which was also a rule from the decision tree in Fig. 3.3. In fact, there is no difference between rules from a decision tree (or a rule induction system) and CARs if we consider only categorical (or discrete) attributes (more on this later). The differences are in the mining processes and the final rule sets. CAR mining finds all rules in data that satisfy the user-specified minimum support (minsup) and minimum confidence (minconf) constraints. A decision tree or a rule induction system finds only a **subset** of the rules (expressed as a tree or a list of rules) for classification.

Example 13: Recall that the decision tree in Fig. 3.3 gives the following three rules:

Own_house = true → Class =Yes   [sup=6/15, conf=6/6]
Own_house = false, Has_job = true → Class=Yes [sup=3/15, conf=3/3]
Own_house = false, Has_job = false → Class=No [sup=6/15, conf=6/6].

However, there are many other rules that exist in data, e.g.,

Age = young, Has_job = true → Class=Yes   [sup=2/15, conf=2/2]
Age = young, Has_job = false → Class=No   [sup=3/15, conf=3/3]
Credit_rating = fair → Class=No   [sup=4/15, conf=4/5]

and many more, if we use minsup = 2/15 = 13.3% and minconf = 70%. ■

In many cases, rules that are not in the decision tree (or a rule list) may be able to perform classification more accurately. Empirical comparisons reported by several researchers show that classification using CARs can perform more accurately on many data sets than decision trees and rule induction systems (see Bibliographic Notes for references).

The complete set of rules from CAR mining is also beneficial from a rule usage point of view. In some applications, the user wants to act on some interesting rules. For example, in an application for finding causes of product problems, more rules are preferred to fewer rules because with more rules, the user is more likely to find rules that indicate causes of the problems. Such rules may not be generated by a decision tree or a rule induction system. A deployed data mining system based on CARs is reported in [352]. We should, however, also bear in mind of the following:

1. Decision tree learning and rule induction do not use the minsup or minconf constraint. Thus, some rules that they find can have very low supports, which, of course, are likely to be pruned because the chance that they overfit the training data is high. Although we can set a low minsup for CAR mining, it may cause combinatorial explosion. In practice, in addition to minsup and minconf, a limit on the total number of rules to be generated may be used to further control the CAR generation process. When the number of generated rules reaches the limit, the algorithm stops. However, with this limit, we may not be able to generate long rules (with many conditions). Recall that the Apriori algorithm works in a level-wise fashion, i.e., short rules are generated before long rules. In some applications, this might not be an issue as short rules are often preferred and are sufficient for classification or for action. Long rules normally have very low supports and tend to overfit the data. However, in some other applications, long rules can be useful.
2. CAR mining does not use continuous (numeric) attributes, while decision trees deal with continuous attributes naturally. Rule induction can use continuous attributes as well. There is still no satisfactory method to deal with such attributes directly in association rule mining. Fortunately, many attribute discretization algorithms exist that can automatically discretize the value range of a continuous attribute into suitable intervals [e.g., 151, 172], which are then considered as discrete values.

Mining Class Association Rules for Classification

There are many techniques that use CARs to build classifiers. Before describing them, let us first discuss some issues related to CAR mining for classification. Since a CAR mining algorithm has been discussed in Sect. 2.5, we will not repeat it here.

Rule Pruning: CAR rules are highly redundant, and many of them are not statistically significant (which can cause overfitting). Rule pruning is thus needed. The idea of pruning CARs is basically the same as that in decision tree building or rule induction. Thus, we will not discuss it further (see [343, 328] for some of the pruning methods).

Multiple Minimum Class Supports: As discussed in Sect. 2.5.3, a single minsup is inadequate for mining CARs because many practical classification data sets have uneven class distributions, i.e., some classes cover a large proportion of the data, while others cover only a very small proportion (which are called **rare** or **infrequent classes**).

Example 14: Suppose we have a dataset with two classes, Y and N. 99% of the data belong to the Y class, and only 1% of the data belong to the N class. If we set minsup = 1.5%, we will not find any rule for class N. To solve the problem, we need to lower down the minsup. Suppose we set minsup = 0.2%. Then, we may find a huge number of overfitting rules for class Y because minsup = 0.2% is too low for class Y. ■

Multiple minimum class supports can be applied to deal with the problem. We can assign a different **minimum class support** $minsup_i$ for each class c_i, i.e., all the rules of class c_i must satisfy $minsup_i$. Alternatively, we can provide one single total minsup, denoted by t_minsup, which is then distributed to each class according to the class distribution:

$$minsup_i = t\_minsup \times sup(c_i) \qquad (12)$$

where $sup(c_i)$ is the support of class c_i in training data. The formula gives frequent classes higher minsups and infrequent classes lower minsups.

Parameter Selection: The parameters used in CAR mining are the minimum supports and the minimum confidences. Note that a different minimum confidence may also be used for each class. However, minimum confidences do not affect the classification much because classifiers tend to use high confidence rules. One minimum confidence is sufficient as long as it is not set too high. To determine the best $minsup_i$ for each class c_i, we can try a range of values to build classifiers and then use a validation set to select the final value. Cross-validation may be used as well.

Data Formats: The algorithm for CAR mining given in Sect. 2.5.2 is for mining transaction data sets. However, many classification data sets are in the table format. As we discussed in Sect. 2.3, a tabular data set can be easily converted to a transaction data set.

Classifier Building

After all CAR rules are found, a classifier is built using the rules. There are many existing methods, which can be grouped into three categories.

Use the Strongest Rule: This is perhaps the simplest strategy. It simply uses CARs directly for classification. For each test instance, it finds the strongest rule that covers the instance. Recall that a rule **covers** an instance if the instance satisfies the conditions of the rule. The class of the strongest rule is then assigned as the class of the test instance. The strength of a rule can be measured in various ways, e.g., based on confidence, χ^2 test, or a combination of both support and confidence values.

Select a Subset of the Rules to Build a Classifier: The representative method of this category is the one used in the CBA system. The method is similar to the sequential covering method, but applied to class association rules with additional enhancements as discussed above.

Let the set of all discovered CARs be S. Let the training data set be D. The basic idea is to select a subset L ($\subseteq S$) of high confidence rules to cover the training data D. The set of selected rules, including a default class, is then used as the classifier. The selection of rules is based on a total order defined on the rules in S.

Definition: Given two rules, r_i and r_j, $r_i \succ r_j$ (also called r_i precedes r_j or r_i has a higher precedence than r_j) if

1. the confidence of r_i is greater than that of r_j, or
2. their confidences are the same, but the support of r_i is greater than that of r_j, or
3. both the confidences and supports of r_i and r_j are the same, but r_i is generated earlier than r_j.

A CBA classifier L is of the form:

$$L = <r_1, r_2, ..., r_k, default\text{-}class>$$

where $r_i \in S$, $r_a \succ r_b$ if $b > a$. In classifying a test case, the first rule that satisfies the case classifies it. If no rule applies to the case, it takes the default class (*default-class*). A simplified version of the algorithm for building such a classifier is given in Fig. 3.13. The classifier is the *RuleList*.

This algorithm can be easily implemented by making one pass through the training data for every rule. However, this is extremely inefficient for large data sets. An efficient algorithm that makes at most two passes over the data is given in [343].

Combine Multiple Rules: Like the first approach, this approach does not take any additional step to build a classifier. At the classification time, for

Algorithm CBA(S, D)

```
1   S = sort(S);                    // sorting is done according to the precedence ≻
2   RuleList = ∅;                   // the rule list classifier
3   for each rule r ∈ S in sequence do
4       if D ≠ ∅ AND r classifies at least one example in D correctly then
5           delete from D all training examples covered by r;
6           add r at the end of RuleList
7       endif
8   endfor
9   add the majority class as the default class at the end of RuleList
```

Fig. 3.13. A simple classifier building algorithm

each test instance, the system first finds the subset of rules that covers the instance. If all the rules in the subset have the same class, the class is assigned to the test instance. If the rules have different classes, the system divides the rules into groups according to their classes, i.e., all rules of the same class are in the same group. The system then compares the aggregated effects of the rule groups and finds the strongest group. The class label of the strongest group is assigned to the test instance. To measure the strength of each rule group, there again can be many possible techniques. For example, the CMAR system uses a weighted χ^2 measure [328].

3.5.2 Class Association Rules as Features

In the above two methods, rules are directly used for classification. In this method, rules are used as features to augment the original data or simply form a new data set, which is then fed to a traditional classification algorithm, e.g., decision trees or the naïve Bayesian method.

To use CARs as features, only the conditional part of each rule is needed, and it is often treated as a Boolean feature/attribute. If a data instance in the original data contains the conditional part, the value of the feature/attribute is set to 1, and otherwise it is set to 0. Several applications of this method have been reported [23, 131, 255, 314]. The reason that this approach is helpful is that CARs capture multi-attribute or multi-item correlations with class labels. Many classification algorithms do not find such correlations (e.g., naïve Bayesian), but they can be quite useful.

3.5.3 Classification Using Normal Association Rules

Not only can class association rules be used for classification, but also normal association rules. For example, association rules are commonly used in e-commerce Web sites for product recommendations, which work as follows: When a customer purchases some products, the system will recommend him/her some other related products based on what he/she has already purchased (see Chap. 12).

Recommendation is essentially a classification or prediction problem. It predicts what a customer is likely to buy. Association rules are naturally applicable to such applications. The classification process is the following:

1. The system first uses previous purchase transactions (the same as market basket transactions) to mine association rules. In this case, there are no fixed classes. Any item can appear on the left-hand side or the right-hand side of a rule. For recommendation purposes, usually only one item appears on the right-hand side of a rule.
2. At the prediction (e.g., recommendation) time, given a transaction (e.g., a set of items already purchased by a customer), all the rules that cover the transaction are selected. The strongest rule is chosen and the item on the right-hand side of the rule (i.e., the consequent) is then the predicted item and recommended to the user. If multiple rules are very strong, multiple items can be recommended.

This method is basically the same as the "**use the strongest rule**" method described in Sect. 3.5.1. Again, the rule strength can be measured in various ways, e.g., confidence, χ^2 test, or a combination of both support and confidence. For example, in [337], the product of support and confidence is used as the rule strength. Clearly, the other two methods discussed in Sect. 3.5.1 can be applied as well.

The key advantage of using association rules for recommendation is that they can predict any item since any item can be the class item on the right-hand side. Traditional classification algorithms only work with a single fixed class attribute, and are not easily applicable to recommendations.

Finally, we note that multiple minimum supports (Sect. 2.4) can be of significant help. Otherwise, **rare items** will never be recommended, which causes the **coverage** problem (see Sect. 12.3.3). It is shown in [389] that using multiple minimum supports can dramatically increase the coverage.

3.6 Naïve Bayesian Classification

Supervised learning can be naturally studied from a probabilistic point of view. The task of classification can be regarded as estimating the class **posterior** probabilities given a test example d, i.e.,

$$\Pr(C = c_j \mid d). \tag{13}$$

We then see which class c_j is more probable. The class with the highest probability is assigned to the example d.

Formally, let A_1, A_2, ..., $A_{|A|}$ be the set of attributes with discrete values in the data set D. Let C be the class attribute with $|C|$ values, c_1, c_2, ..., $c_{|C|}$. Given a test example d with observed attribute values a_1 through $a_{|A|}$, where a_i is a possible value of A_i (or a member of the domain of A_i), i.e.,

$$d = <A_1=a_1, ..., A_{|A|}=a_{|A|}>.$$

The prediction is the class c_j such that $\Pr(C=c_j \mid A_1=a_1, ..., A_{|A|}=a_{|A|})$ is maximal. c_j is called a **maximum *a posteriori*** (MAP) hypothesis.

By Bayes' rule, the above quantity (13) can be expressed as

$$
\begin{aligned}
&\Pr(C = c_j \mid A_1 = a_1,..., A_{|A|} = a_{|A|}) \\
&= \frac{\Pr(A_1 = a_1,..., A_{|A|} = a_{|A|} \mid C = c_j)\Pr(C = c_j)}{\Pr(A_1 = a_1,..., A_{|A|} = a_{|A|})} \\
&= \frac{\Pr(A_1 = a_1,..., A_{|A|} = a_{|A|} \mid C = c_j)\Pr(C = c_j)}{\sum_{k=1}^{|C|}\Pr(A_1 = a_1,..., A_{|A|} = a_{|A|} \mid C = c_k)\Pr(C = c_k)}.
\end{aligned}
\tag{14}
$$

$\Pr(C=c_j)$ is the class **prior** probability of c_j, which can be estimated from the training data. It is simply the fraction of the data in D with class c_j.

If we are only interested in making a classification, $\Pr(A_1=a_1, ..., A_{|A|}=a_{|A|})$ is irrelevant for decision making because it is the same for every class. Thus, only $\Pr(A_1=a_1, ..., A_{|A|}=a_{|A|} \mid C=c_j)$ needs to be computed, which can be written as

$$
\begin{aligned}
&\Pr(A_1=a_1, ..., A_{|A|}=a_{|A|} \mid C=c_j) \\
&= \Pr(A_1=a_1 \mid A_2=a_2, ..., A_{|A|}=a_{|A|}, C=c_j) \times \Pr(A_2=a_2, ..., A_{|A|}=a_{|A|} \mid C=c_j).
\end{aligned}
\tag{15}
$$

Recursively, the second term above (i.e., $\Pr(A_2=a_2, ..., A_{|A|}=a_{|A|} \mid C=c_j)$) can be written in the same way (i.e., $\Pr(A_2=a_2 \mid A_3=a_3 ..., A_{|A|}=a_{|A|}, C=c_j) \times \Pr(A_3=a_3, ..., A_{|A|}=a_{|A|} \mid C=c_j)$), and so on. However, to further our derivation, we need to make an important assumption.

Conditional independence assumption: We assume that all attributes are conditionally independent given the class $C = c_j$. Formally, we assume,

$$\Pr(A_1=a_1 \mid A_2=a_2, ..., A_{|A|}=a_{|A|}, C=c_j) = \Pr(A_1=a_1 \mid C=c_j)
\tag{16}$$

and similarly for A_2 through $A_{|A|}$. We then obtain

$$\Pr(A_1 = a_1,..., A_{|A|} = a_{|A|} \mid C = c_j) = \prod_{i=1}^{|A|}\Pr(A_i = a_i \mid C = c_j)
\tag{17}$$

$$\Pr(C = c_j \mid A_1 = a_1,..., A_{|A|} = a_{|A|})$$

$$= \frac{\Pr(C = c_j)\prod_{i=1}^{|A|}\Pr(A_i = a_i \mid C = c_j)}{\sum_{k=1}^{|C|}\Pr(C = c_k)\prod_{i=1}^{|A|}\Pr(A_i = a_i \mid C = c_k)}. \tag{18}$$

Next, we need to estimate the *prior* probabilities $\Pr(C=c_j)$ and the conditional probabilities $\Pr(A_i=a_i \mid C=c_j)$ from the training data, which are straightforward.

$$\Pr(C = c_j) = \frac{\text{number of examples of class } c_j}{\text{total number of examples in the data set}} \tag{19}$$

$$\Pr(A_i = a_i \mid C = c_j) = \frac{\text{number of examples with } A_i = a_i \text{ and class } c_j}{\text{number of examples of class } c_j}. \tag{20}$$

If we only need a decision on the most probable class for each test instance, we only need the numerator of Equation (18) since the denominator is the same for every class. Thus, given a test case, we compute the following to decide the most probable class for the test case:

$$c = \arg\max_{c_j} \Pr(C = c_j)\prod_{i=1}^{|A|}\Pr(A_i = a_i \mid C = c_j) \tag{21}$$

Example 15: Suppose that we have the training data set in Fig. 3.14, which has two attributes A and B, and the class C. We can compute all the probability values required to learn a naïve Bayesian classifier.

A	B	C
m	b	t
m	s	t
g	q	t
h	s	t
g	q	t
g	q	f
g	s	f
h	b	f
h	q	f
m	b	f

Fig. 3.14. An example of a training data set

$$\Pr(C = t) = 1/2, \qquad \Pr(C=f) = 1/2$$

$\Pr(A=m \mid C=t) = 2/5$	$\Pr(A=g \mid C=t) = 2/5$	$\Pr(A=h \mid C=t) = 1/5$
$\Pr(A=m \mid C=f) = 1/5$	$\Pr(A=g \mid C=f) = 2/5$	$\Pr(A=h \mid C=f) = 2/5$
$\Pr(B=b \mid C=t) = 1/5$	$\Pr(B=s \mid C=t) = 2/5$	$\Pr(B=q \mid C=t) = 2/5$
$\Pr(B=b \mid C=f) = 2/5$	$\Pr(B=s \mid C=f) = 1/5$	$\Pr(B=q \mid C=f) = 2/5$

Now we have a test example:

$$A = m \quad B = q \quad C = ?$$

We want to know its class. Equation (21) is applied. For $C = t$, we have

$$\Pr(C = t)\prod_{j=1}^{2}\Pr(A_j = a_j \mid C = t) = \frac{1}{2}\times\frac{2}{5}\times\frac{2}{5} = \frac{2}{25}.$$

For class $C = f$, we have

$$\Pr(C = f)\prod_{j=1}^{2}\Pr(A_j = a_j \mid C = f) = \frac{1}{2}\times\frac{1}{5}\times\frac{2}{5} = \frac{1}{25}.$$

Since $C = t$ is more probable, t is the predicted class of the test case.   ■

It is easy to see that the probabilities (i.e., $\Pr(C=c_j)$ and $\Pr(A_i=a_i \mid C=c_j)$) required to build a naïve Bayesian classifier can be found in one scan of the data. Thus, the algorithm is linear in the number of training examples, which is one of the great strengths of the naïve Bayes, i.e., it is extremely efficient. In terms of classification accuracy, although the algorithm makes the strong assumption of conditional independence, several researchers have shown that its classification accuracies are surprisingly strong. See experimental comparisons of various techniques in [148, 285, 349].

To learn practical naïve Bayesian classifiers, we still need to address some additional issues: how to handle numeric attributes, zero counts, and missing values. Below, we deal with each of them in turn.

Numeric Attributes: The above formulation of the naïve Bayesian learning assumes that all attributes are categorical. However, most real-life data sets have numeric attributes. Therefore, in order to use the naïve Bayeisan algorithm, each numeric attribute needs to be discretized into intervals. This is the same as for class association rule mining. Existing discretization algorithms in [e.g., 151, 172] can be used.

Zero Counts: It is possible that a particular attribute value in the test set never occurs together with a class in the training set. This is problematic because it will result in a 0 probability, which wipes out all the other probabilities $\Pr(A_i=a_i \mid C=c_j)$ when they are multiplied according to Equation

(21) or Equation (18). A principled solution to this problem is to incorporate a small-sample correction into all probabilities.

Let n_{ij} be the number of examples that have both $A_i = a_i$ and $C = c_j$. Let n_j be the total number of examples with $C=c_j$ in the training data set. The uncorrected estimate of $Pr(A_i=a_i \mid C=c_j)$ is n_{ij}/n_j, and the corrected estimate is

$$Pr(A_i = a_i \mid C = c_j) = \frac{n_{ij} + \lambda}{n_j + \lambda m_i} \tag{22}$$

where m_i is the number of values of attribute A_i (e.g., 2 for a Boolean attribute), and λ is a multiplicative factor, which is commonly set to $\lambda = 1/n$, where n is the total number of examples in the training set D [148, 285]. When $\lambda = 1$, we get the well known **Laplace's law of succession** [204]. The general form of correction (also called **smoothing**) in Equation (22) is called the **Lidstone's law of succession** [330]. Applying the correction $\lambda = 1/n$, the probabilities of Example 15 are revised. For example,

Pr($A=m \mid C=t$) = (2+1/10) / (5 + 3*1/10) = 2.1/5.3 = 0.396
Pr($B=b \mid C=t$) = (1+1/10) / (5 + 3*1/10) = 1.1/5.3 = 0.208.

Missing Values: Missing values are ignored, both in computing the probability estimates in training and in classifying test instances.

3.7 Naïve Bayesian Text Classification

Text classification or categorization is the problem of learning classification models from training documents labeled with pre-defined classes. That learned models are then used to classify future documents. For example, we have a set of news articles of three classes or topics, Sport, Politics, and Science. We want to learn a classifier that is able to classify future news articles into these classes.

Due to the rapid growth of online documents in organizations and on the Web, automated document classification is an important problem. Although the techniques discussed in the previous sections can be applied to text classification, it has been shown that they are not as effective as the methods presented in this section and in the next two sections. In this section, we study a naïve Bayesian learning method that is specifically formulated for texts, which makes use of text specific characteristics. However, the ideas are similar to those in Sect. 3.6. Below, we first present a probabilistic framework for texts, and then study the naïve Bayesian equations for their classification. There are several slight variations of this model. This section is mainly based on the formulation given in [365].

3.7.1 Probabilistic Framework

The naïve Bayesian learning method for text classification is derived based on a probabilistic **generative model**. It assumes that each document is generated by a **parametric distribution** governed by a set of **hidden parameters**. Training data is used to estimate these parameters. The parameters are then applied to classify each test document using Bayes' rule by calculating the **posterior probability** that the distribution associated with a class (represented by the unobserved class variable) would have generated the given document. Classification then becomes a simple matter of selecting the most probable class.

The generative model is based on two assumptions:

1. The data (or the text documents) are generated by a mixture model.
2. There is one-to-one correspondence between mixture components and document classes.

A **mixture model** models the data with a number of statistical distributions. Intuitively, each distribution corresponds to a data cluster and the parameters of the distribution provide a description of the corresponding cluster. Each distribution in a mixture model is also called a **mixture component** (the distribution can be of any kind). Figure 3.15 plots two **probability density functions** of a mixture of two Gaussian distributions that generate a 1-dimensional data set of two classes, one distribution per class, whose parameters (denoted by θ_i) are the mean (μ_i) and the standard deviation (σ_i), i.e., $\theta_i = (\mu_i, \sigma_i)$.

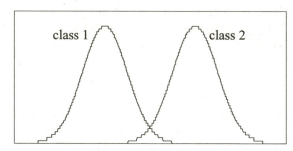

Fig. 3.15. Probability density functions of the two distributions in the mixture model

Let the number of mixture components (or distributions) in a mixture model be K, and the jth distribution have the parameters θ_j. Let Θ be the set of parameters of all components, $\Theta = \{\varphi_1, \varphi_2, ..., \varphi_K, \theta_1, \theta_2, ..., \theta_K\}$, where φ_j is the **mixture weight** (or **mixture probability**) of the mixture component j and θ_j is the set of parameters of component j. The mixture weights are subject to the constraint $\sum_{j=1}^{K} \varphi_j = 1$. The meaning of mixture weights (or probabilities) will be clear below.

Let us see how the mixture model generates a collection of documents. Recall the classes C in our classification problem are $c_1, c_2, \ldots, c_{|C|}$. Since we assume that there is one-to-one correspondence between mixture components and classes, each class corresponds to a mixture component. Thus $|C| = K$, and the jth mixture component can be represented by its corresponding class c_j and is parameterized by θ_j. The mixture weights are **class prior probabilities**, i.e., $\varphi_j = \Pr(c_j|\Theta)$. The mixture model generates each document d_i by:

1. first selecting a mixture component (or class) according to class prior probabilities (i.e., mixture weights), $\varphi_j = \Pr(c_j|\Theta)$;
2. then having this selected mixture component (c_j) generate a document d_i according to its parameters, with distribution $\Pr(d_i|c_j; \Theta)$ or more precisely $\Pr(d_i|c_j; \theta_j)$.

The probability that a document d_i is generated by the mixture model can be written as the sum of total probability over all mixture components. Note that to simplify the notation, we use c_j instead of $C = c_j$ as in the previous section:

$$\Pr(d_i \mid \Theta) = \sum_{j=1}^{|C|} \Pr(c_j \mid \Theta) \Pr(d_i \mid c_j; \Theta). \qquad (23)$$

Since each document is attached with its class label, we can now derive the naïve Bayesian model for text classification. Note that in the above probability expressions, we include Θ to represent their dependency on Θ as we employ a generative model. In an actual implementation, we need not be concerned with Θ, i.e., it can be ignored.

3.7.2   Naïve Bayesian Model

A text document consists of a sequence of sentences, and each sentence consists of a sequence of words. However, due to the complexity of modeling word sequence and their relationships, several assumptions are made in the derivation of the Bayesian classifier. That is also why we call the final classification model, *naïve Bayesian* classification.

Specifically, the naïve Bayesian classification treats each document as a "bag" of words. The generative model makes the following assumptions:

1. Words of a document are generated independently of the context, that is, independently of the other words in the same document given the class label. This is the familiar naïve Bayesian assumption used before.
2. The probability of a word is independent of its position in the document. For example, the probability of seeing the word "student" in the first position of the document is the same as seeing it in any other position. The document length is chosen independent of its class.

With these assumptions, each document can be regarded as generated by a **multinomial distribution**. In other words, each document is drawn from a multinomial distribution of words with as many independent trials as the length of the document. The words are from a given vocabulary $V = \{w_1, w_2, \ldots, w_{|V|}\}$, $|V|$ being the number of words in the vocabulary. To see why this is a multinomial distribution, we give a short introduction to the multinomial distribution.

A **multinomial trial** is a process that can result in any of k outcomes, where $k \geq 2$. Each outcome of a multinomial trial has a probability of occurrence. The probabilities of the k outcomes are denoted by $p_1, p_2, \ldots, p_k$. For example, the rolling of a die is a multinomial trial, with six possible outcomes 1, 2, 3, 4, 5, 6. For a fair die, $p_1 = p_2 = \ldots = p_k = 1/6$.

Now assume n independent trials are conducted, each with the k possible outcomes and the k probabilities, $p_1, p_2, \ldots, p_k$. Let us number the outcomes 1, 2, 3, ..., k. For each outcome, let X_t denote the number of trials that result in that outcome. Then, $X_1, X_2, \ldots, X_k$ are discrete random variables. The collection of $X_1, X_2, \ldots, X_k$ is said to have the **multinomial distribution** with parameters, $n, p_1, p_2, \ldots, p_k$.

In our context, n corresponds to the length of a document, and the outcomes correspond to all the words in the vocabulary V ($k = |V|$). $p_1, p_2, \ldots, p_k$ correspond to the probabilities of occurrence of the words in V in a document, which are $\Pr(w_t|c_j; \Theta)$. X_t is a random variable representing the number of times that word w_t appears in a document. We can thus directly apply the probability function of the multinomial distribution to find the probability of a document given its class (including the probability of document length, $\Pr(|d_i|)$, which is assumed to be independent of class):

$$\Pr(d_i \mid c_j; \Theta) = \Pr(\mid d_i \mid) \mid d_i \mid! \prod_{t=1}^{|V|} \frac{\Pr(w_t \mid c_j; \Theta)^{N_{ti}}}{N_{ti}!} \tag{24}$$

where N_{ti} is the number of times that word w_t occurs in document d_i and

$$\sum_{t=1}^{|V|} N_{ti} = \mid d_i \mid, \text{ and } \sum_{t=1}^{|V|} \Pr(w_t \mid c_j; \Theta) = 1. \tag{25}$$

The parameters θ_j of the generative component for each class c_j are the probabilities of all words w_t in V, written as $\Pr(w_t|c_j; \Theta)$, and the probabilities of document lengths, which are the same for all classes (or mixture components) due to our assumption.

Parameter Estimation: The parameters can be estimated from the training data $D = \{D_1, D_2, ..., D_{|C|}\}$, where D_j is the subset of data for class c_j (recall $|C|$ is the number of classes). The vocabulary V is the set of all distinctive words in D. Note that we do not need to estimate the probability of each document length as it is not used in our final classifier. The estimate of Θ is written as $\hat{\Theta}$. The parameters are estimated based on empirical counts.

The estimated probability of word w_t given class c_j is simply the number of times that w_t occurs in the training data D_j (of class c_j) divided by the total number of word occurrences in the training data for that class:

$$\Pr(w_t \mid c_j; \hat{\Theta}) = \frac{\sum_{i=1}^{|D|} N_{ti} \Pr(c_j \mid d_i)}{\sum_{s=1}^{|V|} \sum_{i=1}^{|D|} N_{si} \Pr(c_j \mid d_i)}. \tag{26}$$

In Equation (26), we do not use D_j explicitly. Instead, we include $\Pr(c_j|d_i)$ to achieve the same effect because $\Pr(c_j|d_i) = 1$ for each document in D_j and $\Pr(c_j|d_i) = 0$ for documents of other classes. Again, N_{ti} is the number of times that word w_t occurs in document d_i.

In order to handle 0 counts for infrequently occurring words that do not appear in the training set, but may appear in the test set, we need to smooth the probability to avoid probabilities of 0 or 1. This is the same problem as in Sect. 3.6. The standard way of doing this is to augment the count of each distinctive word with a small quantity λ ($0 \le \lambda \le 1$) or a fraction of a word in both the numerator and denominator. Thus, any word will have at least a very small probability of occurrence.

$$\Pr(w_t \mid c_j; \hat{\Theta}) = \frac{\lambda + \sum_{i=1}^{|D|} N_{ti} \Pr(c_j \mid d_i)}{\lambda |V| + \sum_{s=1}^{|V|} \sum_{i=1}^{|D|} N_{si} \Pr(c_j \mid d_i)}. \tag{27}$$

This is called the **Lidstone smoothing** (Lidstone's law of succession). When $\lambda = 1$, the smoothing is known as the **Laplace smoothing**. Many experiments have shown that $\lambda < 1$ works better for text classification [7]. The best λ value for a data set can be found through experiments using a validation set or through cross-validation.

Finally, class prior probabilities, which are mixture weights φ_j, can be easily estimated using the training data as well.

$$\Pr(c_j \mid \hat{\Theta}) = \frac{\sum_{i=1}^{|D|} \Pr(c_j \mid d_i)}{|D|}. \tag{28}$$

Classification: Given the estimated parameters, at the classification time, we need to compute the probability of each class c_j for the test document d_i. That is, we compute the probability that a particular mixture component c_j generated the given document d_i. Using Bayes rule and Equations (23), (24), (27), and (28), we have

$$\Pr(c_j \mid d_i; \hat{\Theta}) = \frac{\Pr(c_j \mid \hat{\Theta}) \Pr(d_i \mid c_j; \hat{\Theta})}{\Pr(d_i \mid \hat{\Theta})} \tag{29}$$

$$= \frac{\Pr(c_j \mid \hat{\Theta}) \prod_{k=1}^{|d_i|} \Pr(w_{d_i,k} \mid c_j; \hat{\Theta})}{\sum_{r=1}^{|C|} \Pr(c_r \mid \hat{\Theta}) \prod_{k=1}^{|d_i|} \Pr(w_{d_i,k} \mid c_r; \hat{\Theta})},$$

where $w_{d_i,k}$ is the word in position k of document d_i (which is the same as using w_t and N_{ti}). If the final classifier is to classify each document into a single class, the class with the highest posterior probability is selected:

$$\arg\max_{c_j \in C} \Pr(c_j \mid d_i; \hat{\Theta}). \tag{30}$$

3.7.3 Discussion

Most assumptions made by naïve Bayesian learning are violated in practice. For example, words in a document are clearly not independent of each other. The mixture model assumption of one-to-one correspondence between classes and mixture components may not be true either because a class may contain documents from multiple topics. Despite such violations, researchers have shown that naïve Bayesian learning produces very accurate models.

Naïve Bayesian learning is also very efficient. It scans the training data only once to estimate all the probabilities required for classification. It can be used as an incremental algorithm as well. The model can be updated easily as new data comes in because the probabilities can be conveniently revised. Naïve Bayesian learning is thus widely used for text classification.

The naïve Bayesian formulation presented here is based on a mixture of **multinomial distributions**. There is also a formulation based on **multivariate Bernoulli distributions** in which each word in the vocabulary is a binary feature, i.e., it either appears or does not appear in the document.

Thus, it does not consider the number of times that a word occurs in a document. Experimental comparisons show that multinomial formulation consistently produces more accurate classifiers [365].

3.8 Support Vector Machines

Support vector machines (SVM) is another type of learning system [525], which has many desirable qualities that make it one of most popular algorithms. It not only has a solid theoretical foundation, but also performs classification more accurately than most other algorithms in many applications, especially those applications involving very high dimensional data. For instance, it has been shown by several researchers that SVM is perhaps the most accurate algorithm for text classification. It is also widely used in Web page classification and bioinformatics applications.

In general, SVM is a **linear learning system** that builds two-class classifiers. Let the set of training examples D be

$$\{(\mathbf{x}_1, y_1), (\mathbf{x}_2, y_2), \ldots, (\mathbf{x}_n, y_n)\},$$

where $\mathbf{x}_i = (x_{i1}, x_{i2}, \ldots, x_{ir})$ is a r-dimensional **input vector** in a real-valued space $X \subseteq \mathcal{R}^r$, y_i is its **class label** (output value) and $y_i \in \{1, -1\}$. 1 denotes the positive class and -1 denotes the negative class. Note that we use slightly different notations in this section. For instance, we use y instead of c to represent a class because y is commonly used to represent classes in the SVM literature. Similarly, each data instance is called an **input vector** and denoted by a bold face letter. In the following, we use bold face letters for all vectors.

To build a classifier, SVM finds a linear function of the form

$$f(\mathbf{x}) = \langle \mathbf{w} \cdot \mathbf{x} \rangle + b \tag{31}$$

so that an input vector $\mathbf{x}_i$ is assigned to the positive class if $f(\mathbf{x}_i) \geq 0$, and to the negative class otherwise, i.e.,

$$y_i = \begin{cases} 1 & \text{if} \langle \mathbf{w} \cdot \mathbf{x}_i \rangle + b \geq 0 \\ -1 & \text{if} \langle \mathbf{w} \cdot \mathbf{x}_i \rangle + b < 0 \end{cases} \tag{32}$$

Hence, $f(\mathbf{x})$ is a real-valued function $f: X \subseteq \mathcal{R}^r \rightarrow \mathcal{R}$. $\mathbf{w} = (w_1, w_2, \ldots, w_r) \in \mathcal{R}^r$ is called the **weight vector**. $b \in \mathcal{R}$ is called the **bias**. $\langle \mathbf{w} \cdot \mathbf{x} \rangle$ is the **dot product** of $\mathbf{w}$ and $\mathbf{x}$ (or **Euclidean inner product**). Without using vector notation, Equation (31) can be written as:

$$f(x_1, x_2, \ldots, x_r) = w_1 x_1 + w_2 x_2 + \ldots + w_r x_r + b,$$

where x_i is the variable representing the ith coordinate of the vector **x**. For convenience, we will use the vector notation from now on.

In essence, SVM finds a hyperplane

$$\langle \mathbf{w} \cdot \mathbf{x} \rangle + b = 0 \tag{33}$$

that separates positive and negative training examples. This hyperplane is called the **decision boundary** or **decision surface**.

Geometrically, the hyperplane $\langle \mathbf{w} \cdot \mathbf{x} \rangle + b = 0$ divides the input space into two half spaces: one half for positive examples and the other half for negative examples. Recall that a hyperplane is commonly called **a line** in a 2-dimensional space and **a plane** in a 3-dimensional space.

Fig. 3.16(A) shows an example in a 2-dimensional space. Positive instances (also called positive data points or simply positive points) are represented with small filled rectangles, and negative examples are represented with small empty circles. The thick line in the middle is the decision boundary hyperplane (a line in this case), which separates positive (above the line) and negative (below the line) data points. Equation (31), which is also called the **decision rule** of the SVM classifier, is used to make classification decisions on test instances.

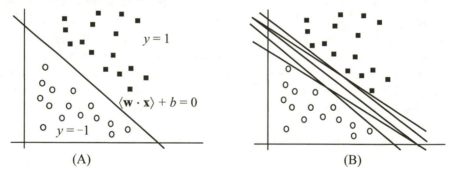

Fig. 3.16. (A) A linearly separable data set and (B) possible decision boundaries

Fig. 3.16(A) raises two interesting questions:

1. There are an infinite number of lines that can separate the positive and negative data points as illustrated by Fig. 3.16(B). Which line should we choose?
2. A hyperplane classifier is only applicable if the positive and negative data can be linearly separated. How can we deal with nonlinear separations or data sets that require nonlinear decision boundaries?

The SVM framework provides good answers to both questions. Briefly, for question 1, SVM chooses the hyperplane that maximizes the margin (the gap) between positive and negative data points, which will be defined formally shortly. For question 2, SVM uses **kernel** functions. Before we dive into the details, we should note that SVM requires numeric data and only builds two-class classifiers. At the end of the section, we will discuss how these limitations may be addressed.

3.8.1 Linear SVM: Separable Case

This sub-section studies the simplest case of linear SVM. It is assumed that the positive and negative data points are linearly separable.

From linear algebra, we know that in $\langle \mathbf{w} \cdot \mathbf{x} \rangle + b = 0$, $\mathbf{w}$ defines a direction perpendicular to the hyperplane (see Fig. 3.17). $\mathbf{w}$ is also called the **normal vector** (or simply **normal**) of the hyperplane. Without changing the normal vector $\mathbf{w}$, varying b moves the hyperplane parallel to itself. Note also that $\langle \mathbf{w} \cdot \mathbf{x} \rangle + b = 0$ has an inherent degree of freedom. We can rescale the hyperplane to $\langle \lambda \mathbf{w} \cdot \mathbf{x} \rangle + \lambda b = 0$ for $\lambda \in \mathfrak{R}^+$ (positive real numbers) without changing the function/hyperplane.

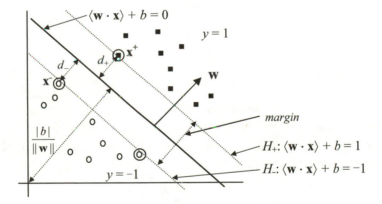

Fig. 3.17. Separating hyperplanes and margin of SVM: Support vectors are circled

Since SVM maximizes the margin between positive and negative data points, let us find the margin. Let d_+ (respectively d_-) be the shortest distance from the separating hyperplane ($\langle \mathbf{w} \cdot \mathbf{x} \rangle + b = 0$) to the closest positive (negative) data point. The **margin** of the separating hyperplane is $d_+ + d_-$. SVM looks for the separating hyperplane with the largest margin, which is also called the **maximal margin hyperplane**, as the final **decision boundary**. The reason for choosing this hyperplane to be the decision boundary is because theoretical results from *structural risk minimization* in computational learning theory show that maximizing the margin minimizes the upper bound of classification errors.

Let us consider a positive data point $(\mathbf{x}^+, 1)$ and a negative $(\mathbf{x}^-, -1)$ that are closest to the hyperplane $\langle \mathbf{w} \cdot \mathbf{x} \rangle + b = 0$. We define two parallel hyperplanes, H_+ and H_-, that pass through $\mathbf{x}^+$ and $\mathbf{x}^-$ respectively. H_+ and H_- are also parallel to $\langle \mathbf{w} \cdot \mathbf{x} \rangle + b = 0$. We can rescale $\mathbf{w}$ and b to obtain

$$H_+: \quad \langle \mathbf{w} \cdot \mathbf{x}^+ \rangle + b = 1 \tag{34}$$

$$H_-: \quad \langle \mathbf{w} \cdot \mathbf{x}^- \rangle + b = -1 \tag{35}$$

such that
$$\langle \mathbf{w} \cdot \mathbf{x}_i \rangle + b \geq 1 \qquad \text{if } y_i = 1$$
$$\langle \mathbf{w} \cdot \mathbf{x}_i \rangle + b \leq -1 \qquad \text{if } y_i = -1,$$

which indicate that no training data fall between hyperplanes H_+ and H_-.

Now let us compute the distance between the two **margin hyperplanes** H_+ and H_-. Their distance is the **margin** $(d_+ + d_-)$. Recall from vector space in linear algebra that the (perpendicular) Euclidean distance from a point $\mathbf{x}_i$ to a hyperplane $\langle \mathbf{w} \cdot \mathbf{x} \rangle + b = 0$ is:

$$\frac{|\langle \mathbf{w} \cdot \mathbf{x}_i \rangle + b|}{\|\mathbf{w}\|}, \tag{36}$$

where $\|\mathbf{w}\|$ is the Euclidean norm of $\mathbf{w}$,

$$\|\mathbf{w}\| = \sqrt{\langle \mathbf{w} \cdot \mathbf{w} \rangle} = \sqrt{w_1^2 + w_2^2 + \ldots + w_r^2} \tag{37}$$

To compute d_+, instead of computing the distance from $\mathbf{x}^+$ to the separating hyperplane $\langle \mathbf{w} \cdot \mathbf{x} \rangle + b = 0$, we pick up any point $\mathbf{x}_s$ on $\langle \mathbf{w} \cdot \mathbf{x} \rangle + b = 0$ and compute the distance from $\mathbf{x}_s$ to $\langle \mathbf{w} \cdot \mathbf{x}^+ \rangle + b = 1$ by applying Equation 36 and noticing that $\langle \mathbf{w} \cdot \mathbf{x}_s \rangle + b = 0$,

$$d_+ = \frac{|\langle \mathbf{w} \cdot \mathbf{x}_s \rangle + b - 1|}{\|\mathbf{w}\|} = \frac{1}{\|\mathbf{w}\|} \tag{38}$$

Likewise, we can compute the distance of $\mathbf{x}_s$ to $\langle \mathbf{w} \cdot \mathbf{x}^+ \rangle + b = -1$ to obtain $d_- = 1/\|\mathbf{w}\|$. Thus, the decision boundary $\langle \mathbf{w} \cdot \mathbf{x} \rangle + b = 0$ lies half way between H_+ and H_-. The margin is thus

$$margin = d_+ + d_- = \frac{2}{\|\mathbf{w}\|} \tag{39}$$

In fact, we can compute the margin in many ways. For example, it can be computed by finding the distances from the origin to the three hyperplanes, or by projecting the vector $(\mathbf{x}_2^- - \mathbf{x}_1^+)$ to the normal vector $\mathbf{w}$.

Since SVM looks for the separating hyperplane that maximizes the margin, this gives us an optimization problem. Since maximizing the margin is the same as minimizing $\|\mathbf{w}\|^2/2 = \langle \mathbf{w} \cdot \mathbf{w} \rangle/2$. We have the following linear separable SVM formulation.

Definition (Linear SVM: Separable Case): Given a set of linearly separable training examples,

$$D = \{(\mathbf{x}_1, y_1), (\mathbf{x}_2, y_2), \ldots, (\mathbf{x}_n, y_n)\},$$

learning is to solve the following constrained minimization problem,

$$\text{Minimize}: \frac{\langle \mathbf{w} \cdot \mathbf{w} \rangle}{2} \tag{40}$$

$$\text{Subject to}: y_i(\langle \mathbf{w} \cdot \mathbf{x}_i \rangle + b) \geq 1, \quad i = 1, 2, \ldots, n$$

Note that the constraint $y_i(\langle \mathbf{w} \cdot \mathbf{x}_i \rangle + b) \geq 1, \quad i = 1, 2, \ldots, n$ summarizes:

$$\langle \mathbf{w} \cdot \mathbf{x}_i \rangle + b \geq 1 \qquad \text{for } y_i = 1$$
$$\langle \mathbf{w} \cdot \mathbf{x}_i \rangle + b \leq -1 \qquad \text{for } y_i = -1.$$

Solving the problem (40) will produce the solutions for $\mathbf{w}$ and b, which in turn give us the maximal margin hyperplane $\langle \mathbf{w} \cdot \mathbf{x} \rangle + b = 0$ with the margin $2/\|\mathbf{w}\|$.

A full description of the solution method requires a significant amount of optimization theory, which is beyond the scope of this book. We will only use those relevant results from optimization without giving formal definitions, theorems or proofs.

Since the objective function is quadratic and convex and the constraints are linear in the parameters $\mathbf{w}$ and b, we can use the standard Lagrange multiplier method to solve it.

Instead of optimizing only the objective function (which is called unconstrained optimization), we need to optimize the Lagrangian of the problem, which considers the constraints at the same time. The need to consider constraints is obvious because they restrict the feasible solutions. Since our inequality constraints are expressed using "≥", the **Lagrangian** is formed by the constraints multiplied by positive Lagrange multipliers and subtracted from the objective function, i.e.,

$$L_P = \frac{1}{2}\langle \mathbf{w} \cdot \mathbf{w} \rangle - \sum_{i=1}^{n} \alpha_i [y_i(\langle \mathbf{w} \cdot \mathbf{x}_i \rangle + b) - 1] \tag{41}$$

where $\alpha_i \geq 0$ are the **Lagrange multipliers**.

The optimization theory says that an optimal solution to (41) must satisfy certain conditions, called **Kuhn–Tucker conditions**, which play a central role in constrained optimization. Here, we give a brief introduction to these conditions. Let the general optimization problem be

$$\text{Minimize} : f(\mathbf{x})$$
$$\text{Subject to} : g_i(\mathbf{x}) \le b_i, \quad i = 1, 2, ..., n \tag{42}$$

where f is the objective function and g_i is a constraint function (which is different from y_i in (40) as y_i is not a function but a class label of 1 or −1). The Lagrangian of (42) is,

$$L_P = f(\mathbf{x}) + \sum_{i=1}^{n} \alpha_i [g_i(\mathbf{x}) - b_i)] \tag{43}$$

An optimal solution to the problem in (42) must satisfy the following **necessary** (but **not sufficient**) conditions:

$$\frac{\partial L_P}{\partial x_j} = 0, \quad j = 1, 2, ..., r \tag{44}$$

$$g_i(\mathbf{x}) - b_i \le 0, \quad i = 1, 2, ..., n \tag{45}$$

$$\alpha_i \ge 0, \quad i = 1, 2, ..., n \tag{46}$$

$$\alpha_i (b_i - g_i(\mathbf{x}_i)) = 0, \quad i = 1, 2, ..., n \tag{47}$$

These conditions are called the **Kuhn–Tucker conditions**. Note that (45) is simply the original set of constraints in (42). The condition (47) is called the **complementarity condition**, which implies that at the solution point,

If $\alpha_i > 0$ then $g_i(\mathbf{x}) = b_i$.
If $g_i(\mathbf{x}) > b_i$ then $\alpha_i = 0$.

These mean that for active constraints, $\alpha_i > 0$, whereas for inactive constraints $\alpha_i = 0$. As we will see later, they give some very desirable properties to SVM.

Let us come back to our problem. For the minimization problem (40), the Kuhn–Tucker conditions are (48)–(52):

$$\frac{\partial L_P}{\partial w_j} = w_j - \sum_{i=1}^{n} y_i \alpha_i x_{ij} = 0, \quad j = 1, 2, ..., r \tag{48}$$

$$\frac{\partial L_P}{\partial b} = -\sum_{i=1}^{n} y_i \alpha_i = 0 \tag{49}$$

$$y_i (\langle \mathbf{w} \cdot \mathbf{x}_i \rangle + b) - 1 \ge 0, \quad i = 1, 2, ..., n \tag{50}$$

$$\alpha_i \geq 0, \quad i = 1, 2, ..., n \tag{51}$$

$$\alpha_i(y_i(\langle \mathbf{w} \cdot \mathbf{x}_i \rangle + b) - 1) = 0, \quad i = 1, 2, ..., n \tag{52}$$

Inequality (50) is the original set of constraints. We also note that although there is a Lagrange multiplier α_i for each training data point, the complementarity condition (52) shows that only those data points on the margin hyperplanes (i.e., H_+ and H_-) can have $\alpha_i > 0$ since for them $y_i(\langle \mathbf{w} \cdot \mathbf{x}_i \rangle + b)$ $- 1 = 0$. These data points are called **support vectors**, which give the name to the algorithm, *support vector machines*. All the other data points have $\alpha_i = 0$.

In general, Kuhn–Tucker conditions are necessary for an optimal solution, but not sufficient. However, for our minimization problem with a convex objective function and a set of linear constraints, the Kuhn–Tucker conditions are both **necessary** and **sufficient** for an optimal solution.

Solving the optimization problem is still a difficult task due to the inequality constraints. However, the Lagrangian treatment of the convex optimization problem leads to an alternative **dual** formulation of the problem, which is easier to solve than the original problem, which is called the **primal** problem (L_P is called the **primal Lagrangian**).

The concept of duality is widely used in the optimization literature. The aim is to provide an alternative formulation of the problem which is more convenient to solve computationally and/or has some theoretical significance. In the context of SVM, the dual problem is not only easy to solve computationally, but also crucial for using **kernel functions** to deal with nonlinear decision boundaries as we do not need to compute $\mathbf{w}$ explicitly (which will be clear later).

Transforming from the primal to its corresponding dual can be done by setting to zero the partial derivatives of the Lagrangian (41) with respect to the **primal variables** (i.e., $\mathbf{w}$ and b), and substituting the resulting relations back into the Lagrangian. This is to simply substitute (48), which is

$$w_j = \sum_{i=1}^{n} y_i \alpha_i x_{ij}, \quad j = 1, 2, ..., r \tag{53}$$

and (49), which is

$$\sum_{i=1}^{n} y_i \alpha_i = 0, \tag{54}$$

into the original Lagrangian (41) to eliminate the primal variables, which gives us the dual objective function (denoted by L_D),

$$L_D = \sum_{i=1}^{n} \alpha_i - \frac{1}{2} \sum_{i,j=1}^{n} y_i y_j \alpha_i \alpha_j \langle \mathbf{x}_i \cdot \mathbf{x}_j \rangle. \tag{55}$$

L_D contains only **dual variables** and must be maximized under the simpler constraints, (48) and (49), and $\alpha_i \geq 0$. Note that (48) is not needed as it has already been substituted into the objective function L_D. Hence, the **dual** of the primal Equation (40) is

$$\text{Maximize: } L_D = \sum_{i=1}^{n} \alpha_i - \frac{1}{2} \sum_{i,j=1}^{n} y_i y_j \alpha_i \alpha_j \langle \mathbf{x}_i \cdot \mathbf{x}_j \rangle.$$

$$\tag{56}$$

$$\text{Subject to: } \sum_{i=1}^{n} y_i \alpha_i = 0$$

$$\alpha_i \geq 0, \quad i = 1, 2, ..., n.$$

This dual formulation is called the **Wolfe dual**. For our convex objective function and linear constraints of the primal, it has the property that the α_i's at the maximum of L_D gives $\mathbf{w}$ and b occurring at the minimum of L_P (the primal).

Solving (56) requires numerical techniques and clever strategies beyond the scope of this book. After solving (56), we obtain the values for α_i, which are used to compute the weight vector $\mathbf{w}$ and the bias b using Equations (48) and (52) respectively. Instead of depending on one support vector ($\alpha_i > 0$) to compute b, in practice all support vectors are used to compute b, and then take their average as the final value for b. This is because the values of α_i are computed numerically and can have numerical errors. Our final **decision boundary (maximal margin hyperplane)** is

$$\langle \mathbf{w} \cdot \mathbf{x} \rangle + b = \sum_{i \in sv} y_i \alpha_i \langle \mathbf{x}_i \cdot \mathbf{x} \rangle + b = 0 \tag{57}$$

where sv is the set of indices of the support vectors in the training data.

Testing: We apply (57) for classification. Given a test instance $\mathbf{z}$, we classify it using the following:

$$sign(\langle \mathbf{w} \cdot \mathbf{z} \rangle + b) = sign\left(\sum_{i \in sv} y_i \alpha_i \langle \mathbf{x}_i \cdot \mathbf{z} \rangle + b \right). \tag{58}$$

If (58) returns 1, then the test instance $\mathbf{z}$ is classified as positive; otherwise, it is classified as negative.

3.8.2   Linear SVM: Non-separable Case

The linear separable case is the ideal situation. In practice, however, the training data is almost always noisy, i.e., containing errors due to various reasons. For example, some examples may be labeled incorrectly. Furthermore, practical problems may have some degree of randomness. Even for two identical input vectors, their labels may be different.

For SVM to be useful, it must allow noise in the training data. However, with noisy data the linear separable SVM will not find a solution because the constraints cannot be satisfied. For example, in Fig. 3.18, there is a negative point (circled) in the positive region, and a positive point in the negative region. Clearly, no solution can be found for this problem.

Recall that the primal for the linear separable case was:

$$\text{Minimize}: \frac{\langle \mathbf{w} \cdot \mathbf{w} \rangle}{2} \tag{59}$$

$$\text{Subject to}: y_i(\langle \mathbf{w} \cdot \mathbf{x}_i \rangle + b) \geq 1, \quad i = 1, 2, ..., n.$$

To allow errors in data, we can relax the margin constraints by introducing **slack** variables, $\xi_i (\geq 0)$ as follows:

$$\langle \mathbf{w} \cdot \mathbf{x}_i \rangle + b \geq 1 - \xi_i \quad \text{for } y_i = 1$$
$$\langle \mathbf{w} \cdot \mathbf{x}_i \rangle + b \leq -1 + \xi_i \quad \text{for } y_i = -1.$$

Thus we have the new constraints:

$$\text{Subject to}: \quad y_i(\langle \mathbf{w} \cdot \mathbf{x}_i \rangle + b) \geq 1 - \xi_i, \, i = 1, 2, ..., n,$$
$$\xi_i \geq 0, \, i = 1, 2, ..., n.$$

The geometric interpretation is shown in Fig. 3.18, which has two error data points $\mathbf{x}_a$ and $\mathbf{x}_b$ (circled) in wrong regions.

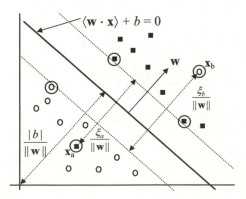

Fig. 3.18. The non-separable case: $\mathbf{x}_a$ and $\mathbf{x}_b$ are error data points

We also need to penalize the errors in the objective function. A natural way is to assign an extra cost for errors to change the objective function to

$$\text{Minimize}: \frac{\langle \mathbf{w} \cdot \mathbf{w} \rangle}{2} + C \left(\sum_{i=1}^{n} \xi_i \right)^k \tag{60}$$

where $C \geq 0$ is a user specified parameter. The resulting optimization problem is still a convex programming problem. $k = 1$ is commonly used, which has the advantage that neither ξ_i nor its Lagrangian multipliers appear in the dual formulation. We only discuss the $k = 1$ case below.

The new optimization problem becomes:

$$\text{Minimize}: \frac{\langle \mathbf{w} \cdot \mathbf{w} \rangle}{2} + C \sum_{i=1}^{n} \xi_i$$

$$\text{Subject to}: y_i(\langle \mathbf{w} \cdot \mathbf{x}_i \rangle + b) \geq 1 - \xi_i, \quad i = 1, 2, ..., n \tag{61}$$

$$\xi_i \geq 0, \quad i = 1, 2, ..., n.$$

This formulation is called the **soft-margin SVM**. The primal Lagrangian (denoted by L_P) of this formulation is as follows

$$L_P = \frac{1}{2} \langle \mathbf{w} \cdot \mathbf{w} \rangle + C \sum_{i=1}^{n} \xi_i - \sum_{i=1}^{n} \alpha_i [y_i(\langle \mathbf{w} \cdot \mathbf{x}_i \rangle + b) - 1 + \xi_i] - \sum_{i=1}^{n} \mu_i \xi_i \tag{62}$$

where $\alpha_i, \mu_i \geq 0$ are the **Lagrange multipliers**. The **Kuhn–Tucker conditions** for optimality are the following:

$$\frac{\partial L_P}{\partial w_j} = w_j - \sum_{i=1}^{n} y_i \alpha_i x_{ij} = 0, \quad j = 1, 2, ..., r \tag{63}$$

$$\frac{\partial L_P}{\partial b} = -\sum_{i=1}^{n} y_i \alpha_i = 0 \tag{64}$$

$$\frac{\partial L_P}{\partial \xi_i} = C - \alpha_i - \mu_i = 0, \quad i = 1, 2, ..., n \tag{65}$$

$$y_i(\langle \mathbf{w} \cdot \mathbf{x}_i \rangle + b) - 1 + \xi_i \geq 0, \quad i = 1, 2, ..., n \tag{66}$$

$$\xi_i \geq 0, \quad i = 1, 2, ..., n \tag{67}$$

$$\alpha_i \geq 0, \quad i = 1, 2, ..., n \tag{68}$$

$$\mu_i \geq 0, \quad i = 1, 2, ..., n \tag{69}$$

$$\alpha_i(y_i(\langle \mathbf{w} \cdot \mathbf{x}_i \rangle + b) - 1 + \xi_i) = 0, \quad i = 1, 2, ..., n \tag{70}$$

$$\mu_i \xi_i = 0, \quad i = 1, 2, ..., n \tag{71}$$

As the linear separable case, we then transform the primal to its dual by setting to zero the partial derivatives of the Lagrangian (62) with respect to the **primal variables** (i.e., $\mathbf{w}$, b and ξ_i), and substituting the resulting relations back into the Lagrangian. That is, we substitute Equations (63), (64) and (65) into the primal Lagrangian (62). From Equation (65), $C - \alpha_i - \mu_i = 0$, we can deduce that $\alpha_i \le C$ because $\mu_i \ge 0$. Thus, the dual of (61) is

$$\text{Maximize: } L_D(\boldsymbol{\alpha}) = \sum_{i=1}^{n} \alpha_i - \frac{1}{2} \sum_{i,j=1}^{n} y_i y_j \alpha_i \alpha_j \langle \mathbf{x}_i \cdot \mathbf{x}_j \rangle \tag{72}$$

$$\text{Subject to: } \sum_{i=1}^{n} y_i \alpha_i = 0$$

$$0 \le \alpha_i \le C, \quad i = 1, 2, ..., n.$$

Interestingly, ξ_i and its Lagrange multipliers μ_i are not in the dual and the objective function is identical to that for the separable case. The only difference is the constraint $\alpha_i \le C$ (inferred from $C - \alpha_i - \mu_i = 0$ and $\mu_i \ge 0$).

The dual problem (72) can also be solved numerically, and the resulting α_i values are then used to compute $\mathbf{w}$ and b. $\mathbf{w}$ is computed using Equation (63) and b is computed using the Kuhn–Tucker complementarity conditions (70) and (71). Since we do not have values for ξ_i, we need to get around it. From Equations (65), (70) and (71), we observe that if $0 < \alpha_i < C$ then both $\xi_i = 0$ and $y_i((\mathbf{w} \cdot \mathbf{x}_i) + b) - 1 + \xi_i) = 0$. Thus, we can use any training data point for which $0 < \alpha_i < C$ and Equation (70) (with $\xi_i = 0$) to compute b:

$$b = \frac{1}{y_i} - \sum_{i=1}^{n} y_i \alpha_i \langle \mathbf{x}_i \cdot \mathbf{x}_j \rangle. \tag{73}$$

Again, due to numerical errors, we can compute all possible b's and then take their average as the final b value.

Note that Equations (65), (70) and (71) in fact tell us more:

$$
\begin{aligned}
\alpha_i = 0 \quad &\Rightarrow \quad y_i((\mathbf{w} \cdot \mathbf{x}_i) + b) \ge 1 \text{ and } \xi_i = 0 \\
0 < \alpha_i < C \quad &\Rightarrow \quad y_i((\mathbf{w} \cdot \mathbf{x}_i) + b) = 1 \text{ and } \xi_i = 0 \\
\alpha_i = C \quad &\Rightarrow \quad y_i((\mathbf{w} \cdot \mathbf{x}_i) + b) \le 1 \text{ and } \xi_i \ge 0
\end{aligned}
\tag{74}
$$

Similar to support vectors for the separable case, (74) shows one of the most important properties of SVM: the solution is sparse in α_i. Most training data points are outside the margin area and their α_i's in the solution are 0. Only those data points that are on the margin (i.e., $y_i((\mathbf{w} \cdot \mathbf{x}_i) + b) = 1$, which are support vectors in the separable case), inside the margin (i.e., α_i

= C and $y_i(\langle \mathbf{w} \cdot \mathbf{x}_i \rangle + b) < 1$), or errors are non-zero. Without this sparsity property, SVM would not be practical for large data sets.

The final decision boundary is (we note that many α_i's are 0)

$$\langle \mathbf{w} \cdot \mathbf{x} \rangle + b = \sum_{i=1}^{n} y_i \alpha_i \langle \mathbf{x}_i \cdot \mathbf{x} \rangle + b = 0. \tag{75}$$

The decision rule for classification (testing) is the same as the separable case, i.e., $sign(\langle \mathbf{w} \cdot \mathbf{x} \rangle + b)$. We notice that for both Equations (75) and (73), $\mathbf{w}$ does not need to be explicitly computed. This is crucial for using kernel functions to handle nonlinear decision boundaries.

Finally, we still have the problem of determining the parameter C. The value of C is usually chosen by trying a range of values on the training set to build multiple classifiers and then to test them on a validation set before selecting the one that gives the best classification result on the validation set. Cross-validation is commonly used as well.

3.8.3   Nonlinear SVM: Kernel Functions

The SVM formulations discussed so far require that positive and negative examples can be linearly separated, i.e., the decision boundary must be a hyperplane. However, for many real-life data sets, the decision boundaries are nonlinear. To deal with nonlinearly separable data, the same formulation and solution techniques as for the linear case are still used. We only transform the input data from its original space into another space (usually of a much higher dimensional space) so that a linear decision boundary can separate positive and negative examples in the transformed space, which is called the **feature space**. The original data space is called the **input space**.

Thus, the basic idea is to map the data in the input space X to a feature space F via a nonlinear mapping ϕ,

$$\phi : X \rightarrow F$$
$$\mathbf{x} \mapsto \phi(\mathbf{x}). \tag{76}$$

After the mapping, the original training data set $\{(\mathbf{x}_1, y_1), (\mathbf{x}_2, y_2), \ldots, (\mathbf{x}_n, y_n)\}$ becomes:

$$\{(\phi(\mathbf{x}_1), y_1), (\phi(\mathbf{x}_2), y_2), \ldots, (\phi(\mathbf{x}_n), y_n)\}. \tag{77}$$

The same linear SVM solution method is then applied to F. Figure 3.19 illustrates the process. In the input space (figure on the left), the training examples cannot be linearly separated. In the transformed feature space (figure on the right), they can be separated linearly.

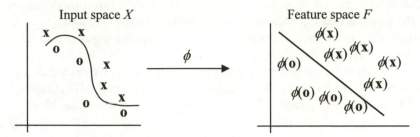

Fig. 3.19. Transformation from the input space to the feature space

With the transformation, the optimization problem in (61) becomes

$$\text{Minimize}: \quad \frac{\langle \mathbf{w} \cdot \mathbf{w} \rangle}{2} + C \sum_{i=1}^{n} \xi_i \tag{78}$$

$$\text{Subject to}: \quad y_i(\langle \mathbf{w} \cdot \phi(\mathbf{x}_i) \rangle + b) \geq 1 - \xi_i, \quad i = 1, 2, ..., n$$

$$\xi_i \geq 0, \quad i = 1, 2, ..., n$$

Its corresponding dual is

$$\text{Maximize}: \quad L_D = \sum_{i=1}^{n} \alpha_i - \frac{1}{2} \sum_{i,j=1}^{n} y_i y_j \alpha_i \alpha_j \langle \phi(\mathbf{x}_i) \cdot \phi(\mathbf{x}_j) \rangle. \tag{79}$$

$$\text{Subject to}: \quad \sum_{i=1}^{n} y_i \alpha_i = 0$$

$$0 \leq \alpha_i \leq C, \quad i = 1, 2, ..., n.$$

The final decision rule for classification (testing) is

$$\sum_{i=1}^{n} y_i \alpha_i \langle \phi(\mathbf{x}_i) \cdot \phi(\mathbf{x}) \rangle + b \tag{80}$$

Example 16: Suppose our input space is 2-dimensional, and we choose the following transformation (mapping):

$$(x_1, x_2) \mapsto (x_1^2, x_2^2, \sqrt{2}x_1 x_2) \tag{81}$$

The training example ((2, 3), −1) in the input space is transformed to the following training example in the feature space:

$$((4, 9, 8.5), -1). \qquad \blacksquare$$

The potential problem with this approach of transforming the input data explicitly to a feature space and then applying the linear SVM is that it may suffer from the curse of dimensionality. The number of dimensions in the feature space can be huge with some useful transformations (see below) even with reasonable numbers of attributes in the input space. This makes it computationally infeasible to handle.

Fortunately, explicit transformations can be avoided if we notice that in the dual representation both the construction of the optimal hyperplane (79) in F and the evaluation of the corresponding decision/classification function (80) only require the evaluation of dot products $\langle \phi(\mathbf{x}) \cdot \phi(\mathbf{z}) \rangle$ and never the mapped vector $\phi(\mathbf{x})$ in its explicit form. This is a crucial point.

Thus, if we have a way to compute the dot product $\langle \phi(\mathbf{x}) \cdot \phi(\mathbf{z}) \rangle$ in the feature space F using the input vectors $\mathbf{x}$ and $\mathbf{z}$ directly, then we would not need to know the feature vector $\phi(\mathbf{x})$ or even the mapping function ϕ itself. In SVM, this is done through the use of **kernel functions**, denoted by K,

$$K(\mathbf{x}, \mathbf{z}) = \langle \phi(\mathbf{x}) \cdot \phi(\mathbf{z}) \rangle, \tag{82}$$

which are exactly the functions for computing dot products in the transformed feature space using input vectors $\mathbf{x}$ and $\mathbf{z}$. An example of a kernel function is the **polynomial kernel**,

$$K(\mathbf{x}, \mathbf{z}) = \langle \mathbf{x} \cdot \mathbf{z} \rangle^d. \tag{83}$$

Example 17: Let us compute this kernel with degree $d = 2$ in a 2-dimensional space. Let $\mathbf{x} = (x_1, x_2)$ and $\mathbf{z} = (z_1, z_2)$.

$$\begin{aligned}
\langle \mathbf{x} \cdot \mathbf{z} \rangle^2 &= (x_1 z_1 + x_2 z_2)^2 \\
&= x_1^2 z_1^2 + 2 x_1 z_1 x_2 z_2 + x_2^2 z_2^2 \\
&= \langle (x_1^2, x_2^2, \sqrt{2} x_1 x_2) \cdot (z_1^2, z_2^2, \sqrt{2} z_1 z_2) \rangle \\
&= \langle \phi(\mathbf{x}) \cdot \phi(\mathbf{z}) \rangle,
\end{aligned} \tag{84}$$

where $\phi(\mathbf{x}) = (x_1^2, x_2^2, \sqrt{2} x_1 x_2)$, which shows that the kernel $\langle \mathbf{x} \cdot \mathbf{z} \rangle^2$ is a dot product in the transformed feature space. The number of dimensions in the feature space is 3. Note that $\phi(\mathbf{x})$ is actually the mapping function used in Example 16. Incidentally, in general the number of dimensions in the feature space for the polynomial kernel function $K(\mathbf{x}, \mathbf{z}) = \langle \mathbf{x} \cdot \mathbf{z} \rangle^d$ is $\binom{r+d-1}{d}$,

which is a huge number even with a reasonable number (r) of attributes in the input space. Fortunately, by using the kernel function in (83), the huge number of dimensions in the feature space does not matter. ∎

The derivation in (84) is only for illustration purposes. We do not need to find the mapping function. We can simply apply the kernel function directly. That is, we replace all the dot products $\langle \phi(\mathbf{x}) \cdot \phi(\mathbf{z}) \rangle$ in (79) and (80) with the kernel function $K(\mathbf{x}, \mathbf{z})$ (e.g., the polynomial kernel in (83)). This strategy of directly using a kernel function to replace dot products in the feature space is called the **kernel trick**. We would never need to explicitly know what ϕ is.

However, the question is, how do we know whether a function is a kernel without performing the derivation such as that in (84)? That is, how do we know that a kernel function is indeed a dot product in some feature space? This question is answered by a theorem called the **Mercer's theorem**, which we will not discuss here. See [118] for details.

It is clear that the idea of kernel generalizes the dot product in the input space. The dot product is also a kernel with the feature map being the identity

$$K(\mathbf{x}, \mathbf{z}) = \langle \mathbf{x} \cdot \mathbf{z} \rangle. \tag{85}$$

Commonly used kernels include

Polynomial: $K(\mathbf{x},\mathbf{z}) = (\langle \mathbf{x} \cdot \mathbf{z} \rangle + \theta)^d$ (86)

Gaussian RBF: $K(\mathbf{x},\mathbf{z}) = e^{-\|\mathbf{x}-\mathbf{z}\|^2/2\sigma}$ (87)

where $\theta \in \mathscr{R}$, $d \in N$, and $\sigma > 0$.

Summary

SVM is a linear learning system that finds the maximal margin decision boundary to separate positive and negative examples. Learning is formulated as a quadratic optimization problem. Nonlinear decision boundaries are found via a transformation of the original data to a much higher dimensional feature space. However, this transformation is never explicitly done. Instead, kernel functions are used to compute dot products required in learning without the need to even know the transformation function.

Due to the separation of the learning algorithm and kernel functions, kernels can be studied independently from the learning algorithm. One can design and experiment with different kernel functions without touching the underlying learning algorithm.

SVM also has some limitations:

1. It works only in real-valued space. For a categorical attribute, we need to convert its categorical values to numeric values. One way to do this is to create an extra binary attribute for each categorical value, and set the attribute value to 1 if the categorical value appears, and 0 otherwise.

2. It allows only two classes, i.e., binary classification. For multiple class classification problems, several strategies can be applied, e.g., one-against-rest, and error-correcting output coding [138].
3. The hyperplane produced by SVM is hard to understand by users. It is difficult to picture where the hyperplane is in a high-dimensional space. The matter is made worse by kernels. Thus, SVM is commonly used in applications that do not required human understanding.

3.9 K-Nearest Neighbor Learning

All the previous learning methods learn some kinds of models from the data, e.g., decision trees, sets of rules, posterior probabilities, and hyperplanes. These learning methods are often called **eager learning** methods as they learn models of the data before testing. In contrast, k-nearest neighbor (kNN) is a **lazy learning** method in the sense that no model is learned from the training data. Learning only occurs when a test example needs to be classified. The idea of kNN is extremely simple and yet quite effective in many applications, e.g., text classification.

It works as follows: Again let D be the training data set. Nothing will be done on the training examples. When a test instance d is presented, the algorithm compares d with every training example in D to compute the similarity or distance between them. The k most similar (closest) examples in D are then selected. This set of examples is called the **k nearest neighbors** of d. d then takes the most frequent class among the k nearest neighbors. Note that $k = 1$ is usually not sufficient for determining the class of d due to noise and outliers in the data. A set of nearest neighbors is needed to accurately decide the class. The general kNN algorithm is given in Fig. 3.20.

Algorithm kNN(D, d, k)
1   Compute the distance between d and every example in D;
2   Choose the k examples in D that are nearest to d, denote the set by P ($\subseteq D$);
3   Assign d the class that is the most frequent class in P (or the majority class).

Fig. 3.20. The k-nearest neighbor algorithm

The key component of a kNN algorithm is the **distance/similarity function**, which is chosen based on applications and the nature of the data. For relational data, the Euclidean distance is commonly used. For text documents, cosine similarity is a popular choice. We will introduce these distance functions and many others in the next chapter.

The number of nearest neighbors k is usually determined by using a validation set, or through cross validation on the training data. That is, a range of k values are tried, and the k value that gives the best accuracy on the validation set (or cross validation) is selected. Figure 3.21 illustrates the importance of choosing the right k.

Example 18: In Fig. 3.21, we have two classes of data, positive (filled squares) and negative (empty circles). If 1-nearest neighbor is used, the test data point ⊕ will be classified as negative, and if 2-nearest neighbors are used, the class cannot be decided. If 3-nearest neighbors are used, the class is positive as two positive examples are in the 3-nearest neighbors.

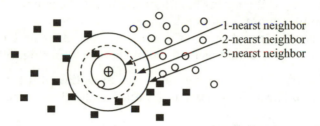

Fig. 3.21. An illustration of k-nearest neighbor classification

Despite its simplicity, researchers have showed that the classification accuracy of kNN can be quite strong and in many cases as accurate as those elaborated methods. For instance, it is showed in [574] that kNN performs equally well as SVM for some text classification tasks. kNN is also very flexible. It can work with any arbitrarily shaped decision boundaries.

kNN is, however, slow at the classification time. Due to the fact that there is no model building, each test instance is compared with every training example at the classification time, which can be quite time consuming especially when the training set D and the test set are large. Another disadvantage is that kNN does not produce an understandable model. It is thus not applicable if an understandable model is required in the application.

3.10 Ensemble of Classifiers

So far, we have studied many individual classifier building techniques. A natural question to ask is: can we build many classifiers and then combine them to produce a better classifier? Yes, in many cases. This section describes two well known ensemble techniques, **bagging** and **boosting**. In both these methods, many classifiers are built and the final classification decision for each test instance is made based on some forms of voting of the committee of classifiers.

3.10.1 Bagging

Given a training set D with n examples and a base learning algorithm, bagging (for *Bootstrap Aggregating*) works as follows [63]:

Training:

1. Create k bootstrap samples S_1, S_2, and S_k. Each sample is produced by drawing n examples at random from D with replacement. Such a sample is called a **bootstrap replicate** of the original training set D. On average, each sample S_i contains 63.2% of the original examples in D, with some examples appearing multiple times.
2. Build a classifier based on each sample S_i. This gives us k classifiers. All the classifiers are built using the same base learning algorithm.

Testing: Classify each test (or new) instance by voting of the k classifiers (equal weights). The majority class is assigned as the class of the instance.

Bagging can improve the accuracy significantly for unstable learning algorithms, i.e., a slight change in the training data resulting in a major change in the output classifier. Decision tree and rule induction methods are examples of unstable learning methods. k-nearest neighbor and naïve Bayesian methods are examples of stable techniques. For stable classifiers, Bagging may sometime degrade the accuracy.

3.10.2 Boosting

Boosting is a family of ensemble techniques, which, like bagging, also manipulates the training examples and produces multiple classifiers to improve the classification accuracy [477]. Here we only describe the popular **AdaBoost** algorithm given in [186]. Unlike bagging, AdaBoost assigns a weight to each training example.

Training: AdaBoost produces a sequence of classifiers (also using the same base learner). Each classifier is dependent on the previous one, and focuses on the previous one's errors. Training examples that are incorrectly classified by the previous classifiers are given higher weights.

Let the original training set D be $\{(\mathbf{x}_1, y_1), (\mathbf{x}_2, y_2), \ldots, (\mathbf{x}_n, y_n)\}$, where $\mathbf{x}_i$ is an input vector, y_i is its class label and $y_i \in Y$ (the set of class labels). With a weight attached to each example, we have, $\{(\mathbf{x}_1, y_1, w_1), (\mathbf{x}_2, y_2, w_2), \ldots, (\mathbf{x}_n, y_n, w_n)\}$, and $\sum_i w_i = 1$. The AdaBoost algorithm is given in Fig. 3.22.

The algorithm builds a sequence of k classifiers (k is specified by the user) using a base learner, called BaseLeaner in line 3. Initially, the weight

AdaBoost(D, Y, BaseLeaner, k)

1.   Initialize $D_1(w_i) \leftarrow 1/n$ for all i;   // initialize the weights
2.   **for** $t = 1$ to k **do**
3. $f_t \leftarrow$ BaseLearner(D_t);   // build a new classifier f_t
4. $e_t \leftarrow \sum_{i:f_t(D_t(\mathbf{x}_i)) \neq y_i} D_t(w_i)$;   // compute the error of f_t
5.   **if** $e_t > \frac{1}{2}$ **then**   // if the error is too large,
6. $k \leftarrow k - 1$;   // remove the iteration and
7.   exit-loop   // exit
8.   **else**
9. $\beta_t \leftarrow e_t / (1 - e_t)$;
10          $D_{t+1}(w_i) \leftarrow D_t(w_i) \times \begin{cases} \beta_t & \text{if } f_t(D_t(\mathbf{x}_i)) = y_i \\ 1 & \text{otherwise} \end{cases}$;   // update the weights
11. $D_{t+1}(w_i) \leftarrow \dfrac{D_{t+1}(w_i)}{\sum_{i=1}^{n} D_{t+1}(w_i)}$ // normalize the weights
12.   **endif**
13. **endfor**
14. $f_{final}(\mathbf{x}) \leftarrow \underset{y \in Y}{\text{argmax}} \sum_{t:f_t(\mathbf{x})=y} \log \dfrac{1}{\beta_t}$ // the final output classifier

Fig. 3.22. The AdaBoost algorithm

for each training example is $1/n$ (line 1). In each iteration, the training data set becomes D_t, which is the same as D, but with different weights. Each iteration builds a new classifier f_t (line 3). The error of f_t is calculated in line 4. If it is too large, delete the iteration and exit (lines 5–7). Lines 9–11 update and normalize the weights for building the next classifier.

Testing: For each test case, the results of the series of classifiers are combined to determine the final class of the test case, which is shown in line 14 of Fig. 3.22 (a weighted voting).

Boosting works better than bagging in most cases as shown in [454]. It also tends to improve performance more when the base learner is unstable.

Bibliographic Notes

Supervised learning has been studied extensively by the machine learning community. The book by Mitchell [385] covers most learning techniques and is easy to read. Duda et al.'s pattern classification book is also a great reference [155]. Most data mining books have one or two chapters on supervised learning, e.g., those by Han and Kamber [218], Hand et al. [221], Tan et al. [512], and Witten and Frank [549].

For decision tree induction, Quinlan's book [453] has all the details and the code of his popular decision tree system C4.5. Other well-known systems include CART by Breiman et al. [62] and CHAD by Kass [270]. Scaling up of decision tree algorithms was also studied in several papers. These algorithms can have the data on disk, and are thus able to run with huge data sets. See [195] for an algorithm and also additional references.

Rule induction algorithms generate rules directly from the data. Well-known systems include AQ by Michalski et al. [381], CN2 by Clark and Niblett [104], FOIL by Quinlan [452], FOCL by Pazzani et al. [438], I-REP by Furnkranz and Widmer [189], and RIPPER by Cohen [106].

Using association rules to build classifiers was proposed by Liu et al. in [343], which also reported the CBA system. CBA selects a small subset of class association rules as the classifier. Other classifier building techniques include combining multiple rules by Li et al. [328], using rules as features by Meretakis and Wüthrich [379], Antonie and Zaiane [23], Deshpande and Karpis [131], Jindal and Liu [255], and Lesh et al. [314], generating a subset of rules by Cong et al. [112, 113], Wang et al. [536], Yin and Han [578], and Zaki and Aggarwal [587]. Other systems include those by Dong et al. [149], Li et al. [319, 320], Yang et al. [570], etc.

The naïve Bayesian classification model described in Sect. 3.6 is based on the papers by Domingos and Pazzani [148], Kohavi et al. [285] and Langley et al [301]. The naïve Bayesian classification for text discussed in Sect. 3.7 is based on the multinomial formulation given by McCallum and Nigam [365]. This model was also used earlier by Lewis and Gale [317], Li and Yamanishi [318], and Nigam et al. [413]. Another formulation of naïve Bayes is based on the multivariate Bernoulli model, which was used in Lewis [316], and Robertson and Sparck-Jones [464].

Support vector machines (SVM) was first introduced by Vapnik and his colleagues in 1992 [59]. Further details were given in his 1995 book [525]. Two other books on SVM and kernel methods are those by Cristianini and Shawe-Taylor [118] and Scholkopf and Smola [479]. The discussion of SVM in this chapter is heavily influenced by Cristianini and Shawe-Taylor's book and the tutorial paper by Burges [74]. Two popular SVM systems are SVM^{Light} (available at http://svmlight.joachims.org/) and LIBSVM (available at http://www.csie.ntu.edu.tw/~cjlin/libsvm/).

Existing classifier ensemble methods include bagging by Breiman [63], boosting by Schapire [477] and Freund and Schapire [186], random forest also by Breiman [65], stacking by Wolpert [552], random trees by Fan [169], and many others.

4 Unsupervised Learning

Supervised learning discovers patterns in the data that relate data attributes to a class attribute. These patterns are then utilized to predict the values of the class attribute of future data instances. These classes indicate some real-world predictive or classification tasks such as determining whether a news article belongs to the category of sports or politics, or whether a patient has a particular disease. However, in some other applications, the data have no class attributes. The user wants to explore the data to find some intrinsic structures in them. Clustering is one technology for finding such structures. It organizes data instances into **similarity groups**, called **clusters** such that the data instances in the same cluster are similar to each other and data instances in different clusters are very different from each other. Clustering is often called **unsupervised learning**, because unlike supervised learning, class values denoting an *a priori* partition or grouping of the data are not given. Note that according to this definition, we can also say that association rule mining is an unsupervised learning task. However, due to historical reasons, clustering is closely associated and even synonymous with unsupervised learning while association rule mining is not. We follow this convention, and describe some main clustering techniques in this chapter.

Clustering has been shown to be one of the most commonly used data analysis techniques. It also has a long history, and has been used in almost every field, e.g., medicine, psychology, botany, sociology, biology, archeology, marketing, insurance, library science, etc. In recent years, due to the rapid increase of online documents and the expansion of the Web, text document clustering too has become a very important task. In Chap. 12, we will also see that clustering is very useful in Web usage mining.

4.1 Basic Concepts

Clustering is the process of organizing data instances into groups whose members are similar in some way. A **cluster** is therefore a collection of data instances which are "similar" to each other and are "dissimilar" to data instances in other clusters. In the clustering literature, a data instance is also called an **object** as the instance may represent an object in the real-world. It is also called a **data point** as it can be seen as a point in an *r*-dimension space, where *r* is the number of attributes in the data.

Fig. 4.1 shows a 2-dimensional data set. We can clearly see three groups of data points. Each group is a cluster. The task of clustering is to find the three clusters hidden in the data. Although it is easy for a human to visually detect clusters in a 2-dimensional or even 3-demensional space, it becomes very hard, if not impossible, to detect clusters visually as the number of dimensions increases. Additionally, in many applications, clusters are not as clear-cut or well separated as the three clusters in Fig. 4.1. Automatic techniques are thus needed for clustering.

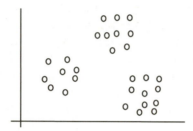

Fig. 4.1. Three natural groups or clusters of data points

After seeing the example in Fig. 4.1, you may ask the question: What is clustering for? To answer it, let us see some application examples from different domains.

Example 1: A company wants to conduct a marketing campaign to promote its products. The most effective strategy is to design a set of personalized marketing materials for each individual customer according to his/her profile and financial situation. However, this is too expensive for a large number of customers. At the other extreme, the company designs only one set of marketing materials to be used for all customers. This one-size-fits-all approach, however, may not be effective. The most cost-effective approach is to segment the customers into a small number of groups according to their similarities and design some targeted marketing materials for each group. This segmentation task is commonly done using clustering algorithms, which **partition** customers into similarity groups. In marketing research, clustering is often called **segmentation**. ∎

Example 2: A company wants to produce and sell T-shirts. Similar to the case above, on one extreme, for each customer it can measure his/her size and have a T-shirt tailor-made for him/her. Obviously, this T-shirt is going to be expensive. On the other extreme, only one size of T-shirts is made.

Since this size may not fit most people, the company might not be able to sell as many T-shirts. Again, the most cost effective way is to group people based on their sizes and make a different generalized size of T-shirts for each group. This is why we see small, medium and large size T-shirts in shopping malls, and seldom see T-shirts with only a single size. The method used to group people according to their sizes is clustering. The process is usually as follows: The T-shirt manufacturer first samples a large number of people and measure their sizes to produce a measurement database. It then clusters the data, which **partitions** the data into some similarity subsets, i.e., clusters. For each cluster, it computes the average of the sizes and then uses the average to mass-produce T-shirts for all people of similar size. ∎

Example 3: Everyday, news agencies around the world generate a large number of news articles. If a Web site wants to collect these news articles to provide an integrated news service, it has to organize the collected articles according to some topic hierarchy. The question is: What should the topics be, and how should they be organized? One possibility is to employ a group of human editors to do the job. However, the manual organization is costly and very time consuming, which makes it unsuitable for news and other time sensitive information. Throwing all the news articles to the readers with no organization is clearly not an option. Although classification is able to classify news articles according to predefined topics, it is not applicable here because classification needs training data, which have to be manually labeled with topic classes. Since news topics change constantly and rapidly, the training data would need to change constantly as well, which is infeasible via manual labeling. Clustering is clearly a solution for this problem because it automatically groups a stream of news articles based on their content similarities. **Hierarchical clustering algorithms** can also organize documents hierarchically, i.e., each topic may contain sub-topics and so on. Topic hierarchies are particularly useful for texts. ∎

The above three examples indicate two types of clustering, **partitional** and **hierarchical**. Indeed, these are the two most important types of clustering approaches. We will study some specific algorithms of these two types of clustering.

Our discussion and examples above also indicate that clustering needs a similarity function to measure how similar two data points (or objects) are, or alternatively a **distance function** to measure the distance between two data points. We will use distance functions in this chapter. The goal of clustering is thus to discover the intrinsic grouping of the input data through the use of a clustering algorithm and a distance function.

Algorithm k-means(k, D)

1   choose k data points as the initial centroids (cluster centers)
2   **repeat**
3   **for** each data point $\mathbf{x} \in D$ do
4   compute the distance from $\mathbf{x}$ to each centroid;
5   assign $\mathbf{x}$ to the closest centroid   // a centroid represents a cluster
6   **endfor**
7   re-compute the centroid using the current cluster memberships
8   **until** the stopping criterion is met

Fig. 4.2. The k-means algorithm

4.2 K-means Clustering

The k-means algorithm is the best known **partitional clustering algorithm.** It is perhaps also the most widely used among all clustering algorithms due to its simplicity and efficiency. Given a set of data points and the required number of k clusters (k is specified by the user), this algorithm iteratively partitions the data into k clusters based on a distance function.

4.2.1 K-means Algorithm

Let the set of data points (or instances) D be

$$\{\mathbf{x}_1, \mathbf{x}_2, \ldots, \mathbf{x}_n\},$$

where $\mathbf{x}_i = (x_{i1}, x_{i2}, \ldots, x_{ir})$ is a vector in a real-valued space $X \subseteq \mathfrak{R}^r$, and r is the number of attributes in the data (or the number of dimensions of the **data space**). The k-means algorithm partitions the given data into k clusters. Each cluster has a cluster **center**, which is also called the cluster **centroid**. The centroid, usually used to represent the cluster, is simply the mean of all the data points in the cluster, which gives the name to the algorithm, i.e., since there are k clusters, thus k means. Figure 4.2 gives the k-means clustering algorithm.

At the beginning, the algorithm randomly selects k data points as the **seed** centroids. It then computes the distance between each seed centroid and every data point. Each data point is assigned to the centroid that is closest to it. A centroid and its data points therefore represent a cluster. Once all the data points in the data are assigned, the centroid for each cluster is re-computed using the data points in the current cluster. This process repeats until a stopping criterion is met. The stopping (or convergence) criterion can be any one of the following:

1. no (or minimum) re-assignments of data points to different clusters.
2. no (or minimum) change of centroids.
3. minimum decrease in the **sum of squared error** (SSE),

$$SSE = \sum_{j=1}^{k} \sum_{\mathbf{x} \in C_j} dist(\mathbf{x}, \mathbf{m}_j)^2,$$ (1)

where k is the number of required clusters, C_j is the jth cluster, $\mathbf{m}_j$ is the centroid of cluster C_j (the mean vector of all the data points in C_j), and $dist(\mathbf{x}, \mathbf{m}_j)$ is the distance between data point $\mathbf{x}$ and centroid $\mathbf{m}_j$.

The k-means algorithm can be used for any application data set where the **mean** can be defined and computed. In **Euclidean space**, the mean of a cluster is computed with:

$$\mathbf{m}_j = \frac{1}{|C_j|} \sum_{\mathbf{x}_i \in C_j} \mathbf{x}_i,$$ (2)

where $|C_j|$ is the number of data points in cluster C_j. The distance from a data point $\mathbf{x}_i$ to a cluster mean (centroid) $\mathbf{m}_j$ is computed with

$$dist(\mathbf{x}_i, \mathbf{m}_j) = \| \mathbf{x}_i - \mathbf{m}_j \|$$ (3)

$$= \sqrt{(x_{i1} - m_{j1})^2 + (x_{i2} - m_{j2})^2 + ... + (x_{ir} - m_{jr})^2}.$$

Example 4: Figure 4.3(A) shows a set of data points in a 2-dimensional space. We want to find 2 clusters from the data, i.e., $k = 2$. First, two data points (each marked with a cross) are randomly selected to be the initial centroids (or seeds) shown in Fig. 4.3(A). The algorithm then goes to the first iteration (the repeat-loop).

Iteration 1: Each data point is assigned to its closest centroid to form 2 clusters. The resulting clusters are given in Fig. 4.3(B). Then the centroids are re-computed based on the data points in the current clusters (Fig. 4.3(C)). This leads to iteration 2.

Iteration 2: Again, each data point is assigned to its closest new centroid to form two new clusters shown in Fig. 4.3(D). The centroids are then re-computed. The new centroids are shown in Fig. 4.3(E).

Iteration 3: The same operations are performed as in the first two iterations. Since there is no re-assignment of data points to different clusters in this iteration, the algorithm ends.

The final clusters are those given in Fig. 4.3(G). The set of data points in each cluster and its centroid are output to the user.

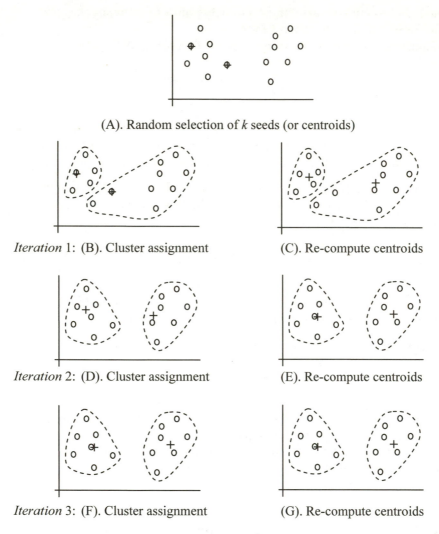

(A). Random selection of k seeds (or centroids)

Iteration 1: (B). Cluster assignment   (C). Re-compute centroids

Iteration 2: (D). Cluster assignment   (E). Re-compute centroids

Iteration 3: (F). Cluster assignment   (G). Re-compute centroids

Fig. 4.3. The working of the k-means algorithm through an example   ■

One problem with the k-means algorithm is that some clusters may become empty during the clustering process since no data point is assigned to
them. Such clusters are called **empty clusters**. To deal with an empty cluster, we can choose a data point as the replacement centroid, e.g., a data
point that is furthest from the centroid of a large cluster. If the sum of the
squared error (SSE) is used as the stopping criterion, the cluster with the
largest squared error may be used to find another centroid.

4.2.2   Disk Version of the K-means Algorithm

The k-means algorithm may be implemented in such a way that it does not need to load the entire data set into the main memory, which is useful for large data sets. Notice that the centroids for the k clusters can be computed incrementally in each iteration because the summation in Equation (2) can be calculated separately first. During the clustering process, the number of data points in each cluster can be counted incrementally as well. This gives us a disk based implementation of the algorithm (Fig. 4.4), which produces exactly the same clusters as that in Fig. 4.2, but with the data on disk. In each for-loop, the algorithm simply scans the data once.

The whole clustering process thus scans the data t times, where t is the number of iterations before convergence, which is usually not very large (< 50). In applications, it is quite common to set a limit on the number of iterations because later iterations typically result in only minor changes to the clusters. Thus, this algorithm may be used to cluster large data sets which cannot be loaded into the main memory. Although there are several special algorithms that scale-up clustering algorithms to large data sets, they all require sophisticated techniques.

Algorithm disk-k-means(k, D)
1   Choose k data points as the initial centriods $\mathbf{m}_j, j = 1, ..., k$;
2   **repeat**
3   initialize $\mathbf{s}_j \leftarrow \mathbf{0}, j = 1, ..., k$;   // $\mathbf{0}$ is a vector with all 0's
4   initialize $n_j \leftarrow 0, j = 1, ..., k$;   // n_j is the number of points in cluster j
5   **for** each data point $\mathbf{x} \in D$ **do**
6           $j \leftarrow \arg\min_{i \in \{1,2,...k\}} dist(\mathbf{x}, \mathbf{m}_i)$;
7   assign $\mathbf{x}$ to the cluster j;
8           $\mathbf{s}_j \leftarrow \mathbf{s}_j + \mathbf{x}$;
9           $n_j \leftarrow n_j + 1$;
10   **endfor**
11      $\mathbf{m}_j \leftarrow \mathbf{s}_j/n_j, j = 1, ..., k$;
12   **until** the stopping criterion is met

Fig. 4.4. A simple disk version of the k-means algorithm

Let us give some explanations of this algorithm. Line 1 does exactly the same thing as the algorithm in Fig. 4.2. Line 3 initializes vector $\mathbf{s}_j$ which is used to incrementally compute the sum in Equation (2) (line 8). Line 4 initializes n_j which records the number of data points assigned to cluster j (line 9). Lines 6 and 7 perform exactly the same tasks as lines 4 and 5 in the original algorithm in Fig. 4.2. Line 11 re-computes the centroids, which are used in the next iteration. Any of the three stopping criteria may be used here. If the sum of squared error is applied, we can modify the algorithm slightly to compute the sum of square error incrementally.

4.2.3   Strengths and Weaknesses

The main strengths of the k-means algorithm are its simplicity and efficiency. It is easy to understand and easy to implement. Its time complexity is $O(tkn)$, where n is the number of data points, k is the number of clusters, and t is the number of iterations. Since both k and t are normally much smaller than n, the k-means algorithm is considered a linear algorithm in the number of data points.

The weaknesses and ways to address them are as follows:

1. The algorithm is only applicable to data sets where the notion of the **mean** is defined. Thus, it is difficult to apply to categorical data sets. There is, however, a variation of the k-means algorithm called **k-modes**, which clusters categorical data. The algorithm uses the mode instead of the mean as the centroid. Assuming that the data instances are described by r categorical attributes, the mode of a cluster C_j is a tuple $\mathbf{m}_j = (m_{j1}, m_{j2}, \ldots, m_{jr})$ where m_{ji} is the most frequent value of the ith attribute of the data instances in cluster C_j. The similarity (or distance) between a data instance and a mode is the number of values that they match (or do not match).

2. The user needs to specify the number of clusters k in advance. In practice, several k values are tried and the one that gives the most desirable result is selected. We will discuss the evaluation of clusters later.

3. The algorithm is sensitive to **outliers**. Outliers are data points that are very far away from other data points. Outliers could be errors in the data recording or some special data points with very different values. For example, in an employee data set, the salary of the Chief-Executive-Officer (CEO) of the company may be considered as an outlier because its value could be many times larger than everyone else. Since the k-means algorithm uses the mean as the centroid of each cluster, outliers may result in undesirable clusters as the following example shows.

Example 5: In Fig. 4.5(A), due to an outlier data point, the two resulting clusters do not reflect the natural groupings in the data. The ideal clusters are shown in Fig. 4.5(B). The outlier should be identified and reported to the user.   ■

There are several methods for dealing with outliers. One simple method is to remove some data points in the clustering process that are much further away from the centroids than other data points. To be safe, we may want to monitor these possible outliers over a few iterations and then decide whether to remove them. It is possible that a very small cluster of data points may be outliers. Usually, a threshold value is used to make the decision.

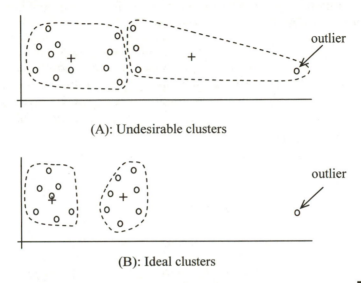

(A): Undesirable clusters

(B): Ideal clusters

Fig. 4.5. Clustering with and without the effect of outliers ∎

Another method is to perform random sampling. Since in sampling we only choose a small subset of the data points, the chance of selecting an outlier is very small. We can use the sample to do a pre-clustering and then assign the rest of the data points to these clusters, which may be done in any of the three ways below:

- Assign each remaining data point to the centroid closest to it. This is the simplest method.
- Use the clusters produced from the sample to perform supervised learning (classification). Each cluster is regarded as a class. The clustered sample is thus treated as the training data for learning. The resulting classifier is then applied to classify the remaining data points into appropriate classes or clusters.
- Use the clusters produced from the sample as seeds to perform **semi-supervised learning**. Semi-supervised learning is a new learning model that learns from a small set of labeled examples (with classes) and a large set of unlabeled examples (without classes). In our case, the clustered sample data are used as the labeled set and the remaining data points are used as the unlabeled set. The results of the learn-

ing naturally cluster all the remaining data points. We will study this technique in the next chapter.

4. The algorithm is sensitive to **initial seeds**, which are the initially selected centroids. Different initial seeds may result in different clusters. Thus, if the sum of squared error is used as the stopping criterion, the algorithm only achieves **local optimal**. The global optimal is computationally infeasible for large data sets.

Example 6: Figure 4.6 shows the clustering process of a 2-dimensional data set. The goal is to find two clusters. The randomly selected initial seeds are marked with crosses in Fig. 4.6(A). Figure 4.6(B) gives the clustering result of the first iteration. Figure 4.6(C) gives the result of the second iteration. Since there is no re-assignment of data points, the algorithm stops.

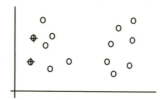

(A). Random selection of seeds (centroids)

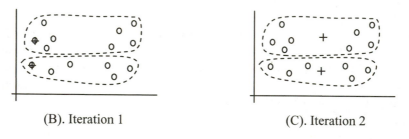

(B). Iteration 1   (C). Iteration 2

Fig. 4.6. Poor initial seeds (centroids)

If the initial seeds are different, we may obtain entirely different clusters as Fig. 4.7 shows. Figure 4.7 uses the same data as Fig. 4.6, but different initial seeds (Fig. 4.7(A)). After two iterations, the algorithm ends, and the final clusters are given in Fig. 4.7(C). These two clusters are more reasonable than the two clusters in Fig. 4.6(C), which indicates that the choice of the initial seeds in Fig. 4.6(A) is poor.

To select good initial seeds, researchers have proposed several methods. One simple method is to first compute the mean **m** (the centroid) of the entire data set (any random data point rather than the mean can be used as well). Then the first seed data point $\mathbf{x}_1$ is selected to be the furthest from the mean $\mathbf{m}$. The second data point $\mathbf{x}_2$ is selected to be the furthest from $\mathbf{x}_1$. Each subsequent data point $\mathbf{x}_i$ is selected such that the sum of distances from $\mathbf{x}_i$ to those already selected data points is the largest. However, if the data has outliers, the method will not work well. To deal with outliers, again, we can randomly select a small sample of the data and perform the same operation on the sample. As we discussed above, since the number of outliers is small, the chance that they show up in the sample is very small.

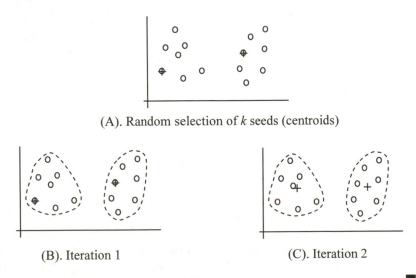

(A). Random selection of k seeds (centroids)

        (B). Iteration 1   (C). Iteration 2

Fig. 4.7. Good initial seeds (centroids)   ■

Another method is to sample the data and use the sample to perform hierarchical clustering, which we will discuss in Sect. 4.4. The centroids of the resulting k clusters are used as the initial seeds.

Yet another approach is to manually select seeds. This may not be a difficult task for text clustering applications because it is easy for human users to read some documents and pick some good seeds. These seeds may help improve the clustering result significantly and also enable the system to produce clusters that meet the user's needs.

5. The k-means algorithm is not suitable for discovering clusters that are not hyper-ellipsoids (or hyper-spheres).

Example 7: Figure 4.8(A) shows a 2-dimensional data set. There are two irregular shaped clusters. However, the two clusters are not hyper-ellipsoids, which means that the k-means algorithm will not be able to find them. Instead, it may find the two clusters shown in Fig. 4.8(B).

The question is: are the two clusters in Fig. 4.8(B) necessarily bad? The answer is no. It depends on the application. It is not true that a clustering algorithm that is able to find arbitrarily shaped clusters is always better. We will discuss this issue in Sect. 4.3.2.

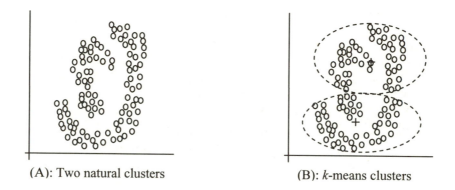

(A): Two natural clusters   (B): k-means clusters

Fig. 4.8. Natural (but irregular) clusters and k-means clusters   ■

Despite these weaknesses, k-means is still the most popular algorithm in practice due to its simplicity, efficiency and the fact that other clustering algorithms have their own lists of weaknesses. There is no clear evidence showing that any other clustering algorithm performs better than the k-means algorithm in general, although it may be more suitable for some specific types of data or applications than k-means. Note also that comparing different clustering algorithms is a very difficult task because unlike supervised learning, nobody knows what the correct clusters are, especially in high dimensional spaces. Although there are several cluster evaluation methods, they all have drawbacks. We will discuss the evaluation issue in Sect. 4.9.

4.3 Representation of Clusters

Once a set of clusters is found, the next task is to find a way to represent the clusters. In some applications, outputting the set of data points that makes up the cluster to the user is sufficient. However, in other applications that involve decision making, the resulting clusters need to be represented in a compact and understandable way, which also facilitates the evaluation of the resulting clusters.

4.3.1 Common Ways of Representing Clusters

There are three main ways to represent clusters:

1. Use the centroid of each cluster to represent the cluster. This is the most popular way. The centroid tells where the center of the cluster is. One may also compute the radius and standard deviation of the cluster to determine the spread in each dimension. The centroid representation alone works well if the clusters are of the hyper-spherical shape. If clusters are elongated or are of other shapes, centroids may not be suitable.
2. Use classification models to represent clusters. In this method, we treat each cluster as a class. That is, all the data points in a cluster are regarded as having the same class label, e.g., the cluster ID. We then run a supervised learning algorithm on the data to find a classification model. For example, we may use the decision tree learning to distinguish the clusters. The resulting tree or set of rules provide an understandable representation of the clusters.

   Figure 4.9 shows a partitioning produced by a decision tree algorithm. The original clustering gave three clusters. Data points in cluster 1 are represented by 1's, data points in cluster 2 are represented by 2's, and data points in cluster 3 are represented by 3's. We can see that the three clusters are separated and each can be represented with a rule.

   $x \le 2 \rightarrow$ cluster 1
   $x > 2, y > 1.5 \rightarrow$ cluster 2
   $x > 2, y \le 1.5 \rightarrow$ cluster 3

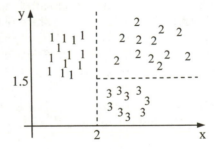

Fig. 4.9. Description of clusters using rules

We make two remarks about this representation method:

- The partitioning in Fig. 4.9 is an ideal case as each cluster is represented by a single rectangle (or rule). However, in most applications, the situation may not be so ideal. A cluster may be split into a few

hyper-rectangles or rules. However, there is usually a dominant or large rule which covers most of the data points in the cluster.
- One can use the set of rules to evaluate the clusters to see whether they conform to some existing domain knowledge or intuition.
3. Use frequent values in each cluster to represent it. This method is mainly for clustering of categorical data (e.g., in the k-modes clustering). It is also the key method used in text clustering, where a small set of frequent words in each cluster is selected to represent the cluster.

4.3.2   Clusters of Arbitrary Shapes

Hyper-elliptical and hyper-spherical clusters are usually easy to represent, using their centroids together with spreads (e.g., standard deviations), rules, or a combination of both. However, other arbitrary shaped clusters, like the natural clusters shown in Fig. 4.8(A), are hard to represent especially in high dimensional spaces.

A common criticism about an algorithm like k-means is that it is not able to find arbitrarily shaped clusters. However, this criticism may not be as bad as it sounds because whether one type of clustering is desirable or not depends on the application. Let us use the natural clusters in Fig. 4.8(A) to discuss this issue together with an artificial application.

Example 8: Assume that the data shown in Fig. 4.8(A) is the measurement data of people's physical sizes. We want to group people based on their sizes into only two groups in order to mass-produce T-shirts of only 2 sizes (say large and small). Even if the measurement data indicate two natural clusters as in Fig. 4.8(A), it is difficult to use the clusters because we need centroids of the clusters to design T-shirts. The clusters in Fig. 4.8(B) are in fact better because they provide us the centroids that are representative of the surrounding data points. If we use the centroids of the two natural clusters as shown in Fig. 4.10 to make T-shirts, it is clearly inappropriate because they are too near to each other in this case. In general, it does not make sense to define the concept of center or centroid for an irregularly shaped cluster.   ∎

Note that clusters of arbitrary shapes can be found by neighborhood search algorithms such as some hierarchical clustering methods (see the next section), and density-based clustering methods [164]. Due to the difficulty of representing an arbitrarily shaped cluster, an algorithm that finds such clusters may only output a list of data points in each cluster, which are not as easy to use. These kinds of clusters are more useful in spatial and image processing applications, but less useful in others.

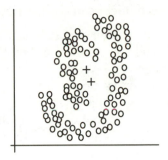

Fig. 4.10. Two natural clusters and their centroids

4.4   Hierarchical Clustering

Hierarchical clustering is another major clustering approach. It has a number of desirable properties which make it popular. It clusters by producing a nested sequence of clusters like a **tree** (also called a **dendrogram**). Singleton clusters (individual data points) are at the bottom of the tree and one root cluster is at the top, which covers all data points. Each internal cluster node contains child cluster nodes. Sibling clusters partition the data points covered by their common parent. Figure 4.11 shows an example.

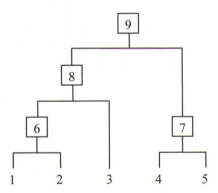

Fig. 4.11. An illustration of hierarchical clustering

At the bottom of the tree, there are 5 clusters (5 data points). At the next level, cluster 6 contains data points 1 and 2, and cluster 7 contains data points 4 and 5. As we move up the tree, we have fewer and fewer clusters. Since the whole clustering tree is stored, the user can choose to view clusters at any level of the tree.

There are two main types of hierarchical clustering methods:

Agglomerative (bottom up) clustering: It builds the dendrogram (tree)
   from the bottom level, and merges the most similar (or nearest) pair of
   clusters at each level to go one level up. The process continues until all
   the data points are merged into a single cluster (i.e., the root cluster).

Divisive (top down) clustering: It starts with all data points in one cluster,
   the root. It then splits the root into a set of child clusters. Each child
   cluster is recursively divided further until only singleton clusters of individual data points remain, i.e., each cluster with only a single point.

Agglomerative methods are much more popular than divisive methods. We
will focus on agglomerative hierarchical clustering. The general agglomerative algorithm is given in Fig. 4.12.

>   **Algorithm** Agglomerative(D)
> 1   Make each data point in the data set D a cluster,
> 2   Compute all pair-wise distances of $x_1, x_2, ..., x_n \in D$;
> 2   **repeat**
> 3   find two clusters that are nearest to each other;
> 4   merge the two clusters form a new cluster c;
> 5   compute the distance from c to all other clusters;
> 12   **until** there is only one cluster left

Fig. 4.12. The agglomerative hierarchical clustering algorithm

Example 9: Figure 4.13 illustrates the working of the algorithm. The data
points are in a 2-dimensional space. Figure 4.13(A) shows the sequence of
nested clusters, and Fig. 4.13(B) gives the dendrogram.   ∎

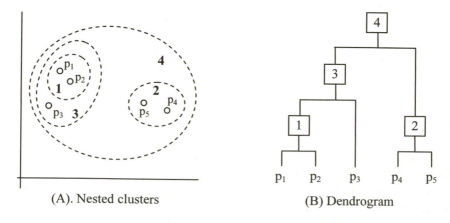

(A). Nested clusters   (B) Dendrogram

Fig. 4.13. The working of an agglomerative hierarchical clustering algorithm

Unlike the k-means algorithm, which uses only the centroids in distance computation, hierarchical clustering may use anyone of several methods to determine the distance between two clusters. We introduce them next.

4.4.1 Single-Link Method

In **single-link** (or **single linkage**) hierarchical clustering, the distance between two clusters is the distance between two closest data points in the two clusters (one data point from each cluster). In other words, the single-link clustering merges the two clusters in each step whose two nearest data points (or members) have the smallest distance, i.e., the two clusters with the **smallest minimum** pair-wise distance. The single-link method is suitable for finding non-elliptical shape clusters. However, it can be sensitive to noise in the data, which may cause the **chain effect** and produce straggly clusters. Figure 4.14 illustrates this situation. The noisy data points (represented with filled circles) in the middle connect two natural clusters and split one of them.

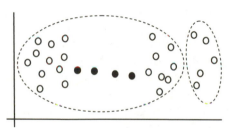

Fig. 4.14. The chain effect of the single-link method

With suitable data structures, single-link hierarchical clustering can be done in $O(n^2)$ time, where n is the number of data points. This is much slower than the k-means method, which performs clustering in linear time.

4.4.2 Complete-Link Method

In **complete-link** (or **complete linkage**) clustering, the distance between two clusters is the **maximum** of all pair-wise distances between the data points in the two clusters. In other words, the complete-link clustering merges the two clusters in each step whose two furthest data points have the smallest distance, i.e., the two clusters with the **smallest maximum** pair-wise distance. Figure 4.15 shows the clusters produced by complete-link clustering using the same data as in Fig. 4.14.

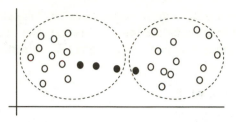

Fig. 4.15. Clustering using the complete-link method

Although the complete-link method does not have the problem of chain effects, it can be sensitive to outliers. Despite this limitation, it has been observed that the complete-link method usually produces better clusters than the single-link method. The worse case time complexity of the complete-link clustering is $O(n^2 \log n)$, where n is the number of data points.

4.4.3 Average-Link Method

This is a compromise between the sensitivity of complete-link clustering to outliers and the tendency of single-link clustering to form long chains that do not correspond to the intuitive notion of clusters as compact, spherical objects. In this method, the distance between two clusters is the average distance of all pair-wise distances between the data points in two clusters. The time complexity of this method is also $O(n^2 \log n)$.

Apart from the above three popular methods, there are several others. The following two methods are also commonly used:

Centroid method: In this method, the distance between two clusters is the distance between their centroids.

Ward's method: In this method, the distance between two clusters is defined as the increase in the sum of squared error (distances) from that of two clusters to that of one merged cluster. Thus, the clusters to be merged in the next step are the ones that will increase the sum the least. Recall that the sum of squared error (SSE) is one of the measures used in the k-means clustering (Equation (1)).

4.4.4. Strengths and Weaknesses

Hierarchical clustering has several advantages compared to the k-means and other partitioning clustering methods. It is able to take any form of distance or similarity function. Moreover, unlike the k-means algorithm which only gives k clusters at the end, the hierarchy of clusters from hierarchical clustering enables the user to explore clusters at any level of detail (or granularity). In many applications, this resulting hierarchy can be very useful in its own right. For example, in text document clustering, the cluster hierarchy may represent a topic hierarchy in the documents.

Some studies have shown that agglomerative hierarchical clustering often produces better clusters than the k-means method. It can also find clusters of arbitrary shapes, e.g., using the single-link method.

Hierarchical clustering also has several weaknesses. As we discussed with the individual methods, the single-link method may suffer from the chain effect, and the complete-link method is sensitive to outliers. The main shortcomings of all hierarchical clustering methods are their computation complexities and space requirements, which are at least quadratic. Compared to the k-means algorithm, this is very inefficient and not practical for large data sets. One can use sampling to deal with the problems. A small sample is taken to do clustering and then the rest of the data points are assigned to each cluster either by distance comparison or by supervised learning (see Sect. 4.3.1). Some **scale-up methods** may also be applied to large data sets. The main idea of the scale-up methods is to find many small clusters first using an efficient algorithm, and then to use the centroids of these small clusters to represent the clusters to perform the final hierarchical clustering (see the BIRCH method in [610]).

4.5   Distance Functions

Distance or similarity functions play a central role in all clustering algorithms. Numerous distance functions have been reported in the literature and used in applications. Different distance functions are also used for different types of attributes (also called **variables**).

4.5.1   Numeric Attributes

The most commonly used distance functions for numeric attributes are the **Euclidean distance** and **Manhattan (city block) distance**. Both distance measures are special cases of a more general distance function called the **Minkowski distance**. We use $dist(\mathbf{x}_i, \mathbf{x}_j)$ to denote the distance between two data points of r dimensions. The Minkowski distance is:

$$dist(\mathbf{x}_i, \mathbf{x}_j) = (\mid x_{i1} - x_{j1} \mid^h + \mid x_{i2} - x_{j2} \mid^h + \ldots + \mid x_{ir} - x_{jr} \mid^h)^{\frac{1}{h}}, \qquad (4)$$

where h is a positive integer.

If $h = 2$, it is the **Euclidean distance**,

$$dist(\mathbf{x}_i, \mathbf{x}_j) = \sqrt{(x_{i1} - x_{j1})^2 + (x_{i2} - x_{j2})^2 + ... + (x_{ir} - x_{jr})^2}. \qquad (5)$$

If $h = 1$, it is the **Manhattan distance**,

$$dist(\mathbf{x}_i, \mathbf{x}_j) = |x_{i1} - x_{j1}| + |x_{i2} - x_{j2}| + ... + |x_{ir} - x_{jr}|. \qquad (6)$$

Other common distance functions include:

Weighted Euclidean distance: A weight is associated with each attribute to express its importance in relation to other attributes.

$$dist(\mathbf{x}_i, \mathbf{x}_j) = \sqrt{w_1(x_{i1} - x_{j1})^2 + w_2(x_{i2} - x_{j2})^2 + ... + w_r(x_{ir} - x_{jr})^2}. \qquad (7)$$

Squared Euclidean distance: the standard Euclidean distance is squared in order to place progressively greater weights on data points that are further apart. The distance is

$$dist(\mathbf{x}_i, \mathbf{x}_j) = (x_{i1} - x_{j1})^2 + (x_{i2} - x_{j2})^2 + ... + (x_{ir} - x_{jr})^2. \qquad (8)$$

Chebychev distance: This distance measure is appropriate in cases where one wants to define two data points as "different" if they are different on any one of the attributes. The Chebychev distance is

$$dist(\mathbf{x}_i, \mathbf{x}_j) = \max(|x_{i1} - x_{j1}|, |x_{i2} - x_{j2}|, ..., |x_{ir} - x_{jr}|). \qquad (9)$$

4.5.2   Binary and Nominal Attributes

The above distance measures are only appropriate for numeric attributes. For **binary** and **nominal** attributes (also called **unordered categorical** attributes), we need different functions. Let us discuss binary attributes first.

A **binary attribute** has two states or values, usually represented by 1 and 0. The two states have no numerical ordering. For example, Gender has two values, male and female, which have no ordering relations but are just different. Existing distance functions for binary attributes are based on the proportion of value matches in two data points. A match means that, for a particular attribute, both data points have the same value. It is easy to use a **confusion matrix** to introduce these measures. Given the ith and jth data points, $\mathbf{x}_i$ and $\mathbf{x}_j$, we can construct the confusion matrix in Fig. 4.16.

To give the distance functions, we further divide binary attributes into **symmetric** and **asymmetric** attributes. For different types of attributes, different distance functions need to be used [271]:

Data point $\mathbf{x}_j$

		1	0	
	1	a	b	$a+b$
Data point $\mathbf{x}_i$	0	c	d	$c+d$
		$a+c$	$b+d$	$a+b+c+d$

a: the number of attributes with the value of 1 for both data points.
b: the number of attributes for which $x_{if} = 1$ and $x_{jf} = 0$, where x_{if} (x_{jf}) is the value of the fth attribute of the data point $\mathbf{x}_i$ ($\mathbf{x}_j$).
c: the number of attributes for which $x_{if} = 0$ and $x_{jf} = 1$.
d: the number of attributes with the value of 0 for both data points.

Fig. 4.16. Confusion matrix of two data points with only binary attributes

Symmetric attributes: A binary attribute is **symmetric** if both of its states (0 and 1) have equal importance, and carry the same weight, e.g., male and female of the attribute Gender. The most commonly used distance function for symmetric attributes is the **simple matching distance**, which is the proportion of mismatches (Equation (10)) of their values. We assume that every attribute in the data set is a symmetric attribute.

$$dist(\mathbf{x}_i, \mathbf{x}_j) = \frac{b+c}{a+b+c+d} \qquad (10)$$

We can also weight some components in Equation (10) according to application needs. For example, we may want mismatches to carry twice the weight of matches, or vice versa:

$$dist(\mathbf{x}_i, \mathbf{x}_j) = \frac{2(b+c)}{a+d+2(b+c)} \qquad (11)$$

$$dist(\mathbf{x}_i, \mathbf{x}_j) = \frac{b+c}{2(a+d)+b+c} \qquad (12)$$

Example 10: Given the following two data points, where each attribute is a symmetric binary attribute,

$\mathbf{x}_1$	1	1	1	0	1	0	0
$\mathbf{x}_2$	0	1	1	0	0	1	0

the distance computed based on the simple matching distance is

$$dist(\mathbf{x}_i, \mathbf{x}_j) = \frac{2+1}{2+2+1+2} = \frac{3}{7} = 0.429. \tag{13}$$

■

Asymmetric attributes: A binary attribute is asymmetric if one of the states is more important or valuable than the other. By convention, we use state 1 to represent the more important state, which is typically the rare or infrequent state. The most commonly used distance measure for asymmetric attributes is the **Jaccard distance**:

$$dist(\mathbf{x}_i, \mathbf{x}_j) = \frac{b+c}{a+b+c}. \tag{14}$$

Similarly, we can vary the Jaccard distance by giving more weight to $(b+c)$ or more weight to a to express different emphases.

$$dist(\mathbf{x}_i, \mathbf{x}_j) = \frac{2(b+c)}{a+2(b+c)}. \tag{15}$$

$$dist(\mathbf{x}_i, \mathbf{x}_j) = \frac{b+c}{2a+b+c}. \tag{16}$$

Note that there is also a **Jaccard coefficient**, which measures similarity (rather than distance) and is defined as $a / (a+b+c)$.

For general **nominal attributes** with more than two states or values, the commonly used distance measure is also based on the simple matching distance. Given two data points $\mathbf{x}_i$ and $\mathbf{x}_j$, let the number of attributes be r, and the number of values that match in $\mathbf{x}_i$ and $\mathbf{x}_j$ be q:

$$dist(\mathbf{x}_i, \mathbf{x}_j) = \frac{r-q}{r}. \tag{17}$$

As that for binary attributes, we can give higher weights to different components in Equation (17) according to different application characteristics.

4.5.3   Text Documents

Although a text document consists of a sequence of sentences and each sentence consists of a sequence of words, a document is usually considered as a "bag" of words in document clustering. The sequence and the position information of words are ignored. Thus a document can be represented as a vector just like a normal data point. However, we use similarity to compare two documents rather than distance. The most commonly used similarity function is the **cosine similarity**. We will study this similarity measure in Sect. 6.2.2 when we discuss information retrieval and Web search.

4.6   Data Standardization

One of the most important steps in data pre-processing for clustering is to standardize the data. For example, using the Euclidean distance, standardization of attributes is highly recommended so that all attributes can have equal impact on the distance computation. This is to avoid obtaining clusters that are dominated by attributes with the largest amounts of variation.

Example 11: In a 2-dimensional data set, the value range of one attribute is from 0 to 1, while the value range of the other attribute is from 0 to 1000. Consider the following pair of data points x_i: (0.1, 20) and x_j: (0.9, 720). The Euclidean distance between the two points is

$$dist(\mathbf{x}_i, \mathbf{x}_j) = \sqrt{(0.9 - 0.1)^2 + (720 - 20)^2} = 700.000457, \tag{18}$$

which is almost completely dominated by (720–20) = 700. To deal with the problem, we standardize the attributes, e.g., to force the attributes to have a common value range. If both attributes are forced to have a scale within the range 0–1, the values 20 and 720 become 0.02 and 0.72. The distance on the first dimension becomes 0.8 and the distance on the second dimension 0.7, which are more equitable. Then, $dist(\mathbf{x}_i, \mathbf{x}_j) = 1.063$. ∎

This example shows that standardizing attributes is important. In fact, different types of attributes require different treatments. We list these treatments below.

Interval-scaled attributes: These are numeric/continuous attributes. Their values are real numbers following a linear scale. Examples of such attributes are age, height, weight, cost, etc. The idea is that intervals keep the same importance through out the scale. For example, the difference in age between 10 and 20 is the same as that between 40 and 50.

There are two main approaches to standardize interval scaled attributes, **range** and **z-score**. The range method divides each value by the range of valid values of the attribute so that the transformed value ranges between 0 and 1. Given the value x_{if} of the fth attribute of the ith data point, the new value $rg(x_{if})$ is,

$$rg(x_{if}) = \frac{x_{if} - \min(f)}{\max(f) - \min(f)}, \tag{19}$$

where min(f) and max(f) are the minimum value and maximum value of attribute f respectively. max(f) − min(f) is the value range of the valid values of attribute f.

The z-score method transforms an attribute value based on the mean and the standard deviation of the attribute. That is, the z-score of the value indicates how far and in what direction the value deviates from the mean of the attribute, expressed in units of the standard deviation of the attribute. The standard deviation of attribute f, denoted by σ_f, is computed with:

$$\sigma_f = \sqrt{\frac{\sum_{i=1}^{n}(x_{if} - \mu_f)^2}{n-1}}, \tag{20}$$

where n is the number of data points in the data set, x_{if} is the same as above, and μ_f is the mean of attribute f, which is computed with:

$$\mu_f = \frac{1}{n}\sum_{i=1}^{n} x_{if}. \tag{21}$$

Given the value x_{if}, its z-score (the new value after transformation) is $z(x_{if})$,

$$z(x_{if}) = \frac{x_{if} - \mu_f}{\sigma_f}. \tag{22}$$

Ratio-Scaled Attributes: These are also numeric attributes taking real values. However, unlike interval-scaled attributes, their scales are not linear. For example, the total amount of microorganisms that evolve in a time t is approximately given by

$$Ae^{Bt},$$

where A and B are some positive constants. This formula is referred to as exponential growth. If we have such attributes in a data set for clustering, we have one of the following two options:

1. Treat it as an interval-scaled attribute. This is often not recommended due to scale distortion.
2. Perform logarithmic transformation to each value, x_{if}, i.e.,

$$\log(x_{if}). \tag{23}$$

After the transformation, the attribute can be treated as an interval-scaled attribute.

Nominal (Unordered Categorical) Attributes: As we discussed in Sect. 4.5.2, the value of such an attribute can take anyone of a set of states (also called categories). The states have no logical or numerical ordering. For example, the attribute *fruit* may have the possible values, Apple, Orange, and Pear, which have no ordering. A **binary attribute** is a special case of a nominal attribute with only two states or values.

Although nominal attributes are not standardized as numeric attributes, it is sometime useful to convert a nominal attribute to a set of binary attributes. Let the number of values of a nominal attribute be v. We can then create v binary attributes to represent them, i.e., one binary attribute for each value. If a data instance for the nominal attribute takes a particular value, the value of its corresponding binary attribute is set to 1, otherwise it is set to 0. The resulting binary attributes can be used as numeric attributes. We will discuss this again in Sect. 4.7.

Example 12: For the nominal attribute *fruit*, we create three binary attributes called, Apple, Orange, and Pear in the new data. If a particular data instance in the original data has Apple as the value for *fruit*, then in the transformed data, we set the value of the attribute Apple to 1, and the values of attributes Orange and Pear to 0.   ■

Ordinal (Ordered Categorical) Attributes: An ordinal attribute is like a nominal attribute, but its values have a numerical ordering. For example, the age attribute may have the values, Young, Middle-Age and Old. The common approach to distance computation is to treat ordinal attributes as interval-scaled attributes and use the same methods as for interval-scaled attributes to standardize the values of ordinal attributes.

4.7   Handling of Mixed Attributes

So far, we have assumed that a data set contains only one type of attributes. However, in practice, a data set may contain mixed attributes. That is, it may contain any subset of the six types of attributes, **interval-scaled**, **symmetric binary**, **asymmetric binary**, **ratio-scaled**, **ordinal** and **nominal** attributes. Clustering a data set involving mixed attributes is a challenging problem.

One way to deal with such a data set is to choose a dominant attribute type and then convert the attributes of other types to this type. For example, if most attributes in a data set are interval-scaled, we can convert ordinal attributes and ratio-scaled attributes to interval-scaled attributes as discussed above. It is also appropriate to treat symmetric binary attributes as interval-scaled attributes. However, it does not make much sense to convert a nominal attribute with more than two values or an asymmetric binary attribute to an interval-scaled attribute, but it is still frequently done in practice by assigning some numbers to them according to some hidden ordering. For instance, in the example of Apple, Orange, and Pear, one may order them according to their prices, and thus make the attribute *fruit* an ordinal attribute or even an interval-scaled attribute. In the previous section, we also saw that a nominal attribute can be converted to a set of (symmetric) binary attributes, which in turn can be regarded as interval-scaled attributes.

Another method of handling mixed attributes is to compute the distance of each attribute of the two data points separately and then combine all the individual distances to produce an overall distance. We describe one such method, which is due to Gower [205] and is also described in [218, 271]. We describe the combination formula first (Equation (24)) and then present the methods to compute individual distances.

$$dist(\mathbf{x}_i, \mathbf{x}_j) = \frac{\sum_{f=1}^{r} \delta_{ij}^{f} d_{ij}^{f}}{\sum_{f=1}^{r} \delta_{ij}^{f}}. \tag{24}$$

This distance value is between 0 and 1. r is the number of attributes in the data set. The indicator δ_{ij}^{f} is 1 if both values x_{if} and x_{jf} for attribute f are non-missing, and it is set to 0 otherwise. It is also set to 0 if attribute f is asymmetric and the match is 0–0. Equation (24) cannot be computed if all δ_{ij}^{f}'s are 0. In such a case, some default value may be used or one of the data points is removed. d_{ij}^{f} is the distance contributed by attribute f, and it is in the range 0–1. If f is a binary or nominal attribute,

$$d_{ij}^{f} = \begin{cases} 1 & if \ x_{if} \neq x_{jf} \\ 0 & otherwise \end{cases} \tag{25}$$

If all the attributes are nominal, Equation (24) reduces to Equation (17). The same is true for symmetric binary attributes, in which we recover the simple matching distance (Equation (10)). When the attributes are all asymmetric, we obtain the Jaccard distance (Equation (14)).

If attribute f is interval-scaled, we use

$$d_{ij}^{f} = \frac{|x_{if} - x_{jf}|}{R_f} \tag{26}$$

where R_f is the value range of attribute f, which is

$$R_f = \max(f) - \min(f) \tag{27}$$

Ordinal attributes and ratio-scaled attributes are handled in the same way after conversion.

If all the attributes are interval-scaled, Equation (24) becomes the Manhattan distance assuming that all attribute values are standardized by dividing their values with the ranges of their corresponding attributes.

4.8   Which Clustering Algorithm to Use?

Clustering research and application has a long history. Over the years, a vast collection of clustering algorithms has been designed. This chapter only introduced several of the main algorithms.

Given an application data set, choosing the "best" clustering algorithm to cluster the data is a challenge. Every clustering algorithm has limitations and works well with only certain data distributions. However, it is very hard, if not impossible, to know what distribution the application data follows. Worse still, the application data set may not fully follow any "ideal" structure or distribution required by the algorithms. Apart from choosing a suitable clustering algorithm from a large collection of available algorithms, deciding how to standardize the data, to choose a suitable distance function and to select other parameter values (e.g., k in the k-means algorithm) are complex as well. Due to these complexities, the common practice is to run several algorithms using different distance functions and parameter settings, and then to carefully analyze and compare the results.

The interpretation of the results should be based on insight into the meaning of the original data together with knowledge of the algorithms used. That is, it is crucial that the user of a clustering algorithm fully understands the algorithm and its limitations. He/she should also have the domain expertise to examine the clustering results. In many cases, generating cluster descriptions using a supervised learning method (e.g., decision tree induction) can be particularly helpful to the analysis and comparison.

4.9   Cluster Evaluation

After a set of clusters is found, we need to assess the goodness of the clusters. Unlike classification, where it is easy to measure accuracy using labeled test data, for clustering nobody knows what the correct clusters are given a data set. Thus, the quality of a clustering is much harder to evaluate. We introduce a few commonly used evaluation methods below.

User Inspection: A panel of experts is asked to inspect the resulting clusters and to score them. Since this process is subjective, we take the average of the scores from all the experts as the final score of the clustering. This manual inspection is obviously a labor intensive and time consuming task. It is subjective as well. However, in most applications, some level of manual inspection is necessary because no other existing evaluation methods are able to guarantee the quality of the final clusters. It should be noted that direct user inspection may be easy for certain types of data, but not for others. For example, user inspection is not hard for text documents because one can read them easily. However, for a relational table with only numbers, staring at the data instances in each cluster makes no sense. The user can only meaningfully study the centroids of the clusters, or rules that characterize the clusters generated by a decision tree algorithm or some other supervised learning methods (see Sect. 4.3.1).

Ground Truth: In this method, classification data sets are used to evaluate clustering algorithms. Recall that a classification data set has several classes, and each data instance/point is labeled with one class. Using such a data set for cluster evaluation, we make the assumption that each class corresponds to a cluster. For example, if a data set has three classes, we assume that it has three clusters, and we request the clustering algorithm to also produce three clusters. After clustering, we compare the cluster memberships with the class memberships to determine how good the clustering is. A variety of measures can be used to assess the clustering quality, e.g., entropy, purity, precision, recall, and F-score.

To facilitate evaluation, a confusion matrix can be constructed from the resulting clusters. From the matrix, various measurements can be computed. Let the set of classes in the data set D be $C = (c_1, c_2, ..., c_k)$. The clustering method also produces k clusters, which partition D into k disjoint subsets, $D_1, D_2, ..., D_k$.

Entropy: For each cluster, we can measure its entropy as follows:

$$entropy(D_i) = -\sum_{j=1}^{k} \Pr_i(c_j) \log_2 \Pr_i(c_j), \tag{28}$$

where $\Pr_i(c_j)$ is the proportion of class c_j data points in cluster i or D_i. The total entropy of the whole clustering (which considers all clusters) is

$$entropy_{total}(D) = \sum_{i=1}^{k} \frac{|D_i|}{|D|} \times entropy(D_i). \tag{29}$$

Purity: This measures the extent that a cluster contains only one class of data. The purity of each cluster is computed with

$$purity(D_i) = \max_j(\Pr_i(c_j)).$$ (30)

The total purity of the whole clustering (considering all clusters) is

$$purity_{total}(D) = \sum_{i=1}^{k} \frac{|D_i|}{|D|} \times purity(D_i).$$ (31)

Precision, recall, and F-score can be computed as well for each cluster based on the class that is the most frequent in the cluster. Note that these measures are based on a single class (see Sect. 3.3.2).

Example 13: Assume we have a text collection D of 900 documents from three topics (or three classes), Science, Sports, and Politics. Each class has 300 documents, and each document is labeled with one of the topics (classes). We use this collection to perform clustering to find three clusters. Class/topic labels are not used in clustering. After clustering, we want to measure the effectiveness of the clustering algorithm.

First, a confusion matrix (Fig. 4.17) is constructed based on the clustering results. From Fig. 4.17, we see that cluster 1 has 250 Science documents, 20 Sports documents, and 10 Politics documents. The entries of the other rows have similar meanings. The last two columns list the entropy and purity values of each cluster and also the total entropy and purity of the whole clustering (last row). We observe that cluster 1, which contains mainly Science documents, is a much better (or purer) cluster than the other two. This fact is also reflected by both their entropy and purity values.

Cluster	Science	Sports	Politics	Entropy	Purity
1	250	20	10	0.589	0.893
2	20	180	80	1.198	0.643
3	30	100	210	1.257	0.617
Total	300	300	300	1.031	0.711

Fig. 4.17. Confusion matrix with entropy and purity values

Obviously, we can use the total entropy or the total purity to compare different clustering results from the same algorithm with different parameter settings or from different algorithms.

Precision and recall may be computed similarly for each cluster. For example, the precision of Science documents in cluster 1 is 0.89. The recall of Science documents in cluster 1 is 0.83. The F-score for Science documents is thus 0.86.   ∎

A final remark about this evaluation method is that although an algorithm may perform well on some labeled data sets, there is no guarantee that it will perform well on the actual application data at hand, which has no class labels. However, the fact that it performs well on some labeled data sets does give us some confidence on the quality of the algorithm. This evaluation method is said to be based on **external data** or **information**.

There are also methods that evaluate clusters based on the **internal information** in the clusters (without using external data with class labels). These methods measure **intra-cluster cohesion** (compactness) and **inter-cluster separation** (isolation). Cohesion measures how near the data points in a cluster are to the cluster centroid. Sum of squared error (SSE) is a commonly used measure. Separation measures how far apart different cluster centroids are from one another. Any distance functions can be used for the purpose. We should note, however, that good values for these measurements do not always mean good clusters. In most applications, expert judgments are still the key. Clustering evaluation remains to be a very difficult problem.

Indirect Evaluation: In some applications, clustering is not the primary task. Instead, it is used to help perform another more important task. Then, we can use the performance on the primary task to determine which clustering method is the best for the task. For instance, in a Web usage mining application, the primary task is to recommend books to online shoppers. If the shoppers can be clustered according to their profiles and their past purchasing history, we may be able to provide better recommendations. A few clustering methods can be tried, and their results are then evaluated based on how well they help with the recommendation task. Of course, here we assume that the recommendation results can be reliably evaluated.

4.10 Discovering Holes and Data Regions

In this section, we wander a little to discuss something related but quite different from the preceding algorithms. We show that unsupervised learning tasks may be performed by using supervised learning techniques [350].

In clustering, data points are grouped into clusters according to their distances (or similarities). However, clusters only represent one aspect of the hidden knowledge in data. Another aspect that we have not studied is the **holes**. If we treat data instances as points in an r-dimensional space, a hole is simply a region in the space that contains no or few data points. The existence of holes is due to the following two reasons:

1. insufficient data in certain areas, and/or
2. certain attribute-value combinations are not possible or seldom occur.

Although clusters are important, holes in the space can be quite useful too. For example, in a disease database we may find that certain symptoms and/or test values do not occur together, or when a certain medicine is used, some test values never go beyond certain ranges. Discovery of such information can be of great importance in medical domains because it could mean the discovery of a cure to a disease or some biological laws.

The technique discussed in this section aims to divide the data space into two types of regions, **data regions** (also called **dense regions**) and **empty regions** (also called **sparse regions**). A data region is an area in the space that contains a concentration of data points and can be regarded as a cluster. An empty region is a **hole**. A supervised learning technique similar to decision tree induction is used to separate the two types of regions. The algorithm (called CLTree for CLuser Tree [350]) works for numeric attributes, but can be extended to discrete or categorical attributes.

Decision tree learning is a popular technique for classifying data of various classes. For a decision tree algorithm to work, we need at least two classes of data. A clustering data set, however, has no class label for each data point. Thus, the technique is not directly applicable. However, the problem can be dealt with by a simple idea.

We can regard each data instance/point in the data set as having a class label Y. We assume that the data space is uniformly distributed with another type of points, called **non-existing points**, which we will label N. With the N points added to the original data space, our problem of partitioning the data space into data regions and empty regions becomes a supervised classification problem. The decision tree algorithm can be adapted to solve the problem. Let us use an example to illustrate the idea.

Example 14: Figure 4.18(A) gives a 2-dimensional space with 24 data (Y) points. Two data regions (clusters) exist in the space. We then add some uniformly distributed N points (represented by "o") to the data space (Fig. 4.18(B)). With the augmented data set, we can run a decision tree algorithm to obtain the partitioning of the space in Fig. 4.18(B). Data regions and empty regions are separated. Each region is a rectangle, which can be expressed as a rule. ∎

The reason that this technique works is that if there are clusters (or dense data regions) in the data space, the data points cannot be uniformly distributed in the entire space. By adding some uniformly distributed N

points, we can isolate data regions because within each data region there are significantly more Y points than N points. The decision tree technique is well known for this partitioning task.

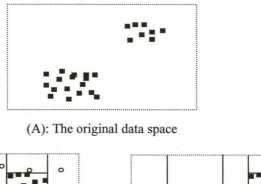

(A): The original data space

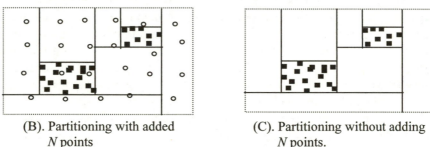

(B). Partitioning with added
N points

(C). Partitioning without adding
N points.

Fig. 4.18. Separating data and empty regions using a decision tree

An interesting question is: can the task be performed without physically adding the N points to the original data? The answer is yes. Physically adding N points increases the size of the data and thus the running time. A more important issue is that it is unlikely that we can have points truly uniformly distributed in a high-dimensional space as we would need an exponential number of them. Fortunately, we do not need to physically add any N points. We can compute them when needed. The CLTree method is able to produce the partitioning in Fig. 4.18(C) with no N points added. The details are quite involved. Interested readers can refer to [350]. This method has some interesting characteristics:

- It provides descriptions or representations of the resulting data regions and empty regions in terms of hyper-rectangles, which can be expressed as rules as we have seen in Sect. 3.2 of Chap. 3 and in Sect. 4.3.1. Many applications require such descriptions, which can be easily interpreted by users.
- It automatically detects outliers, which are data points in empty regions.
- It may not use all attributes in the data just as in decision tree building

for supervised learning. That is, it can automatically determine what attributes are important and what are not. This means that it can perform subspace clustering, i.e., finding clusters that exist in some subspaces (represented by some subsets of the attributes) of the original space.

This method also has limitations. The main limitation is that data regions of irregular shapes are hard to handle since decision tree learning only generates hyper-rectangles (formed by axis-parallel hyper-planes), which are rules. Hence, an irregularly shaped data or empty region may be split into several hyper-rectangles. Post-processing is needed to join them if desired (see [350] for additional details).

Bibliographic Notes

Clustering or unsupervised learning has a long history and a very large body of work. This chapter described only some widely used core algorithms. Most other algorithms are variations or extensions of these methods. For a comprehensive coverage of clustering, please refer to several books dedicated to clustering, e.g., those by Everitt [167], Hartigan [222], Jain and Dubes [252], Kaufman and Rousseeuw [271], and Mirkin [383]. Most data mining texts also have excellent coverage of clustering techniques, e.g., Han and Kamber [218] and Tan et al. [512], which have influenced the writing of this chapter. Below, we review some more recent developments on clustering and give some further readings.

A density-based clustering algorithm based on local data densities was proposed by Ester et al. [164] and Xu et al. [564] for finding clusters of arbitrary shapes. Hinneburg and Keim [239], Sheikholeslami et al. [485] and Wang et al. [538] proposed several grid-based clustering methods which first partition the space into small grids. A popular neural network clustering algorithm is the Self-Organizing Map (SOM) by Kohonen [287]. Fuzzy clustering (e.g., fuzzy c-means) was studied by Bezdek [50] and Dunn [157]. Cheeseman et al. [94] and Moore [396] studied clustering using mixture models. The method assumes that clusters are a mixture of Gaussians and uses the EM algorithm [127] to learn a mixture density. We will see in Chap. 5 that EM based partially supervised learning algorithms are basically clustering methods with some given initial seeds.

Most clustering algorithms work on numeric data. Categorical data and/or transaction data clustering were investigated by Barbará et al. [36], Ganti et al. [193], Gibson et al. [197], Guha et al. [212], Wang et al. [537], etc. A related area in artificial intelligence is the conceptual clustering, which was studied by Fisher [178], Misha et al. [384] and many others.

Many clustering algorithms, e.g., hierarchical clustering algorithms, have high time complexities and are thus not suitable for large data sets. Scaling up such algorithms becomes an important issue for large applications. Several researchers have designed techniques to scale up clustering algorithms, e.g., Bradley et al. [61], Guha et al. [211], Ng and Han [406], and Zhang et al. [610].

In recent years, there were quite a few new developments in clustering. The first one is **subspace clustering**. Traditional clustering algorithms use the whole space to find clusters, but natural clusters may exist in only some sub-spaces. That is, some clusters may only use certain subsets of the attributes. This problem was investigated by Agrawal et al. [8], Aggarwal et al. [4], Aggarwal and Yu [5], Cheng et al. [99], Liu et al. [350], Zaki et al. [590], and many others.

The second new research is **semi-supervised clustering**, which means that the user can provide some initial information to guide the clustering process. For example, the user can select some initial seeds [39] and/or specify some constraints, e.g., **must-link** (two points must be in the same cluster) and **cannot-link** (two points cannot be in the same cluster) [528].

The third is the **spectral clustering**, which emerged from several fields, e.g., VLSI [17] and computer vision [481, 489, 542]. It clusters data points by computing eigenvectors of the similarity matrix. Recently, it was also studied in machine learning and data mining [141, 404, 594].

Yet another new research is **co-clustering**, which simultaneously clusters both rows and columns. This approach was studied by Cheng and Church [100], Dhillon [134], Dhillon et al. [135], and Hartigan [223].

Regarding document and Web page clustering, most implementations are still based on k-means and hierarchical clustering methods, or their variations but using text specific similarity or distance functions. Steinbach et al. [506], and Zhao and Karypis [614, 615] experimented with k-means and agglomerative hierarchical clustering methods and also proposed some improvements. Many researchers also worked on **clustering of search engine results** (or snippets) to organize search results into different topics, e.g., Hearst and Pedersen [233], Kummamuru et al. [294], Leouski and Croft [311], Zamir and Etzioni [591, 592], and Zeng et al. [593].

5 Partially Supervised Learning

With *Wee Sun Lee*

In supervised learning, the learning algorithm uses labeled training examples from every class to generate a classification function. One of the drawbacks of this classic paradigm is that a large number of labeled examples are needed in order to learn accurately. Since labeling is often done manually, it can be very labor intensive and time consuming. In this chapter, we study two **partially supervised learning** tasks. As their names suggest, these two learning tasks do not need full supervision, and thus are able to reduce the labeling effort. The first is the task of **learning from labeled and unlabeled examples**, which is commonly known as **semi-supervised learning**. In this chapter, we also call it **LU learning** (L and U stand for "labeled" and "unlabeled" respectively). In this learning setting, there is a small set of labeled examples of every class, and a large set of unlabeled examples. The objective is to make use of the unlabeled examples to improve learning.

The second is the task of **learning from positive and unlabeled examples**. This problem assumes two-class classification. However, the training data only has a set of labeled positive examples and a set of unlabeled examples, but no labeled negative examples. In this chapter, we also call this problem **PU learning** (P and U stand for "positive" and "unlabeled" respectively). The objective is to build an accurate classifier without labeling any negative examples. We study these two problems in the context of text classification and Web page classification in this chapter. However, the general ideas and the algorithms are also applicable to other kinds of classification tasks.

5.1 Learning from Labeled and Unlabeled Examples

As we described in Chap. 3, the common approach to learning a classification function is to label a set of examples with some pre-defined categories or classes, and then use a learning algorithm to produce a classifier. This classifier is applied to assign classes to future instances (or test data). In the context of text classification and Web page classification, the examples are text documents and Web pages. This approach to building a classifier is called supervised learning because the training documents/pages have been labeled with pre-defined classes.

The main bottleneck of building such a classifier is that a large, often prohibitive, number of labeled training documents are needed to build accurate classifiers. In text classification, the labeling is typically done manually by reading the documents, which is a time consuming task. However, we cannot eliminate labeling completely because without it a machine learning algorithm will not know what the user is interested in. Although unsupervised learning or clustering may help to some extent, clustering does not guarantee to produce the categorization results required by the user. This raises an important question: Can the manual labeling effort be reduced, and can other sources of information be used so that the number of labeled examples required for learning would not be too large?

This section addresses the problem of learning from a small set of **labeled examples** and a large set of **unlabeled examples**, i.e., LU learning. Thus, in this setting only a small set of examples needs to be labeled for each class. However, since a small set of labeled examples is not sufficient for building an accurate classifier, a large number of unlabeled examples are utilized to help. One key point to note is that although the number may be small, every class must have some labeled examples.

In many applications, unlabeled examples are easy to come by. This is especially true for online documents. For example, if we want to build a classifier to classify news articles into different categories or classes, it is fairly easy to collect a huge number of unlabeled news articles from the Web. In fact, in many cases, the new data that need to be classified (which have no class labels) can be used as the unlabeled examples.

The question is: why do the unlabeled data help? In the context of text classification, one reason is that the unlabeled data provide information on the joint probability distribution over words. For example, using only the labeled data we find that documents containing the word "homework" tend to belong to a particular class. If we use this fact to classify the unlabeled documents, we may find that "lecture" co-occurs with "homework" frequently in the unlabeled set. Then, "lecture" may also be an indicative word for the class. Such correlations provide a helpful source of information to increase classification accuracy, especially when the labeled data are scarce.

Several researchers have shown that unlabeled data help learning. That is, under certain conditions using both labeled and unlabeled data in learning is better than using a small set of labeled data alone. Their techniques can thus alleviate the labor-intensive labeling effort. We now study some of these learning techniques, and also discuss their limitations.

5.1.1   EM Algorithm with Naïve Bayesian Classification

One of the LU learning techniques uses the Expectation–Maximization (EM) algorithm [127]. EM is a popular iterative algorithm for maximum likelihood estimation in problems with missing data. The EM algorithm consists of two steps, the **Expectation step** (or **E-step**), and the **Maximization step** (or **M-step**). The E-step basically fills in the missing data based on the current estimation of the parameters. The M-step, which maximizes the likelihood, re-estimates the parameters. This leads to the next iteration of the algorithm, and so on. EM converges to a local minimum when the model parameters stabilize.

The ability of EM to work with missing data is exactly what is needed for learning from labeled and unlabeled examples. The documents in the labeled set (denoted by L) all have class labels (or values). The documents in the unlabeled set (denoted by U) can be regarded as having missing class labels. We can use EM to estimate them based on the current model, i.e., to assign probabilistic class labels to each document d_i in U, i.e., $Pr(c_j|d_i)$. After a number of iterations, all probabilities will converge.

Note that the EM algorithm is not really a specific "algorithm", but is a framework or strategy. It simply runs a base algorithm iteratively. We will use the naïve Bayesian (NB) algorithm discussed in Sect. 3.7 as the base algorithm, and run it iteratively. The parameters that EM estimates in this case are the probability of each word given a class and the class prior probabilities (see Equation (27) and (28) in Sect. 3.7 of Chap. 3).

Although it is quite involved to derive the EM algorithm with the NB classifier, it is fairly straightforward to implement and to apply the algorithm. That is, we use a NB classifier in each iteration of EM, Equation (29) in Chap. 3 for the E-step, and Equations (27) and (28) in Chap. 3 for the M-step. Specifically, we first build a NB classifier f using the labeled examples in L. We then use f to classify the unlabeled examples in U, more accurately to assign a probability to each class for every unlabeled example, i.e., $Pr(c_j|d_i)$, which takes the value in [0, 1] instead of {0, 1}. Some explanations are in order here.

Let the set of classes be $C = \{c_1, c_2, ..., c_{|C|}\}$. Each iteration of EM will assign every example d_i in U a probability distribution on the classes that it may belong to. That is, it assigns d_i the class probabilities of $Pr(c_1|d_i)$, $Pr(c_2|d_i)$, ..., $Pr(c_{|C|}|d_i)$. This is different from the example in the labeled set L, where each document belongs to only a single class c_k, i.e., $Pr(c_k|d_i) = 1$ and $Pr(c_j|d_i) = 0$ for $j \neq k$.

Based on the assignments of $Pr(c_j|d_i)$ to each document in U, a new NB classifier can be constructed. This new classifier can use both the labeled set L and the unlabeled set U as the examples in U now have probabilistic

Algorithm EM(*L*, *U*)
1   Learn an initial naïve Bayesian classifier *f* from only the labeled set *L* (using Equations (27) and (28) in Chap. 3);
2   **repeat**
        // E-Step
3   **for** each example d_i in *U* **do**
4   Using the current classifier *f* to compute $\Pr(c_j|d_i)$ (using Equation (29) in Chap. 3).
5   **end**
        // M-Step
6   learn a new naïve Bayesian classifier *f* from $L \cup U$ by computing $\Pr(c_j)$ and $\Pr(w_t|c_j)$ (using Equations (27) and (28) in Chap. 3).
7   **until** the classifier parameters stabilize
Return the classifier *f* from the last iteration.

Fig. 5.1. The EM algorithm with naïve Bayesian classification

labels, $\Pr(c_j|d_i)$. This leads to the next iteration. The process continues until the classifier parameters ($\Pr(w_t|c_j)$ and $\Pr(c_j)$) no longer change (or have minimum changes).

The EM algorithm with NB classification was proposed for LU learning by Nigam et al. [413]. The algorithm is shown in Fig. 5.1. EM here can also be seen as a clustering method with some initial seeds (labeled data) in each cluster. The class labels of the seeds indicate the class labels of the resulting clusters.

The derivation of the EM algorithm in Fig. 5.1 is quite involved and is given as an appendix at the end of this chapter. Two assumptions are made in the derivation. They are in fact the two mixture model assumptions in Sect. 3.7 of Chap. 3 for deriving the naïve Bayesian classifier for text classification, i.e.,

1. the data is generated by a mixture model, and
2. there is a one-to-one correspondence between mixture components and classes.

It has been shown that the EM algorithm in Fig. 5.1 works well if the two mixture model assumptions for a particular data set are true. Note that although naïve Bayesian classification makes additional assumptions as we discussed in Sect. 3.7 of Chap. 3, it performs surprisingly well despite the obvious violation of the assumptions. The two mixture model assumptions, however, can cause major problems when they do not hold. In many real-life situations, they may be violated. It is often the case that a class (or topic) contains a number of sub-classes (or sub-topics). For example, the class **Sports** may contain documents about different sub-classes of sports, e.g., Baseball, Basketball, Tennis, and Softball. Worse still, a class c_j may even contain documents from completely different topics, e.g., Science, Politics, and Sports. The first assumption above is usually not a problem. The second assumption is critical. If the condition holds, EM works very well and is particularly useful when the labeled set is very small, e.g., fewer than five labeled documents per class. In such cases, every iteration of EM is able to improve the classifier dramatically. However, if the second condition does not hold, the classifier from each iteration can become worse and worse. That is, the unlabeled set hurts learning instead of helping it.

Two methods are proposed to remedy the situation.

Weighting the Unlabeled Data: In LU learning, the labeled set is small, but the unlabeled set is very large. So the EM's parameter estimation is almost completely determined by the unlabeled set after the first iteration. This means that EM essentially performs unsupervised clustering. When the two mixture model assumptions are true, the natural clusters of the data are in correspondence with the class labels. The resulting clusters can be used as the classifier. However, when the assumptions are not true, the clustering can go very wrong, i.e., the clustering may not converge to mixture components corresponding to the given classes, and are therefore detrimental to classification accuracy. In order to reduce the effect of the problem, we can weight down the unlabeled data during parameter estimation (EM iterations). Specifically, we change the computation of $\Pr(w_t|c_j)$ (Equation (27) in Chap. 3) to the following, where the counts of the unlabeled documents are decreased by a factor of μ, $0 \le \mu \le 1$:

$$\Pr(w_t \mid c_j) = \frac{\lambda + \sum_{i=1}^{|D|} \Lambda(i) N_{ti} \Pr(c_j \mid d_i)}{\lambda |V| + \sum_{s=1}^{|V|} \sum_{i=1}^{|D|} \Lambda(i) N_{ti} \Pr(c_j \mid d_i)}, \qquad (1)$$

where

$$\Lambda(i) = \begin{cases} \mu & \text{if } d_i \in U \\ 1 & \text{if } d_i \in L. \end{cases} \qquad (2)$$

When $\mu = 1$, each unlabeled document is weighted the same as a labeled document. When $\mu = 0$, the unlabeled data are not considered. The value of μ can be chosen based on leave-one-out cross-validation accuracy on the labeled training data. The μ value that gives the best result is used.

Finding Mixture Components: Instead of weighting unlabeled data low, we can attack the problem head on, i.e., by finding the mixture components (sub-classes) of the class. For example, the original class Sports may consist of documents from Baseball, Tennis, and Basketball, which are three mixture components (sub-classes or sub-topics) of Sports. Instead of using the original class, we try to find these components and treat each of them as a class. That is, if we can find the three mixture components, we can use them to replace the class Sports. There are several automatic approaches for identifying mixture components. For example, a hierarchical clustering technique was proposed in [111] to find the mixture components, which showed good performances. A simple approach based on leave-one-out cross-validation on the labeled training set was also given in [413].

Manually identifying different components may not be a bad option for text documents because one only needs to read the documents in the labeled set (or some sampled unlabeled documents), which is very small.

5.1.2 Co-Training

Co-training is another approach to learning from labeled and unlabeled examples. This approach assumes that the set of attributes (or features) in the data can be partitioned into two subsets. Each of them is sufficient for learning the target classification function. For example, in Web page classification, one can build a classifier using either the text appearing on the page itself, or the anchor text attached to hyperlinks pointing to the page from other pages on the Web. This means that we can use the same training data to build two classifiers using two subsets of features.

Traditional learning algorithms do not consider this division of features (attributes), or feature redundancy. All the features are pooled together in learning. In some cases, feature selection algorithms are applied to remove redundant features. Co-training exploits this feature division to learn separate classifiers over each of the feature sets, and utilizes the fact that the two classifiers must agree on their labeling of the unlabeled data to do LU learning.

Blum and Mitchell [55] formalize the co-training setting and provide a theoretical guarantee for accurate learning subject to certain assumptions. In the formalization, we have an example (data) space $X = X_1 \times X_2$, where X_1 and X_2 provide two different "views" of the example. That is, each example $\mathbf{x}$ (represented as a vector) is given as a pair $(\mathbf{x}_1, \mathbf{x}_2)$. This simply means that the set of features (or attributes) is partitioned into two subsets. Each "view" or feature subset is sufficient for correct classification. Under some further assumptions, it was proven that co-training algorithms can learn from unlabeled data starting from only a weak classifier built using the small set of labeled training documents.

The first assumption is that the example distribution is **compatible** with the target functions; that is, for most examples, the target classification functions over the feature sets predict the same label. In other words, if f denotes the combined classifier, f_1 denotes the classifier learned from X_1, f_2 denotes the classifier learned from X_2 and c is the actual class label of example $\mathbf{x}$, then $f(\mathbf{x}) = f_1(\mathbf{x}_1) = f_2(\mathbf{x}_2) = c$ for most examples.

The second assumption is that the features in one set of an example are **conditionally independent** of the features in the other set, given the class of the example. In the case of Web page classification, this assumes that the words on a Web page are not related to the words on its incoming hyperlinks, except through the class of the Web page. This is a somewhat unrealistic assumption in practice.

The co-training algorithm explicitly uses the feature split to learn from labeled and unlabeled data. The algorithm is iterative. The main idea is that in each iteration, it learns a classifier from the labeled set L with each subset of the features, and then applies the classifier to classify (or label) the unlabeled examples in U. A number (n_i) of most confidently classified examples in U from each class c_i are added to L. This process ends when U becomes empty (or a fixed number of iterations is reached). In practice, we can set a different n_i for a different class c_i depending on class distributions. For example, if a data set has one third of class 1 examples and two thirds of class 2 examples, we can set $n_1 = 1$ and $n_2 = 2$.

The whole co-training algorithm is shown in Fig. 5.2. Lines 2 and 3 build two classifiers f_1 and f_2 from the two "views" of the data respectively. f_1 and f_2 are then applied to classify the unlabeled examples in U (lines 4 and 5). Some most confidently classified examples are removed from U and added to L. The algorithm then goes to the next iteration.

Algorithm co-training(L, U)
1   **repeat**
2   Learn a classifier f_1 using L based on only $\mathbf{x}_1$ portion of the examples $\mathbf{x}$.
3   Learn a classifier f_2 using L based on only $\mathbf{x}_2$ portion of the examples $\mathbf{x}$.
4   Apply f_1 to classify the examples in U, for each class c_i, pick n_i examples that f_1 most confidently classifies as class c_i, and add them to L.
5   Apply f_2 to classify the examples in U, for each class c_i, pick n_i examples that f_2 most confidently classifies as class c_i, and add them to L.
6   **until** U becomes empty or a fixed number of iterations are reached

Fig. 5.2. A co-training algorithm

When the co-training algorithm ends, it returns two classifiers. At classification time, for each test example the two classifiers are applied separately and their scores are combined to decide the class. For naïve Bayesian classifiers, we multiply the two probability scores, i.e.,

$$Pr(c_j|\mathbf{x}) = Pr(c_j|\mathbf{x}_1)Pr(c_j|\mathbf{x}_2) \tag{3}$$

The key idea of co-training is that classifier f_1 adds examples to the labeled set that are used for learning f_2 based on the X_2 view, and vice versa. Due to the conditional independence assumption, the examples added by f_1 can be considered as new and random examples for learning f_2 based on the X_2 view. Then the learning will progress. The situation is illustrated in Fig. 5.3. This example has classes, positive and negative, and assumes linear separation of the two classes. In the X_1 view (Fig. 5.3(A)), the circled examples are most confident positive and negative examples classified (or labeled) by f_1 in the unlabeled set U. In the X_2 view (Fig. 5.3(B)), these circled examples appear randomly. With these random examples from U added to L, a better f_2 will be learned in the next iteration.

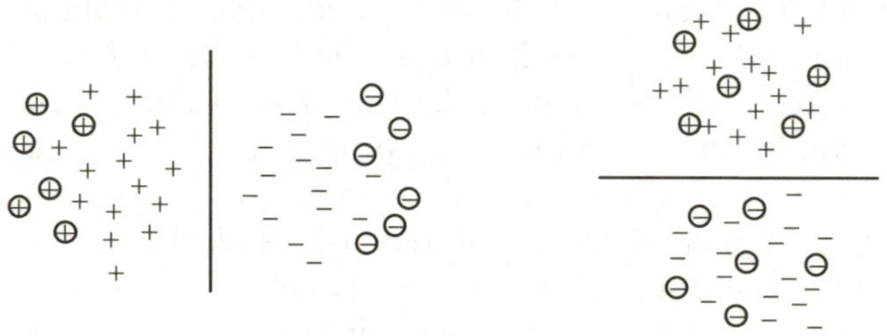

(A) X_1 view: data in U labeled by h_1 (B) X_2 view: the same data

Fig. 5.3. Two views of co-training.

However, if the added examples to L are not random examples in the X_2 space but very similar to the situation in Fig. 5.3(A), then these examples are not informative to learning. That is, if the two subsets of features are correlated given the class or the conditional independence assumption is violated, the added examples will not be random but isolated in a specific region similar to those in Fig. 5.3(A). Then they will not be as useful or informative to learning. Consequently, co-training will not be effective.

In [411], it is shown that co-training produces more accurate classifiers than the EM algorithm presented in the previous section, even for data sets whose feature division does not completely satisfy the strict requirements of compatibility and conditional independence.

5.1.3 Self-Training

Self-training, which is similar to both EM and co-training, is another method for LU learning. It is an incremental algorithm that does not use the split of features. Initially, a classifier (e.g., naïve Bayesian classifier) is trained with the small labeled set considering all features. The classifier is then applied to classify the unlabeled set. Those most confidently classified (or unlabeled) documents of each class, together with their predicted class labels, are added to the labeled set. The classifier is then re-trained and the procedure is repeated. This process iterates until all the unlabeled documents are given class labels. The basic idea of this method is that the classifier uses its own predictions to teach itself.

5.1.4   Transductive Support Vector Machines

Support vector machines (SVM) is one of the most effective methods for text classification. One way to use unlabeled data in training SVM is by selecting the labels of the unlabeled data in such a way that the resulting margin of the classifier is maximized. Training for the purpose of labeling known (unlabeled) test instances is referred to as **transduction**, giving rise to the name **transductive SVM** [526]. An example of how transduction can change the decision boundary is shown in Fig. 5.4. In this example, the old decision boundary, constructed using only labeled data, would have a very small margin on the unlabeled data. By utilizing the unlabeled data in the training process, a classifier that has the largest margin on both the labeled and unlabeled data can be obtained.

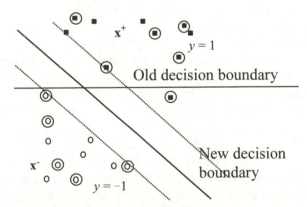

Fig. 5.4. The old decision boundary (before the addition of unlabeled data) and the new decision boundary created by transductive SVM. The unlabeled data are indicated by circles around them

The main difficulty with applying transductive SVM is the computational complexity. When all the labels are observed, training SVM is a convex optimization problem that can be solved efficiently. The problem of assigning labels to unlabeled examples in such a way that the resulting margin of the classifier is maximized can no longer be solved efficiently.

To solve the problem, Joachims [259] used a sub-optimal iterative method that starts by learning a classifier using only the labeled data. The method then treats a subset of unlabeled instances that are most confidently labeled positive by the learned classifier as initial positive examples while the rest of the unlabeled examples are treated as initial negative examples. The number of instances to label as positive can be specified by the user to change the precision–recall trade-off and is maintained throughout the iterations. The method then tries to improve the soft margin cost function by iteratively changing the labels of some of the instances and re-training the SVM. The ratio of positive to negative instances is maintained by selecting one positively labeled instance p and one negatively labeled instance q to change in each iteration. It was shown in [259] that if the two instances are selected such that the slack variables $\xi_p > 0$, $\xi_q > 0$ and $\xi_p + \xi_q > 2$, the soft margin cost function will decreases at each iteration. Further improvements described in [259] include allowing the soft margin error of unlabeled examples to be penalized differently from the soft margin error of the labeled examples and penalizing the soft margin error on the positive unlabeled examples differently from the soft margin error on the negative unlabeled examples. The penalty on the unlabeled examples is also iteratively increased from a small value to the desired value. This may improve the chances of finding a good local optimum as it may be easier to improve the cost function when the penalty is small. The method was applied successfully to text classification problems.

Like other methods of learning from labeled and unlabeled examples, transductive SVM can be sensitive to its assumptions. When the large margin assumption is correct on the dataset, it may improve performance but when the assumption is incorrect, it can decrease performance compared to supervised learning. As an example, the transductive SVM performed poorly using small labeled data sets when separating Project Web pages from other types of university Web pages in [259]. It was conjectured that, with a small number of labeled data, separating the Web pages according to some of the underlying topics of the Web pages may give a larger margin than separating them according to whether the Web pages are Project pages or not.

5.1.5 Graph-Based Methods

Graph-based LU learning methods can be viewed as extensions of **nearest neighbor supervised learning** algorithms that work with both labeled and unlabeled instances. The basic idea in these methods is to treat labeled and unlabeled instances as vertices in a graph where a similarity function is used to define the edge weights between instances. The graph, with similar instances connected by larger weights, is then used to help label the unlabeled instances in such a way that labels of vertices connected by edges with large weights tend to agree with each other. Methods used for constructing the graphs include connecting each instance to its k-nearest neighbors, connecting each instance to other instances within a certain distance δ and using a fully connected graph with an exponentially decreasing similarity function such as the Gaussian function to assign the weights. The assumptions used in these methods are similar to those of the nearest neighbor classifier, that is, near neighbors should have the same labels and we have a good measure of similarity between instances. We discuss three types of graph-based LU learning methods below: **mincut**, **Gaussian fields** and **spectral graph transducer**. All three methods work on binary classification problems but, like the support vector machines, can be used with strategies such as one-against-rest for solving multiple class classification problems.

Mincut: This method was proposed by Blum and Chalwa [54]. A weighted graph $G = (V, E, W)$ is constructed first, where V consists of the labeled and unlabeled instances, E consists of edges between the instances and W is a function on the edges with $W(i, j) = w_{ij}$ denoting the similarity of instances i and j. The vertices associated with labeled instances are then given values from $\{0, 1\}$ consistent with their binary labels. The idea in the mincut algorithm is to find an assignment of values v_i from the set $\{0, 1\}$ to the unlabeled instances in V such that the cost function $\sum_{(i,j)\in E} w_{ij} |v_i - v_j|$ is minimized. The advantage of this formulation is that the problem can be solved in polynomial time even though it is a combinatorial optimization problem. One way to do this is to transform the problem into a max-flow problem (see [116] for a description of the max-flow problem). To do that, we convert the graph into a flow network by introducing a source vertex v_+ and a sink vertex v_-, where the source vertex is connected by edges with infinite capacities to the positive labeled instances while the sink vertex is connected by edges with infinite capacities to the negative labeled instances. The other edge weights in the graph are also treated as edge capacities in the flow network. A cut of the network is a partition of the vertices into two subsets V_+ and V_- such that $v_+ \in V_+$ and $v_- \in V_-$. A minimum cut is a partition that has the smallest sum of capacities in the edges connecting V_+ and V_-. Finding a minimum cut is equivalent to minimizing the function $\sum_{(i,j)\in E} w_{ij} |v_i - v_j|$ since all the vertices are assigned values from $\{0,$

1}. Max-flow algorithms can be used to efficiently find a mincut of the network in time $O(|V|^3)$.

Gaussian Fields: Instead of minimizing $\sum_{(i,j)\in E} w_{ij} |v_i - v_j|$, Zhu et al. [619] proposed minimizing $\sum_{(i,j)\in E} w_{ij}(v_i - v_j)^2$ with the value of the vertices being selected from [0, 1] instead of {0, 1}. The advantage of using this formulation is that it allows the solution to be obtained using linear algebra. Let W be the weight matrix corresponding to the graph,

$$W = \begin{bmatrix} W_{LL} & W_{LU} \\ W_{UL} & W_{UU} \end{bmatrix}, \tag{4}$$

where W_{LL}, W_{LU}, W_{UL} and W_{UU} are sub-matrices with the subscript L denoting labeled instances and the subscript U denoting the unlabeled instances. Let D be a diagonal matrix where $D_{ii} = \sum_j w_{ij}$ is the sum of the entries in row (or column) i. We also form a vector $\mathbf{v}$, consisting of values assigned to the labeled and unlabeled instances. The labeled instances are assigned fixed values in {0, 1} consistent with their labels while the values v_i assigned to the unlabeled instances are chosen to minimize $\sum_{(i,j)\in E} w_{ij}(v_i - v_j)^2$. The solution can be written as

$$\mathbf{v}_U = (D_{UU} - W_{UU})^{-1} W_{UL} \mathbf{v}_L, \tag{5}$$

where $\mathbf{v}_U$ is the part of the vector $\mathbf{v}$ that contains values assigned to the unlabeled instances, $\mathbf{v}_L$ is the part of the vector that contains values assigned to labeled instances and D_{UU} is the sub-matrix of D consisting of sum of entries of rows in W associated with unlabeled instances.

The optimization problem $\sum_{(i,j)\in E} w_{ij}(v_i - v_j)^2$ can also be written in matrix form as $\mathbf{v}^T \Delta \mathbf{v}$ where $\Delta = D - W$ is known as the **combinatorial Laplacian** of the graph. The matrix Δ is known to be positive semidefinite, so it can be viewed as an inverse covariance matrix of a multivariate Gaussian random variable, giving rise to the name Gaussian field.

Spectral Graph Transducer: One potential problem with the mincut formulation is that the mincut cost function tends to prefer unbalanced cuts where the number of instances in either the positive or negative class vastly outnumbers the number of instances in the other class. Unbalanced cuts tend to have a lower cost in the mincut formulation because the number of edges between V_+ and V_- is maximized when the sizes of V_+ and V_- are equal and is small when either one of them is small. For example, if we have n vertices and V_+ contains a single element, then there are potentially $n-1$ edges between V_+ and V_-. In contrast, if V_+ and V_- are the same size, then there are potentially $n^2/4$ edges between the two sets of vertices.

Let cut(V_+, V_-) be the sum of the edge weights connecting V_+ and V_-. To mitigate the effect of preferring unbalanced cut, Joachims [261] proposed to minimize a cost function of normalized cut $\dfrac{cut(V_+, V_-)}{|V_+||V_-|}$, where the cut value is normalized by the number of edges between the two sets. Minimizing this cost function is computationally difficult, so Joachims [261] proposed minimizing a relaxed version of the problem.

Let Δ be the combinatorial Laplacian of the graph. It can be shown that minimizing the normalized cut (with no labeled data) using α and β number of instances (α and β are specified by the user) in the two partitions is equivalent to minimizing $\mathbf{v}^T \Delta \mathbf{v}$ for $v_i \in \{\gamma_+, \gamma_-\}$, where

$$\gamma_+ = \sqrt{\frac{\beta}{\alpha}} \quad \text{and} \quad \gamma_- = -\sqrt{\frac{\alpha}{\beta}}. \tag{6}$$

Instead of using $v_i \in \{\gamma_+, \gamma_-\}$, Joachims [261] proposed to allow v_i to take real values under the constraint $\mathbf{v}^T \mathbf{1} = 0$ and $\mathbf{v}^T \mathbf{v} = n$, where $\mathbf{1}$ is the all one vector. To make sure that the labeled instances are properly classified, a term $(\mathbf{v} - \gamma)^T C(\mathbf{v} - \gamma)$ is added to the cost function, where C is a diagonal matrix with non-zero entries only for labeled instances and γ is the target vector for approximation by $\mathbf{v}$. The components of γ that correspond to positive and negative instances are set to γ_+ and γ_- respectively, while the components of γ that correspond to unlabeled instances do not affect the cost function because their corresponding diagonal entries of C are set to zero. The values of the non-zero entries of C can be set by the user to give different misclassification costs to each instance. This gives the combined optimization problem of

$$\min_v \mathbf{v}^T \Delta \mathbf{v} + c(\mathbf{v} - \gamma)^T C(\mathbf{v} - \gamma) \tag{7}$$

$$s.t. \qquad \mathbf{v}^T \mathbf{1} = 0 \; and \; \mathbf{v}^T \mathbf{v} = n$$

where c gives a trade-off between the cost for the labeled and unlabeled parts. The solution of Equation (7) is obtained using spectral methods.

The Gaussian field method and spectral graph transduction have been applied to the natural language processing problem of word sense disambiguation in [414, 442]. Word sense disambiguation is the problem of assigning appropriate meanings to words (which may have multiple meanings) according to the context that they appear in. Although some improvements are observed, the success of these methods is still limited.

5.1.6 Discussion

We discuss two issues: (1) whether the unlabeled set U is always helpful and (2) the evaluation of LU learning.

Does the Unlabeled Set Always Help? The answer is no. As we have seen, all approaches make strong assumptions. For example, EM makes two mixture model assumptions, and co-training makes the feature split assumption. When the assumptions are true for an application data set, unlabeled data can help learning (even dramatically). When the assumptions are not true, the unlabeled data may harm learning. Automatically detecting bad match of the problem structure with the model assumptions in advance is, however, very hard and remains an open problem.

A related issue is that researchers have not shown that when the labeled data set is sufficiently large, the unlabeled data still help. Manual labeling more text documents may not be as difficult as it seems in some applications, especially when the number of classes is small. In most cases, to label a document one does not even need to read the entire document (if it is long). Typically, the first few sentences can already tell its class. Compounded with the problem of inability to decide whether the unlabeled data indeed help classification, practical applications of LU learning are still limited.

Evaluation: The evaluation of LU learning is commonly done in the same way as traditional classification. However, there is a problem with the availability of sufficient test data. In practice, users always want to have a reasonable guarantee on the predictive accuracy of a classification system before they are willing to use the system. This means that test data sets have to be used to evaluate the system. Existing algorithms for LU learning assume that there is a large set of labeled test data for this purpose. However, this contradicts the LU learning problem statement, which says that the labeled set is very small. If we can ask the user to label more data, then we do not need LU learning because some examples of the test set can be used in training. Evaluation without more labeled data is also an open problem.

One may look at this problem in another way. We first use the classifier generated by LU learning to classify the unlabeled set or a new test set and then sample some classified documents to be checked manually in order to estimate the classification accuracy. If classification is sufficiently accurate, the results of the classifier will be used. Otherwise, improvements need to be made. In this case, additional labeled data obtained during manual inspection can be added to the original labeled set. You see we end up doing more labeling! Hopefully, we do not have to do too much labeling.

5.2   Learning from Positive and Unlabeled Examples

In some applications, the problem is to identify a particular class *P* of documents from a set of mixed documents, which contains documents of class *P* and also other kinds of documents. We call the class of documents that one is interested in the positive class documents, or simply **positive documents**. We call the rest of the documents the negative class documents or simply **negative documents**.

This problem can be seen as a classification problem with two classes, positive and negative. However, there are no labeled negative documents for training. The problem is stated more formally as follows,

Problem Statement: Given a set *P* of *positive documents* that we are interested in, and a set *U* of **unlabeled documents** (*the mixed set*), which contains both *positive documents* and *negative documents*, we want to build a classifier using *P* and *U* that can identify positive documents in *U* or in a separate test set – in other words, we want to accurately classify positive and negative documents in *U* or in the test (or future) data set.

This problem is called **PU learning**. Note that the set *U* can be used in both training and testing because *U* is unlabeled.

The key feature of this problem is that there is no labeled negative document for learning. Traditional supervised learning algorithms are thus not directly applicable because they all require both labeled positive and labeled negative documents to build a classifier. This is also the case for LU learning, although the labeled set for each class may be very small.

5.2.1   Applications of PU Learning

The PU learning problem occurs frequently in Web and text retrieval applications because most of the time the user is only interested in Web pages or text documents of a particular topic. For example, one may be interested in only travel related pages (positive pages). Then all the other types of pages are negative pages. Let us use a concrete example to show the actual setting of a PU learning application.

Example 1: We want to build a repository of data mining research papers. We can start with an initial set of papers from a data mining conference or journal, which are positive examples. We then want to find data mining papers from online journals and conference series in the fields of databases and artificial intelligence. Journals and conferences in these fields all contain some data mining papers. They also contain many other types of papers. The problem is how to extract data mining papers from such conferences and journals, or in other words, how to classify the papers from these sources into data mining papers and non-data mining papers without labeling any negative papers in any source.   ∎

In practical applications, positive documents are usually available because if one has worked on a particular task for some time one should have accumulated many related documents. Even if no positive document is available initially, collecting some from the Web or any other source is relatively easy. One can then use this set to find the same class of documents from other sources without manually labeling any negative documents. PU learning is particularly useful in the following situations:

1. **Learning with multiple unlabeled sets:** In some applications, one needs to find positive documents from a large number of document collections. For example, we want to identify Web pages that sell printers. We can easily obtain a set of positive pages from an online merchant, e.g., amazon.com. Then we want to find printer pages from other merchants. We can crawl each site one by one, and extract printer pages from each site using PU learning. We do not need to manually label negative pages (non-printer pages) from any site.

   Although it may not be hard to label some negative pages from a single site, it is difficult to label for every site. Note that in general the classifier built based on the negative pages from one site s_1 may not be used to classify pages from another site s_2 because the negative pages in s_2 can be very different from the negative pages in s_1. The reason is that although both sites sell printers, the other products that they sell can be quite different. Thus using the classifier built for s_1 to classify pages in s_2 may violate the fundamental assumption of machine learning: the distribution of training examples is identical to the distribution of test examples. As a consequence, we may obtain poor accuracy results.

2. **Learning with unreliable negative examples:** This situation often occurs in experimental sciences. For example, in biology, biologists perform experiments to determine some biological functions. They are often quite confident about positive cases that they have discovered. However, they may not be confident about negative cases because laboratory results can be affected by all kinds of conditions. The negative cases are thus unreliable. It is perhaps more appropriate to treat such negative cases as unlabeled examples than negative examples.

PU learning is also useful for the following seemingly unrelated problems:

Detecting unexpected documents in the test set: In many applications, the test set (or future instance set in practice) contains some classes of documents that are not seen in the training set. For instance, our training set has only two classes, Sports and Science. Then, a learning algorithm will produce a classifier to separate Sports and Science documents. However, in the test set, some other types of documents, e.g., Politics and Religion, may appear, which are called **unexpected documents**. In traditional classification, those Politics and Religion documents in the test set will be classified as either Sports or Science documents, which is clearly inappropriate. In PU learning, we can remove Politics and Religion documents by treating the whole training set as the positive data, and the whole test set as the unlabeled data. A study of this problem is reported in [323].

Detecting outliers: Traditional outlier detection algorithms are mainly based on clustering. During clustering, those data points that are too far away from cluster centroids are considered outliers. PU learning can be applied to outlier detection as follows: A random sample is drawn from the original data. The sample is treated as the set of positive examples, and the remaining data is treated as the set of unlabeled examples. The method given in [323] can then be applied. This strategy may work because since the number of outliers is small, the chance of selecting them during sampling will be extremely small. To make the method more robust, one can run the technique multiple times using multiple random samples.

Before discussing theoretical foundations of PU learning, let us first develop some intuition on why PU learning is possible and why unlabeled data are helpful. Figure 5.5 shows the idea.

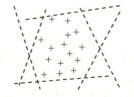 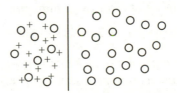

(A) With only positive data   (B) With both positive and unlabeled data

Fig. 5.5. Unlabeled data are helpful

In Fig. 5.5(A), we see only positive documents (data points) represented with +'s. We assume that a linear classifier is sufficient for the classification task. In this case, it is hard to know where to draw the line to separate positive and negative examples because we do not know where the negative examples might be. There are infinite possibilities. However, if the unlabeled data (represented by small circles) are added to the space (Fig. 5.5(B)), it is very clear where the separation line should be. Let us now discuss the theoretical result of PU learning.

5.2.2 Theoretical Foundation

Let $(\mathbf{x}_i, y_i)$ be random variables drawn independently from probability distribution $D_{(\mathbf{x},y)}$ where $y \in \{-1, 1\}$ is the conditional random variable that we wish to estimate given $\mathbf{x}$. $\mathbf{x}_i$ represents a document, and y_i is its class, which can be 1 (positive) or -1 (negative). Let $D_{\mathbf{x}|y=1}$ be the conditional distribution from which the positive examples are independently drawn and let $D_{\mathbf{x}}$ be the marginal distribution from which unlabeled examples are independently drawn. Our objective is to learn a classification function f that can separate positive and negative documents. Since learning is to produce a classifier that has the minimum probability of error, $\Pr(f(\mathbf{x}) \neq y)$, let us rewrite it into a more useful form,

$$\Pr(f(\mathbf{x}) \neq y) = \Pr(f(\mathbf{x})=1 \text{ and } y=-1) + \Pr(f(\mathbf{x})=-1 \text{ and } y=1). \tag{8}$$

The first term can be rewritten as

$$\begin{aligned}
&\Pr(f(\mathbf{x})=1 \text{ and } y=-1) \\
&= \Pr(f(\mathbf{x})=1) - \Pr(f(\mathbf{x})=1 \text{ and } y=1) \\
&= \Pr(f(\mathbf{x})=1) - (\Pr(y=1) - \Pr(f(\mathbf{x})=-1 \text{ and } y=1)).
\end{aligned} \tag{9}$$

Substituting (9) into Equation (8), we obtain

$$\begin{aligned}
&\Pr(f(\mathbf{x}) \neq y) \\
&= \Pr(f(\mathbf{x})=1) - \Pr(y=1) + 2\Pr(f(\mathbf{x})=-1|y=1)\Pr(y=1).
\end{aligned} \tag{10}$$

Since $\Pr(y = 1)$ is constant (although it is unknown), we can minimize the probability of error by minimizing

$$\Pr(f(\mathbf{x})=1) + 2\Pr(f(\mathbf{x})=-1|y=1)\Pr(y=1). \tag{11}$$

If we can hold $\Pr(f(\mathbf{x})=-1|y=1)$ small, then learning is approximately the same as minimizing $\Pr(f(\mathbf{x})=1)$. Holding $\Pr(f(\mathbf{x})=-1|y=1)$ small while minimizing $\Pr(f(\mathbf{x})=1)$ is approximately the same as minimizing $\Pr_u(f(\mathbf{x})=1)$ (on the unlabeled set U) while holding $\Pr_P(f(\mathbf{x})=1) \geq r$ (on the positive set P), where r is the **recall**, i.e., $\Pr(f(\mathbf{x})=1|y=1)$. Note that $(\Pr_P(f(\mathbf{x})=1) \geq r)$ is the same as $(\Pr_P(f(\mathbf{x})=-1) \leq 1-r)$.

Two theorems given by Liu et al. [348] state these formally and show that in both the noiseless case (P has no error) and the noisy case (P contains errors, i.e., some negative documents) reasonably good learning results can be achieved if

- the problem is posed as a **constrained optimization problem** where the algorithm tries to minimize the number of unlabeled examples labeled positive subject to the constraint that the fraction of errors on the positive examples is no more than $1-r$.

Example 2: Figure 5.6 illustrates the constrained optimization problem. Assume that positive and negative documents can be linearly separated. Positive documents are represented with +'s, and unlabeled documents with small circles. Assume also that the positive set has no error and we want the recall r on the positive set to be 100%. Each line in the figure is a possible linear classifier. Every document on the left of each line will be labeled (classified) by the line as positive, and every document on the right will be labeled as negative. Lines 1 and 2 are clearly not solutions because the constraint "the fraction of errors on the positive examples must be no more than $1 - r \, (= 0)$" is violated, although the number of unlabeled examples labeled (classified) as positive is minimized by line 1. Lines 4, 5, and 6 are poor solutions too because the number of unlabeled examples labeled as positive is not minimized by any of them. Line 3 is the optimal solution. Under the constraint that no positive example is labeled negative, line 3 minimizes the number of unlabeled examples labeled as positive. ∎

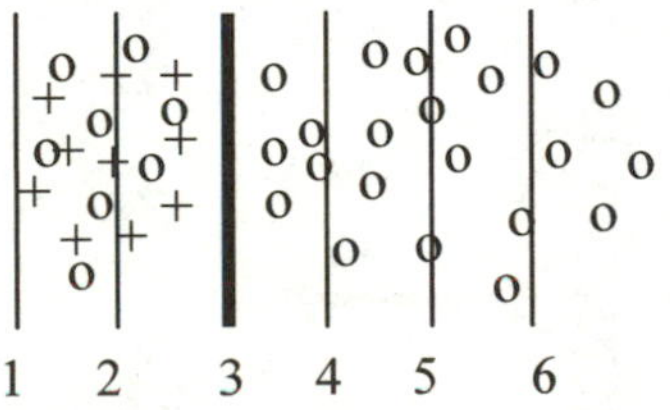

Fig. 5.6. An illustration of the constrained optimization problem

Based on the constrained optimization idea, two kinds of approaches have been proposed to build PU classifiers: the **two-step approach** and the **direct approach**. In the actual learning algorithms, the user may not need to specify a desired recall level r on the positive set because some of these algorithms have their evaluation methods that can automatically determine whether a good solution has been found.

5.2.3   Building Classifiers: Two-Step Approach

As its name suggests the two-step approach works in two steps:

1. Identifying a set of **reliable negative documents** (denoted by RN) from the unlabeled set U.
2. Building a classifier using P, RN and $U - RN$. This step may apply an existing learning algorithm once or iteratively depending on the quality and the size of the RN set.

This two-step approach is illustrated in Fig. 5.7. Here, we assume that step 2 uses an iterative algorithm. In step 1, a set of reliable negative documents (RN) is found from the unlabeled set U, which divides U into two subsets, RN and Q (= $U - RN$). Q is called the **likely positive set**. In step 2, the algorithm iteratively improves the results by adding more documents to RN until a convergence criterion is met. We can see that the process is trying to minimize the number of unlabeled examples labeled positive since Q becomes smaller and smaller while RN becomes larger and larger. In other words, it tries to iteratively increase the number of unlabeled examples that are labeled negative while maintaining the positive examples in P correctly classified. We present several techniques for each step below.

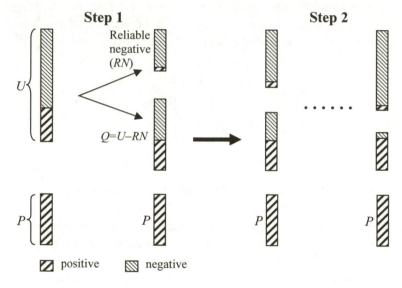

Fig. 5.7. An illustration of the two-step approach

Techniques for Step 1

We introduce four methods to extract reliable negative documents from the unlabeled set U.

Spy Technique: This technique works by sending some "spy" documents from the positive set P to the unlabeled set U. Figure 5.8 gives the algorithm of the technique, which is used in the S-EM system [348]. The algorithm has three sub-steps:

1. It randomly samples a set S of positive documents from P and put them in U (lines 2 and 3). The default sampling ratio of s% is 15% in S-EM.

Algorithm Spy(*P*, *U*)
1.   *RN* ← ∅;
2.   *S* ← *Sample*(*P*, *s*%);
3.   *Us* ← *U* ∪ *S*;
4.   *Ps* ← *P* − *S*;
5.   Assign each document in *Ps* the class label 1;
6.   Assign each document in *Us* the class label −1;
7.   NB(*Us*, *Ps*);   // This produces a NB classifier.
8.   Classify each document in *Us* using the NB classifier;
9.   Determine a probability threshold *t* using *S*;
10. **for** each document *d* ∈ *Us* **do**
11.   **if** its probability Pr(1|*d*) < *t* **then**
12.   *RN* ← *RN* ∪ {*d*};
13.   **endif**
14. **endfor**

Fig. 5.8. The spy technique.

The documents in *S* act as "spy" documents from the positive set to the unlabeled set *U*. Since the spies behave similarly to the unknown positive documents in *U*, they allow the algorithm to infer the behavior of the unknown positive documents in *U*.

2. It randomly samples a set *S* of positive documents from *P* and put them in *U* (lines 2 and 3). The default sampling ratio of *s*% is 15% in S-EM. The documents in *S* act as "spy" documents from the positive set to the unlabeled set *U*. Since the spies behave similarly to the unknown positive documents in *U*, they allow the algorithm to infer the behavior of the unknown positive documents in *U*.

3. It runs the naïve Bayesian (NB) algorithm using the set *P* − *S* as positive and the set *U* ∪ *S* as negative (lines 3–7). The NB classifier is then applied to classify each document *d* in *U* ∪ *S* (or *Us*), i.e., to assign it a probabilistic class label Pr(1|*d*), where 1 represents the positive class.

4. It uses the probabilistic labels of the spies to decide which documents are most likely to be negative. A threshold *t* is employed to make the decision. Those documents in *U* with lower probabilities (Pr(1|*d*)) than *t* are the most likely negative documents, denoted by *RN* (lines 10–14).

We now discuss how to determine *t* using spies (line 9). Let the set of spies be $S = \{s_1, s_2, ..., s_k\}$, and the probabilistic labels assigned to each s_i be Pr(1|s_i). Intuitively, we can use the minimum probability in *S* as the threshold value *t*, i.e., $t = min\{Pr(1|s_1), Pr(1|s_2), ..., Pr(1|s_k)\}$, which means that we want to retrieve all spy documents. In a noiseless case, using the minimum probability is acceptable. However, most real-life document collections have outliers and noise. Using the minimum probability is unreliable. The reason is that the posterior probability $Pr(1|s_i)$ of an outlier document s_i in S could be 0 or smaller than most (or even all) actual negative documents. However, we do not know the noise level of the data. To be safe, the S-EM system uses a large noise level l = 15% as the default. The final classification result is not very sensitive to l as long it is not too small. To determine t, we first sort the documents in S according to their $Pr(1|s_i)$ values. We then use the selected noise level l to decide t: we select t such that l percent of documents in S have probability less than t. Hence, t is not a fixed value. The actual parameter is in fact l.

Note that the reliable negative set RN can also be found through multiple iterations. That is, we run the spy algorithm multiple times. Each time a new random set of spies S is selected from P and a different set of reliable negative documents is obtained, denoted by RN_i. The final set of reliable negative documents is the intersection of all RN_i. This may be a better technique because we do not need to worry that one set of random spies S may not be chosen well, especially when the set P is not large.

1DNF Technique: The 1DNF method (Fig. 5.9) is used in [583]. It first builds a positive feature set PF containing words that occur in the positive set P more frequently than in the unlabeled set U (lines 1–7). Line 1 collects all the words in $U \cup P$ to obtain a vocabulary V. Lines 8–13 try to identify reliable negative documents from U. A document in U that does not contain any feature in PF is regarded as a reliable negative document.

NB Technique: This method is employed in [340]. It simply uses a naïve Bayesian classifier to identify a set of reliable negative documents RN from the unlabeled set U. The algorithm is given in Fig. 5.10.

This method may also be run multiple times. Each time we randomly remove a few documents from P to obtain a different set of reliable negative documents, denoted by RN_i. The final set of reliable negative documents RN is the intersection of all RN_i.

Rocchio technique: This method is employed in [321]. The algorithm is the same as that in Fig. 5.10 except that NB is replaced with Rocchio. The Rocchio classification method is described in Sect. 6.3.

Techniques for Step 2

There are two approaches for this step.

1. Run a learning algorithm (e.g., NB or SVM) using P and RN. The set of documents in $U–RN$ is discarded. This method works well if the reliable negative set RN is sufficiently large and contains mostly negative docu-

Algorithm 1DNF(P, U)

1.   Assume the word feature set be $V = \{w_1,\ldots, w_n\}$, $w_i \in U \cup P$;
2.   Let positive feature set $PF \leftarrow \varnothing$;
3.   **for** each $w_i \in V$ **do**   // $freq(w_i, P)$: number of times
4.   **if** ($freq(w_i, P) / |P| > freq(w_i, U) / |U|$) **then**   // that w_i appears in P
5. $PF \leftarrow PF \cup \{w_i\}$;
6.   **endif**
7.   **endfor**;
8. $RN \leftarrow U$;
9.   **for** each document $d \in U$ **do**
10.   **if** $\exists w_j \, freq(w_j, d) > 0$ and $w_j \in PF$ **then**
11. $RN \leftarrow RN - \{d\}$
12.   **endif**
13.   **endfor**

Fig. 5.9. The 1DNF technique for step 1

1.   Assign each document in P the class label 1;
2.   Assign each document in U the class label -1;
3.   Build a NB classifier using P and U;
4.   Use the classifier to classify U. Those documents in U that are classified as negative form the reliable negative set RN.

Fig. 5.10. The NB method for Step 1

ments. The spy technique, NB and Rocchio in step 1 are often able to produce a sufficiently large set of negative documents. The 1DNF technique may only identify a very small set of negative documents. Then running a learning algorithm will not be able to build a good classifier.

2. Run a learning algorithm iteratively till it converges or some stopping criterion is met. This method is used when the set RN is small.

We will not discuss the first approach as it is straightforward. SVM usually does very well. Below, we introduce two techniques for the second approach, which are based on EM and SVM respectively.

EM Algorithm with Naïve Bayesian Classification: The EM algorithm can be used naturally here [348]. As in LU learning, the Expectation step basically fills in the missing data. In our case, it produces and revises the probabilistic labels of the documents in U–RN (see below). The parameters are estimated in the Maximization step after the missing data are filled. This leads to the next iteration of the algorithm. EM converges when its parameters stabilize. Using NB in each iteration, EM employs the same equations as those used in building a NB classifier (Equation (29) for the Expectation step, and Equations (27) and (28) for the Maximization step).

Algorithm EM(P, U, RN)
1.   Each document in P is assigned the class label 1;
2.   Each document in RN is assigned the class label -1;
3.   Learn an initial NB classifier f from P and RN (using Equations (27) and
         (28) in Chap. 3);
4   **repeat**
         // E-Step
5   **for** each example d_i in U–RN **do**
6   Using the current classifier f to compute $Pr(c_j|d_i)$ using Equation (29)
                 in Chap. 3.
7   **end**
         // M-Step
8   learn a new NB classifier f from P, RN and U–RN by computing $Pr(c_j)$
                 and $Pr(w_t|c_j)$ (using Equations (27) and (28) in Chap. 3).
9   **until** the classifier parameters stabilize
10. Return the classifier f from the last iteration.

Fig. 5.11. EM algorithm with the NB classifier

Algorithm I-SVM(P, RN, Q)
1.   Every document in P is assigned the class label 1;
2.   Every document in RN is assigned the class label -1;
3.   **loop**
4.   Use P and RN to train a SVM classifier f;
5.   Classify Q using f;
6.   Let W be the set of documents in Q that is classified as negative;
7.   **if** $W = \varnothing$ **then** *exit-loop*   // convergence
8.   **else** $Q \leftarrow Q - W$;
9. $RN \leftarrow RN \cup W$;
10.   **endif**;

Fig. 5.12. Running SVM iteratively

The class probability given to each document in U–RN takes the value in $[0, 1]$ instead of $\{0, 1\}$. The algorithm is given in Fig. 5.11.

The EM algorithm here makes the same mixture model assumptions as in LU learning. Thus, it has the same problem of model mismatch. See the discussions in Sect. 5.1.1.

Iterative SVM: In this method, SVM is run iteratively using P, RN and Q ($= U$–RN). The algorithm, called I-SVM, is given in Fig. 5.12. The basic idea is as follows: In each iteration, a new SVM classifier f is constructed from P and RN (line 4). Here RN is regarded as the set of negative examples (line 2). The classifier f is then applied to classify the documents in Q (line 5). The set W of documents in Q that are classified as negative (line

6) is removed from Q (line 8) and added to RN (line 9). The iteration stops when no document in Q is classified as negative, i.e., $W = \varnothing$ (line 7). The final classifier is the result. This method is used in [321, 582, 583].

Finally, we note again that if the first step is able to identify a large number of reliable negative documents from U, running SVM once in step 2 is sufficient. Iterative approaches may not be necessary, which are also less efficient. The Spy, NB and Rocchio methods for step 1 are often able to identify a large number of reliable negative documents. See [340] for an evaluation of various methods based on two benchmark text collections.

Classifier Selection

The iterative methods discussed above produce a new classifier at each iteration. However, the classifier at the convergence may not be the best classifier. In general, each iteration of the algorithm gives a classifier that may potentially be a better classifier than the classifier produced at convergence. This is true for both EM and SVM.

The main problem with EM is that classes and topics may not have one-to-one correspondence. This is the same problem as in LU learning. SVM may also produce poor classifiers at the convergence because SVM is sensitive to noise. If the RN set is not chosen well or in an iteration some positive documents are classified as negative, then the subsequent iterations may produce very poor results. In such cases, it is often better to stop at an earlier iteration. One simple method is to apply the theory directly. That is, each classifier is applied to classify a positive validation set, P_v. If many documents from P_v (e.g., > 5%) are classified as negative, the algorithm should stop (that is a recall of 95%). If the set P is small, the method can also be applied to P directly. A principled method is given in the next subsection, i.e., Equation (14).

5.2.4   Building Classifiers: Direct Approach

We now present a direct approach, called **biased-SVM**. This approach modifies the SVM formulation slightly so that it is suitable for PU learning. Let the set of training examples be $\{(\mathbf{x}_1, y_1), (\mathbf{x}_2, y_2), \ldots, (\mathbf{x}_n, y_n)\}$, where $\mathbf{x}_i$ is an input vector and y_i is its class label, $y_i \in \{1, -1\}$. Assume that the first $k-1$ examples are positive examples P (labeled 1), while the rest are unlabeled examples U, which are treated as negative and labeled -1. Thus, the negative set has errors, i.e., containing positive documents. We consider two cases.

1. **Noiseless case:** There is no error in the positive examples but only in unlabeled examples. The theoretical result in Sect. 5.2.2 states that if the sample size is large enough, minimizing the number of unlabeled examples classified as positive while constraining the positive examples to be correctly classified will give a good classifier. Following the theory, in this noiseless case, we have this following SVM formulation

$$\text{Minimize}: \frac{\langle \mathbf{w} \cdot \mathbf{w} \rangle}{2} + C \sum_{i=k}^{n} \xi_i \tag{12}$$

$$\text{Subject to}: \langle \mathbf{w} \cdot \mathbf{x}_i \rangle + b \geq 1, \quad i = 1, 2, ..., k-1$$

$$-1(\langle \mathbf{w} \cdot \mathbf{x}_i \rangle + b) \geq 1 - \xi_i, \quad i = k, k+1, ..., n$$

$$\xi_i \geq 0, \quad i = k, k+1, ..., n$$

In this formulation, we do not allow any error in the positive set P, which is the first constraint, but allow errors for the negative set (the original unlabeled set), which is the second constraint. Clearly, the formulation follows the theory exactly due to the second term in the objective function. The subscript in the second term starts from k, which is the index of the first unlabeled example. To distinguish this formulation from the classic SVM, we call it the **biased-SVM** [340].

2. **Noisy case:** In practice, the positive set may also contain some errors. Thus, if we allow noise (or error) in positive examples, we have the following soft margin version of the biased-SVM which uses two parameters C_+ and C_- to weigh positive errors and negative errors differently.

$$\text{Minimize}: \frac{\langle \mathbf{w} \cdot \mathbf{w} \rangle}{2} + C_+ \sum_{i=1}^{k-1} \xi_i + C_- \sum_{i=k}^{n} \xi_i \tag{13}$$

$$\text{Subject to}: y_i(\langle \mathbf{w} \cdot \mathbf{x}_i \rangle + b) \geq 1 - \xi_i, \quad i = 1, 2, ..., n$$

$$\xi_i \geq 0, \quad i = 1, 2, ..., n$$

We can vary C_+ and C_- to achieve our objective. Intuitively, we give a bigger value for C_+ and a smaller value for C_- because the unlabeled set, which is assumed to be negative, contains positive data.

We now focus on Equation (13) as it is more realistic in practice. We need to choose values for C_+ and C_-. The common practice is to try a range of values for both C_+ and C_- and use a separate validation set to verify the performance of the resulting classifier. The C_+ and C_- values that give the best classification results on the validation set are selected as the final parameter values for them. Cross-validation is another possible technique for the purpose. Since the need to learn from positive and unlabeled examples often arises in retrieval situations (retrieving positive documents from the unlabeled set), we employ the commonly used F-score as the performance measure, $F = 2pr/(p+r)$, where p is the precision and r is the recall.

Unfortunately it is not clear how to estimate the F-score without labeled negative examples. In [309], Lee and Liu proposed an alternative performance measure to compare different classifiers. It behaves similarly to the F-score but can be estimated directly from the validation set without the need of labeled negative examples. The measure is

$$\frac{r^2}{\Pr(f(\mathbf{x})=1)}, \tag{14}$$

where f is the classifier and $\Pr(f(\mathbf{x})=1)$ is the probability that a document is classified as positive. It is not easy to see why Equation (14) behaves similarly to the F-score, but we can show that $r^2/\Pr(f(\mathbf{x})=1) = pr/\Pr(y=1)$, where $\Pr(y=1)$ is the probability of positive documents. $pr/\Pr(y=1)$ behaves similarly to the F-score in the sense that it is large when both p and r are large and is small when either p or r is small.

We first write recall (r) and precision (p) in terms of probability:

$$r = \Pr(f(\mathbf{x})=1 \mid y=1) \tag{15}$$

$$p = \Pr(y=1 \mid f(\mathbf{x})=1). \tag{16}$$

According to probability theory, we have

$$\Pr(f(\mathbf{x})=1 \mid y=1)\Pr(y=1) = \Pr(y=1 \mid f(\mathbf{x})=1)\Pr(f(\mathbf{x})=1), \tag{17}$$

which can be written as

$$\frac{r}{\Pr(f(\mathbf{x})=1)} = \frac{p}{\Pr(y=1)}. \tag{18}$$

Multiplying both sides by r, we obtain the result:

$$\frac{r^2}{\Pr(f(\mathbf{x})=1)} = \frac{pr}{\Pr(y=1)}. \tag{19}$$

The quantity $r^2/\Pr(f(\mathbf{x})=1)$ can be estimated based on the validation set, which contains both positive and unlabeled documents. r can be estimated using the positive examples in the validation set and $\Pr(f(\mathbf{x}) = 1)$ can be estimated from the whole validation set.

This criterion in fact reflects the theory in Sect. 5.2.2 very well. The quantity is large when r is large and $\Pr(f(\mathbf{x}) = 1)$ is small, which means that the number of unlabeled examples labeled as positive should be small. In [340], it is shown that biased-SVM works better than two-step techniques.

5.2.5 Discussion

Does PU Learning Always Work? Theoretical results show that it should if the positive set and the unlabeled set are sufficiently large [348]. This has been confirmed by many experimental studies. Interested readers can find the detailed results in [340, 348], which we summarize below:

1. PU learning can achieve about the same classification results as fully supervised learning (i.e., both labeled positive and negative examples are available for training) when the positive set and the unlabeled set are sufficiently large. This implies that labeled negative examples do not provide much information for learning. When the positive set is very small, PU learning is poorer than fully supervised learning.

2. For the two-step approaches, using SVM for the second step performs better than EM. SVM needs to be run only once if step 1 can extract a large number of reliable negative documents. Both Spy and Rocchio (and to some extent NB as well) are able to do that. Thus, the iterative method in step 2 is not necessary.

   The generative model of naïve Bayes with EM in the second step can perform very well if the mixture model assumptions hold [348]. However, if the mixture model assumptions do not hold, the classification results can be very poor [340]. Note that SVM is usually called a **discriminative model** (or **classifier**) because it does not make any model assumptions. It simply finds a hyperplane to separate positive and negative examples in the training data.

3. Biased-SVM performs slightly better than the 2-step approaches. However, it is very slow in training because SVM needs to be run a large number of times in order to select the best values for C_+ and C_-.

Evaluation: Unlike LU learning, here we do not even have labeled negative examples, which makes the evaluation difficult. Although Equation (14) and other heuristics allow a system to choose a "better" classifier among a set of classifiers, it is unable to give the actual accuracy, precision or recall of each classifier. Evaluation is an open problem. The results reported in [340, 348] assume that a set of labeled positive and negative test examples is available, which, of source, is unrealistic in practice because the PU learning model states that no labeled negative example is available.

In some cases, the evaluation can be done with some confidence. For example, if the user needs to extract positive documents from many unlabeled sets as in the example of identifying printer pages from multiple Web sites, a PU learning algorithm can be applied to one site and then the user manually checks the classification result to see whether it is satisfactory. If the result is satisfactory, the algorithm can be applied to the rest of the sites without further manual evaluation.

Appendix: *Derivation of EM for Naïve Bayesian Classification*

EM is a method for performing a classical statistical estimation technique called **maximum likelihood estimation**. In maximum likelihood estimation, the aim is to find the model parameter $\hat{\Theta}$ that maximizes the likelihood function $\Pr(D_o; \Theta)$ for observed data D_o. In other words, maximum likelihood estimation aims to select the model that is most likely to have generated the observed data. In many cases, such as in the naïve Bayesian classification model, the maximum likelihood estimator is easy to find and has a closed form solution when all components of the data D are observed. However, the problem becomes difficult when the data D actually consists of an observed component D_o and an unobserved component D_u. In such cases, iterative methods that converge only to a local maximum, such as the EM method, are usually used.

Maximizing the log likelihood function $\log\Pr(D_o; \Theta)$ produces the same solution as maximizing the likelihood function and is easier to handle mathematically. In the presence of unobserved data D_u, the log likelihood function becomes $\log\Pr(D_o;\Theta) = \log\sum_{D_u}\Pr(D_o,D_u;\Theta)$. Instead of maximizing the log likelihood $\log\sum_{D_u}\Pr(D_o,D_u;\Theta)$ directly, at each iteration T, the EM algorithm finds the value Θ that maximizes the **expected complete log likelihood**

$$\sum_{D_u} E(D_u \mid D_o; \Theta^{T-1})\log\Pr(D_o, D_u; \Theta), \tag{20}$$

where Θ^{T-1} is the parameter that was produced in iteration $T-1$. In many cases, such as in the naïve Bayesian model, the expected log likelihood is easy to maximize and has a closed form solution. It can be shown (see [127]) that the log likelihood increases monotonically with each iteration of the EM algorithm.

We now derive the EM update for the naïve Bayesian model. We first consider the complete log likelihood, that is, the log likelihood when all variables are observed. The conditional probability of a document given its class is (see Sect. 3.7.2 in Chap. 3)

$$\Pr(d_i \mid c_j; \Theta) = \Pr(\mid d_i \mid) \mid d_i \mid! \prod_{t=1}^{|V|} \frac{\Pr(w_t \mid c_j; \Theta)^{N_{ti}}}{N_{ti}!} \tag{21}$$

Each document and its class label are assumed to have been sampled independently. Let $c_{(i)}$ be the class label of document i. The likelihood function can hence be written as

$$\prod_{i=1}^{|D|}\Pr(d_i\,|\,c_{(i)};\Theta)\Pr(c_{(i)};\Theta)=\prod_{i=1}^{|D|}\Pr(|\,d_i\,|)\,|\,d_i\,|!\prod_{t=1}^{|V|}\frac{\Pr(w_t\,|\,c_{(i)};\Theta)^{N_{ti}}}{N_{ti}!}\Pr(c_{(i)};\Theta). \quad (22)$$

Taking logs, we have the complete log likelihood function

$$\sum_{i=1}^{|D|}\sum_{t=1}^{|V|}N_{ti}\log\Pr(w_t\,|\,c_{(i)};\Theta)+\sum_{i=1}^{|D|}\log\Pr(c_{(i)};\Theta)+\phi, \quad (23)$$

where ϕ is a constant containing the terms unaffected by Θ. To facilitate the process of taking expectation when some of the class labels are not observed, we introduce indicator variables, h_{ik}, that take the value 1 when document i takes the label k and the value 0 otherwise. The complete log likelihood can be written in the following equivalent form

$$\sum_{i=1}^{|D|}\sum_{t=1}^{|V|}\sum_{k=1}^{|C|}h_{ik}N_{ti}\log\Pr(w_t\,|\,c_k;\Theta)+\sum_{i=1}^{|D|}\sum_{k=1}^{|C|}h_{ik}\log\Pr(c_k;\Theta)+\phi. \quad (24)$$

When some of the labels are not observed, we take the conditional expectation for the unobserved variables h_{ik} with respect to Θ^{T-1} to get the expected complete log likelihood

$$\sum_{i=1}^{|D|}\sum_{t=1}^{|V|}\sum_{k=1}^{|C|}\Pr(c_k\,|\,d_i;\Theta^{T-1})N_{ti}\log\Pr(w_t\,|\,c_k;\Theta)$$
$$+\sum_{i=1}^{|D|}\sum_{k=1}^{|C|}\Pr(c_k\,|\,d_i;\Theta^{T-1})\log\Pr(c_k;\Theta)+\phi, \quad (25)$$

where, for the observed labels $c_{(i)}$, we use the convention that $\Pr(c_k|d_i;\Theta^{T-1})$ takes the value one for $c_k = c_{(i)}$ and zero otherwise. We maximize the expected complete log likelihood subject to the coefficients summing to one using the Lagrange multiplier method. The Lagrangian is

$$\sum_{i=1}^{|D|}\sum_{t=1}^{|V|}\sum_{k=1}^{|C|}\Pr(c_k\,|\,d_i;\Theta^{T-1})N_{ti}\log\Pr(w_t\,|\,c_k;\Theta)$$
$$+\sum_{i=1}^{|D|}\sum_{k=1}^{|C|}\Pr(c_k\,|\,d_i;\Theta^{T-1})\log\Pr(c_k;\Theta)$$
$$+\lambda\left(1-\sum_{k=1}^{|C|}\Pr(c_k;\Theta)\right)+\sum_{t=1}^{|V|}\sum_{k=1}^{|C|}\lambda_{tk}\left(1-\sum_{t=1}^{|V|}\Pr(w_t\,|\,c_k;\Theta)\right)+\phi. \quad (26)$$

Differentiating the Lagrangian with respect to λ, we get $\sum_{k=1}^{|C|}\Pr(c_k;\Theta)=1$. Differentiating with respect to $\Pr(c_k;\Theta)$, we get

$$\sum_{i=1}^{|D|} \Pr(c_k \mid d_i; \Theta^{T-1}) = \lambda \Pr(c_k; \Theta) \qquad for \; k=1,...,|C|. \qquad (27)$$

Summing the left and right-hand side over k and using $\sum_{k=1}^{|C|} \Pr(c_k; \Theta) = 1$,

we get $\lambda = \sum_{i=1}^{|D|} \sum_{k=1}^{|C|} \Pr(c_k \mid d_i; \Theta^{T-1}) = |D|$. Substituting back, we obtain the update equation

$$\Pr(c_j; \Theta^T) = \frac{\sum_{i=1}^{|D|} \Pr(c_j \mid d_i; \Theta^{T-1})}{|D|}. \qquad (28)$$

Working similarly, we can get the update equation for $\Pr(w_t \mid c_j; \Theta^T)$,

$$\Pr(w_t \mid c_j; \Theta^T) = \frac{\sum_{i=1}^{|D|} N_{ti} \Pr(c_j \mid d_i; \Theta^{T-1})}{\sum_{s=1}^{|V|} \sum_{i=1}^{|D|} N_{si} \Pr(c_j \mid d_i; \Theta^{T-1})}. \qquad (29)$$

To handle the 0 count problem (see Sect. 3.7.2 in Chap. 3), we can use Lidstone smoothing (Equation (27) in Chap. 3).

Bibliographic Notes

Learning with labeled and unlabeled examples (LU learning) using naïve Bayes and EM was proposed by Nigam et al. [413]. They also noted the problem of having mixtures of subclasses in the classes and proposed to identify and use such subclasses as a possible solution. A hierarchical clustering technique was also proposed by Cong et al. [111] for handling the mixture of subclasses problem. Castelli and Cover [83] presented a theoretical study of LU learning using mixture models.

Co-training was introduced by Blum and Mitchell [55]. Follow-on works include those by Collins and Singer [109], Goldman and Zhou [201], etc. Generalization error bounds within the Probably Approximately Correct (PAC) framework was given in [121] by Dasgupta et al. Nigam and Ghani [411] examined the importance of feature division in co-training and compared it to the EM algorithm and self-training.

Transduction was proposed by Vapnik [526] as learning when the test instances are known. Joachims described a heuristic algorithm and built a system for performing transduction using SVM [259]. The transductive SVM given in [259] can also be used for induction, i.e. classifying future unseen instances. In contrast, most graph-based methods are more suited

for transduction, i.e. classifying test instances that are known during train-
ing. The graph-based mincut algorithm was introduced by Blum and
Chalwa [54]. The graph-based Gaussian field method was proposed by
Zhu et al. [619] while the spectral graph transducer was proposed by
Joachims [261]. The edited book by Chapelle et al. [93] gives a compre-
hensive coverage of various LU learning algorithms.

On learning from positive and unlabeled examples (or PU learning),
Denis [129] reported a theoretical study of PAC learning in this setting under the statistical query model [272], which basically assumes that the proportion of positive instances in the unlabeled set is known. Letouzey et al.
[315] presented a learning algorithm based on a modified decision tree
method in this model. PU learning is also studied theoretically by Muggleton [398] from the Bayesian framework where the distribution of functions
and examples are assumed known. Liu et al. [348] gives another theoretical study, in which both the noiseless case and the noisy case are considered. It was concluded that a reasonable generalization (learning) can be
achieved if the problem is posed as a constrained optimization problem
(see Sect. 5.2.2). Most existing algorithms for solving the problem are
based on this constrained optimization model.

Over the years, several practical algorithms were proposed. The first
class of algorithms deals with the problem in two steps. These algorithms
include S-EM [348], PEBL [582, 583], and Roc-SVM [321], which have
been studied in this chapter. The second class of algorithm follows the
theoretical result directly. Lee and Liu [309] described a weighted logistic
regression technique. Liu et al. [340] described a biased-SVM technique.
They both require a performance criterion to determine the quality of the
classifier. The criterion is given in [309], which has been presented in Sect.
5.2.4. Liu et al. reported a comprehensive comparison of various techniques in [340]. It was shown that biased-SVM performed better than other
techniques. Some other works on PU learning include those of Agichtein
[6], Barbara et al. [35], Deng et al. [128], Denise et al. [130], Fung, et al.
[188], Li and Liu [322], Zhang and Lee [603], etc.

A closely related work to PU learning is one-class SVM, which uses
only positive examples to build a classifier. This method was proposed by
Scholkopf et al. [478]. Manevitz and Yousef [360] studied text classification using one-class SVM. Li and Liu [321] showed that its accuracy results were poorer than PU learning for text classification. Unlabeled data
does help classification significantly.

6 Information Retrieval and Web Search

Web search needs no introduction. Due to its convenience and the richness of information on the Web, searching the Web is increasingly becoming the dominant information seeking method. People make fewer and fewer trips to libraries, but more and more searches on the Web. In fact, without effective search engines and rich Web contents, writing this book would have been much harder.

Web search has its root in **information retrieval** (or IR for short), a field of study that helps the user find needed information from a large collection of text documents. Traditional IR assumes that the basic information unit is a **document**, and a large collection of documents is available to form the text database. On the Web, the documents are **Web pages**.

Retrieving information simply means finding a set of documents that is relevant to the user query. A ranking of the set of documents is usually also performed according to their relevance scores to the query. The most commonly used query format is a list of **keywords**, which are also called **terms**. IR is different from data retrieval in databases using SQL queries because the data in databases are highly structured and stored in relational tables, while information in text is unstructured. There is no structured query language like SQL for text retrieval.

It is safe to say that Web search is the single most important application of IR. To a great extent, Web search also helped IR. Indeed, the tremendous success of search engines has pushed IR to the center stage. Search is, however, not simply a straightforward application of traditional IR models. It uses some IR results, but it also has its unique techniques and presents many new problems for IR research.

First of all, efficiency is a paramount issue for Web search, but is only secondary in traditional IR systems mainly due to the fact that document collections in most IR systems are not very large. However, the number of pages on the Web is huge. For example, Google indexed more than 8 billion pages when this book was written. Web users also demand very fast responses. No matter how effective an algorithm is, if the retrieval cannot be done efficiently, few people will use it.

Web pages are also quite different from conventional text documents used in traditional IR systems. First, Web pages have **hyperlinks** and **an-**

chor texts, which do not exist in traditional documents (except citations in research publications). Hyperlinks are extremely important for search and play a central role in search ranking algorithms as we will see in the next chapter. Anchor texts associated with hyperlinks too are crucial because a piece of anchor text is often a more accurate description of the page that its hyperlink points to. Second, Web pages are semi-structured. A Web page is not simply a few paragraphs of text like in a traditional document. A Web page has different fields, e.g., title, metadata, body, etc. The information contained in certain fields (e.g., the title field) is more important than in others. Furthermore, the content in a page is typically organized and presented in several structured blocks (of rectangular shapes). Some blocks are important and some are not (e.g., advertisements, privacy policy, copyright notices, etc). Effectively detecting the main content block(s) of a Web page is useful to Web search because terms appearing in such blocks are more important.

Finally, **spamming** is a major issue on the Web, but not a concern for traditional IR. This is so because the rank position of a page returned by a search engine is extremely important. If a page is relevant to a query but is ranked very low (e.g., below top 30), then the user is unlikely to look at the page. If the page sells a product, then this is bad for the business. In order to improve the ranking of some target pages, "illegitimate" means, called spamming, are often used to boost their rank positions. Detecting and fighting Web spam is a critical issue as it can push low quality (even irrelevant) pages to the top of the search rank, which harms the quality of the search results and the user's search experience.

In this chapter, we first study some information retrieval models and methods that are closely related to Web search. We then dive into some Web search specific issues.

6.1 Basic Concepts of Information Retrieval

Information retrieval (IR) is the study of helping users to find information that matches their information needs. Technically, IR studies the acquisition, organization, storage, retrieval, and distribution of information. Historically, IR is about document retrieval, emphasizing document as the basic unit. Fig. 6.1 gives a general architecture of an IR system.

In Figure 6.1, the user with information need issues a query (**user query**) to the **retrieval system** through the **query operations** module. The retrieval module uses the **document index** to retrieve those documents that contain some query terms (such documents are likely to be relevant to the query), compute relevance scores for them, and then rank the retrieved documents according to the scores. The ranked documents are then presented to the user. The **document collection** is also called the **text database**, which is indexed by the **indexer** for efficient retrieval.

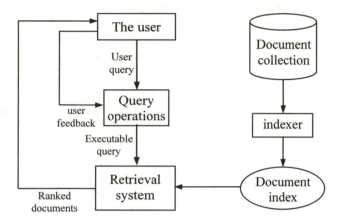

Fig. 6.1. A general IR system architecture

A user query represents the user's information needs, which is in one of the following forms:

1. **Keyword queries:** The user expresses his/her information needs with a list of (at least one) keywords (or **terms**) aiming to find documents that contain some (at least one) or all the query terms. The terms in the list are assumed to be connected with a "soft" version of the logical AND. For example, if one is interested in finding information about Web mining, one may issue the query 'Web mining' to an IR or search engine system. 'Web mining' is retreated as 'Web AND mining'. The retrieval system then finds those likely relevant documents and ranks them suitably to present to the user. Note that a retrieved document does not have to contain all the terms in the query. In some IR systems, the ordering of the words is also significant and will affect the retrieval results.

2. **Boolean queries:** The user can use Boolean operators, AND, OR, and NOT to construct complex queries. Thus, such queries consist of terms and Boolean operators. For example, 'data OR Web' is a Boolean query, which requests documents that contain the word 'data' or 'Web. A page is returned for a Boolean query if the query is logically true in the page (i.e., **exact match**). Although one can write complex Boolean queries using the three operators, users seldom write such queries. Search engines usually support a restricted version of Boolean queries.

3. **Phrase queries:** Such a query consists of a sequence of words that makes up a phrase. Each returned document must contain at least one

instance of the phrase. In a search engine, a phrase query is normally enclosed with double quotes. For example, one can issue the following phrase query (including the double quotes), "Web mining techniques and applications" to find documents that contain the exact phrase.

4. **Proximity queries:** The proximity query is a relaxed version of the phrase query and can be a combination of terms and phrases. Proximity queries seek the query terms within close proximity to each other. The closeness is used as a factor in ranking the returned documents or pages. For example, a document that contains all query terms close together is considered more relevant than a page in which the query terms are far apart. Some systems allow the user to specify the maximum allowed distance between the query terms. Most search engines consider both term proximity and term ordering in retrieval.

5. **Full document queries:** When the query is a full document, the user wants to find other documents that are similar to the query document. Some search engines (e.g., Google) allow the user to issue such a query by providing the URL of a query page. Additionally, in the returned results of a search engine, each snippet may have a link called "more like this" or "similar pages." When the user clicks on the link, a set of pages similar to the page in the snippet is returned.

6. **Natural language questions:** This is the most complex case, and also the ideal case. The user expresses his/her information need as a natural language question. The system then finds the answer. However, such queries are still hard to handle due to the difficulty of natural language understanding. Nevertheless, this is an active research area, called **question answering**. Some search systems are starting to provide question answering services on some specific types of questions, e.g., definition questions, which ask for definitions of technical terms. Definition questions are usually easier to answer because there are strong linguistic patterns indicating definition sentences, e.g., "defined as", "refers to", etc. Definitions can usually be extracted offline [339, 280].

The **query operations** module can range from very simple to very complex. In the simplest case, it does nothing but just pass the query to the retrieval engine after some simple pre-processing, e.g., removal of **stopwords** (words that occur very frequently in text but have little meaning, e.g., "the", "a", "in", etc). We will discuss text pre-processing in Sect. 6.5. In more complex cases, it needs to transform natural language queries into executable queries. It may also accept user feedback and use it to expand and refine the original queries. This is usually called **relevance feedback**, which will be discussed in Sect. 6.3.

The **indexer** is the module that indexes the original raw documents in some data structures to enable efficient retrieval. The result is the **docu-**

ment index. In Sect. 6.6, we study a particular type of indexing scheme, called the **inverted index**, which is used in search engines and most IR systems. An inverted index is easy to build and very efficient to search.

The **retrieval system** computes a relevance score for each indexed document to the query. According to their relevance scores, the documents are ranked and presented to the user. Note that it usually does not compare the user query with every document in the collection, which is too inefficient. Instead, only a small subset of the documents that contains at least one query term is first found from the index and relevance scores with the user query are then computed only for this subset of documents.

6.2 Information Retrieval Models

An IR model governs how a document and a query are represented and how the relevance of a document to a user query is defined. There are four main IR models: Boolean model, vector space model, language model and probabilistic model. The most commonly used models in IR systems and on the Web are the first three models, which we study in this section.

Although these three models represent documents and queries differently, they used the same framework. They all treat each document or query as a **"bag" of words** or **terms**. Term sequence and position in a sentence or a document are ignored. That is, a document is described by a set of distinctive terms. A term is simply a word whose semantics helps remember the document's main themes. We should note that the term here may not be a natural language word in a dictionary. Each term is associated with a weight. Given a collection of documents D, let $V = \{t_1, t_2, ..., t_{|V|}\}$ be the set of distinctive terms in the collection, where t_i is a term. The set V is usually called the **vocabulary** of the collection, and $|V|$ is its size, i.e., the number of terms in V. A weight $w_{ij} > 0$ is associated with each term t_i of a document $\mathbf{d}_j \in D$. For a term that does not appear in document $\mathbf{d}_j$, $w_{ij} = 0$. Each document $\mathbf{d}_j$ is thus represented with a term vector,

$$\mathbf{d}_j = (w_{1j}, w_{2j}, ..., w_{|V|j}),$$

where each weight w_{ij} corresponds to the term $t_i \in V$, and quantifies the level of importance of t_i in document $\mathbf{d}_j$. The sequence of the components (or terms) in the vector is not significant. Note that following the convention of this book, a bold lower case letter is used to represent a vector.

With this vector representation, a collection of documents is simply represented as a relational table (or a matrix). Each term is an attribute, and each weight is an attribute value. In different retrieval models, w_{ij} is computed differently.

6.2.1   Boolean Model

The Boolean model is one of the earliest and simplest information retrieval models. It uses the notion of exact matching to match documents to the user query. Both the query and the retrieval are based on Boolean algebra.

Document Representation: In the Boolean model, documents and queries are represented as sets of terms. That is, each term is only considered present or absent in a document. Using the vector representation of the document above, the weight w_{ij} ($\in \{0, 1\}$) of term t_i in document $\mathbf{d}_j$ is 1 if t_i appears in document $\mathbf{d}_j$, and 0 otherwise, i.e.,

$$w_{ij} = \begin{cases} 1 & \text{if } t_i \text{ appears in } \mathbf{d}_j \\ 0 & \text{otherwise.} \end{cases} \tag{1}$$

Boolean Queries: As we mentioned in Sect. 6.1, query terms are combined logically using the Boolean operators **AND, OR**, and **NOT**, which have their usual semantics in logic. Thus, a Boolean query has a precise semantics. For instance, the query, $((x$ AND $y)$ AND (NOT $z))$ says that a retrieved document must contain both the terms x and y but not z. As another example, the query expression $(x$ OR $y)$ means that at least one of these terms must be in each retrieved document. Here, we assume that x, y and z are terms. In general, they can be Boolean expressions themselves.

Document Retrieval: Given a Boolean query, the system retrieves every document that makes the query logically true. Thus, the retrieval is based on the binary decision criterion, i.e., a document is either relevant or irrelevant. Intuitively, this is called **exact match**. There is no notion of partial match or ranking of the retrieved documents. This is one of the major disadvantages of the Boolean model, which often leads to poor retrieval results. It is quite clear that the frequency of terms and their proximity contribute significantly to the relevance of a document.

Due to this problem, the Boolean model is seldom used alone in practice. Most search engines support some limited forms of Boolean retrieval using explicit **inclusion** and **exclusion operators**. For example, the following query can be issued to Google, 'mining –data +"equipment price"', where + (inclusion) and – (exclusion) are similar to Boolean operators AND and NOT respectively. The operator OR may be supported as well.

6.2.2   Vector Space Model

This model is perhaps the best known and most widely used IR model.

Document Representation

A document in the vector space model is represented as a weight vector, in which each component weight is computed based on some variation of TF or TF-IDF scheme. The weight w_{ij} of term t_i in document $\mathbf{d}_j$ is no longer in $\{0, 1\}$ as in the Boolean model, but can be any number.

Term Frequency (TF) Scheme: In this method, the weight of a term t_i in document $\mathbf{d}_j$ is the number of times that t_i appears in document $\mathbf{d}_j$, denoted by f_{ij}. Normalization may also be applied (see Equation (2)).

The shortcoming of the TF scheme is that it does not consider the situation where a term appears in many documents of the collection. Such a term may not be discriminative.

TF-IDF Scheme: This is the most well known weighting scheme, where TF still stands for the **term frequency** and IDF the **inverse document frequency**. There are several variations of this scheme. Here we only give the most basic one.

Let N be the total number of documents in the system or the collection and df_i be the number of documents in which term t_i appears at least once. Let f_{ij} be the raw frequency count of term t_i in document $\mathbf{d}_j$. Then, the **normalized term frequency** (denoted by tf_{ij}) of t_i in $\mathbf{d}_j$ is given by

$$tf_{ij} = \frac{f_{ij}}{\max\{f_{1j}, f_{2j}, ..., f_{|V|j}\}}, \tag{2}$$

where the maximum is computed over all terms that appear in document $\mathbf{d}_j$. If term t_i does not appear in $\mathbf{d}_j$ then $tf_{ij} = 0$. Recall that $|V|$ is the vocabulary size of the collection.

The inverse document frequency (denoted by idf_i) of term t_i is given by:

$$idf_i = \log \frac{N}{df_i}. \tag{3}$$

The intuition here is that if a term appears in a large number of documents in the collection, it is probably not important or not discriminative. The final TF-IDF term weight is given by:

$$w_{ij} = tf_{ij} \times idf_i. \tag{4}$$

Queries

A query $\mathbf{q}$ is represented in exactly the same way as a document in the document collection. The term weight w_{iq} of each term t_i in $\mathbf{q}$ can also be computed in the same way as in a normal document, or slightly differently. For example, Salton and Buckley [470] suggested the following:

$$w_{iq} = \left(0.5 + \frac{0.5 f_{iq}}{\max\{f_{1q}, f_{2q}, ..., f_{|V|q}\}} \right) \times \log \frac{N}{df_i}. \tag{5}$$

Document Retrieval and Relevance Ranking

It is often difficult to make a binary decision on whether a document is relevant to a given query. Unlike the Boolean model, the vector space model does not make such a decision. Instead, the documents are ranked according to their degrees of relevance to the query. One way to compute the degree of relevance is to calculate the similarity of the query $\mathbf{q}$ to each document $\mathbf{d}_j$ in the document collection D. There are many similarity measures. The most well known one is the **cosine similarity**, which is the cosine of the angle between the query vector $\mathbf{q}$ and the document vector $\mathbf{d}_j$,

$$cosine(\mathbf{d}_j, \mathbf{q}) = \frac{\langle \mathbf{d}_j \bullet \mathbf{q} \rangle}{\|\mathbf{d}_j\| \times \|\mathbf{q}\|} = \frac{\sum_{i=1}^{|V|} w_{ij} \times w_{iq}}{\sqrt{\sum_{i=1}^{|V|} w_{ij}^2} \times \sqrt{\sum_{i=1}^{|V|} w_{iq}^2}}. \tag{6}$$

Cosine similarity is also widely used in text/document clustering.

The dot product of the two vectors is another similarity measure,

$$sim(\mathbf{d}_j, \mathbf{q}) = \langle \mathbf{d}_j \bullet \mathbf{q} \rangle. \tag{7}$$

Ranking of the documents is done using their similarity values. The top ranked documents are regarded as more relevant to the query.

Another way to assess the degree of relevance is to directly compute a relevance score for each document to the query. The **Okapi** method and its variations are popular techniques in this setting. The Okapi retrieval formula given here is based on that in [465, 493]. It has been shown that Okapi variations are more effective than cosine for short query retrieval.

Since it is easier to present the formula directly using the "bag" of words notation of documents than vectors, document $\mathbf{d}_j$ will be denoted by d_j and query $\mathbf{q}$ will be denoted by q. Additional notations are as follows:

t_i is a term
f_{ij} is the raw frequency count of term t_i in document d_j
f_{iq} is the raw frequency count of term t_i in query q
N is the total number of documents in the collection
df_i is the number of documents that contain the term t_i
dl_j is the document length (in bytes) of d_j
$avdl$ is the average document length of the collection

The Okapi relevance score of a document d_j for a query q is:

$$okapi(d_j, q) = \sum_{t_i \in q, d_j} \ln \frac{N - df_i + 0.5}{df_i + 0.5} \times \frac{(k_1 + 1)f_{ij}}{k_1(1 - b + b\frac{dl_j}{avdl}) + f_{ij}} \times \frac{(k_2 + 1)f_{iq}}{k_2 + f_{iq}}, \quad (8)$$

where k_1 (between 1.0-2.0), b (usually 0.75) and k_2 (between 1-1000) are parameters.

Yet another score function is the **pivoted normalization weighting** score function, denoted by *pnw* [493]:

$$pnw(d_j, q) = \sum_{t_i \in q, d_j} \frac{1 + \ln(1 + \ln(f_{ij}))}{(1 - s) + s\frac{dl_j}{avdl}} \times f_{iq} \times \ln \frac{N + 1}{df_i}, \quad (9)$$

where s is a parameter (usually set to 0.2). Note that these are empirical functions based on intuitions and experimental evaluations. There are many variations of these functions used in practice.

6.2.3   Statistical Language Model

Statistical language models (or simply **language models**) are based on probability and have foundations in statistical theory. The basic idea of this approach to retrieval is simple. It first estimates a language model for each document, and then ranks documents by the likelihood of the query given the language model. Similar ideas have previously been used in natural language processing and speech recognition. The formulation and discussion in this section is based on those in [595, 596]. Information retrieval using language models was first proposed by Ponte and Croft [448].

Let the query q be a sequence of terms, $q = q_1q_2...q_m$ and the document collection D be a set of documents, $D = \{d_1, d_2, ..., d_N\}$. In the language modeling approach, we consider the probability of a query q as being "generated" by a probabilistic model based on a document d_j, i.e., $Pr(q|d_j)$. To rank documents in retrieval, we are interested in estimating the posterior probability $Pr(d_j|q)$. Using the Bayes rule, we have

$$Pr(d_j \mid q) = \frac{Pr(q \mid d_j)Pr(d_j)}{Pr(q)} \quad (10)$$

For ranking, $Pr(q)$ is not needed as it is the same for every document. $Pr(d_j)$ is usually considered uniform and thus will not affect ranking. We only need to compute $Pr(q|d_j)$.

The language model used in most existing work is based on unigram, i.e., only individual terms (words) are considered. That is, the model assumes that each term (word) is generated independently, which is essentially a multinomial distribution over words. The general case is the *n*-gram model, where the *n*th term is conditioned on the previous *n*-1 terms.

Based on the multinomial distribution and the unigram model, we have

$$\Pr(q = q_1 q_2 ... q_m \mid d_j) = \prod_{i=1}^{m} \Pr(q_i \mid d_j) = \prod_{i=1}^{|V|} \Pr(t_i \mid d_j)^{f_{iq}}, \tag{11}$$

where f_{iq} is the number of times that term t_i occurs in q, and $\sum_{i=1}^{|V|} \Pr(t_i \mid d_j) = 1$. The retrieval problem is reduced to estimating $\Pr(t_i|d_j)$, which can be the relative frequency,

$$\Pr(t_i \mid d_j) = \frac{f_{ij}}{|d_j|}. \tag{12}$$

Recall that f_{ij} is the number of times that term t_i occurs in document d_j. $|d_j|$ denotes the total number of words in d_j.

However, one problem with this estimation is that a term that does not appear in d_j has the probability of 0, which underestimates the probability of the unseen term in the document. This situation is similar to text classification using the naïve Bayesian model (see Sect. 3.7). A non-zero probability is typically assigned to each unseen term in the document, which is called **smoothing**. Smoothing adjusts the estimates of probabilities to produce more accurate probabilities. The name smoothing comes from the fact that these techniques tend to make distributions more uniform, by adjusting low probabilities such as zero probabilities upward, and high probabilities downward. Not only do smoothing methods aim to prevent zero probabilities, but they also attempt to improve the accuracy of the model as a whole. Traditional additive smoothing is

$$\Pr_{add}(t_i \mid d_j) = \frac{\lambda + f_{ij}}{\lambda |V| + |d_j|}. \tag{13}$$

When $\lambda = 1$, it is the **Laplace smoothing** and when $0 < \lambda < 1$, it is the **Lidstone smoothing**. Many other more sophisticated smoothing methods can be found in [97, 596].

6.3 Relevance Feedback

To improve the retrieval effectiveness, researchers have proposed many techniques. Relevance feedback is one of the effective ones. It is a process where the user identifies some relevant and irrelevant documents in the initial list of retrieved documents, and the system then creates an expanded query by extracting some additional terms from the sample relevant and irrelevant documents for a second round of retrieval. The system may also produce a classification model using the user-identified relevant and irrelevant documents to classify the documents in the document collection into relevant and irrelevant documents. The relevance feedback process may be repeated until the user is satisfied with the retrieved result.

The Rocchio Method

This is one of the early and effective relevance feedback algorithms. It is based on the first approach above. That is, it uses the user-identified relevant and irrelevant documents to expand the original query. The new (or expanded) query is then used to perform retrieval again.

Let the original query vector be $\mathbf{q}$, the set of relevant documents selected by the user be D_r, and the set of irrelevant documents be D_{ir}. The expanded query $\mathbf{q}_e$ is computed as follows,

$$\mathbf{q}_e = \alpha\mathbf{q} + \frac{\beta}{|D_r|}\sum_{\mathbf{d}_r \in D_r}\mathbf{d}_r - \frac{\gamma}{|D_{ir}|}\sum_{\mathbf{d}_{ir} \in D_{ir}}\mathbf{d}_{ir}, \tag{14}$$

where α, β and γ are parameters. Equation (14) simply augments the original query vector $\mathbf{q}$ with additional terms from relevant documents. The original query $\mathbf{q}$ is still needed because it directly reflects the user's information need. Relevant documents are considered more important than irrelevant documents. The subtraction is used to reduce the influence of those terms that are not discriminative (i.e., they appear in both relevant and irrelevant documents), and those terms that appear in irrelevant documents only. The three parameters are set empirically. Note that a slight variation of the algorithm is one without the normalization of $|D_r|$ and $|D_{ir}|$. Both these methods are simple and efficient to compute, and usually produce good results.

Machine Learning Methods

Since we have a set of relevant and irrelevant documents, we can construct a classification model from them. Then the relevance feedback problem becomes a learning problem. Any supervised learning method may be used, e.g., naïve Bayesian classification and SVM. Similarity comparison with the original query is no longer needed.

In fact, a variation of the Rocchio method above, called the **Rocchio classification** method, can be used for this purpose too. Building a Rocchio classifier is done by constructing a prototype vector $\mathbf{c}_i$ for each class i, which is either *relevant* or *irrelevant* in this case (negative elements or components of the vector $\mathbf{c}_i$ are usually set to 0):

$$\mathbf{c}_i = \frac{\alpha}{|D_i|} \sum_{\mathbf{d} \in D_i} \frac{\mathbf{d}}{\|\mathbf{d}\|} - \frac{\beta}{|D - D_i|} \sum_{\mathbf{d} \in D - D_i} \frac{\mathbf{d}}{\|\mathbf{d}\|}, \quad (15)$$

where D_i is the set of documents of class i, and α and β are parameters. Using the TF-IDF term weighting scheme, $\alpha = 16$ and $\beta = 4$ usually work quite well.

In classification, cosine similarity is applied. That is, each test document $\mathbf{d}_t$ is compared with every prototype $\mathbf{c}_i$ based on cosine similarity. $\mathbf{d}_t$ is assigned to the class with the highest similarity value (Fig. 6.2).

Algorithm
1   **for** each class i **do**
2   construct its prototype vector $\mathbf{c}_i$ using Equation (15)
3   **endfor**
4   **for** each test document $\mathbf{d}_t$ **do**
5   the class of $\mathbf{d}_t$ is $\arg\max_i cosine(\mathbf{d}_t, \mathbf{c}_i)$
6   **endfor**

Fig. 6.2. Training and testing of a Rocchio classifier

Apart from the above classic methods, the following learning techniques are also applicable:

Learning from Labeled and Unlabeled Examples (LU Learning): Since the number of user-selected relevant and irrelevant documents may be small, it can be difficult to build an accurate classifier. However, unlabeled examples, i.e., those documents that are not selected by the user, can be utilized to improve learning to produce a more accurate classifier. This fits the LU learning model exactly (see Sect. 5.1). The user-selected relevant and irrelevant documents form the small labeled training set.

Learning from Positive and Unlabeled Examples (PU Learning): The two learning models mentioned above assume that the user can confidently identify both relevant and irrelevant documents. However, in some cases, the user only selects (or clicks) documents that he/she feels relevant based on the title or summary information (e.g., snippets in Web search), which are most likely to be true relevant documents, but does not indicate irrelevant documents. Those documents that are not selected by the user may not be treated as irrelevant because he/she has not seen them. Thus, they can only be regarded as unlabeled documents. This is called **implicit feedback**. In order to learn in this case, we can use PU learning, i.e., learning from positive and unlabeled examples (see Sect. 5.2). We regard the user-selected documents as positive examples, and unselected documents as unlabeled examples. Researchers have experimented with this approach in the Web search context and obtained good results [128].

Using Ranking SVM and Language Models: In the implicit feedback setting, a technique called **ranking SVM** is proposed in [260] to rank the unselected documents based on the selected documents. A language model based approach is also proposed in [487].

Pseudo-Relevance Feedback

Pseudo-relevance feedback is another technique used to improve retrieval effectiveness. Its basic idea is to extract some terms (usually frequent terms) from the top-ranked documents and add them to the original query to form a new query for a second round of retrieval. Again, the process can be repeated until the user is satisfied with the final results. The main difference between this method and the relevance feedback method is that in this method, the user is not involved in the process. The approach simply assumes that the top-ranked documents are likely to be relevant. Through query expansion, some relevant documents missed in the initial round can be retrieved to improve the overall performance. Clearly, the effectiveness of this method relies on the quality of the selected expansion terms.

6.4   Evaluation Measures

Precision and recall measures have been described in Chap. 3 on supervised learning, where each document is classified to a specific class. In IR and Web search, usually no decision is made on whether a document is relevant or irrelevant to a query. Instead, a ranking of the documents is produced for the user. This section studies how to evaluate such rankings.

Again, let the collection of documents in the database be D, and the total number of documents in D be N. Given a user query $\mathbf{q}$, the retrieval algorithm first computes relevance scores for all documents in D and then produce a ranking R_q of the documents based on the relevance scores, i.e.,

$$R_q: \; <\mathbf{d}_1^q, \mathbf{d}_2^q, ..., \mathbf{d}_N^q>, \tag{16}$$

where $\mathbf{d}_1^q \in D$ is the most relevant document to query $\mathbf{q}$ and $\mathbf{d}_N^q \in D$ is the most irrelevant document to query $\mathbf{q}$.

Let $D_q (\subseteq D)$ be the set of actual relevant documents of query $\mathbf{q}$ in D. We can compute the precision and recall values at each $\mathbf{d}_i^q$ in the ranking.

Recall at rank position i or document $\mathbf{d}_i^q$ (denoted by $r(i)$) is the fraction of relevant documents from $\mathbf{d}_1^q$ to $\mathbf{d}_i^q$ in R_q. Let the number of relevant documents from $\mathbf{d}_1^q$ to $\mathbf{d}_i^q$ in R_q be s_i ($\leq |D_q|$) ($|D_q|$ is the size of D_q). Then,

$$r(i) = \frac{s_i}{|D_q|}. \tag{17}$$

Precision at rank position i or document $\mathbf{d}_i^q$ (denoted by $p(i)$) is the fraction of documents from $\mathbf{d}_1^q$ to $\mathbf{d}_i^q$ in R_q that are relevant:

$$p(i) = \frac{s_i}{i} \tag{18}$$

Example 1: We have a document collection D with 20 documents. Given a query $\mathbf{q}$, we know that eight documents are relevant to $\mathbf{q}$. A retrieval algorithm produces the ranking (of all documents in D) shown in Fig. 6.3.

Rank i	+/−	$p(i)$	$r(i)$
1	+	1/1 = 100%	1/8 = 13%
2	+	2/2 = 100%	2/8 = 25%
3	+	3/3 = 100%	3/8 = 38%
4	−	3/4 = 75%	3/8 = 38%
5	+	4/5 = 80%	4/8 = 50%
6	−	4/6 = 67%	4/8 = 50%
7	+	5/7 = 71%	5/8 = 63%
8	−	5/8 = 63%	5/8 = 63%
9	+	6/9 = 67%	6/8 = 75%
10	+	7/10 = 70%	7/8 = 88%
11	−	7/11 = 63%	7/8 = 88%
12	−	7/12 = 58%	7/8 = 88%
13	+	8/13 = 62%	8/8 = 100%
14	−	8/14 = 57%	8/8 = 100%
15	−	8/15 = 53%	8/8 = 100%
16	−	8/16 = 50%	8/8 = 100%
17	−	8/17 = 53%	8/8 = 100%
18	−	8/18 = 44%	8/8 = 100%
19	−	8/19 = 42%	8/8 = 100%
20	−	8/20 = 40%	8/8 = 100%

Fig. 6.3. Precision and recall values at each rank position

In column 1 of Fig. 6.3, 1 represents the highest rank and 20 represents the lowest rank. "+" and "−" in column 2 indicate a relevant document and an irrelevant document respectively. The precision ($p(i)$) and recall ($r(i)$) values at each position i are given in columns 3 and 4. ■

Average Precision: Sometimes we want a single precision to compare different retrieval algorithms on a query **q**. An average precision (p_{avg}) can be computed based on the precision at each relevant document in the ranking,

$$p_{avg} = \frac{\sum_{d_i^q \in D_q} p(i)}{|D_q|}. \tag{19}$$

For the ranking in Fig. 6.3 of Example 1, the average precision is 81%:

$$p_{avg} = \frac{100\% + 100\% + 100\% + 80\% + 71\% + 67\% + 70\% + 62\%}{8} = 81\%. \tag{20}$$

Precision–Recall Curve: Based on the precision and recall values at each rank position, we can draw a precision–recall curve where the x-axis is the recall and the y-axis is the precision. Instead of using the precision and recall at each rank position, the curve is commonly plotted using 11 standard recall levels, 0%, 10%, 20%, ..., 100%.

Since we may not obtain exactly these recall levels in the ranking, interpolation is needed to obtain the precisions at these recall levels, which is done as follows: Let r_i be a recall level, $i \in \{0, 1, 2, ..., 10\}$, and $p(r_i)$ be the precision at the recall level r_i. $p(r_i)$ is computed with

$$p(r_i) = \max_{r_i \le r \le r_{10}} p(r). \tag{21}$$

That is, to interpolate precision at a particular recall level r_i, we take the maximum precision of all recalls between level r_i and level r_{10}.

Example 2: Following Example 1, we obtain the interpolated precisions at all 11 recall levels in the table of Fig. 6.4. The precision-recall curve is shown on the right.

i	$p(r_i)$	r_i
0	100%	0%
1	100%	10%
2	100%	20%
3	100%	30%
4	80%	40%
5	80%	50%
6	71%	60%
7	70%	70%
8	70%	80%
9	62%	90%
10	62%	100%

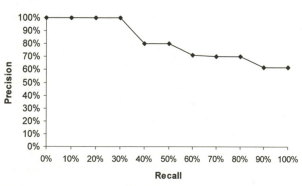

Fig. 6.4. The precision-recall curve

Comparing Different Algorithms: Frequently, we need to compare the retrieval results of different algorithms. We can draw their precision-recall curves together in the same figure for comparison. Figure 6.5 shows the curves of two algorithms on the same query and the same document collection. We observe that the precisions of one algorithm are better than those of the other at low recall levels, but are worse at high recall levels.

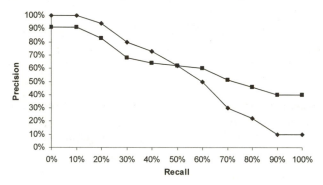

Fig. 6.5. Comparison of two retrieval algorithms based on their precision-recall curves

Evaluation Using Multiple Queries: In most retrieval evaluations, we are interested in the performance of an algorithm on a large number of queries. The overall precision (denoted by $\bar{p}(r_i)$) at each recall level r_i is computed as the average of individual precisions at that recall level, i.e.,

$$\bar{p}(r_i) = \frac{1}{|Q|} \sum_{j=1}^{|Q|} p_j(r_i), \tag{22}$$

where Q is the set of all queries and $p_j(r_i)$ is the precision of query j at the recall level r_i. Using the average precision at each recall level, we can also draw a precision-recall curve.

Although in theory precision and recall do not depend on each other, in practice a high recall is almost always achieved at the expense of precision, and a high precision is achieved at the expense of recall. Thus, precision and recall has a trade-off. Depending on the application, one may want a high precision or a high recall.

One problem with precision and recall measures is that, in many applications, it can be very hard to determine the set of relevant documents D_q for each query **q**. For example, on the Web, D_q is almost impossible to determine because there are simply too many pages to manually inspect. Without D_q, the recall value cannot be computed. In fact, recall does not make much sense for Web search because the user seldom looks at pages ranked below 30. However, precision is critical, and it can be estimated for top ranked documents. Manual inspection of only the top 30 pages is reasonable. The following precision computation is commonly used.

Rank Precision: We compute the precision values at some selected rank positions. For a Web search engine, we usually compute precisions for the top 5, 10, 15, 20, 25 and 30 returned pages (as the user seldom looks at more than 30 pages). We assume that the number of relevant pages is more than 30. Following Example 1, we have $p(5) = 80\%$, $p(10) = 70\%$, $p(15) = 53\%$, and $p(20) = 40\%$.

We should note that precision is not the only measure for evaluating search ranking, reputation or quality of the top ranked pages are also very important as we will see later in this chapter and also in Chap. 7.

F-score: Another often used evaluation measure is the F-score, which we have used in Chap. 3. Here we can compute the F-score at each rank position i. Recall that F-score is the harmonic mean of precision and recall:

$$F(i) = \frac{2}{\dfrac{1}{r(i)} + \dfrac{1}{p(i)}} = \frac{2p(i)r(i)}{p(i) + r(i)}. \tag{23}$$

Finally, the precision and recall **breakeven point** is also a commonly used measure, which we have discussed in Sect. 3.3.2 in Chap. 3.

6.5   Text and Web Page Pre-Processing

Before the documents in a collection are used for retrieval, some preprocessing tasks are usually performed. For traditional text documents (no HTML tags), the tasks are stopword removal, stemming, and handling of digits, hyphens, punctuations, and cases of letters. For Web pages, additional tasks such as HTML tag removal and identification of main content blocks also require careful considerations. We discuss them in this section.

6.5.1   Stopword Removal

Stopwords are frequently occurring and insignificant words in a language that help construct sentences but do not represent any content of the documents. Articles, prepositions and conjunctions and some pronouns are natural candidates. Common stopwords in English include:

a, about, an, are, as, at, be, by, for, from, how, in, is, of, on, or, that, the, these, this, to, was, what, when, where, who, will, with

Such words should be removed before documents are indexed and stored. Stopwords in the query are also removed before retrieval is performed.

6.5.2 Stemming

In many languages, a word has various syntactical forms depending on the contexts that it is used. For example, in English, nouns have plural forms, verbs have gerund forms (by adding "*ing*"), and verbs used in the past tense are different from the present tense. These are considered as syntactic variations of the same root form. Such variations cause low recall for a retrieval system because a relevant document may contain a variation of a query word but not the exact word itself. This problem can be partially dealt with by **stemming**.

Stemming refers to the process of reducing words to their stems or roots. A **stem** is the portion of a word that is left after removing its prefixes and suffixes. In English, most variants of a word are generated by the introduction of suffixes (rather than prefixes). Thus, stemming in English usually means **suffix removal,** or **stripping**. For example, "computer", "computing", and "compute" are reduced to "comput". "walks", "walking" and "walker" are reduced to "walk". Stemming enables different variations of the word to be considered in retrieval, which improves the recall. There are several stemming algorithms, also known as **stemmers**. In English, the most popular stemmer is perhaps the Martin Porter's stemming algorithm [449], which uses a set of rules for stemming.

Over the years, many researchers evaluated the advantages and disadvantages of using stemming. Clearly, stemming increases the recall and reduces the size of the indexing structure. However, it can hurt precision because many irrelevant documents may be considered relevant. For example, both "cop" and "cope" are reduced to the stem "cop". However, if one is looking for documents about police, a document that contains only "cope" is unlikely to be relevant. Although many experiments have been conducted by researchers, there is still no conclusive evidence one way or the other. In practice, one should experiment with the document collection at hand to see whether stemming helps.

6.5.3 Other Pre-Processing Tasks for Text

Digits: Numbers and terms that contain digits are removed in traditional IR systems except some specific types, e.g., dates, times, and other prespecified types expressed with regular expressions. However, in search engines, they are usually indexed.

Hyphens: Breaking hyphens are usually applied to deal with inconsistency of usage. For example, some people use "state-of-the-art", but others use "state of the art". If the hyphens in the first case are removed, we eliminate the inconsistency problem. However, some words may have a hyphen as an integral part of the word, e.g., "Y-21". Thus, in general, the system can follow a general rule (e.g., removing all hyphens) and also have some exceptions. Note that there are two types of removal, i.e., (1) each hyphen is replaced with a space and (2) each hyphen is simply removed without leaving a space so that "state-of-the-art" may be replaced with "state of the art" or "stateoftheart". In some systems both forms are indexed as it is hard to determine which is correct, e.g., if "pre-processing" is converted to "pre processing", then some relevant pages will not be found if the query term is "preprocessing".

Punctuation Marks: Punctuation can be dealt with similarly as hyphens.

Case of Letters: All the letters are usually converted to either the upper or lower case.

6.5.4 Web Page Pre-Processing

We have indicated at the beginning of the section that Web pages are different from traditional text documents. Thus, additional pre-processing is needed. We describe some important ones below.

1. **Identifying different text fields:** In HTML, there are different text fields, e.g., title, metadata, and body. Identifying them allows the retrieval system to treat terms in different fields differently. For example, in search engines terms that appear in the title field of a page are regarded as more important than terms that appear in other fields and are assigned higher weights because the title is usually a concise description of the page. In the body text, those emphasized terms (e.g., under header tags <h1>, <h2>, ..., bold tag <b>, etc.) are also given higher weights.
2. **Identifying anchor text:** Anchor text associated with a hyperlink is treated specially in search engines because the anchor text often represents a more accurate description of the information contained in the page pointed to by its link. In the case that the hyperlink points to an external page (not in the same site), it is especially valuable because it is a summary description of the page given by other people rather than the author/owner of the page, and is thus more trustworthy.
3. **Removing HTML tags:** The removal of HTML tags can be dealt with similarly to punctuation. One issue needs careful consideration, which affects proximity queries and phrase queries. HTML is inherently a vis-

Fig. 6.6. An example of a Web page from Wikipedia

ual presentation language. In a typical commercial page, information is presented in many rectangular blocks (see Fig. 6.6). Simply removing HTML tags may cause problems by joining text that should not be joined. For example, in Fig. 6.6, "cite this article" at the bottom of the left column will join "Main Page" on the right, but they should not be joined. They will cause problems for phrase queries and proximity queries. This problem had not been dealt with satisfactorily by search engines at the time when this book was written.

4. **Identifying main content blocks:** A typical Web page, especially a commercial page, contains a large amount of information that is not part of the main content of the page. For example, it may contain banner ads, navigation bars, copyright notices, etc., which can lead to poor results for search and mining. In Fig. 6.6, the main content block of the page is the block containing "Today's featured article." It is not desirable to index anchor texts of the navigation links as a part of the content of this page. Several researchers have studied the problem of identifying main content blocks. They showed that search and data mining results can be

improved significantly if only the main content blocks are used. We briefly discuss two techniques for finding such blocks in Web pages.

Partitioning based on visual cues: This method uses visual information to help find main content blocks in a page. Visual or rendering information of each HTML element in a page can be obtained from the Web browser. For example, Internet Explorer provides an API that can output the *X* and *Y* coordinates of each element. A machine learning model can then be built based on the location and appearance features for identifying main content blocks of pages. Of course, a large number of training examples need to be manually labeled (see [77, 495] for details).

Tree matching: This method is based on the observation that in most commercial Web sites pages are generated by using some fixed templates. The method thus aims to find such hidden templates. Since HTML has a nested structure, it is thus easy to build a tag tree for each page. **Tree matching** of multiple pages from the same site can be performed to find such templates. In Chap. 9, we will describe a tree matching algorithm for this purpose. Once a template is found, we can identify which blocks are likely to be the main content blocks based on the following observation: the text in main content blocks are usually quite different across different pages of the same template, but the non-main content blocks are often quite similar in different pages. To determine the text similarity of corresponding blocks (which are sub-trees), the **shingle method** described in the next section can be used.

6.5.5 Duplicate Detection

Duplicate documents or pages are not a problem in traditional IR. However, in the context of the Web, it is a significant issue. There are different types of duplication of pages and contents on the Web.

Copying a page is usually called **duplication** or **replication**, and copying an entire site is called **mirroring**. **Duplicate pages** and **mirror sites** are often used to improve efficiency of browsing and file downloading worldwide due to limited bandwidth across different geographic regions and poor or unpredictable network performances. Of course, some duplicate pages are the results of plagiarism. Detecting such pages and sites can reduce the index size and improve search results.

Several methods can be used to find duplicate information. The simplest method is to hash the whole document, e.g., using the MD5 algorithm, or computing an aggregated number (e.g., checksum). However, these methods are only useful for detecting exact duplicates. On the Web, one seldom finds exact duplicates. For example, even different mirror sites may have different URLs, different Web masters, different contact information, different advertisements to suit local needs, etc.

One efficient duplicate detection technique is based on **n-grams** (also called **shingles**). An *n*-gram is simply a consecutive sequence of words of a fixed window size *n*. For example, the sentence, "John went to school with his brother," can be represented with five 3-gram phrases "John went to", "went to school", "to school with", "school with his", and "with his brother". Note that 1-gram is simply the individual words.

Let $S_n(d)$ be the set of distinctive *n*-grams (or shingles) contained in document *d*. Each *n*-gram may be coded with a number or a MD5 hash (which is usually a 32-digit hexadecimal number). Given the *n*-gram representations of the two documents d_1 and d_2, $S_n(d_1)$ and $S_n(d_2)$, the **Jaccard coefficient** can be used to compute the similarity of the two documents,

$$sim(d_1, d_2) = \frac{|S_n(d_1) \cap S_n(d_2)|}{|S_n(d_1) \cup S_n(d_2)|}. \tag{24}$$

A threshold is used to determine whether d_1 and d_2 are likely to be duplicates of each other. For a particular application, the window size *n* and the similarity threshold are chosen through experiments.

6.6 Inverted Index and Its Compression

The basic method of Web search and traditional IR is to find documents that contain the terms in the user query. Given a user query, one option is to scan the document database sequentially to find the documents that contain the query terms. However, this method is obviously impractical for a large collection, such as the Web. Another option is to build some data structures (called **indices**) from the document collection to speed up retrieval or search. There are many index schemes for text [31]. The **inverted index**, which has been shown superior to most other indexing schemes, is a popular one. It is perhaps the most important index method used in search engines. This indexing scheme not only allows efficient retrieval of documents that contain query terms, but also very fast to build.

6.6.1 Inverted Index

In its simplest form, the inverted index of a document collection is basically a data structure that attaches each distinctive term with a list of all documents that contains the term. Thus, in retrieval, it takes constant time to find the documents that contains a query term. Finding documents containing multiple query terms is also easy as we will see later.

Given a set of documents, $D = \{d_1, d_2, ..., d_N\}$, and each document has a unique identifier (ID). An inverted index consists of two parts: a vocabulary V, containing all the distinct terms in the document set, and for each distinct term t_i an **inverted list** of postings. Each **posting** stores the ID (denoted by id_j) of the document d_j that contains term t_i and other pieces of information about term t_i in document d_j. Depending on the need of the retrieval or ranking algorithm, different pieces of information may be included. For example, to support phrase and proximity search, a posting for a term t_i usually consists of the following,

$$<id_j, f_{ij}, [o_1, o_2, ..., o_{|f_{ij}|}]>$$

where id_j is the ID of document d_j that contains the term t_i, f_{ij} is the frequency count of t_i in d_j, and o_k are the offsets (or positions) of term t_i in d_j. Postings of a term are sorted in increasing order based on the id_j's and so are the offsets in each posting (see Example 3). This facilitates compression of the inverted index as we will see in Sect. 6.6.4.

Example 3: We have three documents of id_1, id_2, and id_3:

id_1: Web mining is useful.
      1   2   3   4
id_2: Usage mining applications.
      1   2   3
id_3: Web structure mining studies the Web hyperlink structure.
     1   2   3   4   5   6   7   8

The numbers below each document are the offset position of each word. The vocabulary is the set:

{Web, mining, useful, applications, usage, structure, studies, hyperlink}

Stopwords "is" and "the" have been removed, but no stemming is applied. Figure 6.7 shows two inverted indices.

Applications:	id_2	Applications:	$<id_2, 1, [3]>$
Hyperlink:	id_3	Hyperlink:	$<id_3, 1, [7]>$
Mining:	id_1, id_2, id_3	Mining:	$<id_1, 1, [2]>, <id_2, 1, [2]>, <id_3, 1, [3]>$
Structure:	id_3	Structure:	$<id_3, 2, [2, 8]>$
Studies:	id_3	Studies:	$<id_3, 1, [4]>$
Usage:	id_2	Usage:	$<id_2, 1, [1]>$
Useful:	id_1	Useful:	$<id_1, 1, [4]>$
Web:	id_1, id_3	Web:	$<id_1, 1, [1]>, <id_3, 2, [1, 6]>$

        (A)   (B)

Fig. 6.7. Two inverted indices: a simple version and a more complex version

Figure 6.7(A) is a simple version, where each term is attached with only an inverted list of IDs of the documents that contain the term. Each inverted list in Fig. 6.7(B) is more complex as it contains additional information, i.e., the frequency count of the term and its positions in each document. Note that we use id_i as the document IDs to distinguish them from offsets. In an actual implementation, they may also be positive integers. Note also that a posting can contain other types of information depending on the need of the retrieval or search algorithm (see Sect. 6.8).   ■

6.6.2   Search Using an Inverted Index

Queries are evaluated by first fetching the inverted lists of the query terms, and then processing them to find the documents that contain all (or some) terms. Specifically, given the query terms, searching for relevant documents in the inverted index consists of three main steps:

Step 1 (**vocabulary search**): This step finds each query term in the vocabulary, which gives the inverted list of each term. To speed up the search, the vocabulary usually resides in the main memory. Various indexing methods, e.g., hashing, tries or B-tree, can be used to speed up the search. Lexicographical ordering may also be employed due to its space efficiency. Then the binary search method can be applied. The complexity is $O(\log|V|)$, where $|V|$ is the vocabulary size.

If the query contains only a single term, this step gives all the relevant documents and the algorithm then goes to step 3. If the query contains multiple terms, the algorithm proceeds to step 2.

Step 2 (**results merging**): After the inverted list of each term is found, merging of the lists is performed to find their intersection, i.e., the set of documents containing all query terms. Merging simply traverses all the lists in synchronization to check whether each document contains all query terms. One main heuristic is to use the shortest list as the base to merge with the other longer lists. For each posting in the shortest list, a binary search may be applied to find it in each longer list. Note that partial match (i.e., documents containing only some of the query terms) can be achieved as well in a similar way, which is more useful in practice.

Usually, the whole inverted index cannot fit in memory, so part of it is cached in memory for efficiency. Determining which part to cache involves analysis of query logs to find frequent query terms. The inverted lists of these frequent query terms can be cached in memory.

Step 3 (**Rank score computation**): This step computes a rank (or relevance) score for each document based on a relevance function (e.g., okapi or cosine), which may also consider the phrase and term proximity information. The score is then used in the final ranking.

Example 4: Using the inverted index built in Fig. 6.7(B), we want to search for "web mining" (the query). In step 1, two inverted lists are found:

Mining: $<id_1, 1, [2]>, <id_2, 1, [2]>, <id_3, 1, [3]>$
Web: $<id_1, 1, [1]>, <id_3, 2, [1, 6]>$

In step 2, the algorithm traverses the two lists and finds documents containing both words (documents id_1 and id_3). The word positions are also retrieved. In step 3, we compute the rank scores. Considering the proximity and the sequence of words, we give id_1 a higher rank (or relevance) score than id_3 as "web" and "mining" are next to each other in id_1 and in the same sequence as that in the query. Different search engines may use different algorithms to combine these factors.   ■

6.6.3   Index Construction

The construction of an inverted index is quite simple and can be done efficiently using a trie data structure among many others. The time complexity of the index construction is $O(T)$, where T is the number of all terms (including duplicates) in the document collection (after pre-processing).

For each document, the algorithm scans it sequentially and for each term, it finds the term in the trie. If it is found, the document ID and other information (e.g., the offset of the term) are added to the inverted list of the term. If the term is not found, a new leaf is created to represent the term.

Example 5: Let us build an inverted index for the three documents in Example 3, which are reproduced below for easy reference. Figure 6.8 shows the vocabulary trie and the inverted lists for all terms.

id_1: Web mining is useful.
            1   2   3   4
id_2: Usage mining applications.
            1   2   3
id_3: Web structure mining studies the Web hyperlink structure   ■
            1   2   3   4   5   6   7   8

To build the index efficiently, the trie is usually stored in memory. However, in the context of the Web, the whole index will not fit in the main memory. The following technique can be applied.

We follow the above algorithm to build the index until the memory is full. The partial index I_1 obtained so far is written on the disk. Then, we process the subsequent documents and build the partial index I_2 in memory, and so on. After all documents have been processed, we have k partial indices, I_1, I_2, ..., I_k, on disk. We then merge the partial indices in a hierarchical manner. That is, we first perform pair-wise merges of I_1 and I_2, I_3 and I_4, and so on. This gives us larger indices I_{1-2}, I_{3-4} and so on. After the first level merging is complete, we proceed to the second level merging, i.e., we merge I_{1-2} and I_{3-4}, I_{5-6} and I_{7-8} and so on. This process continues until all the partial indices are merged into a single index. Each merge is fairly straightforward because the vocabulary in each partial index is sorted by the trie construction. The complexity of each merge is thus linear in the number of terms in both partial indices. Since each level needs a linear process of the whole index, the complete merging process takes $O(k\log k)$ time. To reduce the disk space requirement, whenever a new partial index is generated, we can merge it with a previously merged index. That is, when we have I_1 and I_2, we can merge them immediately to produce I_{1-2}, and when I_3 is produced, it is merged with I_{1-2} to produce I_{1-2-3} and so on.

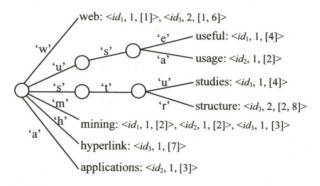

Fig. 6.8. The vocabulary trie and the inverted lists

Instead of using a trie, an alternative method is to use an in-memory hash table (or other data structures) for terms. The algorithm is quite straightforward and will not be discussed further.

On the Web, an important issue is that pages are constantly added, modified or deleted. It may be quite inefficient to modify the main index because a single page change can require updates to a large number of records of the index. One simple solution is to construct two additional indices, one for added pages and one for deleted pages. Modification can be regarded as a deletion and then an addition. Given a user query, it is searched in the main index and also in the two auxiliary indices. Let the pages returned from the search in the main index be D_0, the pages returned from the search in the index of added pages be D_+ and the pages returned from the search in the index of deleted pages be D_-. Then, the final results returned to the user is $(D_0 \cup D_+) - D_-$. When the two auxiliary indices become too large, they can be merged into the main index.

6.6.4   Index Compression

An inverted index can be very large. In order to speed up the search, it should reside in memory as much as possible to avoid disk I/O. Because of this, reducing the index size becomes an important issue. A natural solution to this is **index compression**, which aims to represent the same information with fewer bits or bytes. Using compression, the size of an inverted index can be reduced dramatically. In the lossless compression, the original index can also be reconstructed exactly using the compressed version. Lossless compression methods are the focus of this section.

The inverted index is quite amiable to compression. Since the main space used by an inverted index is for the storage of document IDs and offsets of each term, we thus want to reduce this space requirement. Since all the information is represented with positive integers, we only discuss **integer compression** techniques in this section.

Without compression, on most architectures an integer has a fixed-size representation of four bytes (32 bits). However, few integers need 4 bytes to represent, so a more compact representation (compression) is clearly possible. There are generally two classes of compression schemes for inverted lists: the **variable-bit** scheme and the **variable-byte** scheme.

In the variable-bit (also called **bitwise**) scheme, an integer is represented with an integral number of bits. Well known bitwise methods include **unary coding**, **Elias gamma coding** and **delta coding** [161], and **Golomb coding** [202]. In the variable-byte scheme, an integer is stored in an integral number of bytes, where each byte has 8 bits. A simple bytewise scheme is the variable-byte coding [547]. These coding schemes basically map integers onto self-delimiting binary codewords (bits), i.e., the start bit and the end bit of each integer can be detected with no additional delimiters or markers.

An interesting feature of the inverted index makes compression even more effective. Since document IDs in each inverted list are sorted in increasing order, we can store the difference between any two adjacent document IDs, id_i and id_{i+1}, where $id_{i+1} > id_i$, instead of the actual IDs. This difference is called the **gap** between id_i and id_{i+1}. The gap is a smaller number than id_{i+1} and thus requires fewer bits. In search, if the algorithm linearly traverses each inverted list, document IDs can be recovered easily. Since offsets in each posting are also sorted, they can be stored similarly.

For example, the sorted document IDs are: 4, 10, 300, and 305. They can be represented with gaps, 4, 6, 290 and 5. Given the gap list 4, 6, 290 and 5, it is easy to recover the original document IDs, 4, 10, 300, and 305. We note that for frequent terms (which appear in a large number of documents) the gaps are small and can be encoded with short codes (fewer bits). For infrequent or rare terms, the gaps can be large, but they do not use up much space due to the fact that only a small number of documents contain them. Storing gaps can significantly reduce the index size.

We now discuss each of the coding schemes in detail. Each scheme includes a method for **coding** (or **compression**) and a method for **decoding** (**decompression**).

Unary Coding

Unary coding is simple. It represents a number x with $x-1$ bits of zeros followed by a bit of one. For example, 5 is represented as 00001. The one bit is simply the delimitor. Decoding is also straightforward. This scheme is effective for very small numbers, but wasteful for large numbers. It is thus seldom used alone in practice.

Table 6.1 shows example codes of different coding schemes for 10 decimal integers. Column 2 shows the unary code for each integer.

Table 6.1: Example codes for integers of different coding schemes: Spacing in the Elias, Golomb, and variable-byte codes separates the prefix of the code from the suffix.

Decimal	Unary	Elias Gamma	Elias Delta	Golomb $(b = 3)$	Golomb $(b = 10)$	Variable byte
1	1	1	1	1 10	1 001	0000001 0
2	01	0 10	0 100	1 11	1 010	0000010 0
3	001	0 11	0 101	01 0	1 011	0000011 0
4	0001	00 100	0 1100	01 10	1 100	0000100 0
5	00001	00 101	0 1101	01 11	1 101	0000101 0
6	000001	00 110	0 1110	001 0	1 1100	0000110 0
7	0000001	00 111	0 1111	001 10	1 1101	0000111 0
8	00000001	000 1000	00 100000	001 11	1 1110	0001000 0
9	000000001	000 1001	00 100001	0001 0	1 1111	0001001 0
10	0000000001	000 1010	00 100010	0001 10	01 000	0001010 0

Elias Gamma Coding

Coding: In the Elias gamma coding, a positive integer x is represented by: $1+\lfloor \log_2 x \rfloor$ in unary (i.e., $\lfloor \log_2 x \rfloor$ 0-bits followed by a 1-bit), followed by the binary representation of x without its most significant bit. Note that $1+\lfloor \log_2 x \rfloor$ is simply the number of bits of x in binary. The coding can also be described with the following two steps:

1. Write x in binary.
2. Subtract 1 from the number of bits written in step 1 and prepend that many zeros.

Example 6: The number 9 is represented by 0001001, since $1+\lfloor\log_2 9\rfloor = 4$, or 0001 in unary, and 9 is 001 in binary with the most significant bit removed. Alternatively, we first write 9 in binary, which is 1001 with 4 bits, and then prepend three zeros. In this way, 1 is represented by 1 (in one bit), and 2 is represented by 010. Additional examples are shown in column 3 of Table 6.1. ∎

Decoding: We decode an Elias gamma-coded integer in two steps:

1. Read and count zeroes from the stream until we reach the first one. Call this count of zeroes K.
2. Consider the one that was reached to be the first digit of the integer, with a value of 2^K, read the remaining K bits of the integer.

Example 7: To decompress 0001001, we first read all zero bits from the beginning until we see a bit of 1. We have $K = 3$ zero bits. We then include the 1 bit with the following 3 bits, which give us 1001 (binary for 9). ∎

Gamma coding is efficient for small integers but is not suited to large integers for which the parameterized Golomb code or the Elias delta code is more suitable.

Elias Delta Coding

Elias delta codes are somewhat longer than gamma codes for small integers, but for larger integers such as document numbers in an index of Web pages, the situation is reversed.

Coding: In the Elias delta coding, a positive integer x is stored with the gamma code representation of $1+\lfloor\log_2 x\rfloor$, followed by the binary representation of x less the most significant bit.

Example 8: Let us code the number 9. Since $1+\lfloor\log_2 x\rfloor = 4$, we have its gamma code 00100 for 4. Since 9's binary representation less the most significant bit is 001, we have the delta code of 00100001 for 9. Additional examples are shown in column 4 of Table 6.1. ∎

Decoding: To decode an Elias delta-coded integer x, we first decode the gamma-code part $1+\lfloor\log_2 x\rfloor$ as the magnitude M (the number of bits of x in binary), and then retrieve the binary representation of x less the most significant bit. Specifically, we use the following steps:

1. Read and count zeroes from the stream until you reach the first one. Call this count of zeroes L.
2. Considering the one that was reached to be the first bit of an integer, with a value of 2^L, read the remaining L digits of the integer. This is the

integer M.

3. Put a one in the first place of our final output, representing the value 2^M. Read and append the following M-1 bits.

Example 9: We want to decode 00100001. We can see that $L = 2$ after step 1, and after step 2, we have read and consumed 5 bits. We also obtain $M = 4$ (100 in binary). Finally, we prepend 1 to the M-1 bits (which is 001) to give 1001, which is 9 in binary. ∎

While Elias codes yield acceptable compression and fast decoding, a better performance in both aspects is possible with the Golomb coding.

Golomb Coding

The Golomb coding is a form of parameterized coding in which integers to be coded are stored as values relative to a constant b. Several variations of the original Golomb scheme exist, which save some bits in coding compared to the original scheme. We describe one version here.

Coding: A positive integer x is represented in two parts:

1. The first part is a unary representation of $q+1$, where q is the quotient $\lfloor (x/b) \rfloor$, and
2. The second part is a special binary representation of the remainder $r = x-qb$. Note that there are b possible remainders. For example, if $b = 3$, the possible remainders will be 0, 1, and 2.

The binary representation of a remainder requires $\lfloor \log_2 b \rfloor$ or $\lceil \log_2 b \rceil$ bits. Clearly, it is not possible to write every remainder in $\lfloor \log_2 b \rfloor$ bits in binary. To save space, we want to write the first few remainders using $\lfloor \log_2 b \rfloor$ bits and the rest using $\lceil \log_2 b \rceil$ bits. We must do so such that the decoder knows when $\lfloor \log_2 b \rfloor$ bits are used and when $\lceil \log_2 b \rceil$ bits are used. Let $i = \lfloor \log_2 b \rfloor$. We code the first d remainders using i bits,

$$d = 2^{i+1} - b. \tag{25}$$

It is worth noting that these d remainders are all less than d. The rest of the remainders are coded with $\lceil \log_2 b \rceil$ bits and are all greater than or equal to d. They are coded using a special binary code (also called a **fixed prefix code**) with $\lceil \log_2 b \rceil$ (or $i+1$) bits.

Example 10: For $b = 3$, to code $x = 9$, we have the quotient $q = \lfloor 9/3 \rfloor = 3$. For remainder, we have $i = \lfloor \log_2 3 \rfloor = 1$ and $d = 1$. Note that for $b = 3$, there are three remainders, i.e., 0, 1, and 2, which are coded as 0, 10, and 11 respectively. The remainder for 9 is $r = 9 - 3 \times 3 = 0$. The final code for 9 is 00010. Additional examples for $b = 3$ are shown in column 5 of Table 6.1.

For $b = 10$, to code $x = 9$, we have the quotient $q = \lfloor 9/10 \rfloor = 0$. For remainder, we have $i = \lfloor \log_2 10 \rfloor = 3$ and $d = 6$. Note that for $b = 10$, there are 10 remainders, i.e., 0, 1, 2, ..., 10, which are coded as 000, 001, 010, 011, 100, 101, 1100, 1101, 1110, 1111 respectively. The remainder of 9 is $r = 9 - 0 \times 5 = 9$. The final code for 9 is 11111. Additional examples for $b = 10$ are shown in column 6 of Table 6.1.   ∎

We can see that the first d remainders are standard binary codes, but the rest are not. They are generated using a tree instead. Figure 6.9 shows an example based on $b = 5$. The leaves are the five remainders. The first three remainders (0, 1, 2) are in the standard binary code, and the rest (3 and 4) have an additional bit. It is important to note that the first 2 bits ($i = 2$) of the remainder 3 (the first remainder coded in $i+1$ bits) is 11, which is 3 (i.e., d) in binary. This information is crucial for decoding because it enables the algorithm to know when $i+1$ bits are used. We also notice that d is completely determined by b, which helps decoding.

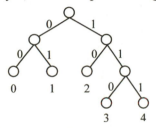

Fig. 6.9. The coding tree for $b = 5$

If b is a power of 2 (called **Golomb–Rice coding**), i.e., $b = 2^k$ for integer $k \geq 0$, every remainder is coded with the same number of bits because $\lfloor \log_2 b \rfloor = \lceil \log_2 b \rceil$. This is also easy to see from Equation (25), i.e., $d = 2^k$.

Decoding: To decode a Golomb-coded integer x, we use the following steps:

1. Decode unary-coded quotient q (the relevant bits are comsumed).
2. Compute $i = \lfloor \log_2 b \rfloor$ and $d = 2^{i+1} - b$.
3. Retrieve the next i bits and assign it to r.
4. If $r \geq d$ then
      retrieve one more bit and append it to r at the end;
      $r = r - d$.
5. Return $x = qb + r$.

Some explanation is in order for step 4. As we discussed above, if $r \geq d$ we need $i+1$ bits to code the remainder. The first line of step 4 retrieves the additional bit and appends it to r. The second line obtains the true value of the remainder r.

Example 11: We want to decode 11111 for $b = 10$. We see that $q = 0$ because there is no zero at the beginning. The first bit is consumed. We know that $i = \lfloor \log_2 10 \rfloor = 3$ and $d = 6$. We then retrieve the next three bits, 111, which is 7 in decimal, and assign it to r (= 111). Since $7 > 6$ (which is d), we retrieve one more bit, which is 1, and r is now 1111 (15 in decimal). The new $r = r - d = 15 - 6 = 9$. Finally, $x = qb + r = 0 + 9 = 9$. ■

Now we discuss the selection of b for each term. For gap compression, Witten et al. [551] reported that a suitable b is

$$b \approx 0.69 \left(\frac{N}{n_t} \right), \tag{26}$$

where N is the total number of documents and n_t is the number of documents that contain term t.

Variable-Byte Coding

Coding: In this method, seven bits in each byte are used to code an integer, with the least significant bit set to 0 in the last byte, or to 1 if further bytes follow. In this way, small integers are represented efficiently. For example, 135 is represented in two bytes, since it lies in the range 2^7 and 2^{14}, as 00000011 00001110. Additional examples are shown in column 6 of Table 6.1.

Decoding: Decoding is performed in two steps:

1. Read all bytes until a byte with the zero last bit is seen.
2. Remove the least significant bit from each byte read so far and concatenate the remaining bits.

For example, 00000011 00001110 is decoded to 00000010000111, which is 135.

Finally, experimental results in [547] show that non-parameterized Elias coding is generally not as space-efficient or as fast as parameterized Golomb coding for retrieval. Gamma coding does not work well. Variable-byte integers are often faster than variable-bit integers, despite having higher storage costs, because fewer CPU operations are required to decode variable-byte integers and they are byte-aligned on disk. A suitable compression technique can allow retrieval to be up to twice as fast than without compression, while the space requirement averages $20\% - 25\%$ of the cost of storing uncompressed integers.

6.7   Latent Semantic Indexing

The retrieval models discussed so far are based on keyword or term matching, i.e., matching terms in the user query with those in the documents. However, many concepts or objects can be described in multiple ways (using different words) due to the context and people's language habits. If a user query uses different words from the words used in a document, the document will not be retrieved although it may be relevant because the document uses some symonyms of the words in the user query. This causes low recall. For example, "picture", "image" and "photo" are **synonyms** in the context of digital cameras. If the user query only has the word "picture", relevant documents that contain "image" or "photo" but not "picture" will not be retrieved.

Latent semantic indexing (LSI), proposed by Deerwester et al. [125], aims to deal with this problem through the identification of statistical associations of terms. It is assumed that there is some underlying latent semantic structure in the data that is partially obscured by the randomness of word choice. It then uses a statistical technique, called **singular value decomposition** (SVD) [203], to estimate this latent structure, and to remove the "noise". The results of this decomposition are descriptions of terms and documents based on the latent semantic structure derived from SVD. This structure is also called the **hidden "concept" space**, which associates syntactically different but semantically similar terms and documents. These transformed terms and documents in the "concept" space are then used in retrieval, not the original terms or documents. Furthermore, the query is also transformed into the "concept" space before retrieval.

Let D be the text collection, the number of distinctive words in D be m and the number of documents in D be n. LSI starts with an $m \times n$ term-document matrix A. Each row of A represents a term and each column represents a document. The matrix may be computed in various ways, e.g., using term frequency or TF-IDF values. We use term frequency as an example in this section. Thus, each entry or cell of the matrix A, denoted by A_{ij}, is the number of times that term i occurs in document j.

6.7.1   Singular Value Decomposition

What SVD does is to factor matrix A (a $m \times n$ matrix) into the product of three matrices, i.e.,

$$A = U\Sigma V^T,$$ (27)

where

U is a $m \times r$ matrix and its columns, called **right singular vectors**, are eigenvectors associated with the r non-zero eigenvalues of AA^T. Furthermore, the columns of U are unit orthogonal vectors, i.e., $U^T U = I$ (identity matrix).

V is an $n \times r$ matrix and its columns, called **right singular vectors**, are eigenvectors associated with the r non-zero eigenvalues of $A^T A$. The columns of V are also unit orthogonal vectors, i.e., $V^T V = I$.

Σ is a $r \times r$ diagonal matrix, $\Sigma = \text{diag}(\sigma_1, \sigma_2, \ldots, \sigma_r)$, $\sigma_i > 0$. $\sigma_1, \sigma_2, \ldots,$ and σ_r, called **singular values**, are the non-negative square roots of the r (non-zero) eigenvalues of AA^T. They are arranged in decreasing order, i.e., $\sigma_1 \geq \sigma_2 \geq \ldots \geq \sigma_r > 0$.

We note that initially U is in fact an $m \times m$ matrix and V an $n \times n$ matrix and Σ an $m \times n$ diagonal matrix. Σ's diagonal consists of nonnegative eigenvalues of AA^T or $A^T A$. However, due to zero eigenvalues, Σ has zero-valued rows and columns. Matrix multiplication tells us that those zero-valued rows and columns from Σ can be dropped. Then, the last $m-r$ columns in U and the last $n-r$ columns in V can also be dropped.

m is the number of row (terms) in A, representing the number of terms.

n is the number of columns in A, representing the number of documents.

r is the **rank** of A, $r \leq \min(m, n)$.

The singular value decomposition of A always exists and is unique up to

1. allowable permutations of columns of U and V and elements of Σ leaving it still diagonal; that is, columns i and j of Σ may be interchanged *iff* row i and j of Σ are interchanged, and columns i and j of U and V are interchanged.
2. sign (+/–) flip in U and V.

An important feature of SVD is that we can delete some insignificant dimensions in the transformed (or "concept") space to optimally (in the least square sense) approximate matrix A. The significance of the dimensions is indicated by the magnitudes of the singular values in Σ, which are already sorted. In the context of information retrieval, the insignificant dimensions may represent "noise" in the data, and should be removed. Let us use only the k largest singular values in Σ and set the remaining small ones to zero. The approximated matrix of A is denoted by A_k. We can also reduce the size of the matrices Σ, U and V by deleting the last $r-k$ rows and columns from Σ, the last $r-k$ columns in U and the last $r-k$ columns in V. We then obtain

$$A_k = U_k \Sigma_k V_k^T, \tag{28}$$

which means that we use the k-largest singular triplets to approximate the original (and somewhat "noisy") term-document matrix A. The new space is called the **k-concept space**. Figure 6.10 shows the original matrices and the reduced matrices schematically.

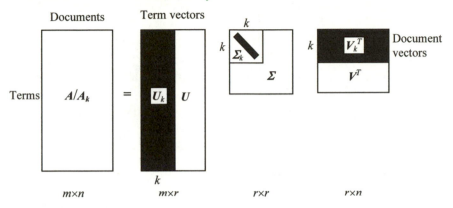

Fig. 6.10. The schematic representation of A and A_k

It is critical that the LSI method does not re-construct the original term-document matrix A perfectly. The truncated SVD captures most of the important underlying structures in the association of terms and documents, yet at the same time removes the noise or variability in word usage that plagues keyword matching retrieval methods.

Intuitive Idea of LSI: The intuition of LSI is that SVD rotates the axes of m-dimensional space of A such that the first axis runs along the largest variation (variance) among the documents, the second axis runs along the second largest variation (variance) and so on. Figure 6.11 shows an example.

The original x-y space is mapped to the x'-y' space generated by SVD. We can see that x and y are clearly correlated. In our retrieval context, each data point represents a document and each axis (x or y) in the original space represents a term. Hence, the two terms are correlated or co-occur frequently. In the SVD, the direction of x' in which the data has the largest variation is represented by the first column vector of U, and the direction of y' is represented by the second column vector of U. ΣV^T represents the documents in the transformed "concept" space. The singular values in Σ are simply scaling factors.

We observe that y' direction is insignificant, and may represent some "noise", so we can remove it. Then, every data point (document) is projected to x'. We have an outlier document $\mathbf{d}_i$ that contains term x, but not term y. However, if it is projected to x', it becomes closer to other points.

Let us see what happens if we have a query $\mathbf{q}$ represented with a star in Fig. 6.11, which contains only a single term "y". Using the traditional exact term matching, $\mathbf{d}_i$ is not relevant because "y" does not appear in $\mathbf{d}_i$. However, in the new space after projection, they are quite close or similar.

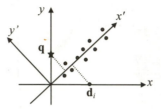

Fig. 6.11. Intuition of the LSI.

6.7.2   Query and Retrieval

Given a user query $\mathbf{q}$ (represented by a column vector as those in A), it is first converted into a document in the k-concept space, denoted by $\mathbf{q}_k$. This transformation is necessary because SVD has transformed the original documents into the k-concept space and stored them in V_k. The idea is that $\mathbf{q}$ is treated as a new document in the original space represented as a column in A, and then mapped to $\mathbf{q}_k$ (a row vector) as an additional document (or column) in V_k^T. From Equation (28), it is easy to see that

$$\mathbf{q} = U_k \Sigma_k \mathbf{q}_k^{\ T}. \tag{29}$$

Since the columns in U are unit orthogonal vectors, $U_k^T U_k = I$. Thus,

$$U_k^{\ T} \mathbf{q} = \Sigma_k \mathbf{q}_k^{\ T}. \tag{30}$$

As the inverse of a diagonal matrix is still a diagonal matrix, and each entry on the diagonal is $1/\sigma_i$ ($1 \leq i \leq k$), if it is multiplied on both sides of Equation (30), we obtain,

$$\Sigma_k^{\ -1} U_k^{\ T} \mathbf{q} = \mathbf{q}_k^{\ T}. \tag{31}$$

Finally, we get the following (notice that the transpose of a diagonal matrix is itself),

$$\mathbf{q}_k = \mathbf{q}^T U_k \Sigma_k^{\ -1}. \tag{32}$$

For retrieval, we simply compare $\mathbf{q}_k$ with each document (row) in V_k using a similarity measure, e.g., the cosine similarity. Recall that each row of V_k (or each column of V_k^T) corresponds to a document (column) in A. This method has been used traditionally.

Alternatively, since $\Sigma_k V_k^T$ (not V_k^T) represents the documents in the transformed k-concept space, we can compare the similarity of the query document in the transformed space, which is $\Sigma_k \mathbf{q}_k^T$, and each transformed document in $\Sigma_k V_k^T$ for retrieval. The difference between the two methods is obvious. This latter method considers scaling effects of the singular values in Σ_k, but the former does not. However, it is not clear which method performs better as I know of no reported study on this alternative method.

6.7.3   An Example

Example 12: We will use the example in [125] to illustrate the process. The document collection has the following nine documents. The first five documents are related to human computer interaction, and the last four documents are related to graphs. To reduce the size of the problem, only the underlined terms are used in our computation.

c_1:   Human machine interface for Lab ABC computer applications
c_2:   A survey of user opinion of computer system response time
c_3:   The EPS user interface management system
c_4:   System and human system engineering testing of EPS
c_5:   Relation of user-perceived response time to error measurement
m_1:   The generation of random, binary, unordered trees
m_2:   The intersection graph of paths in trees
m_3:   Graph minors IV: Widths of trees and well-quasi-ordering
m_4:   Graph minors: A survey

The term-document matrix A is given below, which is a 9×12 matrix.

	c_1	c_2	c_3	c_4	c_5	m_1	m_2	m_3	m_4	
	1	0	0	1	0	0	0	0	0	human
	1	0	1	0	0	0	0	0	0	interface
	1	1	0	0	0	0	0	0	0	computer
	0	1	1	0	1	0	0	0	0	user
	0	1	1	2	0	0	0	0	0	system
$A =$	0	1	0	0	1	0	0	0	0	response
	0	1	0	0	1	0	0	0	0	time
	0	0	1	1	0	0	0	0	0	EPS
	0	1	0	0	0	0	0	0	1	survey
	0	0	0	0	0	1	1	1	0	trees
	0	0	0	0	0	0	1	1	1	graph
	0	0	0	0	0	0	0	1	1	minors

After performing SVD, we obtain three matrices, U, Σ and V^T, which are given below. Singular values on the diagonal of Σ are in decreasing order.

$$U = \begin{pmatrix}
0.22 & -0.11 & 0.29 & -0.41 & -0.11 & -0.34 & 0.52 & -0.06 & -0.41 \\
0.20 & -0.07 & 0.14 & -0.55 & 0.28 & 0.50 & -0.07 & -0.01 & -0.11 \\
0.24 & 0.04 & -0.16 & -0.59 & -0.11 & -0.25 & -0.30 & 0.06 & 0.49 \\
0.40 & 0.06 & -0.34 & 0.10 & 0.33 & 0.38 & 0.00 & 0.00 & 0.01 \\
0.64 & -0.17 & 0.36 & 0.33 & -0.16 & -0.21 & -0.17 & 0.03 & 0.27 \\
0.27 & 0.11 & -0.43 & 0.07 & 0.08 & -0.17 & 0.28 & -0.02 & -0.05 \\
0.27 & 0.11 & -0.43 & 0.07 & 0.08 & -0.17 & 0.28 & -0.02 & -0.05 \\
0.30 & -0.14 & 0.33 & 0.19 & 0.11 & 0.27 & 0.03 & -0.02 & -0.17 \\
0.21 & 0.27 & -0.18 & -0.03 & -0.54 & 0.08 & -0.47 & -0.04 & -0.58 \\
0.01 & 0.49 & 0.23 & 0.03 & 0.59 & -0.39 & -0.29 & 0.25 & -0.23 \\
0.04 & 0.62 & 0.22 & 0.00 & -0.07 & 0.11 & 0.16 & -0.68 & 0.23 \\
0.03 & 0.45 & 0.14 & -0.01 & -0.30 & 0.28 & 0.34 & 0.68 & 0.18
\end{pmatrix}$$

$$\Sigma = \begin{pmatrix}
3.34 & 0 & 0 & 0 & 0 & 0 & 0 & 0 & 0 \\
0 & 2.54 & 0 & 0 & 0 & 0 & 0 & 0 & 0 \\
0 & 0 & 2.35 & 0 & 0 & 0 & 0 & 0 & 0 \\
0 & 0 & 0 & 1.64 & 0 & 0 & 0 & 0 & 0 \\
0 & 0 & 0 & 0 & 1.50 & 0 & 0 & 0 & 0 \\
0 & 0 & 0 & 0 & 0 & 1.31 & 0 & 0 & 0 \\
0 & 0 & 0 & 0 & 0 & 0 & 0.85 & 0 & 0 \\
0 & 0 & 0 & 0 & 0 & 0 & 0 & 0.56 & 0 \\
0 & 0 & 0 & 0 & 0 & 0 & 0 & 0 & 0.36
\end{pmatrix}$$

$$V^T = \begin{pmatrix}
0.20 & -0.06 & 0.11 & -0.95 & 0.05 & -0.08 & 0.18 & -0.01 & -0.06 \\
0.61 & 0.17 & -0.50 & -0.03 & -0.21 & -0.26 & -0.43 & 0.05 & 0.24 \\
0.46 & -0.13 & 0.21 & 0.04 & 0.38 & 0.72 & -0.24 & 0.01 & 0.02 \\
0.54 & -0.23 & 0.57 & 0.27 & -0.21 & -0.37 & 0.26 & -0.02 & -0.08 \\
0.28 & 0.11 & -0.51 & 0.15 & 0.33 & 0.03 & 0.67 & -0.06 & -0.26 \\
0.00 & 0.19 & 0.10 & 0.02 & 0.39 & -0.30 & -0.34 & 0.45 & -0.62 \\
0.01 & 0.44 & 0.19 & 0.02 & 0.35 & -0.21 & -0.15 & -0.76 & 0.02 \\
0.02 & 0.62 & 0.25 & 0.01 & 0.15 & 0.00 & 0.25 & 0.45 & 0.52 \\
0.08 & 0.53 & 0.08 & -0.03 & -0.60 & 0.36 & 0.04 & -0.07 & -0.45
\end{pmatrix}$$

Now let us choose only two largest singular values from Σ, i.e., $k = 2$. Thus, the concept space has only two dimensions. The other two matrices are also truncated accordingly. We obtain the 3 matrix U_k, Σ_k and V_k^T:

$$\begin{array}{ccc}
U_k & \Sigma_k & V_k^T
\end{array}$$

$$A_k = \begin{pmatrix}
0.22 & -0.11 \\
0.20 & -0.07 \\
0.24 & 0.04 \\
0.40 & 0.06 \\
0.64 & -0.17 \\
0.27 & 0.11 \\
0.27 & 0.11 \\
0.30 & -0.14 \\
0.21 & 0.27 \\
0.01 & 0.49 \\
0.04 & 0.62 \\
0.03 & 0.45
\end{pmatrix} \begin{bmatrix} 3.34 & 0 \\ 0 & 2.54 \end{bmatrix} \begin{bmatrix} 0.20 & 0.61 & 0.46 & 0.54 & 0.28 & 0.00 & 0.02 & 0.02 & 0.08 \\ -0.06 & 0.17 & -0.13 & -0.23 & 0.11 & 0.19 & 0.44 & 0.62 & 0.53 \end{bmatrix}$$

Now we issue a search query $\mathbf{q}$, "user interface", to find relevant documents. The transformed query document $\mathbf{q}_k$ of query $\mathbf{q}$ in the k-concept space is computed below using Equation (32), which is (0.179 -0.004).

$$\mathbf{q}_k = \begin{pmatrix} 0 \\ 1 \\ 0 \\ 1 \\ 0 \\ 0 \\ 0 \\ 0 \\ 0 \\ 0 \\ 0 \\ 0 \end{pmatrix}^T \begin{pmatrix} 0.22 & -0.11 \\ 0.20 & -0.07 \\ 0.24 & 0.04 \\ 0.40 & 0.06 \\ 0.64 & -0.17 \\ 0.27 & 0.11 \\ 0.27 & 0.11 \\ 0.30 & -0.14 \\ 0.21 & 0.27 \\ 0.01 & 0.49 \\ 0.04 & 0.62 \\ 0.03 & 0.45 \end{pmatrix} \begin{bmatrix} 3.34 & 0 \\ 0 & 2.54 \end{bmatrix}^{-1} = (0.179 \ -0.004)$$

$\mathbf{q}_k$ is then compared with every document vector in V_k using the cosine similarity. The similarity values are as follows:

c_1:   0.964
c_2:   0.957
c_3:   0.968
c_4:   0.928
c_5:   0.922
m_1:   −0.022
m_2:   0.023
m_3:   0.010
m_4:   0.127

We obtain the final ranking of (c_3, c_1, c_2, c_4, c_5, m_4, m_2, m_3, m_1).   ■

6.7.4   Discussion

LSI has been shown to perform better than traditional keywords based methods. The main drawback is the time complexity of the SVD, which is $O(m^2n)$. It is thus difficult to use for a large document collection such as the Web. Another drawback is that the concept space is not interpretable as its description consists of all numbers with little semantic meaning.

Determining the optimal number of dimensions k of the concept space is also a major difficulty. There is no general consensus for an optimal number of dimensions. The original paper [125] of LSI suggests 50–350 dimensions. In practice, the value of k needs to be determined based on the specific document collection via trial and error, which is a very time consuming process due to the high time complexity of the SVD.

To close this section, one can imagine that association rules may be able to approximate the results of LSI and avoid its shortcomings. Association rules represent term correlations or co-occurrences. Association rule mining has two advantages. First, its mining algorithm is very efficient. Since we may only need rules with 2-3 terms, which are sufficient for practical purposes, the mining algorithm only needs to scan the document collection 2-3 times. Second, rules are easy to understand. However, little research has been done in this direction so far.

6.8 Web Search

We now put it all together and describe the working of a search engine. Since it is difficult to know the internal details of a commercial search engine, most contents in this section are based on research papers, especially the early Google paper [68]. Due to the efficiency problem, latent semantic indexing is probably not used in Web search yet. Current search algorithms are still mainly based on the vector space model and term matching.

A search engine starts with the crawling of pages on the Web. The crawled pages are then parsed, indexed, and stored. At the query time, the index is used for efficient retrieval. We will not discuss crawling here. Its details can be found in Chap. 8. The subsequent operations of a search engine are described below:

Parsing: A parser is used to parse the input HTML page, which produces a stream of tokens or terms to be indexed. The parser can be constructed using a lexical analyzer generator such as YACC and Flex (which is from the GNU project). Some pre-processing tasks described in Sect. 6.5 may also be performed before or after parsing.

Indexing: This step produces an inverted index, which can be done using any of the methods described in Sect. 6.6. For retrieval efficiency, a search engine may build multiple inverted indices. For example, since the titles and anchor texts are often very accurate descriptions of the pages, a small inverted index may be constructed based on the terms appeared in them alone. Note that here the anchor text is for indexing the page that its link points to, not the page containing it. A full index is then built based on all the text in each page, including anchor texts (a piece of anchor text is indexed both for the page that contains it, and for the page that its link points to). In searching, the algorithm may search in the small index first and then the full index. If a sufficient number of relevant pages are found in the small index, the system may not search in the full index.

Searching and Ranking: Given a user query, searching involves the following steps:

1. pre-processing the query terms using some of the methods described in Sect. 6.5, e.g., stopword removal and stemming;
2. finding pages that contain all (or most of) the query terms in the inverted index;
3. ranking the pages and returning them to the user.

The ranking algorithm is the heart of a search engine. However, little is known about the algorithms used in commercial search engines. We give a general description based on the algorithm in the early Google system.

As we discussed earlier, traditional IR uses cosine similarity values or any other related measures to rank documents. These measures only consider the content of each document. For the Web, such content based methods are not sufficient. The problem is that on the Web there are too many relevant documents for almost any query. For example, using "web mining" as the query, the search engine Google estimated that there were 46,500,000 relevant pages. Clearly, there is no way that any user will look at this huge number of pages. Therefore, the issue is how to rank the pages and present the user the "best" pages at the top.

An important ranking factor on the Web is the quality of the pages, which was hardly studied in traditional IR because most documents used in IR evaluations are from reliable sources. However, on the Web, anyone can publish almost anything, so there is no quality control. Although a page may be 100% relevant, it may not be a quality page due to several reasons. For example, the author may not be an expert of the query topic, the information given in the page may be unreliable or biased, etc.

However, the Web does have an important mechanism, the hyperlinks (links), that can be used to assess the quality of each page to some extent. A link from page x to page y is an implicit conveyance of authority of page x to page y. That is, the author of page x believes that page y contains quality or **authoritative information**. One can also regard the fact that page x points to page y as a vote of page x for page y. This democratic nature of the Web can be exploited to assess the quality of each page. In general, the more votes a page receives, the more likely it is a **quality page**. The actual algorithms are more involved than simply counting the number of votes or links pointing to a page (called **in-links**). We will describe the algorithms in the next chapter. **PageRank** is the most well known such algorithm (see Sect. 7.3). It makes use of the link structure of Web pages to compute a quality or reputation score for each page. Thus, a Web page can be evaluated based on both its content factors and its reputation. Content-based evaluation depends on two kinds of information:

Occurrence Type: There are several types of occurrences of query terms in a page:

Title: a query term occurs in the title field of the page.

Anchor text: a query term occurs in the anchor text of a page pointing to the current page being evaluated.

URL: a query term occurs in the URL of the page. Many URL addresses contain some descriptions of the page. For example, a page on Web mining may have the URL http://www.domain.edu/Web-mining.html.

Body: a query term occurs in the body field of the page. In this case, the prominence of each term is considered. Prominence means whether the term is emphasized in the text with a large font, or bold and/or italic tags. Different prominence levels can be used in a system. Note that anchor texts in the page can be treated as plain texts for the evaluation of the page.

Count: The number of occurrences of a term of each type. For example, a query term may appear in the title field of the page 2 times. Then, the title count for the term is 2.

Position: This is the position of each term in each type of occurrence. The information is used in proximity evaluation involving multiple query terms. Query terms that are near to each other are better than those that are far apart. Furthermore, query terms appearing in the page in the same sequence as they are in the query are also better.

For the computation of the content based score (also called the **IR score**), each occurrence type is given an associated weight. All **type weights** form a fixed vector. Each raw term count is converted to a **count weight**, and all count weights also form a vector.

The quality or reputation of a page is usually computed based on the link structure of Web pages, which we will study in Chap. 7. Here, we assume that a reputation score has been computed for each page.

Let us now look at two kinds of queries, **single word queries** and **multi-word queries**. A single word query is the simplest case with only a single term. After obtaining the pages containing the term from the inverted index, we compute the dot product of the **type weight vector** and the **count weight vector** of each page, which gives us the IR score of the page. The IR score of each page is then combined with its **reputation score** to produce the final score of the page.

For a multi-word query, the situation is similar, but more complex since there is now the issue of considering term proximity and ordering. Let us simplify the problem by ignoring the term ordering in the page. Clearly, terms that occur close to each other in a page should be weighted higher than those that occur far apart. Thus multiple occurrences of terms need to be matched so that nearby terms are identified. For every matched set, a proximity value is calculated, which is based on how far apart the terms are in the page. Counts are also computed for every type and proximity. Each type and proximity pair has a type-proximity-weight. The counts are converted into count-weights. The dot product of the count-weights and the type-proximity-weights gives an IR score to the page. Term ordering can be considered similarly and included in the IR score, which is then combined with the page reputation score to produce the final rank score.

6.9 Meta-Search and Combining Multiple Rankings

In the last section, we described how an individual search engine works. We now discuss how several search engines can be used together to produce a **meta-search engine**, which is a search system that does not have its own database of Web pages. Instead, it answers the user query by combining the results of some other search engines which normally have their databases of Web pages. Figure 6.12 shows a meta-search architecture.

After receiving a query from the user through the **search interface**, the meta-search engine submits the query to the underlying search engines (called its **component search engines**). The returned results from all these search engines are then combined (**fused** or **merged**) and sent to the user.

A meta-search engine has some intuitive appeals. First of all, it increases the search coverage of the Web. The Web is a huge information source, and each individual search engine may only cover a small portion of it. If we use only one search engine, we will never see those relevant pages that are not covered by the search engine.

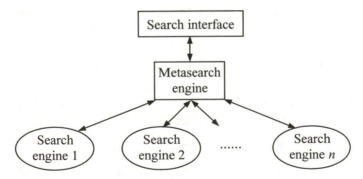

Fig. 6.12. A meta-search architecture

Meta-search may also improve the search effectiveness. Each component search engine has its ranking algorithm to rank relevant pages, which is often biased, i.e., it works well for certain types of pages or queries but not for others. By combining the results from multiple search engines, their biases can be reduced and thus the search precision can be improved.

The key operation in meta-search is to combine the ranked results from the component search engines to produce a single ranking. The first task is to identify whether two pages from different search engines are the same, which facilitates combination and **duplicate removal**. Without downloading the full pages (which is too time consuming), this process is not simple due to aliases, symbolic links, redirections, etc. Typically, several heuristics are used for the purpose, e.g., comparing domain names of URLs, titles of the pages, etc.

The second task is to combine the ranked results from individual search engines to produce a single ranking, i.e., to fuse individual rankings. There are two main classes of meta-search combination (or fusion) algorithms: ones that use similarity scores returned by each component system and ones that do not. Some search engines return a similarity score (with the query) for each returned page, which can be used to produce a better combined ranking. We discuss these two classes of algorithms below.

It is worth noting that the first class of algorithms can also be used to combine scores from different similarity functions in a single IR system or in a single search engine. Indeed, the algorithms below were originally proposed for this purpose. It is likely that search engines already use some such techniques (or their variations) within their ranking mechanisms because a ranking algorithm needs to consider multiple factors.

6.9.1 Combination Using Similarity Scores

Let the set of candidate documents to be ranked be $D = \{d_1, d_2, ..., d_N\}$. There are k underlying systems (component search engines or ranking techniques). The ranking from system or technique i gives document d_j the similarity score, s_{ij}. Some popular and simple combination methods are defined by Fox and Shaw in [184].

CombMIN: The combined similarity score for each document d_j is the minimum of the similarities from all underlying search engine systems:

$$\text{CombMIN}(d_j) = \min(s_{1j}, s_{2j}, ..., s_{kj}). \tag{33}$$

CombMAX: The combined similarity score for each document d_j is the maximum of the similarities from all underlying search engine systems:

$$\text{CombMAX}(d_j) = \max(s_{1j}, s_{2j}, ..., s_{kj}). \tag{34}$$

CombSUM: The combined similarity score for each document d_j is the sum of the similarities from all underlying search engine systems.

$$\text{CombSUM}(d_j) = \sum_{i=1}^{k} s_{ij}. \tag{35}$$

CombANZ: It is defined as

$$\text{CombANZ}(d_j) = \frac{\text{CombSUM}(d_j)}{r_j}, \tag{36}$$

where r_j is the number of non-zero similarities, or the number of systems that retrieved d_j.

CombMNZ: It is defined as

$$\text{CombMNZ}(d_j) = \text{CombSUM}(d_j) \times r_j \tag{37}$$

where r_j is the number of non-zero similarities, or the number of systems that retrieved d_j.

It is a common practice to normalize the similarity scores from each ranking using the maximum score before combination. Researchers have shown that, in general, CombSUM and CombMNZ perform better. CombMNZ outperforms CombSUM slightly in most cases.

6.9.2 Combination Using Rank Positions

We now discuss some popular rank combination methods that use only rank positions of each search engine. In fact, there is a field of study called the **social choice theory** [273] that studies voting algorithms as techniques to make group or social decisions (choices). The algorithms discussed below are based on voting in elections.

In 1770 Jean-Charles de Borda proposed "election by order of merit". Each voter announces a (linear) preference order on the candidates. For each voter, the top candidate receives n points (if there are n candidates in the election), the second candidate receives $n-1$ points, and so on. The points from all voters are summed up to give the final points for each candidate. If there are candidates left unranked by a voter, the remaining points are divided evenly among the unranked candidates. The candidate with the most points wins. This method is called the **Borda ranking**.

An alternative method was proposed by Marquis de Condorcet in 1785. The **Condorcet ranking** algorithm is a majoritarian method where the winner of the election is the candidate(s) that beats each of the other candidates in a pair-wise comparison. If a candidate is not ranked by a voter, the candidate loses to all other ranked candidates. All unranked candidates tie with one another.

Yet another simple method, called the **reciprocal ranking**, sums one over the rank of each candidate across all voters. For each voter, the top candidate has the score of 1, the second ranked candidate has the score of 1/2, and the third ranked candidate has the score of 1/3 and so on. If a candidate is not ranked by a voter, it is skipped in the computation for this voter. The candidates are then ranked according to their final total scores. This rank strategy gives much higher weight than Borda ranking to candidates that are near the top of a list.

Example 13: We use an example in the context of meta-search to illustrate the working of these methods. Consider a meta-search system with five underlying search engine systems, which have ranked four candidate documents or pages, a, b, c, and d as follows:

system 1: a, b, c, d
system 2: b, a, d, c
system 3: c, b, a, d
system 4: c, b, d
system 5: c, b

Let us denote the score of each candidate x by Score(x).

Borda Ranking: The score for each page is as follows:

Score(a) = 4 + 3 + 2 + 1 + 1.5 = 11.5
Score(b) = 3 + 4 + 3 + 3 + 3 = 16
Score(c) = 2 + 1 + 4 + 4 + 4 = 15
Score(d) = 1 + 2 + 1 + 2 + 1.5 = 7.5

Thus the final ranking is: b, c, a, d.

Condorcet Ranking: We first build an $n \times n$ matrix for all pair-wise comparisons, where n is the number of pages. Each non-diagonal entry (i, j) of the matrix shows the number of wins, loses, and ties of page i over page j, respectively. For our example, the matrix is as follows:

	a	b	c	d
a	-	1:4:0	2:3:0	3:1:1
b	4:1:0	-	2:3:0	5:0:0
c	3:2:0	3:2:0	-	4:1:0
d	1:3:1	0:5:0	1:4:0	-

Fig. 6.13. The pair-wise comparison matrix for the four candidate pages

After the matrix is constructed, pair-wise winners are determined, which produces a win, lose and tie table. Each pair in Fig. 6.13 is compared, and the winner receives one point in its "win" column and the loser receives one point in its "lose" column. For a pair-wise tie, both receive one point in the "tie" column. For example, for page *a*, it only beats *d* because *a* is ranked ahead of *d* three times out of 5 ranks (Fig. 6.13). The win, lose and tie table for Fig. 6.13 is given in Fig. 6.14 below.

	win	lose	tie
a	1	2	0
b	2	1	0
c	3	0	0
d	0	3	0

Fig. 6.14. The win, lose and tie table for the comparison matrix in Fig. 6.13

To rank the pages, we use their win and lose values. If the number of wins that a page *i* has is higher than another page *j*, then *i* wins over *j*. If their win property is equal, we consider their lose scores, and the page which has a lower lose score wins. If both their win and lose scores are the same, then the pages are tied. The final ranks of the tied pages are randomly assigned. Clearly *c* is the Condorcet winner in our example. The final ranking is: *c*, *b*, *a*, *d*.

Reciprocal Ranking:

Score(a) = 1 + 1/2 + 1/3 = 1.83
Score(b) = 1/2 + 1 + 1/2 + 1/2 + 1/2 = 3
Score(c) = 1/3 + 1/4 + 1 + 1 + 1 = 3.55
Score(d) = 1/4 + 1/3 + 1/4 + 1/3 = 1.17

The final ranking is: *c*, *b*, *a*, *d*.   ∎

6.10 Web Spamming

Web search has become very important in the information age. Increased exposure of pages on the Web can result in significant financial gains and/or fames for organizations and individuals. The rank positions of Web pages in search are perhaps the single most important indicator of such exposures of pages. If a user searches for information that is relevant to your pages but your pages are ranked low by search engines, then the user may not see the pages because one seldom clicks a large number of returned pages. This is not acceptable for businesses, organizations, and even individuals. Thus, it has become very important to understand search engine ranking algorithms and to present the information in one's pages in such a way that the pages will be ranked high when terms related to the contents of the pages are searched. Unfortunately, this also results in **spamming**, which refers to human activities that deliberately mislead search engines to rank some pages higher than they deserve.

There is a gray area between spamming and legitimate page optimization. It is difficult to define precisely what are justifiable and unjustifiable actions aimed at boosting the importance and consequently the rank positions of one's pages.

Assume that, given a user query, each page on the Web can be assigned an information value. All the pages are then ranked according to their information values. Spamming refers to actions that do not increase the information value of a page, but dramatically increase its rank position by misleading search algorithms to rank it high. Due to the fact that search engine algorithms do not understand the content of each page, they use syntactic or surface features to assess the information value of the page. Spammers exploit this weakness to boost the ranks of their pages.

Spamming is annoying for users because it makes it harder to find truly useful information and leads to frustrating search experiences. Spamming is also bad for search engines because spam pages consume crawling bandwidth, pollute the Web, and distort search ranking.

There are in fact many companies that are in the business of helping others improve their page ranking. These companies are called **Search Engine Optimization** (SEO) companies, and their businesses are thriving. Some SEO activities are ethical and some, which generate spam, are not.

As we mentioned earlier, search algorithms consider both content based factors and reputation based factors in scoring each page. In this section, we briefly describe some spam methods that exploit these factors. The section is mainly based on [214] by Gyongyi and Garcia-Molina.

6.10.1 Content Spamming

Most search engines use variations of TF-IDF based measures to assess the relevance of a page to a user query. Content-based spamming methods basically tailor the contents of the text fields in HTML pages to make spam pages more relevant to some queries. Since TF-IDF is computed based on terms, **content spamming** is also called **term spamming**. Term spamming can be placed in any text field:

Title: Since search engines usually give higher weights to terms in the title of a page due to the importance of the title to a page, it is thus common to spam the title.

Meta-Tags: The HTML meta-tags in the page header enable the owner to include some meta information of the page, e.g., author, abstract, keywords, content language, etc. However, meta-tags are very heavily spammed. Search engines now give terms within these tags very low weights or completely ignore their contents.

Body: Clearly spam terms can be placed within the page body to boost the page ranking.

Anchor Text: As we discussed in Sect. 6.8, the anchor text of a hyperlink is considered very important by search engines. It is indexed for the page containing it and also for the page that it points to, so anchor text spam affects the ranking of both pages.

URL: Some search engines break down the URL of a page into terms and consider them in ranking. Thus, spammers can include spam terms in the URL. For example, a URL may be http://www.xxx.com/cheap-MP3-player-case-battery.html

There are two main term spam techniques, which simply create synthetic contents containing spam terms.

1. **Repeating some important terms:** This method increases the TF scores of the repeated terms in a document and thus increases the relevance of the document to these terms. Since plain repetition can be easily detected by search engines, the spam terms can be weaven into some sentences, which may be copied from some other sources. That is, the spam terms are randomly placed in these sentences. For example, if a spammer wants to repeat the word "mining", it may add it randomly in an unrelated (or related) sentence, e.g., "the picture mining quality of this camera mining is amazing," instead of repeating it many times consecutively (next to each other), which is easy to detect.

2. **Dumping of many unrelated terms:** This method is used to make the page relevant to a large number of queries. In order to create the spam content quickly, the spammer may simply copy sentences from related pages on the Web and glue them together.

   Advertisers may also take advantage of some frequently searched terms on the Web and put them in the target pages so that when users search for the frequently search terms, the target pages become relevant. For example, to advertise cruise liners or cruise holiday packages, spammers put "Tom Cruise" in their advertising pages as "Tom Cruise" is a popular film actor in USA and is searched very frequently.

6.10.2 Link Spamming

Since hyperlinks play an important role in determining the reputation score of a page, spammers also spam on hyperlinks.

Out-Link Spamming: It is quite easy to add out-links in one's pages pointing to some **authoritative pages** to boost the hub cores of one's pages. A page is a **hub page** if it points to many authoritative (or quality) pages. The concepts of authority and hub will be formally studied in the next chapter (Sect. 7.4). To create massive out-links, spammers may use a technique called **directory cloning**. There are many directories, e.g., Yahoo!, DMOZ Open Directory, on the Web which contain a large number of links to other Web pages that are organized according to some pre-specified topic hierarchies. Spammers simply replicate a large portion of a directory in the spam page to create a massive out-link structure quickly.

In-Link Spamming: In-link spamming is harder to achieve because it is not easy to add hyperlinks on the Web pages of others. Spammers typically use one or more of the following techniques.

1. *Creating a honey pot*: If a page wants to have a high reputation/quality score, it needs quality pages pointing to it (see Sect. 7.3 in the next chapter). This method basically tries to create some important pages that contain links to target spam pages. For example, the spammer can create a set of pages that contains some very useful information, e.g., glossary of Web mining terms, or Java FAQ and help pages. The honey pots attract people pointing to them because they contain useful information, and consequently have high reputation scores (high quality pages). Such honey pots contain (hidden) links to target spam pages that the spammers want to promote. This strategy can significantly boost the spam pages.

2. *Adding links to Web directories*: Many Web directories allow the user to submit URLs. Spammers can submit the URLs of spam pages at multiple directory sites. Since directory pages often have high quality (or authority) and hub scores, they can boost reputation scores of spam pages significantly.

3. *Posting links to the user-generated content* (reviews, forum discussions, blogs, etc): There are numerous sites on the Web that allow the user to freely post messages, which are called the **user-generated content**. Spammers can add links pointing to their pages to the seemly innocent messages that they post.

4. *Participating in link exchange*: In this case, many spammers form a group and set up a link exchange scheme so that their sites point to each other in order to promote the pages of all the sites.

5. *Creating own spam farm*: In this case, the spammer needs to control a large number of sites. Then, any link structure can be created to boost the ranking of target spam pages.

6.10.3 Hiding Techniques

In most situations, spammer wants to conceal or to hide the spamming sentences, terms and links so that the Web users do not see them. They can use a number of techniques.

Content Hiding: Spam items are made invisible. One simple method is to make the spam terms the same color as the background color. For example, one may use the following for hiding,

```
<body background = white>
   <font color = white> spam items</font>
   ...
</body>
```

To hide a hyperlink, one can also use a very small image and a blank image. For example, one may use

```
<a href = target.html"><img src="blank.gif"> </a>
```

A spammer can also use scripts to hide some of the visual elements on the page, for instance, by setting the visible HTML style attribute to false.

Cloaking: Spam Web servers return a HTML document to the user and a different document to a Web crawler. In this way, the spammer can present the Web user with the intended content and send a spam page to the search engine for indexing.

Spam Web servers can identify Web crawlers in one of the two ways:

1. It maintains a list of IP addresses of search engines and identifies search engine crawlers by matching IP addresses.
2. It identifies Web browsers based on the **user–agent field** in the HTTP request message. For instance, the user–agent name of the following HTTP request message is the one used by the Microsoft Internet Explorer 6 browser:

```
GET /pub/WWW/TheProject.html HTTP/1.1
Host: www.w3.org
User-Agent: Mozilla/4.0 (compatible; MSIE 6.0; Windows NT 5.1)
```

User–agent names are not standard, so it is up to the requesting application what to include in the corresponding message field. However, search engine crawlers usually identify themselves by names distinct from normal Web browsers in order to allow well-intended, and legitimate optimization. For example, some sites serve search engines a version of their pages that is free of navigation links and advertisements.

Redirection: Spammers can also hide a spammed page by automatically redirecting the browser to another URL as soon as the page is loaded. Thus, the spammed page is given to the search engine for indexing (which the user will never see), and the target page is presented to the Web user through redirection. One way to achieve redirection is to use the "refresh" meta-tag, and set the refresh time to zero. Another way is to use scripts.

6.10.4 Combating Spam

Some spamming activities, like redirection using refresh meta-tag, are easy to detect. However, redirections by using scripts are hard to identify because search engine crawlers do not execute scripts. To prevent cloaking, a search engine crawler may identify itself as a regular Web browser.

Using the terms of anchor texts of links that point to a page to index the page is able to fight content spam to some extent because anchor texts from other pages are more trustworthy. This method was originally proposed to index pages that were not fetched by search engine crawlers [364]. It is now a general technique used by search engines as we have seen in Sect. 6.8, i.e., search engines give terms in such anchor texts higher weights. In fact, the terms near a piece of anchor text also offer good editorial judgment about the target page.

The PageRank algorithm [68] is able to combat content spam to a certain degree as it is based on links that point to the target pages, and the pages that point to the target pages need to be reputable or with high PageRank scores as well (see Chap. 7). However, it does not deal with the in-link based spamming methods discussed above.

Instead of combating each individual type of spam, a method (called TrustRank) is proposed in [216] to combat all kinds of spamming methods at the same time. It takes advantage of the approximate isolation of reputable and non-spam pages, i.e., reputable Web pages seldom pointing to spam pages, and spam pages often link to many reputable pages (in an attempt to improve their hub scores). Link analysis methods are used to separate reputable pages and any form of spam without dealing with each spam technique individually.

Combating spam can also be seen as a classification problem, i.e., predicting whether a page is a spam page or not. One can use any supervised learning algorithm to train a spam classifier. The key issue is to design features used in learning. The following are some example features used in [417] to detect content spam.

1. Number of words in the page: A spam page tends to contain more words than a non-spam page so as to cover a large number of popular words.

2. Average word length: The mean word length for English prose is about 5 letters. Average word length of synthetic content is often different.
3. Number of words in the page title: Since search engines usually give extra weights to terms appearing in page titles, spammers often put many keywords in the titles of the spam pages.
4. Fraction of visible content: Spam pages often hide spam terms by making them invisible to the user.

Other features used include the amount of anchor text, compressibility, fraction of page drawn from globally popular words, independent n-gram likelihoods, conditional n-gram likelihoods, etc. Details can be found in [417]. Its spam detection classifier gave very good results. Testing on 2364 spam pages and 14806 non-spam pages (17170 pages in total), the classifier was able to correctly identify 2,037 (86.2%) of the 2364 spam pages, while misidentifying only 526 spam and non-spam pages.

Another interesting technique for fighting spam is to partition each Web page into different blocks using techniques discussed in Sect. 6.5. Each block is given an importance level automatically. To combat link spam, links in less important blocks are given lower transition probabilities to be used in the PageRank computation. The original PageRank algorithm assigns every link in a page an equal transition probability (see Sect. 7.3). The non-uniform probability assignment results in lower PageRank scores for pages pointed to by links in less important blocks. This method is effective because in the link exchange scheme and the honey pot scheme, the spam links are usually placed in unimportant blocks of the page, e.g., at the bottom of the page. The technique may also be used to fight term spam in a similar way, i.e., giving terms in less important blocks much lower weights in rank score computation. This method is proposed in [78].

However, sophisticated spam is still hard to detect. Combating spam is an on-going process. Once search engines are able to detect certain types of spam, spammers invent more sophisticated spamming methods.

Bibliographic Notes

Information retrieval (IR) is a major research field. This chapter only gives a brief introduction to some commonly used models and techniques. There are several text books that have a comprehensive coverage of the field, e.g., those by Baeza-Yates and Ribeiro-Neto [31], Grossman and Frieder [209], Salton and McGill [471], van Rijsbergen (an online book at http://www.dcs.gla.ac.uk/Keith/Preface.html), Witten et al. [551], and Yu and Meng [581].

A similar chapter in the book by Chakrabarti [85] also discusses many Web specific issues and has influenced the writing of this chapter. Below, we discuss some further readings related to Web search and mining.

On index compression, Elias coding was introduced by Elias [161] and Golomb coding was introduced by Golomb [202]. Their applications to index compression was studied by several researchers, e.g., Witten et al. [551], Bell et al. [45], Moffat et al. [392], and Williams and Zobel [547]. Wikipedia is a great source of information on this topic as well.

Latent semantic index (LSI) was introduced by Deerwester et al. [125], which uses the singular value decomposition technique (SVD) [203]. Additional information about LSI and/or SVD can be found in [48, 581, 288]. Telcordia Technologies, where LSI was developed, maintains a LSI page at http://lsi.research.telcordia.com/ with more references.

On Web page pre-processing, the focus has been on identifying the main content blocks of each page because a typical Web page contains a large amount of noise, which can adversely affect the search or mining accuracy. Several researchers have attempted the task, e.g., Bar-Yossef et al. [38], Debnath et al. [124], Gibson, et al. [199], Li et al. [324], Lin and Ho [336], Ma et al. [355], Ramaswamy et al. [456], Song et al. [495], Yi et al. [576], Yin and Lee [579], etc.

Although search is probably the biggest application on the Web, little is known about the actual implementation of a search engine except some principal ideas. Sect. 6.8 is largely based on the Google paper by Brin and Page [68], and bits and pieces in various other sources. Over the years, a large number of researchers have studied Web search. More recent studies on various aspects of search can be found in [37, 79, 89, 262, 289, 297, 451, 460, 508, 567, 569, 611].

For metasearch, the combination methods in Sect. 6.9.1 were proposed by Fox and Shaw [184]. Aslam and Montague [28], Montague and Aslam [394], and Nuray and Can [418] provide good descriptions of Borda ranking and Condorcet ranking. In addition to ranking, Meng et al. [378] discussed many other metasearch issues.

On Web spam, Gyongyi and Garcia-Molina gave an excellent taxonomy of different types of spam [214]. The TrustRank algorithm is also due to them [216]. An improvement to TrustRank was proposed by Wu et al. [557]. General link spam detection was studied by Adali et al. [1], Amitay et al. [19], Baeza-Yates et al. [30], Gyongyi and Garcia-Molina [215], Wu and Davison [555], Zhang et al. [604], etc. Content spam detection was studied by Fetterly et al. [176, 177], and Ntoulas et al. [417]. A cloaking detection algorithm is reported in [556].

7 Link Analysis

Early search engines retrieved relevant pages for the user based primarily on the content similarity of the user query and the indexed pages of the search engines. The retrieval and ranking algorithms were simply direct implementation of those from information retrieval. Starting from 1996, it became clear that content similarity alone was no longer sufficient for search due to two reasons. First, the number of Web pages grew rapidly during the middle to late 1990s. Given any query, the number of relevant pages can be huge. For example, given the search query "classification technique", the Google search engine estimates that there are about 10 million relevant pages. This abundance of information causes a major problem for ranking, i.e., how to choose only 30–40 pages and rank them suitably to present to the user. Second, content similarity methods are easily spammed. A page owner can repeat some important words and add many remotely related words in his/her pages to boost the rankings of the pages and/or to make the pages relevant to a large number of possible queries.

Starting from around 1996, researchers in academia and search engine companies began to work on the problem. They resort to hyperlinks. Unlike text documents used in traditional information retrieval, which are often considered independent of one another (i.e., with no explicit relationships or links among them except in citation analysis), Web pages are connected through hyperlinks, which carry important information. Some hyperlinks are used to organize a large amount of information at the same Web site, and thus only point to pages in the same site. Other hyperlinks point to pages in other Web sites. Such out-going hyperlinks often indicate an implicit conveyance of authority to the pages being pointed to. Therefore, those pages that are pointed to by many other pages are likely to contain authoritative or quality information. Such linkages should obviously be used in page evaluation and ranking in search engines.

During the period of 1997-1998, two most influential hyperlink based search algorithms PageRank [68, 422] and HITS [281] were designed. PageRank is the algorithm that powers the successful search engine Google. Both PageRank and HITS were originated from **social network analysis** [540]. They both exploit the hyperlink structure of the Web to rank pages according to their levels of "prestige" or "authority". We will study these algorithms in this chapter. We should also note that hyperlink-based page evaluation and ranking is not the only method used by search engines. As we discussed in Chap. 6, contents and many other factors are also considered in producing the final ranking presented to the user.

Apart from search ranking, hyperlinks are also useful for finding Web communities. A Web community is a cluster of densely linked pages representing a group of people with a common interest. Beyond explicit hyperlinks on the Web, links in other contexts are useful too, e.g., for discovering communities of named entities (e.g., people and organizations) in free text documents, and for analyzing social phenomena in emails. This chapter will introduce some of the current algorithms.

7.1 Social Network Analysis

Social network is the study of social entities (people in an organization, called **actors**), and their interactions and relationships. The interactions and relationships can be represented with a network or graph, where each vertex (or node) represents an actor and each link represents a relationship. From the network we can study the properties of its structure, and the role, position and prestige of each social actor. We can also find various kinds of sub-graphs, e.g., **communities** formed by groups of actors.

Social network analysis is useful for the Web because the Web is essentially a virtual society, and thus a virtual social network, where each page can be regarded as a social actor and each hyperlink as a relationship. Many of the results from social networks can be adapted and extended for use in the Web context. The ideas from social network analysis are indeed instrumental to the success of Web search engines.

In this section, we introduce two types of social network analysis, **centrality** and **prestige**, which are closely related to hyperlink analysis and search on the Web. Both centrality and prestige are measures of degree of prominence of an actor in a social network. We introduce them below. For a more complete treatment of the topics, please refer to the authoritative text by Wasserman and Faust [540].

7.1.1 Centrality

Important or prominent actors are those that are linked or involved with other actors extensively. In the context of an organization, a person with extensive contacts (links) or communications with many other people in the organization is considered more important than a person with relatively fewer contacts. The links can also be called **ties**. A **central actor** is one involved in many ties. Fig. 7.1 shows a simple example using an undirected graph. Each node in the social network is an actor and each link indicates that the actors on the two ends of the link communicate with each other. Intuitively, we see that the actor i is the most central actor because he/she can communicate with most other actors.

Fig. 7.1. An example of a social network

There are different types of links or involvements between actors. Thus, several types of centrality are defined on undirected and directed graphs. We discuss three popular types below.

Degree Centrality

Central actors are the most active actors that have most links or ties with other actors. Let the total number of actors in the network be n.

Undirected Graph: In an undirected graph, the **degree centrality** of an actor i (denoted by $C_D(i)$) is simply the node degree (the number of edges) of the actor node, denoted by $d(i)$, normalized with the maximum degree, $n-1$.

$$C_D(i) = \frac{d(i)}{n-1}.$$

(1)

The value of this measure ranges between 0 and 1 as $n-1$ is the maximum value of $d(i)$.

Directed Graph: In this case, we need to distinguish **in-links** of actor i (links pointing to i), and **out-links** (links pointing out from i). The degree centrality is defined based on only the out-degree (the number of out-links or edges), $d_o(i)$.

$$C'_D(i) = \frac{d_o(i)}{n-1}.$$

(2)

Closeness Centrality

This view of centrality is based on the closeness or distance. The basic idea is that an actor x_i is central if it can easily interact with all other actors. That is, its distance to all other actors is short. Thus, we can use the shortest distance to compute this measure. Let the shortest distance from actor i to actor j be $d(i,j)$ (measured as the number of links in a shortest path).

Undirected Graph: The closeness centrality $C_C(i)$ of actor i is defined as

$$C_C(i) = \frac{n-1}{\sum_{j=1}^{n} d(i,j)}. \tag{3}$$

The value of this measure also ranges between 0 and 1 as $n-1$ is the minimum value of the denominator, which is the sum of the shortest distances from i to all other actors. Note that this equation is only meaningful for a connected graph.

Directed Graph: The same equation can be used for a directed graph. The distance computation needs to consider directions of links or edges.

Betweenness Centrality

If two non-adjacent actors j and k want to interact and actor i is on the path between j and k, then i may have some control over their interactions. Betweenness measures this control of i over other pairs of actors. Thus, if i is on the paths of many such interactions, then i is an important actor.

Undirected Graph: Let p_{jk} be the number of shortest paths between actors j and k. The betweenness of an actor i is defined as the number of shortest paths that pass i (denoted by $p_{jk}(i)$, $j \neq i$ and $k \neq i$) normalized by the total number of shortest paths of all pairs of actors not including i:

$$C_B(i) = \sum_{j<k} \frac{p_{jk}(i)}{p_{jk}}. \tag{4}$$

Note that there may be multiple shortest paths between actor j and actor k. Some pass i and some do not. We assume that all paths are equally likely to be used. $C_B(i)$ has a minimum of 0, attained when i falls on no shortest path. Its maximum is $(n-1)(n-2)/2$, which is the number of pairs of actors not including i.

In the network of Fig. 7.2, actor 1 is the most central actor. It lies on all 15 shortest paths linking the other 6 actors. $C_B(1)$ has the maximum value of 15, and $C_B(2) = C_B(3) = C_B(4) = C_B(5) = C_B(6) = C_B(7) = 0$.

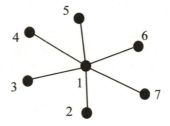

Fig. 7.2. An example of a network illustrating the betweenness centrality

If we are to ensure that the value range is between 0 and 1, we can normal-ize it with $(n-1)(n-2)/2$, which is the maximum value of $C_B(i)$. The stan-dardized betweenness of actor i is defined as

$$C'_B(i) = \frac{2\sum_{j<k} \dfrac{P_{jk}(i)}{P_{jk}}}{(n-1)(n-2)}. \tag{5}$$

Unlike the closeness measure, the betweenness can be computed even if the graph is not connected.

Directed Graph: The same equation can be used but must be multiplied by 2 because there are now $(n-1)(n-2)$ pairs considering a path from j to k is different from a path from k to j. Likewise, p_{jk} must consider paths from both directions.

7.1.2 Prestige

Prestige is a more refined measure of prominence of an actor than central-ity as we will see below. We need to distinguish between ties sent (out-links) and ties received (in-links). A prestigious actor is defined as one who is object of extensive ties as a recipient. In other words, to compute the prestige of an actor, we only look at the ties (links) directed or pointed to the actor (in-links). Hence, the prestige cannot be computed unless the relation is directional or the graph is directed. The main difference between the concepts of centrality and prestige is that centrality focuses on out-links while prestige focuses on in-links. We define three prestige measures. The third prestige measure (i.e., **rank prestige**) forms the basis of most Web page link analysis algorithms, including PageRank and HITS.

Degree Prestige

Based on the definition of the prestige, it is clear that an actor is prestigious if it receives many in-links or nominations. Thus, the simplest measure of prestige of an actor i (denoted by $P_D(i)$) is its in-degree.

$$P_D(i) = \frac{d_I(i)}{n-1}, \tag{6}$$

where $d_I(i)$ is the in-degree of i (the number of in-links of i) and n is the total number of actors in the network. As in the degree centrality, dividing by $n-1$ standardizes the prestige value to the range from 0 and 1. The maximum prestige value is 1 when every other actor links to or chooses actor i.

Proximity Prestige

The degree index of prestige of an actor i only considers the actors that are adjacent to i. The proximity prestige generalizes it by considering both the actors directly and indirectly linked to actor i. That is, we consider every actor j that can reach i, i.e., there is a directed path from j to i.

Let I_i be the set of actors that can reach actor i, which is also called the **influence domain** of actor i. The **proximity** is defined as closeness or distance of other actors to i. Let $d(j, i)$ denote the shortest path distance from actor j to actor i. Each link has the unit distance. To compute the proximity prestige, we use the average distance, which is

$$\frac{\sum_{j \in I_i} d(j,i)}{|I_i|}, \tag{7}$$

where $|I_i|$ is the size of the set I_i. If we look at the ratio or proportion of actors who can reach i to the average distance that these actors are from i, we obtain the proximity prestige, which has the value range of $[0, 1]$:

$$P_P(i) = \frac{|I_i|/(n-1)}{\sum_{j \in I_i} d(j,i) \Big/ |I_i|}, \tag{8}$$

where $|I_i|/(n-1)$ is the proportion of actors that can reach actor i. In one extreme, every actor can reach actor i, which gives $|I_i|/(n-1) = 1$. The denominator is 1 if every actor is adjacent to i. Then, $P_P(i) = 1$. On the other extreme, no actor can reach actor i. Then $|I_i| = 0$, and $P_P(i) = 0$.

Rank Prestige

The above two prestige measures are based on in-degrees and distances. However, an important factor that has not been considered is the **prominence** of individual actors who do the "voting" or "choosing." In the real world, a person i chosen by an important person is more prestigious than chosen by a less important person. For example, a company CEO voting for a person is much more important than a worker voting for the person. If one's circle of influence is full of prestigious actors, then one's own prestige is also high. Thus one's prestige is affected by the ranks or statuses of the involved actors. Based on this intuition, the rank prestige $P_R(i)$ is defined as a linear combination of links that point to i:

$$P_R(i) = A_{1i}P_R(1) + A_{2i}P_R(2) + ... + A_{ni}P_R(n), \tag{9}$$

where $A_{ji} = 1$ if j points to i, and 0 otherwise. This equation says that an actor's rank prestige is a function of the ranks of the actors who vote or choose the actor, which makes perfect sense.

Since we have n equations for n actors, we can write them in the matrix notation. We use P to represent the vector that contains all the rank prestige values, i.e., $P = (P_R(1), P_R(2), ..., P_R(n))^T$ (T means **matrix transpose**). P is represented as a column vector. We use matrix A (where $A_{ij} = 1$ if i points to j, and 0 otherwise) to represent the adjacency matrix of the network or graph. As a notational convention, we use bold italic letters to represent matrices. We then have

$$P = A^T P. \tag{10}$$

This equation is precisely the characteristic equation used for finding the **eigensystem** of the matrix A^T. P is an **eigenvector** of A^T.

This equation and the idea behind it turn out to be very useful in Web search. Indeed, the most well known ranking algorithms for Web search, PageRank and HITS, are directly related to this equation. Sect. 7.3 and 7.4 will focus on these two algorithms and describe how to solve the equation to obtain the prestige value of each actor (or each page on the Web).

7.2   Co-Citation and Bibliographic Coupling

Another area of research concerned with links is the **citation analysis** of scholarly publications. A scholarly publication usually cites related prior work to acknowledge the origins of some ideas in the publication and to compare the new proposal with existing work. Citation analysis is an area of bibliometric research, which studies citations to establish the relationships between authors and their work.

When a publication (also called a paper) cites another publication, a relationship is established between the publications. Citation analysis uses these relationships (links) to perform various types of analysis. A citation can represent many types of links, such as links between authors, publications, journals and conferences, and fields, or even between countries. We will discuss two specific types of citation analysis, **co-citation** and **bibliographic coupling**. The HITS algorithm of Sect. 7.4 is related to these two types of analysis.

7.2.1   Co-Citation

Co-citation is used to measure the similarity of two documents. If papers i and j are both cited by paper k, then they may be said to be related in some sense to one another, even they do not directly cite each other. Figure 7.3 shows that papers i and j are co-cited by paper k. If papers i and j are cited together by many papers, it means that i and j have a strong relationship or similarity. The more papers they are cited by, the stronger their relationship is.

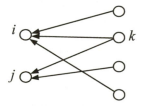

Fig. 7.3. Paper i and paper j are co-cited by paper k

Let L be the citation matrix. Each cell of the matrix is defined as follows: $L_{ij} = 1$ if paper i cites paper j, and 0 otherwise. **Co-citation** (denoted by C_{ij}) is a similarity measure defined as the number of papers that co-cite i and j, and is computed with

$$C_{ij} = \sum_{k=1}^{n} L_{ki}L_{kj}, \tag{11}$$

where n is the total number of papers. C_{ii} is naturally the number of papers that cite i. A square matrix C can be formed with C_{ij}, and it is called the **co-citation matrix**. Co-citation is symmetric, $C_{ij} = C_{ji}$, and is commonly used as a similarity measure of two papers in clustering to group papers of similar topics together.

7.2.2 Bibliographic Coupling

Bibliographic coupling operates on a similar principle, but in a way it is the mirror image of co-citation. Bibliographic coupling links papers that cite the same articles so that if papers i and j both cite paper k, they may be said to be related, even though they do not directly cite each other. The more papers they both cite, the stronger their similarity is. Figure 7.4 shows both papers i and j citing (referencing) paper k.

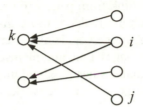

Fig. 7.4. Both paper i and paper j cite paper k

We use B_{ij} to represent the number of papers that are cited by both papers i and j:

$$B_{ij} = \sum_{k=1}^{n} L_{ik} L_{jk}. \tag{12}$$

B_{ii} is naturally the number of references (in the reference list) of paper i. A square matrix B can be formed with B_{ij}, and it is called the **bibliographic coupling matrix**. Bibliographic coupling is also symmetric and is regarded as a similarity measure of two papers in clustering.

We will see later that two important types of pages on the Web, **hubs** and **authorities**, found by the HITS algorithm are directly related to co-citation and bibliographic coupling matrices.

7.3 PageRank

The year 1998 was an important year for Web link analysis and Web search. Both the PageRank and the HITS algorithms were reported in that year. HITS was presented by Jon Kleinberg in January, 1998 at the *Ninth Annual ACM-SIAM Symposium on Discrete Algorithms*. PageRank was presented by Sergey Brin and Larry Page at the *Seventh International World Wide Web Conference (WWW7)* in April, 1998. Based on the algorithm, they built the search engine Google. The main ideas of PageRank and HITS are really quite similar. However, it is their dissimilarity that made a huge difference as we will see later. Since that year, PageRank has emerged as the dominant link analysis model for Web search, partly due to its query-independent evaluation of Web pages and its ability to combat spamming, and partly due to Google's business success. In this section, we focus on PageRank. In the next section, we discuss HITS. A detailed study of these algorithms can also be found in [304].

PageRank relies on the democratic nature of the Web by using its vast link structure as an indicator of an individual page's quality. In essence, PageRank interprets a hyperlink from page x to page y as a vote, by page x, for page y. However, PageRank looks at more than just the sheer number of votes, or links that a page receives. It also analyzes the page that casts the vote. Votes casted by pages that are themselves "important" weigh more heavily and help to make other pages more "important." This is exactly the idea of **rank prestige** in social networks (see Sect. 7.1.2).

7.3.1   PageRank Algorithm

PageRank is a static ranking of Web pages in the sense that a PageRank value is computed for each page off-line and it does not depend on search queries. Since PageRank is based on the measure of prestige in social networks, the PageRank value of each page can be regarded as its prestige. We now derive the PageRank formula. Let us first state some main concepts again in the Web context.

In-links of page i: These are the hyperlinks that point to page i from other pages. Usually, hyperlinks from the same site are not considered.
Out-links of page i: These are the hyperlinks that point out to other pages from page i. Usually, links to pages of the same site are not considered.

From the perspective of prestige, we use the following to derive the PageRank algorithm.

1. A hyperlink from a page pointing to another page is an implicit conveyance of authority to the target page. Thus, the more in-links that a page i receives, the more prestige the page i has.
2. Pages that point to page i also have their own prestige scores. A page with a higher prestige score pointing to i is more important than a page with a lower prestige score pointing to i. In other words, a page is important if it is pointed to by other important pages.

According to rank prestige in social networks, the importance of page i (i's PageRank score) is determined by summing up the PageRank scores of all pages that point to i. Since a page may point to many other pages, its prestige score should be shared among all the pages that it points to. Notice the difference from rank prestige, where the prestige score is not shared.

To formulate the above ideas, we treat the Web as a directed graph $G = (V, E)$, where V is the set of vertices or nodes, i.e., the set of all pages, and E is the set of directed edges in the graph, i.e., hyperlinks. Let the total number of pages on the Web be n (i.e., $n = |V|$). The PageRank score of the page i (denoted by $P(i)$) is defined by:

$$P(i) = \sum_{(j,i)\in E} \frac{P(j)}{O_j}, \tag{13}$$

where O_j is the number of out-links of page j. Mathematically, we have a system of n linear equations (13) with n unknowns. We can use a matrix to represent all the equations. Let P be a n-dimensional column vector of PageRank values, i.e.,

$$P = (P(1), P(2), \ldots, P(n))^T.$$

Let A be the adjacency matrix of our graph with

$$A_{ij} = \begin{cases} \dfrac{1}{O_i} & \text{if}(i, j) \in E \\ 0 & \text{otherwise} \end{cases} \tag{14}$$

We can write the system of n equations with (similar to Equation 10)

$$P = A^T P. \tag{15}$$

This is the characteristic equation of the **eigensystem**, where the solution to P is an **eigenvector** with the corresponding **eigenvalue** of 1. Since this is a circular definition, an iterative algorithm is used to solve it. It turns out that if *some conditions* are satisfied (which will be described shortly), 1 is the largest **eigenvalue** and the PageRank vector P is the **principal eigenvector**. A well known mathematical technique called **power iteration** can be used to find P.

However, the problem is that Equation (15) does not quite suffice because the Web graph does not meet the conditions. To introduce these conditions and the enhanced equation, let us derive the same Equation (15) based on the **Markov chain** [207].

In the Markov chain model, each Web page or node in the Web graph is regarded as a state. A hyperlink is a transition, which leads from one state to another state with a probability. Thus, this framework models Web surfing as a stochastic process. It models a Web surfer randomly surfing the Web as a state transition in the Markov chain. Recall that we used O_i to denote the number of out-links of a node i. Each transition probability is $1/O_i$ if we assume the Web surfer will click the hyperlinks in the page i uniformly at random, the "back" button on the browser is not used and the surfer does not type in an URL. Let A be the state transition probability matrix, a square matrix of the following format,

$$A = \begin{pmatrix} A_{11} & A_{12} & \cdot & \cdot & \cdot & A_{1n} \\ A_{21} & A_{22} & \cdot & \cdot & \cdot & A_{2n} \\ \cdot & \cdot & & & & \cdot \\ \cdot & \cdot & & & & \cdot \\ \cdot & \cdot & & & & \cdot \\ A_{n1} & A_{n2} & \cdot & \cdot & \cdot & A_{nn} \end{pmatrix}$$

A_{ij} represents the transition probability that the surfer in state i (page i) will move to state j (page j). A_{ij} is defined exactly as in Equation (14).

Given an **initial probability distribution** vector that a surfer is at each state (or page) $p_0 = (p_0(1), p_0(2), ..., p_0(n))^T$ (a column vector) and an $n \times n$ **transition probability matrix** A, we have

$$\sum_{i=1}^{n} p_0(i) = 1 \tag{16}$$

$$\sum_{j=1}^{n} A_{ij} = 1. \tag{17}$$

Equation (17) is not quite true for some Web pages because they have no out-links. If the matrix A satisfies Equation (17), we say that A is the **stochastic matrix** of a Markov chain. Let us assume A is a stochastic matrix for the time being and deal with it not being that later.

In a Markov chain, a question of common interest is: Given the initial probability distribution p_0 at the beginning, what is the probability that m steps/transitions later that the Markov chain will be at each state j? We can determine the probability that the system (or the **random surfer**) is in state j after 1 step (1 state transition) by using the following reasoning:

$$p_1(j) = \sum_{i=1}^{n} A_{ij}(1)p_0(i), \tag{18}$$

where $A_{ij}(1)$ is the probability of going from i to j after 1 transition, and $A_{ij}(1) = A_{ij}$. We can write it with a matrix:

$$p_1 = A^T p_0 . \tag{19}$$

In general, the probability distribution after k steps/transitions is:

$$p_k = A^T p_{k-1} . \tag{20}$$

Equation (20) looks very similar to Equation (15). We are getting there.

By the Ergodic Theorem of Markov chains [207], a finite Markov chain defined by the **stochastic transition matrix** A has a unique **stationary probability distribution** if A is **irreducible** and **aperiodic**. These mathematical terms will be defined as we go along.

The stationary probability distribution means that after a series of transitions p_k will converge to a steady-state probability vector π regardless of the choice of the initial probability vector p_0, i.e.,

$$\lim_{k \to \infty} p_k = \pi . \tag{21}$$

When we reach the steady-state, we have $p_k = p_{k+1} = \pi$, and thus $\pi = A^T \pi$. π is the **principal eigenvector** of A^T with **eigenvalue** of 1. In PageRank, π is used as the PageRank vector P. Thus, we again obtain Equation (15), which is re-produced here as Equation (22):

$$P = A^T P . \tag{22}$$

Using the stationary probability distribution π as the PageRank vector is reasonable and quite intuitive because it reflects the long-run probabilities that a random surfer will visit the pages. A page has a high prestige if the probability of visiting it is high.

Now let us come back to the real Web context and see whether the above conditions are satisfied, i.e., whether A is a stochastic matrix and whether it is irreducible and aperiodic. In fact, none of them is satisfied. Hence, we need to extend the ideal-case Equation (22) to produce the "actual PageRank model". Let us look at each condition below.

First of all, A is not a **stochastic (transition) matrix**. A stochastic matrix is the transition matrix for a finite Markov chain whose entries in each row are non-negative real numbers and sum to 1 (i.e., Equation 17). This requires that every Web page must have at least one out-link. This is not true on the Web because many pages have no out-links, which are reflected in transition matrix A by some rows of complete 0's. Such pages are called the **dangling pages** (nodes).

Example 1: Figure 7.5 shows an example of a hyperlink graph.

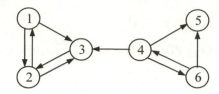

Fig. 7.5. An example of a hyperlink graph

If we assume that the Web surfer will click the hyperlinks in a page uniformly at random, we have the following transition probability matrix:

$$A = \begin{pmatrix} 0 & 1/2 & 1/2 & 0 & 0 & 0 \\ 1/2 & 0 & 1/2 & 0 & 0 & 0 \\ 0 & 1 & 0 & 0 & 0 & 0 \\ 0 & 0 & 1/3 & 0 & 1/3 & 1/3 \\ 0 & 0 & 0 & 0 & 0 & 0 \\ 0 & 0 & 0 & 1/2 & 1/2 & 0 \end{pmatrix}. \tag{23}$$

For example $A_{12} = A_{13} = 1/2$ because node 1 has two out-links. We can see that A is not a stochstic matrix because the fifth row is all 0's, i.e., page 5 is a dangling page. ∎

We can fix this problem in several ways in order to convert A to a stochastic transition matrix. We describe only two ways here:

1. Remove those pages with no out-links from the system during the PageRank computation as these pages do not affect the ranking of any other page directly. Out-links from other pages pointing to these pages are also removed. After PageRanks are computed, these pages and hyperlinks pointing to them can be added in. Their PageRanks are easy to calculate based on Equation (22). Note that the transition probabilities of those pages with removed links will be slightly affected but not significantly. This method is suggested in [68].
2. Add a complete set of outgoing links from each such page i to all the pages on the Web. Thus the transition probability of going from i to every page is $1/n$ assuming uniform probability distribution. That is, we replace each row containing all 0's with e/n, where e is n-dimensional vector of all 1's.

If we use the second method to make A a stochastic matrix by adding a link from page 5 to every page, we obtain

$$\overline{A} = \begin{pmatrix} 0 & 1/2 & 1/2 & 0 & 0 & 0 \\ 1/2 & 0 & 1/2 & 0 & 0 & 0 \\ 0 & 1 & 0 & 0 & 0 & 0 \\ 0 & 0 & 1/3 & 0 & 1/3 & 1/3 \\ 1/6 & 1/6 & 1/6 & 1/6 & 1/6 & 1/6 \\ 0 & 0 & 0 & 1/2 & 1/2 & 0 \end{pmatrix}. \tag{24}$$

Below, we assume that either one of the above is done to make A a stochastic matrix.

Second, A is not **irreducible**. Irreducible means that the Web graph G is strongly connected.

Definition (strongly connected): A directed graph $G = (V, E)$ is **strongly connected** if and only if, for each pair of nodes $u, v \in V$, there is a path from u to v.

A general Web graph represented by A is not irreducible because for some pair of nodes u and v, there is no path from u to v. For example, in Fig. 7.5, there is no directed path from node 3 to node 4. The adjustment in Equation (24) is not enough to ensure irreducibility. That is, in $\overline{A}$, there is still no directed path from node 3 to node 4. This problem and the next problem can be dealt with using a single strategy (to be described shortly).

Finally, A is not **aperiodic**. A state i in a Markov chain being periodic means that there exists a directed cycle that the chain has to traverse.

Definition (aperiodic): A state i is **periodic** with period $k > 1$ if k is the smallest number such that all paths leading from state i back to state i have a length that is a multiple of k. If a state is not periodic (i.e., $k = 1$), it is **aperiodic**. A Markov chain is **aperiodic** if all states are aperiodic.

Example 2: Figure 7.6 shows a periodic Markov chain with $k = 3$. The transition matrix is given on the left. Each state in this chain has a period of 3. For example, if we start from state 1, to come back to state 1 the only path is 1-2-3-1 for some number of times, say h. Thus any return to state 1 will take $3h$ transitions. In the Web, there could be many such cases. ∎

$$A = \begin{bmatrix} 0 & 1 & 0 \\ 0 & 0 & 1 \\ 1 & 0 & 0 \end{bmatrix}$$

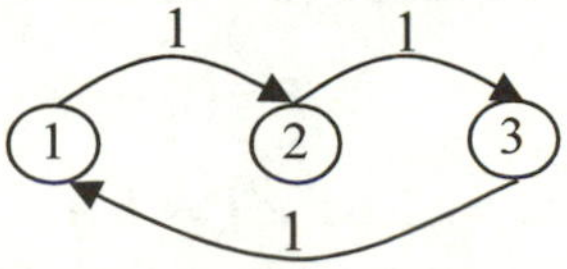

Fig. 7.6. A periodic Markov chain with $k = 3$.

It is easy to deal with the above two problems with a single strategy.

- We add a link from each page to every page and give each link a small transition probability controlled by a parameter d.

The augmented transition matrix becomes **irreducible** because it is clearly strongly connected. It is also **aperiodic** because the situation in Fig. 7.6 no longer exists as we now have paths of all possible lengths from state i back to state i. That is, the random surfer does not have to traverse a fixed cycle for any state. After this augmentation, we obtain an improved PageRank model. In this model, at a page, the random surfer has two options:

1. With probability d, he randomly chooses an out-link to follow.
2. With probability $1-d$, he jumps to a random page without a link.

Equation (25) gives the improved model,

$$P = \left((1-d)\frac{E}{n} + dA^T \right) P \tag{25}$$

where E is ee^T (e is a column vector of all 1's) and thus E is a $n \times n$ square matrix of all 1's. $1/n$ is the probability of jumping to a particular page. n is the total number of nodes in the Web graph. Note that Equation (25) assumes that A has already been made a stochastic matrix.

Example 3: If we follow our example in Fig. 7.5 and Equation (24) (we use $\overline{A}$ for A here), the augmented transition matrix is

$$(1-d)\frac{E}{n} + dA^T = \begin{pmatrix} 1/60 & 7/15 & 1/60 & 1/60 & 1/6 & 1/60 \\ 7/15 & 1/60 & 11/12 & 1/60 & 1/6 & 1/60 \\ 7/15 & 7/15 & 1/60 & 19/60 & 1/6 & 1/60 \\ 1/60 & 1/60 & 1/60 & 1/60 & 1/6 & 7/15 \\ 1/60 & 1/60 & 1/60 & 19/60 & 1/6 & 7/15 \\ 1/60 & 1/60 & 1/60 & 19/60 & 1/6 & 1/60 \end{pmatrix} \tag{26}$$

$(1-d)E/n + dA^T$ is a **stochastic matrix** (but transposed). It is also **irreducible** and **aperiodic** as we discussed above. Here we use $d = 0.9$.

If we scale Equation (25) so that $e^T P = n$, we obtain

$$P = (1-d)e + dA^T P . \tag{27}$$

Before scaling, we have $e^T P = 1$ (i.e., $P(1) + P(2) + \ldots + P(n) = 1$ if we recall that P is the stationary probability vector π of the Markov chain). The scaling is equivalent to multiplying n on both sides of Equation (25).

This gives us the PageRank formula for each page i as follows:

$$P(i) = (1-d) + d\sum_{j=1}^{n} A_{ji}P(j), \qquad (28)$$

which is equivalent to the formula given in the PageRank papers [68, 422]:

$$P(i) = (1-d) + d\sum_{(j,i)\in E} \frac{P(j)}{O_j}. \qquad (29)$$

The parameter d is called the **damping factor** which can be set to between 0 and 1. $d = 0.85$ is used in [68, 422].

The computation of PageRank values of the Web pages can be done using the well known **power iteration method** [203], which produces the principal eigenvector with the eigenvalue of 1. The algorithm is simple, and is given in Fig. 7.7. One can start with any initial assignments of PageRank values. The iteration ends when the PageRank values do not change much or converge. In Fig. 7.7, the iteration ends after the 1-norm of the residual vector is less than a pre-specified threshold ε. Note that the 1-norm for a vector is simply the sum of all the components.

PageRank-Iterate(G)
    $P_0 \leftarrow e/n$
    $k \leftarrow 1$
    repeat
        $P_k \leftarrow (1-d)e + dA^T P_{k-1}$;
        $k \leftarrow k+1$;
    until $||P_k - P_{k-1}||_1 < \varepsilon$
    return P_k

Fig. 7.7. The power iteration method for PageRank

Since we are only interested in the ranking of the pages, the actual convergence may not be necessary. Thus, fewer iterations are needed. In [68], it is reported that on a database of 322 million links the algorithm converges to an acceptable tolerance in roughly 52 iterations.

7.3.2 Strengths and Weaknesses of PageRank

The main advantage of PageRank is its ability to fight spam. A page is important if the pages pointing to it are important. Since it is not easy for Web page owner to add in-links into his/her page from other important pages, it is thus not easy to influence PageRank. Nevertheless, there are reported ways to influence PageRank. Recognizing and fighting spam is an important issue in Web search.

Another major advantage of PageRank is that it is a global measure and is query independent. That is, the PageRank values of all the pages on the Web are computed and saved off-line rather than at the query time. At the query time, only a lookup is needed to find the value to be integrated with other strategies to rank the pages. It is thus very efficient at query time. Both these two advantages contributed greatly to Google's success.

The main criticism is also the query-independence nature of PageRank. It could not distinguish between pages that are authoritative in general and pages that are authoritative on the query topic. Google may have other ways to deal with the problem, which we do not know due to the proprietary nature of Google. Another criticism is that PageRank does not consider time. Let us give some explanation to this.

7.3.3   Timed PageRank

The Web is a dynamic environment, and it changes constantly. Quality pages in the past may not be quality pages now or in the future. Thus, search has a temporal dimension. An algorithm called **TimedPageRank** given in [326, 585] adds the temporal dimension to PageRank. The motivations are:

1. Users are often interested in the latest information. Apart from pages that contain well-established facts and classics which do not change significantly over time, most contents on the Web change constantly. New pages or contents are added, and ideally, outdated contents and pages are deleted. However, in practice many outdated pages and links are not deleted. This causes problems for Web search because such outdated pages may still be ranked very high.
2. PageRank favors pages that have many in-links. To some extent, we can say that it favors older pages because they have existed on the Web for a long time and thus have accumulated many in-links. Then the problem is that new pages which are of high quality and also give the up-to-date information will not be assigned high scores and consequently will not be ranked high because they have fewer or no in-links. It is thus difficult for users to find the latest information on the Web based on PageRank.

The idea of TimedPageRank is simple. Instead of using a constant damping factor d as the parameter in PageRank, TimedPageRank uses a function of time $f(t)$ $(0 \le f(t) \le 1)$, where t is the difference between the current time and the time when the page was last updated. $f(t)$ returns a probability that the Web surfer will follow an actual link on the page. $1-f(t)$ returns the probability that the surfer will jump to a random page. Thus, at a particular page i, the Web surfer has two options:

1. With probability $f(t_i)$, he randomly chooses an out-going link to follow.
2. With probability $1-f(t_i)$, he jumps to a random page without a link.

The intuition here is that if the page was last updated (or created) a long time ago, the pages that it cites (points to) are even older and are probably out of date. Then the $1-f(t)$ value for such a page should be large, which means that the surfer will have a high probability of jumping to a random page. If a page is new, then its $1-f(t)$ value should be small, which means that the surfer will have a high probability to follow an out-link of the page and a small probability of jumping to a random page.

For a complete new page in a Web site, which does not have any in-links at all, the method given in [326] uses the average TimedPageRank value of the past pages in the Web site.

Finally, we note again that the link-based ranking is not the only strategy used in a search engine. Many other information retrieval methods, heuristics and empirical parameters are also employed. However, their details are not published. We also note that PageRank is not the only link-based static and global ranking algorithm. All major search engines, such as *Yahoo*! and MSN, have their own algorithms but are unpublished.

7.4 HITS

HITS stands for **Hypertext Induced Topic Search** [281]. Unlike PageRank which is a static ranking algorithm, HITS is search query dependent. When the user issues a search query, HITS first expands the list of relevant pages returned by a search engine and then produces two rankings of the expanded set of pages, **authority ranking** and **hub ranking**.

An **authority** is a page with many in-links. The idea is that the page may have good or authoritative content on some topic and thus many people trust it and link to it. A **hub** is a page with many out-links. The page serves as an organizer of the information on a particular topic and points to many good authority pages on the topic. When a user comes to this hub page, he/she will find many useful links which take him/her to good content pages on the topic. Figure 7.8 shows an authority page and a hub page.

The key idea of HITS is that a good hub points to many good authorities and a good authority is pointed to by many good hubs. Thus, authorities and hubs have a **mutual reinforcement** relationship. Figure 7.9 shows a set of densely linked authorities and hubs (a **bipartite sub-graph**).

Below, we first present the HITS algorithm, and also make a connection between HITS and co-citation and bibliographic coupling in bibliometric research. We then discuss the strengths and weaknesses of HITS, and describe some possible ways to deal with its weaknesses.

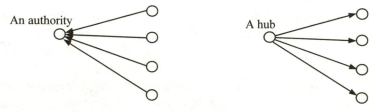

Fig. 7.8. An authority page and a hub page

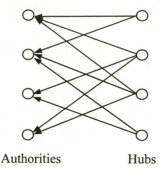

Authorities   Hubs

Fig. 7.9. A densely linked set of authorities and hubs

7.4.1 HITS Algorithm

Before describing the HITS algorithm, let us first describe how HITS collects pages to be ranked. Given a broad search query, q, HITS collects a set of pages as follows:

1. It sends the query q to a search engine system. It then collects t ($t = 200$ is used in the HITS paper) highest ranked pages, which assume to be highly relevant to the search query. This set is called the **root** set W.
2. It then grows W by including any page pointed to by a page in W and any page that points to a page in W. This gives a larger set called S. However, this set can be very large. The algorithm restricts its size by allowing each page in W to bring at most k pages ($k = 50$ is used in the HITS paper) pointing to it into S. The set S is called the **base set**.

HITS then works on the pages in S, and assigns every page in S an **authority score** and a **hub score**. Let the number of pages to be studied be n. We again use $G = (V, E)$ to denote the (directed) link graph of S. V is the set of pages (or nodes) and E is the set of directed edges (or links). We use L to denote the adjacency matrix of the graph.

$$L_{ij} = \begin{cases} 1 & \text{if} (i, j) \in E \\ 0 & \text{otherwise} \end{cases} \tag{30}$$

Let the authority score of the page i be $a(i)$, and the hub score of page i be $h(i)$. The mutual reinforcing relationship of the two scores is represented as follows:

$$a(i) = \sum_{(j,i) \in E} h(j) \tag{31}$$

$$h(i) = \sum_{(i,j) \in E} a(j) \tag{32}$$

Writing them in the matrix form, we use a to denote the column vector with all the authority scores, $a = (a(1), a(2), ..., a(n))^T$, and use h to denote the column vector with all the hub scores, $h = (h(1), h(2), ..., h(n))^T$,

$$a = L^T h \tag{33}$$

$$h = La \tag{34}$$

The computation of authority scores and hub scores is basically the same as the computation of the PageRank scores using the power iteration method. If we use a_k and h_k to denote authority and hub scores at the kth iteration, the iterative processes for generating the final solutions are

$$a_k = L^T L a_{k-1} \tag{35}$$

$$h_k = L L^T h_{k-1} \tag{36}$$

starting with

$$a_0 = h_0 = (1, 1, ..., 1). \tag{37}$$

Note that Equation (35) (or Equation 36) does not use the hub (or authority) vector due to substitutions of Equation (33) and Equation (34).

After each iteration, the values are also normalized (to keep them small) so that

HITS-Iterate(G)

$a_0 \leftarrow h_0 \leftarrow (1, 1, \ldots, 1)$;

$k \leftarrow 1$

Repeat

$\quad a_k \leftarrow L^T L a_{k-1}$;

$\quad h_k \leftarrow L L^T h_{k-1}$;

$\quad a_k \leftarrow a_k / \|a_k\|_1$;   // normalization

$\quad h_k \leftarrow h_k / \|h_k\|_1$;   // normalization

$\quad k \leftarrow k + 1$;

until $\|a_k - a_{k-1}\|_1 < \varepsilon_a$ and $\|h_k - h_{k-1}\|_1 < \varepsilon_h$;

return a_k and h_k

Fig. 7.10. The HITS algorithm based on power iteration

$$\sum_{i=1}^{n} a(i) = 1 \qquad\qquad (38)$$

$$\sum_{i=1}^{n} h(i) = 1 \qquad\qquad (39)$$

The power iteration algorithm for HITS is given in Fig. 7.10. The iteration ends after the 1-norms of the residual vectors are less than some thresholds ε_a and ε_h. Hence, the algorithm finds the principal eigenvectors at "equilibrium" as in PageRank. The pages with large authority and hub scores are better authorities and hubs respectively. HITS will select a few top ranked pages as authorities and hubs, and return them to the user.

Although HITS will always converge, there is a problem with uniqueness of limiting (converged) authority and hub vectors. It is shown that for certain types of graphs, different initializations to the power method produce different final authority and hub vectors. Some results can be inconsistent or wrong. Farahat et al. [171] gave several examples. The heart of the problem is that there are repeated dominant (principal) eigenvalues (several eigenvalues are the same and are dominant eigenvalues), which are caused by the problem that $L^T L$ (respectively $L L^T$) is reducible [303]. The first PageRank solution (Equation 22) has the same problem. However, the PageRank inventors found a way to get around the problem. A modification similar to PageRank may be applied to HITS.

7.4.2 Finding Other Eigenvectors

The HITS algorithm given in Fig. 7.10 finds the principal eigenvectors, which in a sense represent the most densely connected authorities and hubs in the graph G defined by a query. However, in some cases, we may also be interested in finding several densely linked collections of hubs and authorities among the same base set of pages. Each of such collections could potentially be relevant to the query topic, but they could be well-separated from one another in the graph G for a variety of reasons. For example,

1. The query string may be ambiguous with several very different meanings, e.g., "jaguar", which could be a cat or a car.
2. The query string may represent a topic that may arise as a term in the multiple communities, e.g. "classification".
3. The query string may refer to a highly polarized issue, involving groups that are not likely to link to one another, e.g. "abortion".

In each of these examples, the relevant pages can be naturally grouped into several clusters, also called **communities**. In general, the top ranked authorities and hubs represent the major cluster (or **community**). The smaller clusters (or communities), which are also represented by bipartite subgraphs as that in Fig. 7.9, can be found by computing non-principal eigenvectors. Non-principal eigenvectors are calculated in a similar way to power iteration using methods such as **orthogonal iteration** and **QR iteration**. We will not discuss the details of these methods. Interested readers can refer to the book by Golub and Van Loan [203].

7.4.3 Relationships with Co-Citation and Bibliographic Coupling

Authority pages and hub pages have their matches in the bibliometric citation context. An authority page is like an influential research paper (publication) which is cited by many subsequent papers. A hub page is like a survey paper which cites many other papers (including those influential papers). It is no surprise that there is a connection between authority and hub, and co-citation and bibliographic coupling.

Recall that co-citation of pages i and j, denoted by C_{ij}, is computed as

$$C_{ij} = \sum_{k=1}^{n} L_{ki} L_{kj} = (L^T L)_{ij}. \tag{40}$$

This shows that the authority matrix $(L^T L)$ of HITS is in fact the co-citation matrix C in the Web context. Likewise, recall that bibliographic coupling of two pages i and j, denoted by B_{ij}, is computed as

$$B_{ij} = \sum_{k=1}^{n} L_{ik} L_{jk} = (LL^T)_{ij},$$
(41)

which shows that the hub matrix (LL^T) of HITS is the bibliographic coupling matrix B in the Web context.

7.4.4   Strengths and Weaknesses of HITS

The main strength of HITS [281] is its ability to rank pages according to the query topic, which may be able to provide more relevant authority and hub pages. The ranking may also be combined with information retrieval based rankings. However, HITS has several disadvantages.

- First of all, it does not have the anti-spam capability of PageRank. It is quite easy to influence HITS by adding out-links from one's own page to point to many good authorities. This boosts the hub score of the page. Because hub and authority scores are interdependent, it in turn also increases the authority score of the page.
- Another problem of HITS is topic drift. In expanding the root set, it can easily collect many pages (including authority pages and hub pages) which have nothing to do the search topic because out-links of a page may not point to pages that are relevant to the topic and in-links to pages in the root set may be irrelevant as well because people put hyperlinks for all kinds of reasons, including spamming.
- The query time evaluation is also a major drawback. Getting the root set, expanding it and then performing eigenvector computation are all time consuming operations.

Over the years, many researchers tried to deal with these problems. We briefly discuss some of them below.

It was reported by several researchers in [52, 310, 405] that small changes to the Web graph topology can significantly change the final authority and hub vectors. Minor perturbations have little effect on PageRank, which is more stable than HITS. This is essentially due to the random jump step of PageRank. Ng et al. [405] proposed a method by introducing the same random jump step to HITS (by jumping to the base set uniformly at random with probability d), and showed that it could improve the stability of HITS significantly. Lempel and Moran [310] proposed SALSA, *a stochastic algorithm for link structure analysis*. SALSA combines some features of both PageRank and HITS to improve the authority and hub computation. It casts the problem as two Markov chains, an authority

Markov chain and a hub Markov chain. SALSA is less susceptible to spam since the coupling between hub and authority scores is much less strict.

Bharat and Henzinger [52] proposed a simple method to fight two site nepotistic links. That means that a set of pages on one host points to a single page on a second host. This drives up the hub scores of the pages on the first host and the authority score of the page on the second host. A similar thing can be done for hubs. These links may be authored by the same person and thus are regarded as "nepotistic" links to drive up the ranking of the target pages. [52] suggests weighting the links to deal with this problem. That is, if there are k edges from documents on a first host to a single document on a second host we give each edge an **authority weight** of $1/k$. If there are l edges from a single page on a first host to a set of pages on a second host, we give each edge a **hub weight** of $1/l$. These weights are used in the authority and hub computation. There are much more sophisticated spam techniques now involving more than two sites.

Regarding the topic drifting of HITS, existing fixes are mainly based on content similarity comparison during the expansion of the root set. In [88], if an expanded page is too different from the pages in the root set in terms of content similarity (based on cosine similarity), it is discarded. The remaining links are also weighted according to similarity. [88] proposes a method that uses the similarity between the anchor text of a link and the search topic to weight the link (instead of giving each link 1 as in HITS). [84] goes further to segment the page based on the DOM (**Document Object Model**) tree structure to identify the blocks or subtrees that are more related to the query topic instead of regarding the whole page as relevant to the search query. This is a good way to deal with multi-topic pages, which are abundant on the Web. A recent work on this is block-based link analysis [78], which segments each Web page into different blocks. Each block is given a different importance value according to its location in the page and other information. The importance value is then used to weight the links in the HITS (and also PageRank) computation. This will reduce the impact of unimportant links, which usually cause topic drifting and may even be a link spam.

7.5 Community Discovery

Intuitively, a community is simply a group of entities (e.g., people or organizations) that shares a common interest or is involved in an activity or event. In Sect. 7.4.2, we showed that the HITS algorithm can be used to find communities. The communities are represented by dense bipartite subgraphs. We now describe several other community finding algorithms.

Apart from the Web, communities also exist in emails and text documents. This section describes two community finding algorithms for the Web, one community finding algorithm for emails, and one community finding algorithm for text documents.

There are many reasons for discovering communities. For example, in the context of the Web, Kumar et al. [293] listed three reasons:

1. Communities provide valuable and possibly the most reliable, timely, and up-to-date information resources for a user interested in them.
2. They represent the sociology of the Web: studying them gives insights into the evolution of the Web.
3. They enable target advertising at a very precise level.

7.5.1 Problem Definition

Definition (community): Given a finite set of **entities** $S = \{s_1, s_2, ..., s_n\}$ of the same **type**, a **community** is a pair $C = (T, G)$, where T is the **community theme** and $G \subseteq S$ is the set of all entities in S that shares the theme T. If $s_i \in G$, s_i is said to be a **member** of the community C.

Some remarks about this definition are in order:

- A theme defines a community. That is, given a theme T, the set of members of the community is uniquely determined. Thus, two communities are equal if they have the same theme.
- A theme can be defined arbitrarily. For example, it can be an event (e.g., a sport event or a scandal) or a concept (e.g., Web mining).
- An entity s_i in S can be in any number of communities. That is, communities may overlap, or multiple communities may share members.
- The entities in S are of the same type. For example, this definition does not allow people and organizations to be in the same community.
- By no means does this definition cover every aspect of communities in the real world. For example, it does not consider the temporal dimension of communities. Usually a community exists within a specific period of time. Similarly, an entity may belong to a community during some time periods.
- This is a conceptual definition. In practice, different community mining algorithms have their own operational definitions which usually depend on how communities manifest themselves in the given data (which we will discuss shortly). Furthermore, the algorithms may not be able to discover all the members of a community or its precise theme.

Communities may also have hierarchical structures.

Definition (sub-community, super-community, and sub-theme): A community (T, G) may have a set of **sub-communities** $\{(T_1, G_1), ..., (T_m, G_m)\}$, where T_i is a **sub-theme** of T and $G_i \subseteq G$. (T, G) is also called a **super-community** of (T_i, G_i). In the same way, each sub-community (T_i, G_i) can be further decomposed, which gives us a **community hierarchy**.

Community Manifestation in Data: Given a data set, which can be a set of Web pages, a collection of emails, or a set of text documents, we want to find communities of entities in the data. However, the data itself usually does not explicitly give us the themes or the entities (community members) associated with the themes. The system needs to discover the hidden community structures. Thus, the first issue that we need to know is how communities manifest themselves. From such manifested evidences, the system can discover possible communities. Different types of data may have different forms of manifestation. We give three examples.

Web Pages:

1. Hyperlinks: A group of content creators sharing a common interest is usually inter-connected through hyperlinks. That is, members in a community are more likely to be connected among themselves than outside the community.
2. Content words: Web pages of a community usually contain words that are related to the community theme.

Emails:

1. Email exchange between entities: Members of a community are more likely to communicate with one another.
2. Content words: Email contents of a community also contain words related to the theme of the community.

Text documents:

1. Co-occurrence of entities: Members of a community are more likely to appear together in the same sentence and/or the same document.
2. Content words: Words in sentences indicate the community theme.

Clearly, the key form of manifestation of a community is that its members are linked in some way. The associated text often contains words that are indicative of the community theme.

Objective of Community Discovery: Given a data set containing entities, we want to discover hidden communities of the entities. For each community, we want to find the theme and its members. The theme is usually represented with a set of keywords.

7.5.2 Bipartite Core Communities

HITS finds dense bipartite graph communities based on broad topic queries. The question is whether it is possible to find all such communities efficiently from the crawl of the whole Web without using eigenvector computation which is relatively inefficient. Kumar et al. [293] presented a
technique for finding bipartite cores, which are defined as follows.

Recall that the node set of a bipartite graph can be partitioned into two
subsets, which we denote as set F and set C. A **bipartite core** is a complete bipartite sub-graph with at least i nodes in F and at least j nodes in C.
A complete bipartite graph on node sets F and C contains all possible
edges between the vertices of F and the vertices of C. Note that edges
within F or within C are allowed here to suit the Web context, which deviate from the traditional definition of a complete bipartite graph. Intuitively,
the core is a small (i, j)-sized complete bipartite sub-graph of the community, which contains some core members of the community but not all.

The cores that we seek are *directed*, i.e., there is a set of i pages all of
which link to a set of j pages, while no assumption is made of links out of
the latter set of j pages. Intuitively, the former is the set of pages created
by members of the community, pointing to what they believe are the most
valuable pages for that community. For this reason we will refer to the i
pages that contain the links as **fans**, and the j pages that are referenced as
centers (as in community centers). Fans are like specialized *hubs*, and centers are like *authorities*. Figure 7.11 shows an example of a bipartite core.

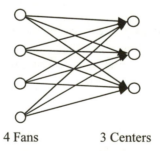

4 Fans   3 Centers

Fig. 7.11. A (4, 3) bipartite core

In Fig. 7.11, each fan page links to every center page. Since there are
four fans and three centers, this is called a (4, 3) bipartite core. Such a core
almost certainly represents a Web community, but a community may have
multiple bipartite cores.

Given a large number of pages crawled from the Web, which is represented as a graph, the procedure for finding bipartite cores consists of two
major steps: pruning and core generation.

Step 1: Pruning

We describe two types of pruning to remove those unqualified pages to be fans or centers. There are also other pruning methods given in [293].

1. **Pruning by in-degree:** we can delete all pages that are very highly referenced (linked) on the Web, such as homepages of Web portals (e.g., Yahoo!, AOL, etc). These pages are referenced for a variety of reasons, having little to do with any single emerging community, and they can be safely deleted. That is, we delete pages with the number of in-links great than k, which is determined empirically ($k = 50$ in [293]).
2. **Iterative pruning of fans and centers:** If we are interested in finding (i, j) cores, clearly any potential fan with an out-degree smaller than j can be pruned and the associated edges deleted from the graph. Similarly, any potential center with an in-degree smaller than i can be pruned and the corresponding edges deleted from the graph. This process can be done iteratively: when a fan gets pruned, some of the centers that it points to may have their in-degrees fall below the threshold i and qualify for pruning as a result. Similarly, when a center gets pruned, a fan that points to it could have its out-degree fall below its threshold of j and qualify for pruning.

Step 2: Generating all (i, j) Cores

After pruning, the remaining pages are used to discover cores. The method works as follows: Fixing j, we start with all $(1, j)$ cores. This is simply the set of all vertices with out-degree at least j. We then construct all $(2, j)$ cores by checking every fan which also points to any center in a $(1, j)$ core. All $(3, j)$ cores can be found in the same fashion by checking every fan which points to any center in a $(2, j)$ core, and so on. The idea is similar to the Apriori algorithm for association rule mining (see Chap. 2) as every proper subset of the fans in any (i, j) core forms a core of smaller size.

Based on the algorithm, Kumar et al. found a large number of topic coherent cores from a crawl of the Web [293]. We note that this algorithm only finds the core pages of the communities, not all members (pages). It also does not find the themes of the communities or their hierarchical organizations.

7.5.3 Maximum Flow Communities

Bipartite cores are usually very small and do not represent full communities. In this section, we define and find maximum flow communities based on the work of Flake et al. [180]. The algorithm requires the user to give a set of seed pages, which are examples of the community that the user wishes to find.

Given a Web link graph $G = (V, E)$, a maximum flow community is defined as a collection $C \subset V$ of Web pages such that each member page $u \in C$ has more hyperlinks (in either direction) within the community C than outside of the community $V\text{-}C$. Identifying such a community is intractable in the general case because it can be mapped into a family of NP-complete graph partition problems. Thus, we need to approximate and recast it into a framework with less stringent conditions based on the network flow model from operations research, specifically the maximum flow model.

The maximum flow model can be stated as follows: We are given a graph $G = (V, E)$, where each edge (u, v) is thought of as having a positive capacity $c(u, v)$ that limits the quantity of a product that may be shipped through the edge. In such a situation, it is often desirable to have the maximum amount of flow from a starting point s (called the **source**) and a terminal point t (called the **sink**). Intuitively, the maximum flow of the graph is determined by the bottleneck edges. For example, given the graph in Fig. 7.12 with the source s and the sink t, if every edge has the unit capacity, the bottleneck edges are W-X and Y-Z.

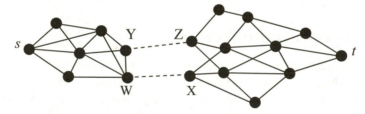

Fig. 7.12. A simple flow network.

The **Max Flow-Min Cut** theorem of Ford and Fulkerson [181] proves that the maximum flow of a network is identical to the minimum cut that separates s and t. Many polynomial time algorithms exist for solving the $s\text{-}t$ maximum flow problem. If Fig. 7.12 is a Web link graph, it is natural to cut the edges W-X and Y-Z to produce two Web communities.

The basic idea of the approach in [180] is as follows: It starts with a set S of *seed* pages, which are example pages of the community that the user wishes to find. The system then crawls the Web to find more pages using the seed pages. A maximum flow algorithm is then applied to separate the community C involving the seed pages and the other pages. These steps may need to be repeated in order to find the desired community. Figure 7.13 gives the algorithm.

Algorithm Find-Community (*S*)
    while number of iteration is less than desired **do**
        build $G = (V, E)$ by doing a fixed depth crawl starting from *S*;
        $k = |S|$;
        $C = $ Max-Flow-Community(G, S, k);
        rank all $v \in C$ by the number of edges in *C*;
        add the highest ranked non-seed vertices to *S*
    end-while
    return all $v \in V$ still connected to the source *s*

Procedure Max-Flow-Community(G, S, k)
    create artificial vertices, *s* and *t* and add to *V*;   // *V* is the vertex set of *G*.
    for all $v \in S$ **do**
        add (s, v) to *E* with $c(s, v) = \infty$ // *E* is the edge set of *G*.
    endfor
    for all $(u, v) \in E, u \neq s$ **do**
        $c(u, v) = k$;
        if $(v, u) \notin E$ **then**
            add (v, u) to *E* with $c(v, u) = k$
        endif
    endfor
    for all $v \in V, v \notin S \cup \{s, t\}$ **do**
        add (v, t) to *E* with $c(v, t) = 1$
    endfor
    Max-Flow(G, s, t);
    return all $v \in V$ still connected to *s*.

Fig. 7.13. The algorithm for mining maximum flow communities

The algorithm Find-Community is the control program. It takes a set *S* of seed Web pages as input, and crawls to a fixed depth including in-links as well as out-links (with in-links found by querying a search engine). It then applies the procedure Max-Flow-Community to the induced graph *G* from the crawl. After a community *C* is found, it ranks the pages in the community by the number of edges that each has inside of the community. Some highest ranked non-seed pages are added to the seed set. This is to create a big seed set for the next iteration in order to crawl more pages. The algorithm then iterates the procedure. Note that the first iteration may only identify a very small community. However, when new seeds are added, increasingly larger communities are identified. Heuristics are used to decide when to stop.

The procedure Max-Flow-Community finds the actual community from *G*. Since a Web graph has no source and sink, it first augments the web graph by adding an artificial source, *s*, with infinite capacity edges routed to all seed vertices in *S*; making each pre-existing edge bidirectional and assigning each edge a constant capacity *k*. It then adds an artificial sink *t* and routes all vertices except the source, the sink, and the seed vertices to *t* with unit capacity. After augmenting the web graph, a residual flow graph is produced by a maximum flow procedure (Max-Flow()). All vertices accessible from *s* through non-zero positive edges form the desired result. The value *k* is heuristically chosen to be the size of the set *S* to ensure that after the artificial source and sink are added to the original graph, the same cuts will be produced as the original graph (see the proof in [179]). Figure 7.14 shows the community finding process.

Finally, we note that this algorithm does not find the theme of the community or the community hierarchy (i.e., sub-communities and so on).

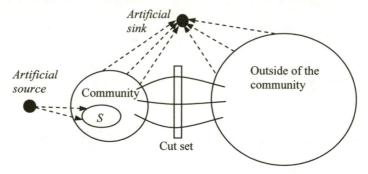

Fig. 7.14. Schematic representation of the community finding process

7.5.4 Email Communities Based on Betweenness

Email has become the predominant means of communication in the information age. It has been established as an indicator of collaboration and knowledge (or information) exchange. Email exchanges provide plenty of data on personal communication for the discovery of shared interests and relationships between people, which were hard to discover previously.

It is fairly straightforward to construct a graph based on email data. People are the vertices and the edges are added between people who corresponded through email. Usually, the edge between two people is added if a minimum number of messages passed between them. The minimum number is controlled by a threshold, which can be tuned.

To analyze an email graph or network, one can make use of all the centrality measures and prestige measures discussed in Sect. 7.1. We now focus on community finding only.

We are interested in people communities, which are subsets of vertices that are related. One way to identify communities is by partitioning the graph into discrete clusters such that there are few edges lying between the clusters. This definition is similar to that of the maximum flow community. **Betweenness** in social networks is a natural measure for identifying those edges in between clusters or communities [523]. The idea is that inter-community links, which are few, have high betweenness values, while the intra-community edges have low betweenness values. However, the betweenness discussed in Sect. 7.1 is evaluated on each person in the network. Here, we need to evaluate the betweenness of each edge. The idea is basically the same and Equation (4) can be used here without normalization because we only find communities in a single graph. The betweenness of an edge is simply the number of shortest paths that pass it.

If the graph is not connected, we identify communities from each connected component. Given a connected graph, the method works iteratively in two steps (Fig. 7.15):

repeat
    Compute the betweenness of each edge in the remaining graph;
    Remove the edge with the highest betweenness
until the graph is suitably partitioned.

Fig. 7.15. Community finding using the betweenness measure.

Since the removal of an edge can strongly affect the betweenness of many other edges, we need to repeatedly re-compute the betweenness of all edges. The idea of the method is very similar to the minimum-cut method discussed in Sect. 7.5.3.

The stopping criteria can be designed according to applications. In general, we consider that the smallest community is a triangle. The algorithm should stop producing more unconnected components if there is no way to generate triangle communities. A component of five or fewer vertices cannot consist of two viable communities. The smallest such component is six, which has two triangles connected by one edge, see Fig. 7.16. If any discovered community does not have a triangle, it may not be considered as a community. Clearly, other stopping criteria can be used.

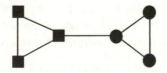

Fig. 7.16. The smallest possible graph of two viable communities.

7.5.5 Overlapping Communities of Named Entities

Most community discovery algorithms are based on graph partitioning, which means that an entity can belong to only a single community. However, in real life, a person can be in multiple communities (see the definition in Sect. 7.5.1). For example, he/she can be in the community of his/her family, the community of his/her colleagues and the community of his/her friends. A heuristic technique is presented in [325] for finding overlapping communities of entities in text documents.

In the Web or email context, there are explicit links connecting entities and forming communities. In free text documents, no explicit links exist. Then the question is: what constitutes a link between two entities in text documents? As we indicated earlier, one simple technique is to regard two entities as being linked if they co-occur in the same sentence. This method is reasonable because if two people are mentioned in a sentence there is usually a relationship between them.

The objective is to find entity communities from a text corpus, which could be a set of given documents or the returned pages from a search engine using a given entity as the search query. An entity here refers to the name of a person or an organization.

The algorithm in [325] consists of four steps:

1. Building a link graph: The algorithm first parses each document. For each sentence, it identifies named entities contained in the sentence. If a sentence has more than one named entities, these entities are pair-wise linked. The keywords in the sentence are attached to the linked pairs to form their textual contents. All the other sentences are discarded.
2. Finding all triangles: The algorithm then finds all triangles, which are the basic building blocks of communities. A triangle consists of three entities bound together. The reason for using triangles is that it has been observed by researchers that a community expands predominantly by triangles sharing a common edge.
3. Finding community cores: It next finds community cores. A community core is a group of tightly bound triangles, which are relaxed complete sub-graphs (or cliques). Intuitively, a core consists of a set of tightly connected members of a community.
4. Clustering around community cores: For those triangles and also entity pairs that are not in any core, they are assigned to cores according to their textual content similarities with the discovered cores.

It is clear that in this algorithm a single entity can appear in multiple communities because an entity can appear in multiple triangles. To finish off, the algorithm also ranks the entities in each community according to degree centrality. Keywords associated with the edges of each community are also ranked. The top keywords are assumed to represent the theme of the community. The technique has been applied to find communities of political figures and celebrities from Web documents with promising results.

Bibliographic Notes

Social network analysis has a relative long history. A large number of interesting problems and algorithms were studied in the past 60 years. The book by Wasserman and Faust [540] is an authoritative text of the field. Co-citation [494] and bibliographic coupling [275] are from bibliometrics, which is a type of research method used in library and information science. The book edited by Borgman [58] is a good source of information on both the research and applications of bibliometrics.

The use of social network analysis in the Web context (also called link analysis) started with the PageRank algorithm proposed by Brin and Page [68] and Page et al. [422], and the HITS algorithm proposed by Kleinberg [281]. PageRank is also the algorithm that powers the Google search engine. Due to several weaknesses of HITS, many researchers have tried to improve it. Various enhancements were reported by Lempel and Moran [310], Bharat and Henzinger [52], Chakrabarti et al. [88], Cai et al. [78], etc. The book by Langville and Meyer [304] contains in-depth analyses of PageRank, HITS and many enhancements to HITS. Other works related to Web link analysis include those in [98, 226, 266, 368] on improving the PageRank computation, in [168] on searching workspace Web, in [103, 182, 183, 416] on the evolution of the Web and the search engine influence on the Web, in [140, 142, 410, 516] on other link based models, in [34, 440, 370, 371] on Web graph and its characteristics, in [37, 51, 235] on sampling of Web pages, and in [32, 425, 585] on the temporal dimension of Web search.

On community discovery, HITS can find some communities by computing non-principal eigenvectors [198, 281]. Kumar et al. [293] proposed the algorithm for finding bipartite cores. Flake et al. [179] introduced the maximum flow community mining. Ino et al. [249] presented a more strict definition of communities. Tyler et al. [523] gave the method for finding email communities based on betweenness. The algorithm for finding overlapping communities of named entities from texts was given by Li et al. [325]. More recent developments on communities and social networks on the Web can be found in [16, 21, 137, 158, 200, 518, 519, 561, 618].

8 Web Crawling

By *Filippo Menczer*

Web **crawlers**, also known as **spiders** or **robots**, are programs that automatically download Web pages. Since information on the Web is scattered among billions of pages served by millions of servers around the globe, users who browse the Web can follow hyperlinks to access information, virtually moving from one page to the next. A crawler can visit many sites to collect information that can be analyzed and mined in a central location, either online (as it is downloaded) or off-line (after it is stored).

Were the Web a static collection of pages, we would have little long term use for crawling. Once all the pages are fetched and saved in a repository, we are done. However, the Web is a dynamic entity evolving at rapid rates. Hence there is a continuous need for crawlers to help applications stay current as pages and links are added, deleted, moved or modified.

There are many applications for Web crawlers. One is business intelligence, whereby organizations collect information about their competitors and potential collaborators. Another use is to monitor Web sites and pages of interest, so that a user or community can be notified when new information appears in certain places. There are also malicious applications of crawlers, for example, that harvest email addresses to be used by spammers or collect personal information to be used in phishing and other identity theft attacks. The most widespread use of crawlers is, however, in support of search engines. In fact, crawlers are the main consumers of Internet bandwidth. They collect pages for search engines to build their indexes. Well known search engines such as Google, *Yahoo!* and MSN run very efficient **universal** crawlers designed to gather all pages irrespective of their content. Other crawlers, sometimes called **preferential crawlers**, are more targeted. They attempt to download only pages of certain types or topics.

This chapter introduces the main concepts, algorithms and data structures behind Web crawlers. After discussing the implementation issues that all crawlers have to address, we describe different types of crawlers: **universal**, **focused**, and **topical**. We also discuss some of the **ethical issues** around crawlers. Finally, we peek at possible future uses of crawlers in support of alternative models where crawling and searching activities are distributed among a large community of users connected by a dynamic and adaptive peer network.

8.1 A Basic Crawler Algorithm

In its simplest form, a crawler starts from a set of **seed** pages (URLs) and then uses the links within them to fetch other pages. The links in these pages are, in turn, extracted and the corresponding pages are visited. The process repeats until a sufficient number of pages are visited or some other objective is achieved. This simple description hides many delicate issues related to network connections, spider traps, URL canonicalization, page parsing, and crawling ethics. In fact, Google founders Sergey Brin and Lawrence Page, in their seminal paper, identified the Web crawler as the most sophisticated yet fragile component of a search engine [68].

Figure 8.1 shows the flow of a basic **sequential** crawler. Such a crawler fetches one page at a time, making inefficient use of its resources. Later in the chapter we discuss how efficiency can be improved by the use of multiple processes, threads, and asynchronous access to resources. The crawler maintains a list of unvisited URLs called the **frontier**. The list is initialized with seed URLs which may be provided by the user or another program. In each iteration of its main loop, the crawler picks the next URL from the frontier, fetches the page corresponding to the URL through HTTP, parses the retrieved page to extract its URLs, adds newly discovered URLs to the frontier, and stores the page (or other extracted information, possibly index terms) in a local disk repository. The crawling process may be terminated when a certain number of pages have been crawled. The crawler may also be forced to stop if the frontier becomes empty, although this rarely happens in practice due to the high average number of links (on the order of ten out-links per page across the Web).

A crawler is, in essence, a graph search algorithm. The Web can be seen as a large graph with pages as its nodes and hyperlinks as its edges. A crawler starts from a few of the nodes (seeds) and then follows the edges to reach other nodes. The process of fetching a page and extracting the links within it is analogous to expanding a node in graph search.

The frontier is the main data structure, which contains the URLs of unvisited pages. Typical crawlers attempt to store the frontier in the main memory for efficiency. Based on the declining price of memory and the spread of 64-bit processors, quite a large frontier size is feasible. Yet the crawler designer must decide which URLs have low priority and thus get discarded when the frontier is filled up. Note that given some maximum size, the frontier will fill up quickly due to the high fan-out of pages. Even more importantly, the crawler algorithm must specify the order in which new URLs are extracted from the frontier to be visited. These mechanisms determine the graph search algorithm implemented by the crawler.

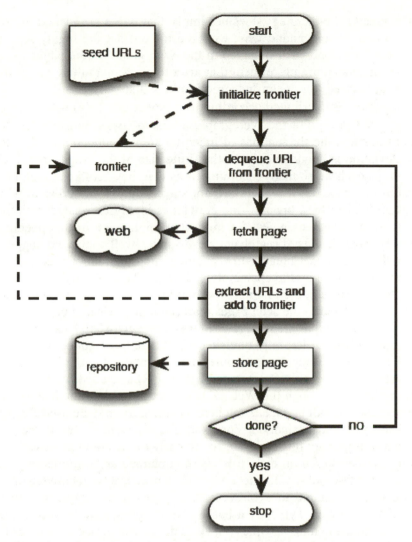

Fig. 8.1. Flow chart of a basic sequential crawler. The main data operations are shown on the left, with dashed arrows.

8.1.1 Breadth-First Crawlers

The frontier may be implemented as a first-in-first-out (FIFO) queue, corresponding to a **breadth-first** crawler. The URL to crawl next comes from the head of the queue and new URLs are added to the tail of the queue. Once the frontier reaches its maximum size, the breadth-first crawler can add to the queue only one unvisited URL from each new page crawled.

The breadth-first strategy does not imply that pages are visited in "random" order. To understand why, we have to consider the highly skewed, long-tailed distribution of indegree in the Web graph. Some pages have a number of links pointing to them that are orders of magnitude larger than the mean. Indeed, the mean indegree is not statistically significant when the indegree k is distributed according to a power law $\Pr(k) \sim k^{-\gamma}$ with exponent $\gamma < 3$ [437]. For the Web graph, this is the case, with $\gamma \approx 2.1$ [69]. This means that the fluctuations of indegree are unbounded, i.e., the standard deviation is bounded only by the finite size of the graph. Intuitively, popular pages have so many incoming links that they act like attractors for breadth-first crawlers. It is therefore not surprising that the order in which pages are visited by a breadth-first crawler is highly correlated with their PageRank or indegree values. An important implication of this phenomenon is an intrinsic **bias** of search engines to index well connected pages.

Another reason that breadth-first crawlers are not "random" is that they are greatly affected by the choice of seed pages. Topical locality measures indicate that pages in the link neighborhood of a seed page are much more likely to be related to the seed pages than randomly selected pages. These and other types of bias are important to **universal crawlers** (Sect. 8.3).

As mentioned earlier, only unvisited URLs are to be added to the frontier. This requires some data structure to be maintained with visited URLs. The **crawl history** is a time-stamped list of URLs fetched by the crawler tracking its path through the Web. A URL is entered into the history only after the corresponding page is fetched. This history may be used for post-crawl analysis and evaluation. For example, we want to see if the most relevant or important resources are found early in the crawl process. While history may be stored on disk, it is also maintained as an in-memory data structure for fast look-up, to check whether a page has been crawled or not. This check is required to avoid revisiting pages or wasting space in the limited-size frontier. Typically a hash table is appropriate to obtain quick URL insertion and look-up times ($O(1)$). The look-up process assumes that one can identify two URLs effectively pointing to the same page. This introduces the need for canonical URLs (see Sect. 8.2).

Another important detail is the need to prevent duplicate URLs from being added to the frontier. A separate hash table can be maintained to store the frontier URLs for fast look-up to check whether a URL is already in it.

8.1.2 Preferential Crawlers

A different crawling strategy is obtained if the frontier is implemented as a **priority queue** rather than a FIFO queue. Typically, **preferential crawl-**

ers assign each unvisited link a priority based on an estimate of the value of the linked page. The estimate can be based on topological properties (e.g., the indegree of the target page), content properties (e.g., the similarity between a user query and the source page), or any other combination of measurable features. For example, the goal of a topical crawler is to follow edges that are expected to lead to portions of the Web graph that are relevant to a user-selected topic. The choice of seeds is even more important in this case than for breadth-first crawlers. We will discuss various preferential crawling algorithms in Sects. 8.4 and 8.5. For now let us simply assume that some function exists to assign a priority value or score to each unvisited URL. If pages are visited in the order specified by the priority values in the frontier, then we have a **best-first** crawler.

The priority queue may be a dynamic array that is always kept sorted by URL scores. At each step, the best URL is picked from the head of the queue. Once the corresponding page is fetched, the URLs extracted from it must, in turn, be scored. They are then added to the frontier in such a manner that the sorting order of the priority queue is maintained. As for breadth-first, best-first crawlers also need to avoid duplicate URLs in the frontier. Keeping a separate hash table for look-up is an efficient way to achieve this. The time complexity of inserting a URL into the priority queue is $O(\log F)$, where F is the frontier size (looking up the hash requires constant time). To dequeue a URL, it must first be removed from the priority queue ($O(\log F)$) and then from the hash table (again $O(1)$). Thus the parallel use of the two data structures yields a logarithmic total cost per URL. Once the frontier's maximum size is reached, only the best URLs are kept; the frontier must be pruned after each new set of links is added.

8.2   Implementation Issues

8.2.1   Fetching

To fetch pages, a crawler acts as a Web client; it sends an HTTP request to the server hosting the page and reads the response. The client needs to timeout connections to prevent spending unnecessary time waiting for responses from slow servers or reading huge pages. In fact, it is typical to restrict downloads to only the first 10-100 KB of data for each page. The client parses the response headers for status codes and redirections. Redirect loops are to be detected and broken by storing URLs from a redirection chain in a hash table and halting if the same URL is encountered twice. One may also parse and store the last-modified header to determine the age of the document, although this information is known to be unreliable. Error-checking and exception handling is important during the page fetching process since the same code must deal with potentially millions of remote servers. In addition, it may be beneficial to collect statistics on timeouts and status codes to identify problems or automatically adjust timeout values. Programming languages such as Java, Python and Perl provide simple programmatic interfaces for fetching pages from the Web. However, one must be careful in using high-level interfaces where it may be harder to detect lower-level problems. For example, a robust crawler in Perl should use the Socket module to send HTTP requests rather than the higher-level LWP library (the World-Wide Web library for Perl). The latter does not allow fine control of connection timeouts.

8.2.2 Parsing

Once (or while) a page is downloaded, the crawler parses its content, i.e., the HTTP payload, and extracts information both to support the crawler's master application (e.g., indexing the page if the crawler supports a search engine) and to allow the crawler to keep running (extracting links to be added to the frontier). Parsing may imply simple URL extraction from hyperlinks, or more involved analysis of the HTML code. The Document Object Model (DOM) establishes the structure of an HTML page as a tag tree, as illustrated in Fig. 8.2. HTML parsers build the tree in a depth-first manner, as the HTML source code of a page is scanned linearly.

Unlike program code, which must compile correctly or else will fail with a syntax error, correctness of HTML code tends to be laxly enforced by browsers. Even when HTML standards call for strict interpretation, de facto standards imposed by browser implementations are very forgiving. This, together with the huge population of non-expert authors generating Web pages, imposes significant complexity on a crawler's HTML parser. Many pages are published with missing required tags, tags improperly nested, missing close tags, misspelled or missing attribute names and values, missing quotes around attribute values, unescaped special characters, and so on. As an example, the double quotes character in HTML is reserved for tag syntax and thus is forbidden in text. The special HTML entity " is to be used in its place. However, only a small number of authors are aware of this, and a large fraction of Web pages contains this illegal character. Just like browsers, crawlers must be forgiving in these cases; they cannot afford to discard many important pages as a strict parser would do. A wise preprocessing step taken by robust crawlers is to apply a tool such as **tidy** (www.w3.org/People/Raggett/tidy) to clean up the HTML content prior to parsing. To add to the complexity, there are many coexisting HTML and XHTML reference versions. However, if the crawler only needs to extract links within a page and/or the text in the page, simpler parsers may suffice. The HTML parsers available in high-level languages such as Java and Perl are becoming increasingly sophisticated and robust.

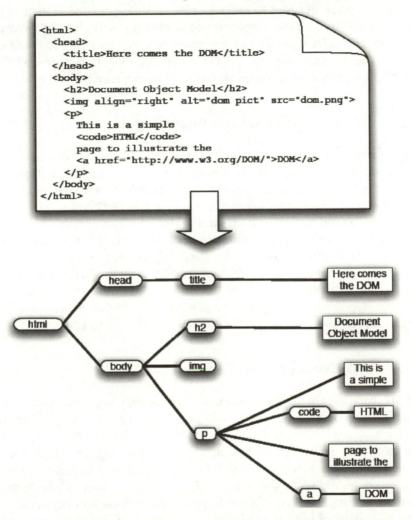

```
<html>
  <head>
    <title>Here comes the DOM</title>
  </head>
  <body>
    <h2>Document Object Model</h2>
    <img align="right" alt="dom pict" src="dom.png">
    <p>
      This is a simple
      <code>HTML</code>
      page to illustrate the
      <a href="http://www.w3.org/DOM/">DOM</a>
    </p>
  </body>
</html>
```

Fig. 8.2. Illustration of the DOM (or tag) tree built from a simple HTML page. Internal nodes (shown as ovals) represent HTML tags, with the <html> tag as the root. Leaf nodes (shown as rectangles) correspond to text chunks.

A growing portion of Web pages are written in formats other than HTML. Crawlers supporting large-scale search engines routinely parse and index documents in many open and proprietary formats such as plain text, PDF, Microsoft Word and Microsoft PowerPoint. Depending on the application of the crawler, this may or may not be required. Some formats present particular difficulties as they are written exclusively for human interaction and thus are especially unfriendly to crawlers. For instance, some commercial sites use graphic animations in Flash; these are difficult for a crawler to parse in order to extract links and their textual content. Other examples include image maps and pages making heavy use of Javascript for interaction. New challenges are going to come as new standards such as Scalable Vector Graphics (SVG), Asynchronous Javascript and XML (AJAX), and other XML-based languages gain popularity.

8.2.3   Stopword Removal and Stemming

When parsing a Web page to extract the content or to score new URLs suggested by the page, it is often helpful to remove so-called **stopwords**, i.e., terms such as articles and conjunctions, which are so common that they hinder the discrimination of pages on the basis of content.

Another useful technique is **stemming**, by which morphological variants of terms are conflated into common roots (stems). In a topical crawler where a link is scored based on the similarity between its source page and the query, stemming both the page and the query helps improve the matches between the two sets and the accuracy of the scoring function.

Both stop-word removal and stemming are standard techniques in information retrieval, and are discussed in greater detail in Chap. 6.

8.2.4   Link Extraction and Canonicalization

HTML parsers provide the functionality to identify tags and associated attribute-value pairs in a given Web page. In order to extract hyperlink URLs from a page, we can use a parser to find anchor (<a>) tags and grab the values of the associated href attributes. However, the URLs thus obtained need to be further processed. First, filtering may be necessary to exclude certain file types that are not to be crawled. This can be achieved with white lists (e.g., only follow links to text/html content pages) or black lists (e.g., discard links to PDF files). The identification of a file type may rely on file extensions. However, they are often unreliable and sometimes missing altogether. We cannot afford to download a document and then decide whether we want it or not. A compromise is to send an HTTP HEAD request and inspect the content-type response header, which is usually a more reliable label.

Another type of filtering has to do with the static or dynamic nature of pages. A dynamic page (e.g., generated by a CGI script) may indicate a query interface for a database or some other application in which a crawler may not be interested. In the early days of the Web, such pages were few and easily recognizable, e.g., by matching URLs against the /cgi-bin/ directory name for CGI scripts, or against the special characters [?=&] used in CGI query strings. However, the use of dynamic content has become much more common; it is used in a variety of sites for content that is perfectly indexable. Most importantly, its dynamic nature is very difficult to recognize via URL inspection. For these reasons, most crawlers no longer make such distinction between static and dynamic content. While a crawler normally would not create query URLs autonomously (unless it is designed to probe the so-called **deep** or **hidden Web,** which contain databases with query interfaces), it will happily crawl URLs hard-coded in HTML source of parsed pages. In other words, if a URL is found in a Web page, it is fair game. There is one important exception to this strategy, the spider trap, which is discussed below.

Before links can be added to the frontier, **relative URLs** must be converted to **absolute URLs**. For example, the relative URL news/today.html in the page http://www.somehost.com/index.html is to be transformed into the absolute form http://www.somehost.com/news/today.html. There are various rules to convert relative URLs into absolute ones. A relative URL can be expressed as a relative or absolute path relative to the Web server's document root directory. The **base URL** may be specified by an HTTP header or a meta-tag within an HTML page, or not at all–in the latter case the directory of the hyperlink's source page is used as a base URL.

Converting relative URLs is just one of many steps that make up the **canonicalization** process, i.e., the conversion of a URL into a **canonical form**. The definition of canonical form is somewhat arbitrary, so that different crawlers may specify different rules. For example, one crawler may always specify the port number within the URL (e.g., http://www.somehost.com:80/), while another may specify the port number only when it is not 80 (the default HTTP port). As long as the canonical form is applied consistently by a crawler, such distinctions are inconsequential. Some programming languages such as Perl provide modules to manage URLs, including methods for absolute/relative conversion and canonicalization. However, several canonicalization steps require the application of heuristic rules, and off-the-shelf tools typically do not provide such functionalities. A crawler may also need to use heuristics to detect when two URLs point to the same page in order to minimize the likelihood that the same page is fetched multiple times. Table 8.1 lists the steps typically employed to canonicalize a URL.

Table 8.1. Some transformations to convert URLs to canonical forms. Stars indicate heuristic rules, where there is a tradeoff between the risk of altering the semantics of the URL (if a wrong guess is made) and the risk of missing duplicate URLs (if no transformation is applied) for the same target

Description and transformation	Example and canonical form
Default port number Remove	http://cs.indiana.edu:80/ http://cs.indiana.edu/
Root directory Add trailing slash	http://cs.indiana.edu http://cs.indiana.edu/
Guessed directory* Add trailing slash	http://cs.indiana.edu/People http://cs.indiana.edu/People/
Fragment Remove	http://cs.indiana.edu/faq.html#3 http://cs.indiana.edu/faq.html
Current or parent directory Resolve path	http://cs.indiana.edu/a/./../b/ http://cs.indiana.edu/b/
Default filename* Remove	http://cs.indiana.edu/index.html http://cs.indiana.edu/
Needlessly encoded characters Decode	http://cs.indiana.edu/%7Efil/ http://cs.indiana.edu/~fil/
Disallowed characters Encode	http://cs.indiana.edu/My File.htm http://cs.indiana.edu/My%20File.htm
Mixed/upper-case host names Lower-case	http://CS.INDIANA.EDU/People/ http://cs.indiana.edu/People/

8.2.5 Spider Traps

A crawler must be aware of **spider traps**. These are Web sites where the URLs of dynamically created links are modified based on the sequence of actions taken by the browsing user (or crawler). Some e-commerce sites such as Amazon.com may use URLs to encode which sequence of products each user views. This way, each time a user clicks a link, the server can log detailed information on the user's shopping behavior for later analysis. As an illustration, consider a dynamic page for product x, whose URL path is /x and that contains a link to product y. The URL path for this link would be /x/y to indicate that the user is going from page x to page y. Now suppose the page for y has a link back to product x. The dynamically created URL path for this link would be /x/y/x, so that the crawler would think this is a new page when in fact it is an already visited page with a new URL. As a side effect of a spider trap, the server may create an entry in a database every time the user (or crawler) clicks on certain dynamic links. An example might be a blog or message board where users can post comments. These situations create sites that appear infinite to a crawler, because the more links are followed, the more new URLs are created. However these new "dummy" links do not lead to existing or new content, but simply to dynamically created form pages, or to pages that have already been visited. Thus a crawler could go on crawling inside the spider trap forever without actually fetching any new content.

In practice spider traps are not only harmful to the crawler, which wastes bandwidth and disk space to download and store duplicate or useless data. They may be equally harmful to the server sites. Not only does the server waste its bandwidth, the side effect of a crawler caught in a spider trap may also be filling a server-side database with bogus entries. The database may eventually become filled to capacity, and the site may be disabled as a result. This is a type of **denial of service** attack carried out unwittingly by the crawler.

In some cases a spider trap needs the client to send a cookie set by the server for the dynamic URLs to be generated. So the problem is prevented if the crawler avoids accepting or sending any cookies. However, in most cases a more proactive approach is necessary to defend a crawler against spider traps. Since the dummy URLs often become larger and larger in size as the crawler becomes entangled in a spider trap, one common heuristic approach to tackle such traps is by limiting the URL sizes to some maximum number of characters, say 256. If a longer URL is encountered, the crawler should simply ignore it. Another way is by limiting the number of pages that the crawler requests from a given domain. The code associated with the frontier can make sure that every consecutive sequence of, say, 100 URLs fetched by the crawler contains at most one URL from each fully qualified host name. This approach is also germane to the issue of crawler etiquette, discussed later.

8.2.6   Page Repository

Once a page is fetched, it may be stored/indexed for the master application (e.g., a search engine). In its simplest form a **page repository** may store the crawled pages as separate files. In this case each page must map to a unique file name. One way to do this is to map each page's URL to a compact string using some hashing function with low probability of collisions, e.g., MD5. The resulting hash value is used as a (hopefully) unique file name. The shortcoming of this approach is that a large scale crawler would incur significant time and disk space overhead from the operating system to manage a very large number of small individual files.

A more efficient solution is to combine many pages into one file. A naïve approach is to simply concatenate some number of pages (say 1,000) into each file, with some special markup to separate and identify the pages within the file. This requires a separate look-up table to map URLs to file names and IDs within each file. A better method is to use a database to store the pages, indexed by (canonical) URLs. Since traditional RDBMSs impose high overhead, embedded databases such as the open-source Berkeley DB are typically preferred for fast access. Many high-level languages such as Java and Perl provide simple APIs to manage Berkeley DB files, for example as tied associative arrays. This way the storage management operations become nearly transparent to the crawler code, which can treat the page repository as an in-memory data structure.

8.2.7 Concurrency

A crawler consumes three main resources: network, CPU, and disk. Each is a bottleneck with limits imposed by bandwidth, CPU speed, and disk seek/transfer times. The simple sequential crawler described in Sect. 8.1 makes a very inefficient use of these resources because at any given time two of them are idle while the crawler attends to the third.

The most straightforward way to speed-up a crawler is through concurrent processes or threads. Multiprocessing may be somewhat easier to implement than multithreading depending on the programming language and platform, but it may also incur a higher overhead due to the involvement of the operating system in the management (creation and destruction) of child processes. Whether threads or processes are used, a concurrent crawler may follow a standard parallel computing model [292] as illustrated in Fig. 8.3. Basically each thread or process works as an independent crawler, except for the fact that access to the shared data structures (mainly the frontier, and possibly the page repository) must be synchronized. In particular a frontier manager is responsible for locking and unlocking the frontier data structures so that only one process or thread can write to them at one time. Note that both enqueueing and dequeuing are write operations. Additionally, the frontier manager would maintain and synchronize access to other shared data structures such as the crawl history for fast look-up of visited URLs.

It is a bit more complicated for a concurrent crawler to deal with an empty frontier than for a sequential crawler. An empty frontier no longer implies that the crawler has reached a dead-end, because other processes may be fetching pages and adding new URLs in the near future. The process or thread manager may deal with such a situation by sending a temporary sleep signal to processes that report an empty frontier. The process manager needs to keep track of the number of sleeping processes; when all the processes are asleep, the crawler must halt.

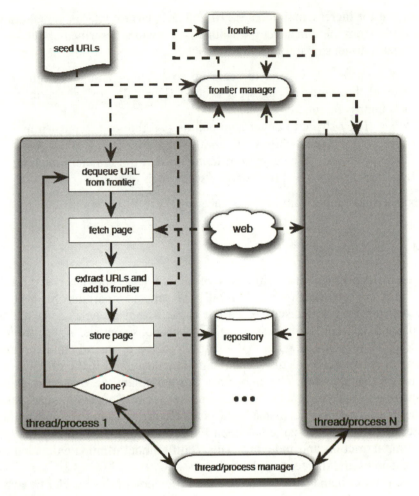

Fig. 8.3. Architecture of a concurrent crawler

The concurrent design can easily speed-up a crawler by a factor of 5 or 10. The concurrent architecture however does not scale up to the performance needs of a commercial search engine. We discuss in Sect. 8.3 further steps that can be taken to achieve more scalable crawlers.

8.3   Universal Crawlers

General purpose search engines use Web crawlers to maintain their indices [25], amortizing the cost of crawling and indexing over the millions of queries received between successive index updates (though indexers are designed for incremental updates [101]. These large-scale universal crawlers differ from the concurrent breadth-first crawlers described above along two major dimensions:

1. *Performance*: They need to scale up to fetching and processing hundreds of thousands of pages per second. This calls for several architectural improvements.
2. *Policy*: They strive to cover as much as possible of the most important pages on the Web, while maintaining their index as fresh as possible. These goals are, of course, conflicting so that the crawlers must be designed to achieve good tradeoffs between their objectives.

Next we discuss the main issues in meeting these requirements.

8.3.1   Scalability

Figure 8.4 illustrates the architecture of a large-scale crawler, based on the accounts in the literature [68, 85, 238]. The most important change from the concurrent model discussed earlier is the use of asynchronous sockets in place of threads or processes with synchronous sockets. Asynchronous sockets are non-blocking, so that a single process or thread can keep hundreds of network connections open simultaneously and make efficient use of network bandwidth. Not only does this eliminate the overhead due to managing threads or processes, it also makes locking access to shared data structures unnecessary. Instead, the sockets are polled to monitor their states. When an entire page has been fetched into memory, it is processed for link extraction and indexing. This "pull" model eliminates contention for resources and the need for locks.

The frontier manager can improve the efficiency of the crawler by maintaining several parallel queues, where the URLs in each queue refer to a single server. In addition to spreading the load across many servers within any short time interval, this approach allows to keep connections with servers alive over many page requests, thus minimizing the overhead of TCP opening and closing handshakes.

The crawler needs to resolve host names in URLs to IP addresses. The connections to the Domain Name System (DNS) servers for this purpose are one of the major bottlenecks of a naïve crawler, which opens a new TCP connection to the DNS server for each URL. To address this bottleneck, the crawler can take several steps. First, it can use UDP instead of TCP as the transport protocol for DNS requests. While UDP does not guarantee delivery of packets and a request can occasionally be dropped, this is rare. On the other hand, UDP incurs no connection overhead with a

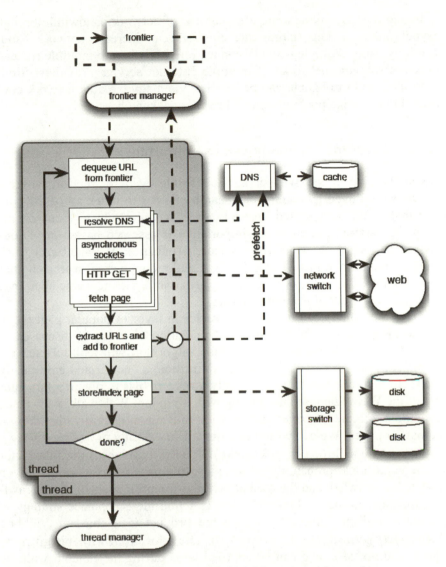

Fig. 8.4. High-level architecture of a scalable universal crawler

significant speed-up over TCP. Second, the DNS server should employ a large, persistent, and fast (in-memory) cache. Finally, the pre-fetching of DNS requests can be carried out when links are extracted from a page. In addition to being added to the frontier, the URLs can be scanned for host names to be sent to the DNS server. This way, when a URL is later ready to be fetched, the host IP address is likely to be found in the DNS cache, obviating the need to propagate the request through the DNS tree.

In addition to making more efficient use of network bandwidth through asynchronous sockets, large-scale crawlers can increase network bandwidth by using multiple network connections switched to multiple routers, thus utilizing the networks of multiple Internet service providers. Similarly, disk I/O throughput can be boosted via a storage area network connected to a storage pool through a fibre channel switch.

8.3.2   Coverage vs. Freshness vs. Importance

Given the size of the Web, it is not feasible even for the largest-scale crawlers employed by commercial search engines to index all of the content that could be accessed. Instead, search engines aim to focus on the most "important" pages, where importance is assessed based on various factors such as link popularity measures (indegree or PageRank) [102, 234]. At the time of this writing the three major commercial search engines report index sizes in the order of 10^{10} pages, while the indexable Web may be at least an order of magnitude larger.

The simplest strategy to bias the crawler in favor of popular pages is to **do nothing** – given the long-tailed distribution of indegree discussed in Sect. 8.1, a simple breadth-first crawling algorithm will tend to fetch the pages with the highest PageRank by definition, as confirmed empirically [401]. In fact, one would have to apply a reverse bias to obtain a fair sample of the Web. Suppose that starting with a random Web walk, we wanted a random sample of pages drawn with uniform probability distribution across all pages. We can write the posterior probability of adding a page p to the sample as $Pr(accept(p)|crawl(p)) \cdot Pr(crawl(p))$ where the first factor is the conditional probability of accepting the page into the sample given that it was crawled, and the second factor is the prior probability of crawling the page in the random walk. We can find the acceptance strategy to obtain a uniform sample by setting the product to a constant, yielding $Pr(accept(p)|crawl(p)) \sim 1/Pr(crawl(p))$. The prior $Pr(crawl(p))$ is given by the PageRank of p, and can be approximated during the random walk by the frequency $f(p)$ that the crawler has encountered a link to p. So therefore, each visited page p should be accepted with probability proportional to $1/f(p)$. Empirical tests on a simulated Web graph validate that this strategy yields a sample of the graph that is statistically representative of the original [235].

The goal to cover as many pages as possible (among the most important ones) is in conflict with the need to maintain a **fresh** index. Because of the highly dynamic nature of the Web, with pages being added, deleted, and modified all the time, it is necessary for a crawler to revisit pages already in the index in order to keep the index up-to-date. Many studies have been conducted to analyze the dynamics of the Web, i.e., the statistical properties of the processes leading to change in Web structure and content 66, 101, 152, 177, 416]. They all indicate that the Web changes at very rapid rates. While early studies relied on the values reported by Web servers in the last-modified HTTP header, recently there is consensus that this information has little reliability. The most recent and exhaustive study at the time of this writing [416] reports that while new pages are created at a rate of about 8% per week, only about 62% of the content of these pages is really new because pages are often copied from existing ones. The link structure of the Web is more dynamic, with about 25% new links created per week. Once created, pages tend to change little so that most of the changes observed in the Web are due to additions and deletions rather than modifications. Finally, there is an agreement on the observation that the degree of change of a page is a better predictor of future change than the frequency of change [177, 416]. This suggests that crawler revisit strategies based on frequency of change [25, 101] may not be the most appropriate for achieving a good tradeoff between coverage and freshness.

8.4 Focused Crawlers

Rather than crawling pages from the entire Web, we may want to crawl only pages in certain categories. One applications of such a preferential crawler would be to maintain a Web taxonomy such as the Yahoo! Directory (dir.yahoo.com) or the volunteer-based **Open Directory Project** (ODP, dmoz.org). Suppose you are the ODP editor for a certain category; you may wish to launch such a crawler from an initial seed set of pages relevant to that category, and see if any new pages discovered should be added to the directory, either directly under the category in question or one of its subcategories. A **focused crawler** attempts to bias the crawler towards pages in certain categories in which the user is interested.

Chakrabarti et al. [87] proposed a focused crawler based on a classifier. The idea is to first build a text classifier using labeled example pages from, say, the ODP. Then the classifier would guide the crawler by preferentially selecting from the frontier those pages that appear most likely to belong to the categories of interest, according to the classifier's prediction. To train the classifier, example pages are drawn from various categories in the taxonomy as shown in Fig. 8.5. The classification algorithm used was the naïve Bayesian method (see Chap. 3). For each category c in the taxonomy we can build a Bayesian classifier to compute the probability $\Pr(c|p)$ that a crawled page p belongs to c (by definition, $\Pr(top|p) = 1$ for the *top* or *root*

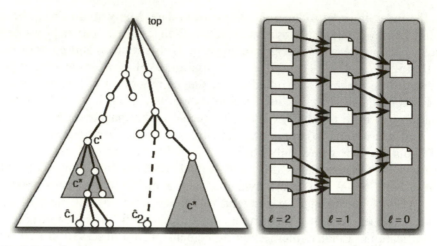

Fig. 8.5. Left: A taxonomy supporting a focused crawler. The areas in gray represent the categories of interest c^*. A crawler with hard focus would add to the frontier the links extracted from a page classified in the leaf category $\hat{c}_1$ because its ancestor category c' is of interest to the user, while the links from a page classified in $\hat{c}_2$ would be discarded. Right: A context graph with $L = 3$ layers constructed to train a context focused crawler from the target set in layer $\ell = 0$.

category). The user can select a set c^* of categories of interest. Each crawled page is assigned a relevance score.

$$R(p) = \sum_{c \in c^*} \cdot \Pr(c \mid p). \tag{1}$$

Two strategies were explored. In the "soft" focused strategy, the crawler uses the score $R(p)$ of each crawled page p as a priority value for all unvisited URLs extracted from p. The URLs are then added to the frontier, which is treated as a priority queue (see Sect. 8.1.2). In the "hard" focused strategy, for a crawled page p, the classifier first finds the leaf category $\hat{c}(p)$ in the taxonomy most likely to include p:

$$\hat{c}(p) = \arg \max_{c: \nexists c' \subset c} \Pr(c \mid p). \tag{2}$$

If an ancestor of $\hat{c}(p)$ is a focus category, i.e., $\exists c': \hat{c}(p) \subset c' \wedge c' \in c^*$, then the URLs from the crawled page p are added to the frontier. Otherwise they are discarded. The idea is illustrated in Fig. 8.5 (left). For example, imagine a crawler focused on **soccer** ($c' = $ **soccer** $\in c^*$) visits a page p in the FIFA World Cup Germany 2006 site. If the classifier correctly assigns p to the leaf category $\hat{c}$=Sports/Soccer/Competitions/World_Cup/2006, the links extracted from p are added to the frontier because 2006 is a sub-category of Sports/Soccer ($\hat{c} \subset$ **soccer**). The soft and hard focus strategies worked equally well in experiments.

Another element of the focused crawler is the use of a distiller. The distiller applies a modified version of the HITS algorithm [282] to find topical hubs. These hubs provide links to authoritative sources on a focus category. The distiller is activated at various times during the crawl and some of the top hubs are added to the frontier.

Context-Focused Crawlers are another type of focused crawlers. They also use naïve Bayesian classifiers as a guide, but in this case the classifiers are trained to estimate the link distance between a crawled page and a set of relevant target pages [139]. To see why this might work, imagine looking for information on "machine learning." One might go to the home pages of computer science departments and from there to faculty pages, which may then lead to relevant pages and papers. A department home page, however, may not contain the keywords "machine learning." A typical focused or best-first crawler would give such a page a low priority and possibly never follow its links. However, if the crawler could estimate that pages about machine learning are only two links away from a page containing the keywords "computer science department," then it would give the department home page a higher priority.

The context-focused crawler is trained using a **context graph** with L layers (Fig. 8.5 right). The seed (target) pages form the layer 0 of the graph. The pages corresponding to the in-links to the seed pages are in layer 1. The in-links to the layer 1 pages make up the layer 2, and so on. The in-links to any page can be obtained by submitting a link: query to a search engine. The seed pages in layer 0 (and possibly those in layer 1) are then concatenated into a single large document, and the top few terms according to the TF-IDF weighting scheme (see Chap. 6) are selected as the vocabulary (feature space) to be used for classification. A naïve Bayesian classifier is built for each layer in the context graph. A prior probability $\Pr(\ell) = 1/L$ is assigned to each layer. All the pages in a layer are used to compute $\Pr(t|\ell)$, the probability of occurrence of a term t given the layer (class) ℓ. At the crawling time, these are used to compute $\Pr(p|\ell)$ for each crawled page p. The posterior probability $\Pr(\ell|p)$ of p belonging to layer ℓ can then be computed for each layer from Bayes' rule. The layer ℓ^* with highest posterior probability wins:

$$\ell^*(p) = \arg\max_{\ell} \Pr(\ell \mid p). \tag{3}$$

If $\Pr(\ell^*|p)$ is less than a threshold, p is classified into the "other" class, which represents pages that do not have a good fit with any of the layers in the context graph. If $\Pr(\ell^*|p)$ exceeds the threshold, p is classified into ℓ^*.

The set of classifiers corresponding to the context graph provides a mechanism to estimate the link distance of a crawled page from a relevant page. If the mechanism works, the computer science department page in our example will get classified into layer 2. The crawler maintains a separate priority queue for each layer, containing the links extracted from visited pages classified in that layer. Each queue is sorted by the scores $\Pr(\ell|p)$. The next URL to crawl is taken from the non-empty queue with the smallest ℓ. So the crawler gives precedence to links that appear to be closest to relevant targets. It is shown in [139] that the context-focused crawler outperforms the standard focused crawler in experiments.

While the majority of focused crawlers in the literature have employed the naïve Bayesian method as the classification algorithm to score unvisited URLs, an extensive study with hundreds of topics has provided strong evidence that classifiers based on SVM or neural networks can yield significant improvements in the quality of the crawled pages [433].

8.5 Topical Crawlers

For many preferential crawling tasks, labeled (positive and negative) examples of pages are not available in sufficient numbers to train a focused crawler before the crawl starts. Instead, we typically have a small set of seed pages and a description of a topic of interest to a user or user community. The topic can consist of one or more example pages (possibly the seeds) or even a short query. Preferential crawlers that start with only such information are often called **topical crawlers** [85, 102, 377]. They do not have text classifiers to guide crawling.

Even without the luxury of a text classifier, a topical crawler can be smart about preferentially exploring regions of the Web that appear relevant to the target topic by comparing features collected from visited pages with cues in the topic description.

To illustrate a topical crawler with its advantages and limitations, let us consider the MySpiders applet (myspiders.informatics.indiana.edu). Figure 8.6 shows a screenshot of this application. The applet is designed to demonstrate two topical crawling algorithms, **best-N-first** and **InfoSpiders**, both discussed below [431].

MySpiders is interactive in that a user submits a query just like one would do with a search engine, and the results are then shown in a window. However, unlike a search engine, this application has no index to search for results. Instead the Web is crawled in real time. As pages deemed relevant are crawled, they are displayed in a list that is kept sorted by a user-selected criterion: score or recency. The score is simply the content (cosine) similarity between a page and the query (see Chap. 6); the recency of a page is estimated by the last-modified header, if returned by the server (as noted earlier this is not a very reliable estimate).

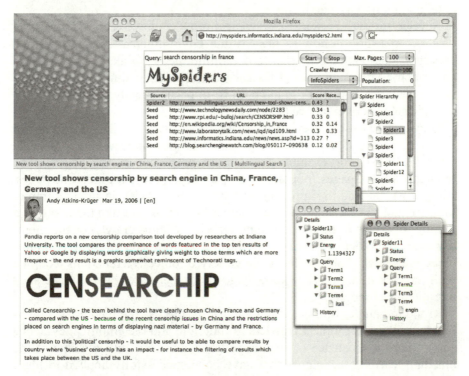

Fig. 8.6. Screenshot of the MySpiders applet in action. In this example the user has launched a population of crawlers with the query "search censorship in france" using the InfoSpiders algorithm. The crawler reports some seed pages obtained from a search engine, but also a relevant blog page (bottom left) that was not returned by the search engine. This page was found by one of the agents, called Spider2, crawling autonomously from one of the seeds. We can see that Spider2 spawned a new agent, Spider13, who started crawling for pages also containing the term "italy." Another agent, Spider5, spawned two agents one of which, Spider11, identified and internalized the relevant term "engine."

One of the advantages of topic crawling is that all hits are fresh by definition. No stale results are returned by the crawler because the pages are visited at query time. This makes this type of crawlers suitable for applications that look for very recently posted documents, which a search engine may not have indexed yet. On the down side, the search is slow compared to a traditional search engine because the user has to wait while the crawler fetches and analyzes pages. If the user's client machine (where the applet runs) has limited bandwidth, e.g., a dial-up Internet connection, the wait is likely infeasible. Another disadvantage is that the ranking algorithms cannot take advantage of global prestige measures, such as PageRank, available to a traditional search engine.

Several research issues around topical crawlers have received attention. One key question is how to identify the environmental signals to which crawlers should attend in order to determine the best links to follow. Rich cues such as the markup and lexical (text) signals within Web pages, as well as features of the link graph built from pages already seen, are all reasonable sources of evidence to exploit.

Crawlers can use the evidence available to them in different ways, for example more or less greedily. The goals of the application also provide crucial context. For example the desired properties of the pages to be fetched (similar pages, popular pages, authoritative pages, recent pages, and so on) can lead to significant differences in crawler design and implementation. The task could be constrained by parameters like the maximum number of pages to be fetched (long crawls vs. short crawls) or the memory available. A crawling task can thus be viewed as a constrained multiobjective search problem. The wide variety of objective functions, coupled with the lack of appropriate knowledge about the search space, make such a problem challenging.

In the remainder of this section we briefly discuss the theoretical conditions necessary for topical crawlers to function, and the empirical evidence supporting the existence of such conditions. Then we review some of the machine learning techniques that have been successfully applied to identify and exploit useful cues for topical crawlers.

8.5.1   Topical Locality and Cues

The central assumption behind topical crawlers is that Web pages contain reliable cues about each other's content. This is a necessary condition for designing a crawler that has a better-than-random chance to preferentially visit pages relevant with respect to a given topic. Indeed, if no estimates could be made about unvisited pages, then all we could do is a random walk through the Web graph, or an exhaustive search (using breadth-first or depth-first search algorithms). Fortunately, crawling algorithms can use cues from words and hyperlinks, associated respectively with a **lexical** and a **link topology**. In the former, two pages are close to each other if they have similar textual content; in the latter, if there is a short path between them (we will see what "short" means).

Lexical metrics are text similarity measures derived from the vector space model (see Chap. 6). The **cluster hypothesis** behind this model is that a document lexically close to a relevant document (with respect to the given query) is also relevant with high probability [461].

Link metrics typically look at hyperlinks as directed edges in a graph, but a path can also be defined in an undirected sense, in which case two pages have a short link distance between them if they are co-cited or co-referenced, even if there is no directed path between them. Links are a very rich source of topical information about Web pages.

From a crawler's perspective, there are two central questions:

1. **link-content conjecture**: whether two pages that link to each other are more likely to be lexically similar to each other, compared to two randomly selected pages;
2. **link-cluster conjecture**: whether two pages that link to each other are more likely to be semantically related to each other, compared to two randomly selected pages.

A first answer to the link-content conjecture was obtained by computing the cosine similarity between linked and random pairs of pages, showing that the similarity is an order of magnitude higher in the former case [123]. The same study also showed that the anchor text tends to be a good (similar) description of the target page.

The link-content conjecture can be generalized by looking at the decay in content similarity as a function of link distance from a source page. This decay was measured by launching an exhaustive breadth-first crawl from seed sets of 100 topics in the Yahoo! directory [372]. Let us use the cosine similarity measure $\sigma(p_1, p_2)$ between pages p_1 and p_2 (see Chap. 6). We can measure the link distance $\delta_1(p_1, p_2)$ along the shortest directed path from p_1 and p_2, revealed by the breadth-first crawl. Both distances $\delta_1(q, p)$ and similarities $\sigma(q, p)$ were averaged for each topic q over all pages p in the crawl set P_d^q for each depth d:

$$\delta(q,d) \equiv \left\langle \delta_1(q,p) \right\rangle_{P_d^p} = \frac{1}{N_d^q} \sum_{i=1}^{d} i(N_i^q - N_{i-1}^q) \tag{4}$$

$$\sigma(q,d) \equiv \left\langle \sigma(q,p) \right\rangle_{P_d^p} = \frac{1}{N_d^q} \sum_{p \in P_d^p} \sigma(q,p) \tag{5}$$

where N_d^q is the size of the cumulative page set $P_d^q = \{p \mid \delta_1(q, p) \leq d\}$. The crawlers were stopped at depth $d = 3$, yielding 3000 data points

$$\{(p, d): q \in \{1, \ldots, 100\}, d \in \{1, 2, 3\}\}.$$

These points were then used for fitting an exponential decay model:

$$\sigma(\delta) \approx \sigma_\infty + (1 - \sigma_\infty) e^{-\alpha_1 \delta^{\alpha_2}} \tag{6}$$

where σ_∞ is the noise level in similarity, measured empirically by averaging across random pairs of pages. The parameters α_1 and α_2 are set by fitting the data. This was done for pages in various top-level domains, and the resulting similarity decay curves are plotted in Fig. 8.7.

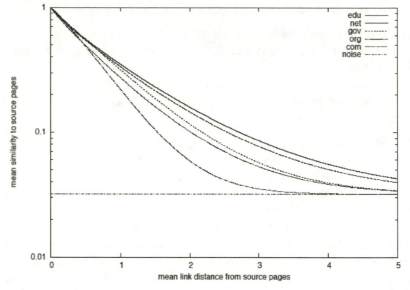

Fig. 8.7. Illustration of the link-content conjecture. The curves plot, for each top-level domain, the decay in mean cosine similarity between pages as a function of their mean directed link distance, obtained by fitting data from 100 exhaustive breadth-first crawls starting from the 100 Yahoo! directory topics [372].

The curves provide us with a rough estimate of how far in link space one can make inferences about lexical content. We see that a weak signal is still present three links away from the starting pages for all but the .com domain, and even further for the .edu domain. Such heterogeneity is not surprising – academic pages are written carefully to convey information and proper pointers, while business sites often do not link to related sites because of competition. Therefore a topical crawler in the commercial domain would have a harder task, other things being equal. A solution may be to use undirected links. More specifically, if a crawler can obtain in-links to good pages (by querying a search engine), it can use **co-citation** to detect hubs. If a page links to several good pages, it is probably a good hub and all its out-links should be given high priority. This strategy, related to the so-called **sibling locality** [3], has been used in focused crawlers [87] and in topical crawlers for business intelligence [432]. In addition to co-citation, one could look at **bibliographic coupling**: if several good pages link to a certain page, that target is likely to be a good authority so it and its in-links should be given high priority. Fig. 8.8 illustrates various ways in which crawlers can exploit co-citation and bibliographic coupling.

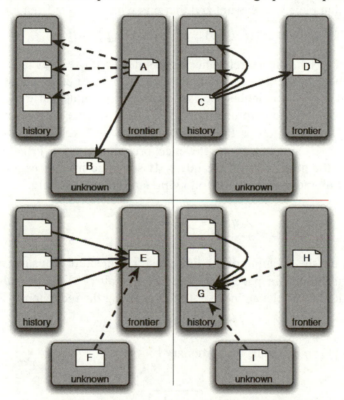

Fig. 8.8. Crawling techniques exploiting co-citation (top) and bibliographic coupling (bottom). Dashed edges represent in-links, which require access to a search engine or connectivity server. Page A is a good hub, so it should be given high priority; once fetched, page B linked by it can be discovered and placed in the frontier with high priority since it is likely to be a good authority. Page C is also a good hub, so D should be given high priority. Page E is a good authority, so it should be given high priority. Its URL can also be used to discover F, which may be a good hub and should be placed in the frontier. G is also a good authority, so H should be given high priority and I should be placed in the frontier.

The link-cluster conjecture, also known as **linkage locality** [87], states that one can infer the meaning of a page by looking at its neighbors. This is actually more important than inferring lexical content, since the latter is only relevant insofar as it is correlated with the semantic content of pages. The same exhaustive crawl data used to validate the link-content conjecture can also be used to explore the link-cluster conjecture, namely the extent to which relevance is preserved within link space neighborhoods and the decay in expected relevance as one browses away from a relevant page [372]. The link-cluster conjecture can be simply formulated in terms of the conditional probability that a page p is relevant with respect to some query q, given that page r is relevant and that p is within d links from r:

$$R_q(d) \equiv \Pr(rel_q(p) \mid rel_q(r) \wedge \delta_1(r,p) \leq d] \tag{7}$$

where $rel_q()$ is a binary relevance assessment with respect to q. In other words a page has a higher than random probability of being about a certain topic if it is in the neighborhood of other pages about that topic. $R_q(d)$ is the posterior relevance probability given the evidence of a relevant page nearby. The conjecture is then represented by the likelihood ratio $\lambda(q, d)$ between $R_q(d)$ and the prior relevance probability $G_q \equiv \Pr(rel_q(p))$, also known as the **generality** of the query. If semantic inferences are possible within a link radius d, then the following condition must hold:

$$\lambda(q,d) \equiv \frac{R_q(d)}{G_q} > 1. \tag{8}$$

To illustrate the meaning of the link-cluster conjecture, consider a random crawler searching for pages about a topic q. Call $\eta_q(t)$ the probability that the crawler hits a relevant page at time t. Solving the recursion

$$\eta_q(t+1) = \eta_q(t)R_q(1) + (1 - \eta_q(t))G_q \tag{9}$$

for $\eta_q(t+1) = \eta_q(t)$ yields the stationary hit rate

$$\eta_q^* = \frac{G_q}{1 + G_q - R_q(1)}. \tag{10}$$

The link-cluster conjecture is a necessary and sufficient condition for such a crawler to have a better than chance hit rate:

$$\eta_q^* > G_q \Leftrightarrow \lambda(q,1) > 1. \tag{11}$$

Figure 8.9 plots the mean likelihood ratio $\lambda(q, d)$ versus the mean link distance $\delta(q, d)$ obtained by fitting an exponential decay function

$$\lambda(\delta) \approx 1 + \alpha_3 e^{-\alpha_4 \delta^{\alpha_5}} \tag{12}$$

to the same 300 data points $\{(q, d)\}$. Note that this three-parameter model is more complex than the one used to validate the link-content conjecture, because $\lambda(\delta = 0)$ must also be estimated from the data ($\lambda(q, 0) = 1/G_q$). The fitted curve reveals that being within a radius of three links from a relevant page increases the relevance probability by a factor $\lambda(q, d) \gg 1$. This is very reassuring for the design of topical crawlers. It also suggests that crawlers should attempt to remain within a few links from some relevant source. In this range hyperlinks create detectable signals about lexical and semantic content, despite the Web's apparent lack of structure.

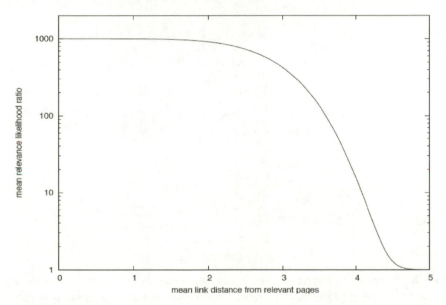

Fig. 8.9. Illustration of the link-cluster conjecture. The curve plots the decay in mean likelihood ratio as a function of mean directed link distance from a relevant page, obtained by fitting data from 100 exhaustive breadth-first crawls starting from as many Yahoo! directory topics [372].

The link-content and link-cluster conjectures can be further developed by looking at the **correlation** between content-based, link-based, and semantic-based similarity measures. Using the ODP as a ground truth, we can express the **semantic similarity** between any two pages in the taxonomy [373, 359] and see how it can be approximated by content and link similarity measures. For content one can consider for example cosine similarity based on **TF** or **TF-IDF** term weights. For link similarity one can similarly represent a page as a bag of links (in-links, out-links, or both/undirected) and then apply a **Jaccard coefficient** or a **cosine similarity**. Figure 8.10 shows, for various topical domains from the ODP, the correlation between semantic similarity and two representative content and link similarity measures. We observe significant heterogeneity in the correlations, suggesting that topical crawlers have an easier job in some topics (e.g., "news") than others (e.g., "games"). Another observation is that in some topical domains (e.g., "home") textual content is a more reliable signal, while in others (e.g., "computers") links are more helpful.

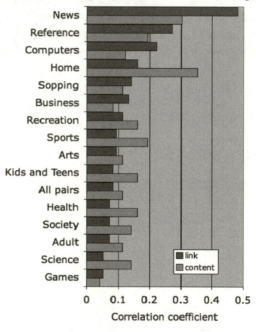

Fig. 8.10. Pearson correlation coefficients between the semantic similarity extracted from ODP [359] and two representative content and link similarity measures. The correlations are measured using a stratified sample of 150,000 URLs from the ODP, for a total of 4 billion pairs [373]. Content similarity is cosine with TF weights, and link similarity is the Jaccard coefficient with undirected links.

8.5.2 Best-First Variations

The majority of crawling algorithms in the literature are variations of the best-first scheme described in Sect. 8.1.2. The difference is in the heuristics that they use to score unvisited URLs. A very simple instance is the case where each URL is queued into the frontier with priority given by the content similarity between the topic description and the page from which the URL was extracted. Content similarity can be measured with the stan-

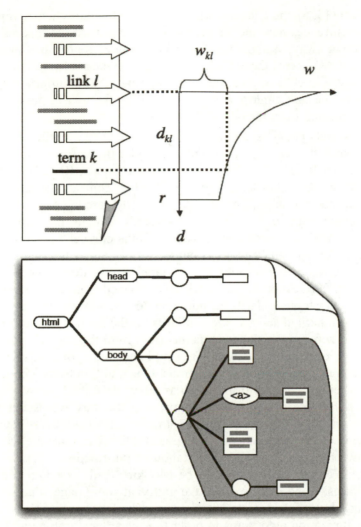

Fig. 8.11. Link context from distance-weighted window (top) and from the DOM tree (bottom).

dard cosine similarity, using TF or TFIDF term weights (in the latter case the crawler must have global or topic-contextual term frequency information available). This simple crawler is also known as **naïve best-first**.

Many variations of the naïve best-first crawlers are possible. Some give more importance to certain HTML markups, such as the title, or to text segments marked by special tags, such as headers. Other techniques focus on determining the most appropriate textual context to score a link. One alternative to using the entire page or just the anchor text as context, used by

InfoSpiders [375] and Clever [86], is a weighted window where topic keywords occurrences near the anchor count more toward the link score than
those farther away, as shown in Fig. 8.11. Another approach is to consider
the tag (DOM) tree of the HTML page [85]. The idea is to walk up the tree
from the link anchor toward the root, stopping at an appropriate aggregation node. The link context is then obtained by the text in the tag subtree
rooted at the aggregation node (Fig. 8.11).

SharkSearch [237] is an improved version of the earlier FishSearch
crawler [60]. It uses a similarity measure like the one used in the naïve
best-first crawler as a first step for scoring unvisited URLs. The similarity
is computed for anchor text, a fixed-width link context, the entire source
page, and ancestor pages. The ancestors of a URL are the pages that appear
on the crawl path to the URL. SharkSearch, like its predecessor Fish-
Search, maintains a depth bound. That is, if the crawler finds unimportant
pages on a crawl path it stops crawling further along that path. To this
end, each URL in the frontier is associated with a depth and a potential
score. The score of an unvisited URL is obtained from a linear combination of anchor text similarity, window context similarity, and an inherited
score. The inherited score is the similarity of the source page to the topic,
unless it is zero, in which case it is inherited from the source's parent (and
recursively from its ancestors). The implementation of SharkSearch requires to preset three similarity coefficients in addition to the depth bound.
This crawler does not perform as well as others described below.

Rather than (or in addition to) improving the way we assign priority
scores to unvisited URLs, we can also improve on a naïve best-first crawler
by altering the priority scheme. A classic trade-off in machine learning is
that between exploration and exploitation of information. A crawler is no
different: it can greedily pursue the best-looking leads based on noisy quality estimates, or be more explorative and visit some pages that seem less
promising, but might lead to better pages. The latter approach is taken in
many optimization algorithms in order to escape local optima and reach a
global optimum with some probability. As it turns out, the same strategy is
also advantageous for topical crawlers. Visiting some URLs with lower
priority leads to a better overall quality of the crawler pages than strictly
following the best-first order. This is demonstrated by **best-N-first**, a
crawling algorithm that picks N URLs at a time from the frontier (the top N
by priority score) and fetches them all. Once all N pages are visited, the
newly extracted URLs are merge-sorted into the priority queue, and the
cycle is repeated. The best-N-first crawler with $N = 256$ is a very strong
competitor, outperforming most of the other topical crawlers in the literature [434, 377]. Figure 8.12 shows a comparison with two crawlers dis-

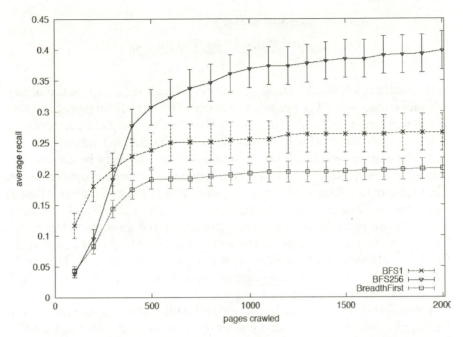

Fig. 8.12. Performance of best-*N*-first crawler with *N* = 256 (BFS256) compared with a naïve best-first crawler (BFS1) and a breadth-first crawler. Recall refers to sets of relevant pages that the crawlers are supposed to discover; averages and error bars are computed across 100 crawls from as many ODP topics.

cussed thus far. Note that a concurrent implementation of a best-first crawler with *N* threads or processes is equivalent to a best-*N*-first crawler.

8.5.3 Adaptation

All the crawlers discussed thus far use a static strategy both to evaluate unvisited URLs and to manage the frontier. Thus they do not learn from experience or adapt to the context of a particular topic in the course of the crawl. In this section we describe a number of machine learning techniques that have been incorporated into **adaptive topical crawlers**.

The **intelligent crawler** uses a statistical model for learning to assign priorities to the URLs in the frontier, considering Bayesian interest factors derived from different features [3]. For example, imagine that the crawler is supposed to find pages about soccer and that 40% of links with the keyword football in the anchor text lead to relevant pages, versus a background or prior frequency of only 2% of crawled pages being relevant. Then the crawler assigns an interest factor

$$\lambda(soccer, football \in anchor)$$

$$= \frac{\Pr[rel_{soccer}(p)\,|\,football \in anchor(p)]}{\Pr[rel_{soccer}(p)]} = 20 \tag{13}$$

to the feature "keyword football in anchor." Recall $rel_{soccer}(p)$ is the binary relevance score (0 or 1) of page p to soccer. The interest factors are treated as independent sources of evidence, or likelihoods. They are combined by a linear combination of log-likelihoods, with user-defined weight parameters. The features employed by the intelligent crawler may be diverse, depending on the particular crawling task. They may include tokens extracted from candidate URLs, source page content and links, co-citation (sibling) relationships, and/or other characteristics of the visited and unvisited URLs. As more evidence is accumulated and stored throughout the crawl, the interest factors are recalculated and the priorities updated, so that the frontier is always sorted according to the most recent estimates. Thus intelligent crawlers adapt to the content and link structure of the Web neighborhoods being explored.

The original focused crawlers described earlier also use machine learning, in particular a classifier that guides the crawler. However the classifier is trained before the crawl is launched, and no learning occurs during the crawl. Therefore we do not consider it an adaptive crawler. However, in a later "accelerated" version of the focused crawler [85], an online learning apprentice was added to the system; the original (baseline) classifier then acts as a critic, providing the apprentice with training examples for learning to classify outgoing links from the features of the pages from which they are extracted. Suppose page p_1 is fetched and contains a link to page p_2. Later, p_2 is fetched and the baseline classifier assigns it to a relevant class. This information is passed to the apprentice, which uses the labeled example ("the link from p_1 to p_2 is good") to learn to classify the link to p_2 as good based on the textual features in the context of the anchor within p_1. Future links with a similar context should be given high priority. Conversely, if p_2 is deemed irrelevant by the baseline classifier, the apprentice learns to predict ("bad link") when it encounters a link with a similar context in the future. The features used to train the apprentice were textual tokens associated with a link context based on the DOM tree, and the learning algorithm used by the apprentice was a naïve Bayesian classifier. This approach led to a significant reduction in the number of irrelevant pages fetched by the focused crawler.

While the accelerated focused crawler is not a topical crawler because it still needs labeled examples to train the baseline classifier prior to the crawl, the idea of training an apprentice online during the crawl can be applied in topical crawlers as well. Indeed this is a type of **reinforcement learning** technique employed in several crawlers, using different features and/or different learning algorithms for the apprentice. In reinforcement learning [263] we have a network where nodes are states and directed links are actions. An action $a \in A$ (think "anchor") moves an agent from a state $p \in P$ (think "page") to another state according to a transition function $L: P \times A \rightarrow P$. Thus an adaptive crawler is seen as an agent moving from page to page. Actions are rewarded according to a function $r: P \times A \rightarrow \mathfrak{R}$. We want to learn a policy mapping states to actions, $\pi: P \rightarrow A$, that maximizes future reward discounted over time:

$$V^{\pi}(p_0) = \sum_{t=0}^{\infty} \gamma^t r(p_t, a_t) \tag{14}$$

where we follow action (link) $a_t = \pi(p_t)$ from state (page) p_t at each time step t. The parameter γ determines how future rewards are discounted ($0 \leq \gamma < 1$). If $\gamma = 0$, the reinforcement learning policy is the greedy one employed by the naïve best-first crawler. To learn an optimal policy, we define the value of selecting action a from state p, and following the optimal policy thereafter:

$$Q(p, a) = r(p, a) + \gamma V^*[L(p, a)] \tag{15}$$

where V^* is the value function of the optimal policy $\pi^*(p) = \text{argmax}_a Q(p, a)$. The question then becomes how to estimate the function Q, i.e., to assign a value to a link a based on the context information in page p from which the link is extracted. However, the actions available to the crawler are not limited to the links from the last page visited; any of the actions corresponding to the URLs in the frontier are available. Furthermore, there is no reason why the Q value of a link should be a function of a particular source page; if links to the same target page are extracted from multiple source pages, the estimated values of the anchors can be combined, for example $Q(u) = \max_{\{(p, a): L(p, a) = u\}} Q(p, a)$. This way Q values can be computed not for links (anchors), but for target pages (URLs); the state and action spaces are thus greatly reduced, basically collapsing all visited pages into a single degenerate state and all links to their target URLs. The policy π reduces to the simple selection of the URL in the frontier with the maximum Q value.

One way to calculate Q values is via a naïve Bayesian classifier. This method was found to work well compared to a breadth-first crawler for the tasks of crawling computer science research papers and company directory information [366, 459]. In this case, the classifier was trained off-line rather than online while crawling, using labeled examples as in the focused crawler. Training the classifier to predict future reward ($\gamma > 0$) was better than only using immediate reward ($\gamma = 0$). For future reward the authors use a heavy discount $\gamma = 0.5$, arguing that it is optimal to be greedy in selecting URLs from the frontier, so that one can crawl toward the nearest relevant page. This assumes that all relevant targets are within reach. So there is no reason to delay reward. However, as discussed earlier, a crawler typically deals with noisy data, so the classifier's Q estimates are not entirely reliable; more importantly, a typical crawler cannot possibly cover the entire search space. These factors suggest that it may be advantageous to occasionally give up some immediate reward in order to explore other directions, potentially leading to pockets or relevant pages unreachable by a greedy crawler (see Fig. 8.12).

Using a previously trained classifier to compute Q values for URLs in the frontier means that supervised learning is combined with reinforcement learning. As for focused crawlers, labeled examples must be available prior to the start of the crawl. This may be possible in tasks such as the collection of research articles, but is not a realistic assumption for typical topical crawlers. An adaptive crawling algorithm that actually uses reinforcement learning while crawling online, without any supervised learning, is **InfoSpiders**. This crawler employs various machine learning techniques to exploit various types of Web regularities. InfoSpiders are inspired by artificial life models in which a population of agents lives, learn, evolve, and die in an environment. Individual agents learn during their lifetimes based on their experiences, with the environment playing the role of a critic, providing rewards and penalties for actions. Agents may also reproduce, giving rise to new agents similar to them, and die. The environment is the Web, the actions consist of following links and visiting pages and the text and link features of pages are the signals that agents can internalize into their learned and evolved behaviors. Feedback from the environment consists of a finite energy resource, necessary for survival. Each action has an energy cost, which may be fixed or proportional to, say, the size of a fetched page or the latency of a page download [126]. Energy is gained from visiting new pages relevant to the query topic. A cache prevents an agent from accumulating energy by visiting the same page multiple times. In the recent version of InfoSpiders, each agent maintains its own frontier of unvisited URLs [377]. The agents can be implemented as concurrent processes/threads, with non-contentious access to their local frontiers. Fig. 8.13 illustrates the representation and flow of an individual agent.

The adaptive representation of each InfoSpiders agent consists of a list of keywords (initialized with the topic description) and a neural net used to evaluate new links. Each input unit of the neural net receives a count of the frequency with which the keyword occurs in the vicinity of each link, weighted to give more importance to keywords occurring near the anchor and maximum for the anchor text (Fig. 8.11). The neural net has a single output unit whose activation is used as a Q value (score) for each link u in input. The agent's neural net learns to predict the Q value of the link's target URL u given the inputs from the link's source page p. The reward function $r(u)$ is the cosine similarity between the agent's terms and the target page u. The future discounted optimal value $\gamma V^{*}(u)$ is approximated using the highest neural net prediction among the links subsequently extracted from u. This procedure is similar to the reinforcement learning algorithm described above, except that the neural net replaces the naïve Bayesian classifier. The neural net is trained by the back-propagation algorithm [469]. This mechanism is called **connectionist reinforcement learning** [335]. While the neural net can in principle model nonlinear relationships between term frequencies and pages, in practice we have used a simple perceptron whose prediction is a linear combination of the keyword weights. Such a learning technique provides each InfoSpiders agent with the capability to adapt its own link-following behavior in the course of a crawl by associating relevance estimates with particular patterns of keyword frequencies around links.

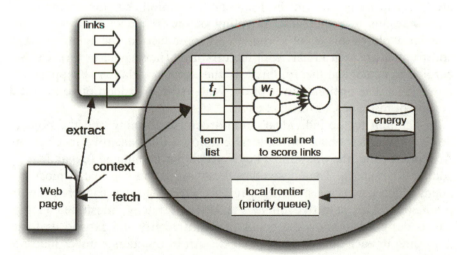

Fig. 8.13. A single InfoSpiders agent. The link context is the weighted window as shown in Fig. 8.11: for each newly extracted URL and for each term in the agent's term list, this produces a weight that is fed into the neural network, whose output is stored as the link's priority score in the frontier.

The neural net's link scores are combined with estimates based on the cosine similarity between the agent's keyword list and the entire source page. A parameter α ($0 \le \alpha \le 1$) regulates the relative importance given to the estimates based on the neural net versus the source page. Based on the combined score σ the agent uses a stochastic selector to pick one of the links in the frontier with probability

$$\Pr(u) = \frac{e^{\beta\sigma(u)}}{\sum_{u' \in \phi} e^{\beta\sigma(u')}} \qquad (16)$$

where u is a URL in the local frontier ϕ. Parameter β regulates the greediness of the link selector. Its value can be fixed or evolved with the agent.

After a new page u has been fetched, the agent receives an energy payoff proportional to the difference between the reward $r(u)$ and the cost charged for the download. An agent dies when it runs out of energy. The energy level is also used to determine whether or not the agent should reproduce after visiting a page. An agent reproduces when the energy level passes a fixed threshold. The reproduction is meant to bias the search toward areas with pages relevant to the topic. Topical locality suggests that if an agent visits a few relevant pages in rapid sequence, more relevant pages are likely to be nearby (in the frontier). To exploit this, the accumulated energy results in a short-term doubling of the frequency with which the crawler explores this agent's frontier. At reproduction, the agent's energy and frontier are split in half with the offspring (new agent or thread). According to ecological theory, this way the agent population is supposed to move toward an **optimal cover** of the Web graph in proportion to the local density of resources, or relevant pages.

In addition to the individual's reinforcement learning and the population's evolutionary bias, InfoSpiders employ a third adaptive mechanism. At reproduction, the offspring's keyword vector is mutated (expanded) by adding a new term. The chosen term/keyword is the one that is most frequent in the parent's last visited page, i.e., the page that triggered the reproduction. This **selective query expansion** strategy, illustrated in Fig. 8.6, is designed to allow the population to diversify and expand its focus according to each agent's local context. An InfoSpiders crawler incorporating all of these adaptive techniques has been shown to outperform various versions of naïve best-first crawlers (Fig. 8.14) when visiting a sufficiently large number of pages (more than 10,000) so that the agents have time to adapt [377, 504].

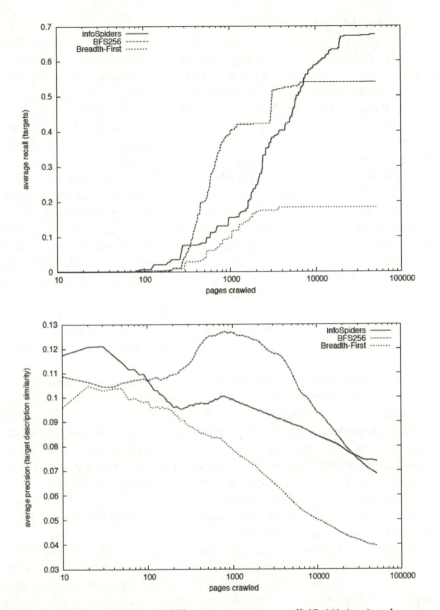

Fig. 8.14. Performance plots [377]: average target recall $\langle R_T(t) \rangle$ (top) and average precision $\langle P_D(t) \rangle$ (similarity to topic description, bottom). The averages are calculated over 10 ODP topics. After 50,000 pages crawled, one tailed t-tests reveal that both BFS256 and InfoSpiders outperform the breadth-first crawler on both performance metrics. InfoSpiders outperform BFS256 on recall, while the difference in precision is not statistically significant.

8.6   Evaluation

Given the goal of building a "good" crawler, a critical question is how to evaluate crawlers so that one can reliably compare two crawling algorithms and conclude that one is "better" than the other. Since a crawler is usually designed to support some application, e.g., a search engine, it can be indirectly evaluated through the application it supports. However, attribution is problematic; if a search engine works better than another (assuming that were easy to determine!), how can we attribute this difference in performance to the underlying crawling algorithms as opposed to the ranking or indexing schemes? Thus it is desirable to evaluate crawlers directly.

Often crawler evaluation has been carried out by comparing a few crawling algorithms on a limited number of queries/tasks without considering the statistical significance. Such anecdotal results, while important, do not suffice for thorough performance comparisons. As the Web crawling field has matured, a need has emerged for evaluating and comparing disparate crawling strategies on common tasks through well-defined performance measures. Let us review the elements of such an evaluation framework, which can be applied to topical as well as focused crawlers.

A comparison between crawlers must be unbiased and must allow one to measure statistically significant differences. This requires a sufficient number of crawl runs over different topics, as well as sound methodologies that consider the temporal nature of crawler outputs. Significant challenges in evaluation include the general unavailability of relevant sets for particular topics or queries. Unfortunately, meaningful experiments involving real users for assessing the relevance of pages as they are crawled are extremely problematic. In order to obtain a reasonable notion of crawl effectiveness one would have to recruit a very large number of subjects, each of whom would have to judge a very large number of pages. Furthermore, crawls against the live Web pose serious time constraints and would be overly burdensome to the subjects.

To circumvent these problems, crawler evaluation typically relies on defining measures for automatically estimating page relevance and quality. The crawler literature reveals several performance measures used for these purposes. A page may be considered relevant if it contains some or all of the keywords in the topic/query. The frequency with which the keywords appear on the page may also be considered [102]. While the topic of interest to the user is often expressed as a short query, a longer description may be available in some cases. Similarity between the short or long description and each crawled page may be used to judge the page's relevance [237, 376, 504]. The pages used as the crawl's seed URLs may be combined together into a single document, and the cosine similarity between this document and a crawled page may serve as the page's relevance score [18]. A classifier may be trained to identify relevant pages. The training may be done using seed pages or other pre-specified relevant pages as positive examples. The trained classifier then provides boolean or continuous relevance scores for each of the crawled pages [87, 139]. Note that if the same classifier, or a classifier trained on the same labeled examples, is used both to guide a (focused) crawler and to evaluate it, the evaluation is not unbiased. Clearly the evaluating classifier would be biased in favor of crawled pages. To partially address this issue, an evaluation classifier may be trained on a different set than the crawling classifier. Ideally the training sets should be disjoint. At a minimum the training set used for evaluation must be extended with examples not available to the crawler [433]. Another approach is to start N different crawlers from the same seeds and let them run until each crawler gathers P pages. All of the $N{\times}P$ pages collected from the crawlers are ranked against the topic query/description using a retrieval algorithm such as cosine. The rank provided by the retrieval system for each page is then used as a relevance score. Finally, one may use algorithms, such as PageRank or HITS, that provide authority or popularity estimates for each crawled page. A simpler method would be to use just the number of in-links to the crawled page to derive similar information [18, 102]. Many variations of link-based methods using topical weights may be applied to measure the topical quality of pages [52, 88].

Once each page is assessed, a method is needed to summarize the performance of a crawler across a set of crawled pages. Given a particular measure of page relevance and/or importance we can summarize the performance of the crawler with metrics that are analogous to the information retrieval notions of **precision** and **recall** (see Chap. 6). Lacking well-defined relevant sets, the classic boolean relevance is replaced by one of the scores outlined above. A few precision-like measures are found in the literature. In case we have boolean relevance scores, we could measure the rate at which "good" pages are found; if 100 relevant pages are found in the first 500 pages crawled, we have an acquisition rate or **harvest rate** of 20% at 500 pages [3]. If the relevance scores are continuous (e.g., from cosine similarity or a trained classifier) they can be averaged over the crawled pages. The average relevance, as shown in Fig. 8.14, may be computed over the progress of the crawl [376]. Sometimes running averages are calculated over a window of a number of pages, e.g., the last 50 pages from a current crawl point [87]. Another measure from information retrieval that has been applied to crawler evaluation is **search length** [375], defined as the number of pages (or the number of irrelevant pages) crawled before a certain percentage of the relevant pages are found. Search length is akin to the reciprocal of precision for a preset level of recall.

Recall-like measures would require normalization by the number of relevant pages. Since this number is unknown for Web crawling tasks, it might appear that recall cannot be applied to crawlers. However, even if unknown, the size of the relevant set is a constant. Therefore, it can be disregarded as a scaling factor when comparing two crawling algorithms on the same topical query. One can simply sum the quality or relevance estimates (obtained by one of the methods outlined above) over the course of a crawl, and obtain a total relevance as shown in Fig. 8.14.

It is possible to design crawling experiments so that a set of relevant target pages is known by the experimenter. Then precision and recall can be calculated from the fraction of these relevant targets that are discovered by the crawler, rather than based on relevance estimates. One way to obtain a set of relevant pages is from a public directory such as the ODP. This way one can leverage the classification already carried out by the volunteer editors of the directory. The experimenter can select as topics a set of categories from the ODP, whose distance from the root of the ODP taxonomy can be determined so as to obtain topics with generality/specificity appropriate for the crawling task [377, 504]. Figure 8.5 (left) illustrates how subtrees rooted at a chosen category can be used to harvest a set of relevant target pages. If a page is classified in a subtopic of a target topic, it can be considered relevant with respect to the target topic.

If a set of known relevant target pages is used to measure the performance of a topical crawler, these same pages cannot be used as seeds for the crawl. Two approaches have been proposed to obtain suitable seed pages. One is to perform a **back-crawl** from the target pages [504]. By submitting link: queries to a search engine API, one can obtain a list of pages linking to each given target; the process can be repeated from these parent pages to find "grandparent" pages, and so on until a desired link distance is reached. The greater the link distance, the harder the task is for the crawler to locate the relevant targets from these ancestor seed pages. The procedure has the desired property that directed paths are guaranteed to exist from any seed page to some relevant targets. Given the potentially large fan-in of pages, sampling is likely required at each stage of the back-crawl to obtain a suitable number of seeds. The process is similar to the construction of a context graph, as shown in Fig. 8.5 (right). A second approach is to split the set of known relevant pages into two sets; one set can be used as seeds, the other as targets. While there is no guarantee that the targets are reachable from the seeds, this approach is significantly simpler because no back-crawl is necessary. Another advantage is that each of the two relevant subsets can be used in turn as seeds and targets. In this way, one can measure the overlap between the pages crawled starting from the two disjoint sets.

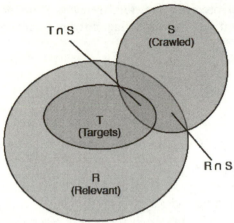

Fig. 8.15. Illustration of precision and recall measures based on known relevant target pages and underlying independence assumption/requirement.

A large overlap is interpreted as **robustness** of the crawler in covering relevant portions of the Web [87, 85].

The use of known relevant pages as proxies for unknown relevant sets implies an important assumption, which we can illustrate by the Venn diagram in Fig. 8.15. Here S is a set of crawled pages and T is the set of known relevant target pages, a subset of the relevant set R. Let us consider the measure of recall. Using T as if it were the relevant set means that we are estimating the recall $|R \cap S| / |R|$ by $|T \cap S| / |T|$. This approximation only holds if T is a representative, unbiased sample of R independent of the crawl process. While the crawler attempts to cover as much as possible of R, it should not have any information about how pages in T are sampled from R. If T and S are not independent, the measure is biased and unreliable. For example if a page had a higher chance of being selected in T because it was in S, or vice versa, then the recall would be overestimated. The same independence assumption holds for precision-like measures, where we estimate $|R \cap S| / |S|$ by $|T \cap S| / |S|$. A consequence of the independence requirement is that if the ODP is used to obtain T, the experimenter must prevent the crawler from accessing the ODP. This would bias the results because, once a relevant ODP category page is found, all of the relevant target pages can be reached by the crawler in a short breadth-first sweep. Preventing access to the ODP may pose a challenge because so many ODP mirrors exist on the Web. They may not be known by the experimenter, and not trivial to detect.

To summarize, crawler performance measures [504] can be characterized along two dimensions: the source of relevance assessments (target pages vs. similarity to their descriptions) and the normalization factor (average relevance, or precision, vs. total relevance, or recall). Using target pages as the relevant sets we can define crawler precision and recall as follows:

$$R_T(t,\theta) = \frac{|S_t \cap T_\theta|}{|T_\theta|} \tag{17}$$

$$P_T(t,\theta) = \frac{|S_t \cap T_\theta|}{|S_t|} \tag{18}$$

where S_t is the set of pages crawled at time t (t can be wall clock time, network latency, number of pages visited, number of bytes downloaded, and so on). T_θ is the relevant target set, where θ represents the parameters used to select the relevant target pages. This could include for example the depth of ODP category subtrees used to extract topic-relevant pages. Analogously we can define crawler precision and recall based on similarity to target descriptions:

$$R_D(t,\theta) = \frac{\sum_{p \in S_t} \sigma(p,D_\theta)}{|T_\theta|} \tag{19}$$

$$P_D(t,\theta) = \frac{\sum_{p \in S_t} \sigma(p,D_\theta)}{|S_t|} \tag{20}$$

where D_θ is the textual description of the target pages, selected with parameters θ, and σ is a text-based similarity function, e.g., cosine similarity (see Chap. 6). Figure 8.14 shows two examples of performance plots for three different crawlers discussed earlier in this chapter. The two plots depict R_T and P_D as a function of pages crawled. InfoSpiders and the BFS256 crawler are found to outperform the breadth-first crawler. InfoSpiders gain a slight edge in recall once the agents have had an opportunity to adapt. This evaluation involves each of the three crawlers visiting 50,000 pages for each of 10 topics, for a total of 1.5 million pages.

Another set of evaluation criteria can be obtained by scaling or normalizing any of the above performance measures by the critical resources used by a crawler. This way, one can compare crawling algorithms by way of performance/cost analysis. For example, with limited network bandwidth one may see latency as a major bottleneck for a crawling task. The time spent by a crawler on network I/O can be monitored and applied as a scaling factor to normalize precision or recall. Using such a measure, a crawler designed to preferentially visit short pages, or pages from fast servers [126], would outperform one that can locate pages of equal or even better quality but less efficiently.

8.7 Crawler Ethics and Conflicts

Crawlers, especially when efficient, can put a significant strain on the resources of Web servers, mainly on their network bandwidth. A crawler that sends many page requests to a server in rapid succession, say ten or more per second, is considered impolite. The reason is that the server would be so busy responding to the crawler that its service to other requests, including those from human browsing interactively, would deteriorate. In the extreme case a server inundated with requests from an aggressive crawler would become unable to respond to other requests, resulting in an effective denial of service attack by the crawler.

To prevent such incidents, it is essential for a crawler to put in place measures to distribute its requests across many servers, and to prevent any one server (fully qualified host name) from receiving requests at more than some reasonably set maximum rate (say, one request every few seconds). In a concurrent crawler, this task can be carried out by the frontier manager, when URLs are dequeued and passed to individual threads or processes. This practice not only is required by politeness toward servers, but also has the additional benefits of limiting the impact of spider traps and not overloading the server, which will respond slowly.

Preventing server overload is just one of a number of policies required of ethical Web agents [160]. Such policies are often collectively referred to as **crawler etiquette**. Another requirement is to disclose the nature of the crawler using the User-Agent HTTP header. The value of this header should include not only a name and version number of the crawler, but also a pointer to where Web administrators may find information about the crawler. Often a Web site is created for this purpose and its URL is included in the User-Agent field. Another piece of useful information is the email contact to be specified in the From header.

Finally, crawler etiquette requires compliance with the **Robot Exclusion Protocol**. This is a de facto standard providing a way for Web server administrators to communicate which files may not be accessed by a crawler. This is accomplished via an optional file named robots.txt in the root directory of the Web server (e.g., http://www.somehost.com/robots.txt). The file provides access policies for different crawlers, identified by the User-agent field. For any user-agent value (or the default "*") a number of Disallow entries identify directory subtrees to be avoided. Compliant crawlers must fetch and parse a server's robots.txt file before sending requests to that server. For example, the following policy in robots.txt:

```
User-agent: *
Disallow: /
```

directs any crawler to stay away from the entire server. Some high-level languages such as Perl provide modules to parse robots.txt files. It is wise for a crawler to cache the access policies of recently visited servers, so that the robots.txt file need not be fetched and parsed every time a request is sent to the same server. Additionally, Web authors can indicate if a page may or may not be indexed, cached, or mined by a crawler using a special HTML meta-tag. Crawlers need to fetch a page in order to parse this tag, therefore this approach is not widely used. More details on the robot exclusion protocols can be found at http://www.robotstxt.org/wc/robots.html.

When discussing the interactions between information providers and search engines or other applications that rely on Web crawlers, confusion sometime arises between the ethical, technical, and legal ramifications of the Robot Exclusion Protocol. Compliance with the protocol is an ethical issue, and non-compliant crawlers can justifiably be shunned by the Web community. However, compliance is voluntary, and a robots.txt file cannot enforce it. Servers can, however, block access to a client based on its IP address. Thus it is likely that a crawler which does not comply with the Exclusion Protocol and does not follow proper etiquette will be quickly blocked by many servers. Crawlers may disguise themselves as browsers by sending a browser's identifying string in the User-Agent header. This way a server administrator may not immediately detect lack of compliance with the Exclusion Protocol, but an aggressive request profile is likely to reveal the true nature of the crawler. To avoid detection, some mischievous crawlers send requests at low and randomized rates. While such behaviors may be reprehensible, they are not illegal – at least not at the time of this writing. Nonetheless, there have been cases of businesses bringing lawsuits against search organizations for not complying with the Robot Exclusion Protocol. In a recent lawsuit involving the Internet Archive's WayBack Machine (www.archive.org), a plaintiff not only attributed legal weight to the Exclusion Protocol, but also expected that a newly added robots.txt policy should have retroactive value!

Deception does not occur only by crawlers against servers. Some servers also attempt to deceive crawlers. For example, Web administrators may attempt to improve the ranking of their pages in a search engine by providing different content depending on whether a request originates from a browser or a search engine crawler, as determined by inspecting the request's User-Agent header. This technique, called **cloaking**, is frowned upon by search engines, which remove sites from their indices when such abuses are detected. For more information about Web spam, see Chap. 6.

One of the most serious challenges for crawlers originates from the rising popularity of pay-per-click advertising. If a crawler is not to follow advertising links, it needs to have a robust detection algorithm to discriminate ads from other links. A bad crawler may also pretend to be a genuine user who clicks on the advertising links in order to collect more money from merchants for the hosts of advertising links.

The above examples suggest a view of the Web as a new playground for artificial intelligence (AI). Crawlers need to become increasingly sophisticated to prevent insidious forms of spam from polluting and exploiting the Web environment. Malicious crawlers are also becoming smarter in their efforts, not only to spam but also to steal personal information and in general to deceive people and crawlers for illicit gains. One chapter of this arms race has been the development of CAPTCHAs [14], graphics-based inverse Turing tests automatically generated by server sites to keep out malicious crawlers. Maybe a stronger AI will be a positive outcome of crawler evolution; maybe a less usable Web will be a hefty price to pay.

Interestingly, the gap between humans and crawlers may be narrowing from both sides. While crawlers become smarter, some humans are dumbing down their content to make it more accessible to crawlers. For example some online news providers use simpler titles than can be easily classified and interpreted by a crawler as opposed or in addition to witty titles that can only be understood by humans.

Another gap that is getting narrower is the distinction between browsers and crawlers, with a growing gray area between the two. A business may wish to disallow crawlers from its site if it provides a service by which it wants to entice human users to visit the site, say to make a profit via ads on the site. A competitor crawling the information and mirroring it on its own site, with different ads, is a clear violator not only of the Robot Exclusion Protocol but also possibly of copyright law. What about an individual user who wants to access the information but automatically hide the ads? There are many browser extensions that allow users to perform all kinds of tasks that deviate from the classic browsing activity, including hiding ads, altering the appearance and content of pages, adding and deleting links, adding functionality to pages, pre-fetching pages, and so on. Such extensions have some of the functionalities of crawlers. Should they identify themselves through the User-Agent header as distinct from the browser with which they are integrated? Should a server be allowed to exclude them? And should they comply with such exclusion policies? These too are questions about ethical crawler behaviors that remain open for the moment.

8.8 Some New Developments

The typical use of (universal) crawlers thus far has been for creating and maintaining indexes for general purpose search engines. However a more diverse use of (topical) crawlers is emerging both for client and server based applications. Topical crawlers are becoming important tools to support applications such as specialized Web portals (a.k.a. "vertical" search engines), **live crawling**, and competitive intelligence.

Another characteristic of the way in which crawlers have been used by search engines up to now is the one-directional relationship between users, search engines, and crawlers. Users are consumers of information provided by search engines, search engines are consumers of information provided by crawlers, and crawlers are consumers of information provided by users (authors). This one-directional loop does not allow, for example, information to flow from a search engine (say, the queries submitted by users) to a crawler. It is likely that commercial search engines will soon leverage the huge amounts of data collected from their users to focus their crawlers on the topics most important to the searching public. To investigate this idea in the context of a vertical search engine, a system was built in which the crawler and the search engine engage in a symbiotic relationship [430]. The crawler feeds the search engine which in turn helps the crawler. It was found that such a symbiosis can help the system learn about a community's interests and serve such a community with better focus.

As discussed in Sect. 8.3, universal crawlers have to somehow focus on the most "important" pages given the impossibility to cover the entire Web and keep a fresh index of it. This has led to the use of global prestige measures such as PageRank to bias universal crawlers, either explicitly [102, 234] or implicitly through the long-tailed structure of the Web graph [401]. An important problem with these approaches is that the focus is dictated by popularity among "average" users and disregards the heterogeneity of user interests. A page about a mathematical theorem may appear quite uninteresting to the average user, if one compares it to a page about a pop star using indegree or PageRank as a popularity measure. Yet the math page may be highly relevant and important to a small community of users (mathematicians). Future crawlers will have to learn to discriminate between low-quality pages and high-quality pages that are relevant to very small communities.

Social networks have recently received much attention among Web users as vehicles to capture commonalities of interests and to share relevant information. We are witnessing an explosion of social and collaborative engines in which user recommendations, opinions, and annotations are aggregated and shared. Mechanisms include tagging (e.g., del.icio.us and flickr.com), ratings (e.g., stumbleupon.com), voting (e.g., digg.com), and hierarchical similarity (GiveALink.org) [363]. One key advantage of social systems is that they empower humans rather than depending on crawlers to discover relevant resources. Further, the aggregation of user recommendations gives rise to a natural notion of trust. Crawlers could be designed to expand the utility of information collected through social systems. For example it would be straightforward to obtain seed URLs relevant to specific communities of all sizes. Crawlers would then explore the Web for other resources in the neighborhood of these seed pages, exploiting topical locality to locate and index other pages relevant to those communities.

Social networks can emerge not only by mining a central repository of user-provided resources, but also by connecting hosts associated with individual users or communities scattered across the Internet. Imagine a user creating its own micro-search engine by employing a personalized topical crawler, seeded for example with a set of bookmarked pages. Desktop search applications make it easy to also share certain local files, if so desired. Can federations of such micro-engine agents emerge on the basis of mutual interests? Peer-to-peer (P2P) networks are beginning to be seen as robust architectures ideal for brokering among individual needs and catering to communities [354].

Adaptive peer-based search systems driven by simple distributed adaptive query routing algorithms can spontaneously organize into networks with efficient communication and with emerging clusters capturing semantic locality. Specifically, in a P2P search application called 6Search (6S), each peer crawls the Web in a focused way, guided by its user's information context. Each peer submits and responds to queries to/from its neighbors. This search process has no centralized control. Peers depend on local adaptive routing algorithms to dynamically change the topology of the peer network and search for the best neighbors to answer their queries. Machine learning techniques are being explored to improve local adaptive routing. Validation of the 6S framework and network via simulations with 70–500 model users based on actual Web crawls has yielded encouraging preliminary results. The network topology rapidly converges from a random network to a **small-world** network, with clusters emerging to match user communities with shared interests [15]. Additionally the quality of the results is significantly better than obtained by centralized search engines built with equivalent resources, and comparable with the results from much larger search engines such as Google [553, 554].

The integration of effective personalized/topical crawlers with adaptive query routing algorithms is the key to the success of peer-based social search systems. Many synergies may be exploited in this integration by leveraging contextual information about the local peer that is readily available to the crawler, as well as information about the peer's neighbors that can be mined through the stream of queries and results routed through the local peer. An open-source prototype of 6S enabling sharing of bookmarks, one-click crawling, and distributed collaborative search is available (http://homer.informatics.indiana.edu/~nan/6S/). If successful, this kind of application could create a new paradigm for crawling and searching where universal crawlers and search engines are complemented with swarms of personal crawlers and micro-engines tuned to the specialized information needs of individual users and dynamic self-organized social networks.

Bibliographic Notes

General ideas and techniques about crawling can be found in [68, 85, 263], but little is known about implementation details of commercial crawlers. Focused crawling discussed in this chapter is based on [87, 85, 139]. Literature on topical crawling algorithms is extensive [e.g., 3, 60, 102, 126, 237, 366, 369, 375, 377, 432, 434, 459]. Topical crawlers have been used for building focused repositories, automating resource discovery, and supporting software agents. For example, topical crawlers are used to collect papers for building scientific literature digital libraries such as CiteSeer and Google Scholar [308, 366, 550]. Applications of topical crawlers to business and competitive intelligence are discussed in [432], and biomedical applications in [503]. Controversial applications to harvest personal information for spam and phishing purposes are illustrated in [251].

On best-first crawlers, various methods have been used to determine an appropriate textual context in which to evaluate and score unvisited links. Using the anchor text is one strategy [123]. Another strategy is to use windows of a fixed size, e.g., 50 words around the anchor, in place of/in addition to the anchor text [237]. The weighted window used by InfoSpiders [375] yields a weight for each link, which is then fed to a neural network to score each link. In the tag (DOM) tree approach [85], using the parent node of the anchor as aggregation node worked well in a business intelligence crawling task [432]. There is a tradeoff analogous to that between precision and recall when we consider the optimal size of a link context: small contexts (e.g., anchor text) have the highest average similarities to the target page, but also highest chance to miss important cues about the target. Larger contexts (e.g., parent or grand-parent aggregator node) have lower average similarities to the target, but lower chance to miss all the keywords in the target. This suggests a greedy optimization scheme: climb the DOM tree from the anchor until sufficient terms are present in the link context [429]. This approach outperformed both the fixed-window method (with optimal window size) and the DOM tree method with a fixed aggregator depth (anchor, parent, or grandparent).

Early versions of InfoSpiders were described in [369, 374, 375, 376]. Certain aspects of evolutionary computation have also been used in other topical crawlers such as the itsy bitsy spider [96]. Another adaptive mechanism for topical crawlers inspired by natural processes is ant colony optimization [194]. The idea is that a population of agents leaves a trail of pheromone along the paths that lead to relevant pages, gradually biasing the crawl toward promising portions of the Web graph. A more extensive review of adaptive topical crawling algorithms can be found in [380].

9 Structured Data Extraction: Wrapper Generation

Web information extraction is the problem of extracting target information items from Web pages. There are two general problems: extracting information from natural language text and extracting structured data from Web pages. This chapter focuses on extracting structured data. A program for extracting such data is usually called a **wrapper**. Extracting information from text is studied mainly in the natural language processing community.

Structured data on the Web are typically data records retrieved from underlying databases and displayed in Web pages following some fixed templates. In this chapter, we still call them **data records**. Extracting such data records is useful because it enables us to obtain and integrate data from multiple sources (Web sites and pages) to provide value-added services, e.g., customizable Web information gathering, comparative shopping, meta-search, etc. With more and more companies and organizations disseminating information on the Web, the ability to extract such data from Web pages is becoming increasingly important. At the time of writing this book, there are several companies working on extracting products sold online, product reviews, job postings, research publications, forum discussions, statistics data tables, news articles, search results, etc.

Researchers and Internet companies started to work on the extraction problem from the middle of 1990s. There are three main approaches:

1. **Manual approach:** By observing a Web page and its source code, the human programmer finds some patterns and then writes a program to extract the target data. To make the process simpler for programmers, several pattern specification languages and user interfaces have been built. However, this approach is not scalable to a large number of sites.
2. **Wrapper induction:** This is the supervised learning approach, and is semi-automatic. The work started around 1995-1996. In this approach, a set of extraction rules is learned from a collection of manually labeled pages or data records. The rules are then employed to extract target data items from other similarly formatted pages.
3. **Automatic extraction:** This is the unsupervised approach started around 1998. Given a single or multiple pages, it automatically finds

patterns or grammars from them for data extraction. Since this approach eliminates the manual labeling effort, it can scale up data extraction to a huge number of sites and pages.

The first approach will not be discussed further. This chapter focuses on the last two approaches. Sects. 9.2 and 9.3 study supervised wrapper learning, and the rest of the chapter studies automatic extraction.

9.1 Preliminaries

To start our discussion, let us see some real pages that contain structured data that we want to extract. We then develop a Web data model and a HTML mark-up encoding scheme for the data model. Data extraction is simply the reverse engineering task. That is, given the HTML mark-up encoded data (i.e., Web pages), the extraction system recovers the original data model and extracts data from the encoded data records.

9.1.1 Two Types of Data Rich Pages

There are mainly two types of data rich pages. Data in such pages are usually retrieved from underlying databases and displayed on the Web following some fixed **templates**. This task is often done by computer programs.

1. **List pages:** Each of such pages contains several lists of objects. Figure 9.1 shows such a page, which has two lists of products. From a layout point of view, we see two **data regions** (one horizontal and one vertical). Within each region, the data records are formatted using the same template. The templates used in the two regions are different.
2. **Detail pages:** Such a page focuses on a single object. For example, in Fig. 9.2, the page focuses on the product "iPod Video 30GB, Black". That is, it contains all the details of the product, name, image, price and other purchasing information, product description, customer rating, etc.

Note that when we say that a page focuses on a particular object (or lists of objects), we do not mean that the page contains no other information. In fact, it almost certainly contains other information. For example, in the page for "iPod Video 30GB, Black" (Fig. 9.2), there are some related products on the right-hand side, company information at the top, and copyright notices, terms and conditions, privacy statements at the bottom, etc. They are not shown in Fig. 9.2 as we want the main part of the product clearly eligible. For list pages, it is often easy to use some heuristics to identify the main data regions, but for detail pages, it is harder.

Fig. 9.1. A segment of a list page with two data regions

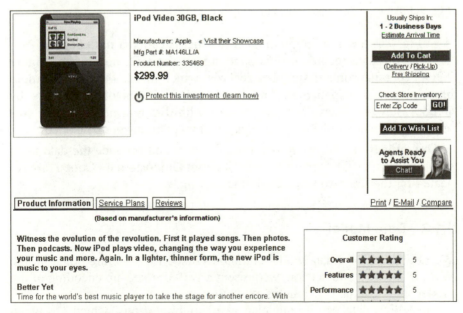

Fig. 9.2. A segment of a detail page

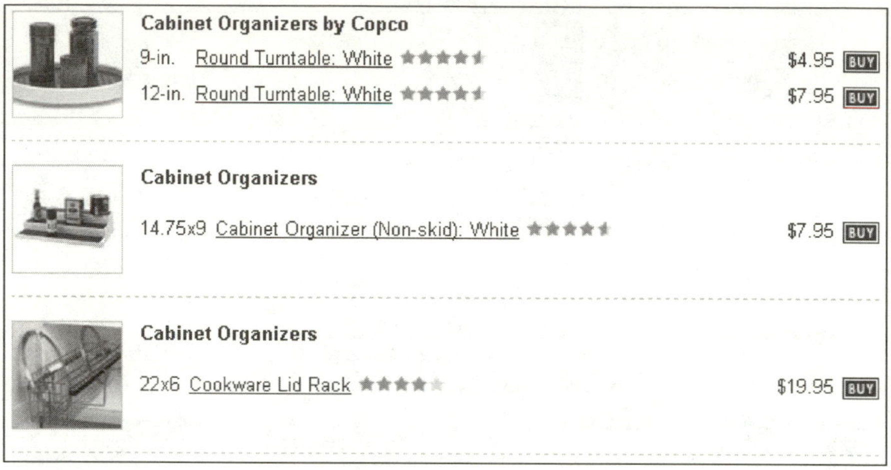

(A) An example of a nested data record

image 1	Cabinet Organizers by Copco	9-in.	Round Turntable: White	*****	$4.95
image 1	Cabinet Organizers by Copco	12-in.	Round Turntable: White	*****	$7.95
image 2	Cabinet Organizers	14.75x9	Cabinet Organizer (Non-skid): White	*****	$7.95
image 3	Cabinet Organizers	22x6	Cookware Lid Rack	****	$19.95

(B) Extraction results

Fig. 9.3. An example input page and output data table

In Fig. 9.1, the description of each product is called a **data record**. Notice that the data records in this page are all flat with no nesting. Figure 9.3(A) contains some nested data records, which makes the problem more interesting and also harder. The first product, "Cabinet Organizers by Copco," has two sizes (9-in. and 12-in.) with different prices. These two organizers are not at the same level as "Cabinet Organizers by Copco".

Our objective: We want to extract the data and produce the data table given in Fig. 9.3(B). "image 1" and "Cabinet Organizers by Copco" are repeated for the first two rows due to the nesting.

9.1.2   Data Model

We now describe a data model commonly used for structured data on the Web. In the next sub-section, we present a HTML mark-up encoding of the model and the data, which helps extraction.

Most Web data can be modeled as **nested relations**, which are typed objects allowing nested sets and tuples. The types are defined as follows:

- There is a set of **basic types**, $B = \{B_1, B_2, ..., B_k\}$. Each B_i is an atomic type, and its domain, denoted by $dom(B_i)$, is a set of constants;
- If $T_1, T_2, ..., T_n$ are basic or set types, then $[T_1, T_2, ..., T_n]$ is a **tuple type** with the domain $dom([T_1, T_2, ..., T_n]) = \{[v_1, v_2, ..., v_n] \mid v_i \in dom(T_i)\}$;
- If T is a tuple type, then $\{T\}$ is a **set type** with the domain $dom(\{T\})$ being the power set of $dom(T)$.

A basic type B_i is analogous to the type of an attribute in relational databases, e.g., string and int. In the context of the Web, B_i is usually a text string, image-file, etc. The example in Fig. 9.4 shows a nested tuple type **product**, with attributes

- *name* (of type *string*),
- *image* (of type *image-file*), and
- *differentSizes* (a *set* type), consisting of a set of tuples with attributes:
  - *size* (of type *string*), and
  - *price* (of type *string*).

```
product [ name:          string;
          image:         image-file;
          differentSizes: {[ size:   string;
                             price:  string; ]}]
```

Fig. 9.4. An example nested type

We can also define flat tuple and set types:

- If $T_1, T_2, ..., T_n$ are basic types, then $[T_1, T_2, ..., T_n]$ is a **flat tuple type**;
- If T is a flat tuple type, then $\{T\}$ is a **flat set type**.

Classic **flat relations** are of flat set types. Nested relations are of arbitrary set types. Types can be represented as trees.

- A basic type B_i is a leaf tree or node;
- A tuple type $[T_1, T_2, ..., T_n]$ is a tree rooted at a **tuple node** with n subtrees, one for each T_i;
- A set type $\{T\}$ is a tree rooted at a **set node** with one sub-tree.

An **instance** of a type T is simply an element of $dom(T)$. Clearly, instances can be represented as trees as well:

- An instance (constant) of a basic type is a leaf tree;
- A tuple instance $[v_1, v_2, ..., v_n]$ forms a tree rooted at a tuple node with n children or sub-trees representing attribute values $v_1, v_2, ..., v_n$;
- A set instance $\{e_1, e_2, ..., e_n\}$ forms a set node with n children or sub-trees representing the set elements $e_1, e_2, ..., $ and e_n.

An instance of a tuple type (also known as a **tuple instance**) is usually called a **data record** in the data extraction research. An instance of a set type (also known as a **set instance**) is usually called a **list** as in an actual Web page the data records in the set are presented in a particular order. An instance of a flat tuple type is called a **flat data record** (no nested lists), and an instance of a **flat set type** is called **a list of flat data records**.

We note that attribute names are not included in the type tree. We next introduce a **labeling** of a type tree, which is defined recursively:

- If a set node is labeled φ, then its child is labeled $\varphi.0$, a tuple node;
- If a tuple node is labeled φ, then its n children are labeled $\varphi.1$, ..., $\varphi.n$.

We can think of labels as abstract names for types or attributes. For example, in Fig. 9.4 the top level tuple type is "product", its three children are attributes: product.name, product.image, and product.differentSizes. $\varphi.0$ labels a tuple node without a name of two attributes, "size" and "price".

9.1.3   HTML Mark-Up Encoding of Data Instances

In a Web page, the data is encoded or formatted with HTML mark-up tags. This sub-section discusses the encoding of data instances in the above abstract data model using HTML tags.

Web pages are written in HTML consisting of plain texts, tags and links to image, audio and video files, and other pages. Most HTML tags work in pairs. Each pair consists of an **open tag** and a **close tag** indicated by < > and </> respectively. Within each corresponding tag-pair, there can be other pairs of tags, resulting in nested structures. Thus, HTML tags can naturally encode nested data. We note the following:

1. There are no designated tags for each type as HTML was not designed as a data encoding language. Any HTML tag can be used for any type.
2. For a tuple type, values (**data items**) of different attributes are usually encoded differently to distinguish them and to highlight important items.
3. A tuple may be partitioned into several groups or sub-tuples. Each group covers a disjoint subset of attributes and may be encoded differently.

Based on these characteristics of the HTML language, the HTML mark-up encoding of instances is defined recursively below. We encode based on the type tree, where each node of the tree is associated with an encoding function, which will encode (or mark-up) all the instances of the type in the same way. We will use the tuple type and its attributes explicitly because values of different attributes in the tuple type are typically encoded differently. We use $T.i$ to represent a value instance of the tuple type T and attribute i. We use *enc* to denote an abstract encoding function.

- For a leaf node of a basic type labeled φ, an instance c is encoded with

  $enc(\phi.c) = OPEN\text{-}TAGS\ c\ CLOSE\text{-}TAGS$

  where *OPEN-TAGS* is a sequence of open HTML tags, and *CLOSE-TAGS* is the sequence of corresponding close HTML tags. The number of tags is greater than or equal to 0.

- For a tuple node labeled φ of n children or attributes, $[\varphi.1, ..., \varphi.n]$, the attributes are first partitioned into h (≥ 1) groups $<\varphi.1, ..., \varphi.e>$, $<\varphi.(e+1),..., \varphi.g> ... <\varphi.(k+1), ..., \varphi.n>$ and an instance $[v_1, ..., v_n]$ of the tuple node is encoded with

$$enc(\varphi.[v_1, ..., v_n]) = OPEN\text{-}TAGS_1\ enc(v_1) ... enc(v_e)\ CLOSE\text{-}TAGS_1$$
$$OPEN\text{-}TAGS_2\ enc(v_{e+1})...enc(v_g)\ CLOSE\text{-}TAGS_2$$
$$...$$
$$OPEN\text{-}TAGS_h\ enc(v_{k+1})...enc(v_n)\ CLOSE\text{-}TAGS_h$$

  where *OPEN-TAGS$_i$* is a sequence of open HTML tags, and *CLOSE-TAGS$_i$* is the sequence of corresponding close tags. The number of tags is greater than or equal to 0.

- For a set node labeled φ, an non-empty set instance $\{e_1, e_2, ..., e_n\}$ is encoded with

  $$enc(\varphi.\{e_1, ..., e_n\}) = OPEN\text{-}TAGS\ enc(e_{j_1})...enc(e_{j_n})\ CLOSE\text{-}TAGS,$$

  where *OPEN-TAGS* is a sequence of open HTML tags, and *CLOSE-TAGS* is the sequence of corresponding close HTML tags. The number of tags is greater than or equal to 0. The set elements are ordered based on an ordering function $<$. With ordering, a set instance is called a **list**. An empty set instance is encoded with *OPEN-TAGS CLOSE-TAGS*.

By no means does this mark-up encoding cover all cases in Web pages. In fact, each group of a tuple type can be further divided. Anyway, you get the idea. We should also note that in an actual Web page the encoding is usually done not only by HTML tags, but also by words and punctuation marks. For example, in Fig. 9.5, if we are interested in extracting the addresses and the area codes, the punctuation marks are useful.

Restaurant Name: **Good Noodles**

- 205 Willow, *Glen*, Phone 1-773-366-1987
- 25 Oak, *Forest*, Phone (800) 234-7903
- 324 Halsted St., *Chicago*, Phone 1-800-996-5023
- 700 Lake St., *Oak Park*, Phone: (708) 798-0008

Fig. 9.5. Words and punctuation marks are also used in data encoding

9.2   Wrapper Induction

We are now ready to study the first approach to data extraction, namely **wrapper induction**, which is based on supervised learning. A wrapper induction system learns data extraction rules from a set of labeled training examples. Labeling is usually done manually, which simply involves marking the data items in the training pages/examples that the user wants to extract. The learned rules are then applied to extract target data from other pages with the same mark-up encoding or the same template.

The algorithm discussed in this section is based on the Stalker system [399]. Related work includes WIEN [296], Softmealy [244], WL^2 [108], the system in [250], etc. The next section describes a different learning approach, which is based on the IDE system given in [599].

Stalker models the Web data as nested relations. Let us model the restaurant page in Fig. 9.5. It has four addresses in four different cities. The type tree of the data is given in Fig. 9.6 (the country code is omitted). For each type, we also added an intuitive label. The wrapper uses a tree structure based on this to facilitate extraction rule learning and data extraction.

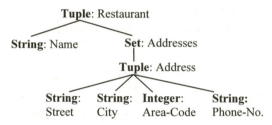

Fig. 9.6. Type tree of the restaurant page in Fig. 9.5

Below, we first introduce the data extraction process, and then describe the learning algorithm for generating extraction rules.

9.2.1   Extraction from a Page

A Web page can be seen as a sequence of **tokens** S (e.g., words, numbers and HTML tags). The extraction is done using a tree structure called the *EC* tree (**embedded catalog tree**), which models the data embedding in a HTML page. The *EC* tree is based on the type tree above. The root of the tree is the document containing the whole token sequence S of the page, and the content of each child node is a subsequence of the sequence of its parent node. To extract a node of interest, the wrapper uses the *EC* description of the page and a set of extraction rules. Figure 9.7 shows the *EC* tree of the page in Fig. 9.5. Note that we use **LIST** here because the set of addresses are already ordered in a page. For an extraction task, the *EC* tree for a data source is specified by the user (not discovered by the system).

Fig. 9.7. The EC tree of the HTML page in Fig. 9.5

For each node in the tree, the wrapper identifies or extracts the content of the node from its parent, which contains the sequence of tokens of all its children. Each extraction is done using two rules, **the start rule** and **the end rule**. The start rule identifies the beginning of the node and the end rule identifies the end of the node. This strategy is applicable to both leaf nodes (which represent data items) and list nodes. For a list node, **list iteration rules** are needed to break the list into individual data records (tuple instances). To extract items from the data records, data extraction rules are applied to each record. All the rules are learned during wrapper induction, which will be discussed in Sect. 9.2.2. Given the *EC* tree and the rules, any node can be extracted by following the tree path *P* from the root to the node by extracting each node in *P* from its parent.

The extraction rules are based on the idea of **landmarks**. Each landmark is a sequence of consecutive tokens and is used to locate the beginning or the end of a target item. Let us use the example in Fig. 9.5 to introduce extraction rules and the extraction process based on the *EC* tree (Fig. 9.7). Figure 9.8 shows the HTML source code of the page in Fig. 9.5.

```
1:  <p> Restaurant Name: <b>Good Noodles</b><br><br>
2:  <li> 205 Willow, <i>Glen</i>, Phone 1-<i>773</i>-366-1987</li>
3:  <li> 25 Oak, <i>Forest</i>, Phone (800) 234-7903 </li>
4:  <li> 324 Halsted St., <i>Chicago</i>, Phone 1-<i>800</i>-996-5023 </li>
5:  <li> 700 Lake St., <i>Oak Park</i>, Phone: (708) 798-0008 </li>  </p>
```

Fig. 9.8. The HTML source of the page in Fig. 9.5

Let us try to extract the restaurant name "Good Noodles". The following rule can be used to identify the beginning of the name:

R1: *SkipTo(<b>)*

This rule means that the system should start from the beginning of the page and skip all the tokens until it sees the first <b> tag. <b> is a landmark.

Obviously, *SkipTo*(:) and *SkipTo*(<i>) will not work. According to the *EC* tree in Fig. 9.7, **R1** is applied to the parent of node *name*, which is the root node. The root node contains the token sequence of the whole page.

Similarly, to identify the end of the restaurant name, we can use:

R2:   *SkipTo*(</b>)

R2 is applied from the end of the page toward the beginning. **R1** is called the **start rule** and **R2** is called the **end rule**.

Note that a rule may not be unique. For example, we can also use the following rules (and many more) to identify the beginning of the name:

R3:   *SkiptTo*(Name _Punctuation_ _HtmlTag_)

or   **R4**:   *SkiptTo*(Name) *SkipTo*(<b>)

R3 means that we skip everything till the word "Name" followed by a punctuation symbol and then a HTML tag. In this case, "Name _Punctuation_ _HtmlTag_" together is a landmark. *_Punctuation_* and *_HtmlTag_* are called **wildcards**. A wildcard represents a class of tokens. For example, *_HtmlTag_* represents any HTML tag, i.e., any HTML tag matches the wildcard *_HtmlTag_*. **R4** means that we skip everything till the word "Name" and then again skip everything till the tag <b>. Since wrapper induction algorithms find simple rules first, **R1** will be produced.

Now, suppose that we also want to extract each area code. The wrapper needs to perform the following steps:

1. Identify the entire list of addresses. We can use the start rule *SkipTo*(

), and the end rule *SkipTo*(</p>).
2. Iterate through the list (lines 2-5 in Fig. 9.8) to break it into four individual records. To identify the beginning of each address, the wrapper can start from the first token of the parent and repeatedly apply the start rule *SkipTo*(<li>) to the content of the list. Each successive identification of the beginning of an address starts from where the previous one ends. Similarly, to identify the end of each address, it starts from the last token of its parent and repeatedly apply the end rule *SkipTo*(</li>).

Once each address record is identified or extracted, we can extract the area code in it. Due to variations in the format of area codes (some are in italic and some are not), we need to use disjunctions. In this case, the disjunctive start and the end rules are respectively **R5** and **R6**:

R5: either *SkipTo*(( )   **R6: either** *SkipTo*( ))
       or *SkipTo*(-<i>)   **or** *SkipTo*(</i>)

In a disjunctive rule, the disjuncts are applied sequentially until a disjunct can identify the target node.

E1:   205 Willow, \<i>Glen\</i>, Phone 1-\<i>773\</i>-366-1987
E2:   25 Oak, \<i>Forest\</i>, Phone (800) 234-7903
E3:   324 Halsted St., \<i>Chicago\</i>, Phone 1-\<i>800\</i>-996-5023
E4:   700 Lake St., \<i>Oak Park\</i>, Phone: (708) 798-0008

Fig. 9.9. Training examples: four addresses with labeled area codes

Finally, we summarize the data extraction features of Stalker.

1. Extraction is done hierarchically based on the *EC* tree, which enables extraction of items at any level of the hierarchy.
2. The extraction of each node is independent of its siblings. No contextual or ordering information of siblings is used in extraction or rule learning.
3. Each extraction is done using two rules, the start rule and the end rule. Each rule consists of an ordered list of disjuncts (could be one).

9.2.2   Learning Extraction Rules

We now present the wrapper learning algorithm for generating extraction rules. The basic idea is as follows: To generate the start rule for a node in the EC tree, some *prefix* tokens or their wildcards of the node are identified as the landmarks that can uniquely identify the beginning of the node. To generate the end rule for a node, some *suffix* tokens or their wildcards of the node are identified as the landmarks. The rule generation process for the start rule and the end rule is basically the same. Their applications are also similar except that to apply a start rule the system starts by consuming the first token in the sequence of the parent and goes towards the last token, while for an end rule the system starts from the last token in the sequence of the parent and goes towards the first. Without loss of generality, in this section, we will discuss only the generation of start rules.

For rule learning, the user first marks or labels the target items that need to be extracted in a few training examples. For instance, we have the examples in Fig. 9.8, which are addresses from the page in Fig. 9.5. Suppose we want to generate rules to extract the area code from each address. The area codes are labeled (marked) as in Fig. 9.9. A graphic user interface can make the labeling process very easy.

Given a set of labeled training examples *E*, the learning algorithm should generate extraction rules that extract all the target items (also called **positive items**) without extracting any other items (called **negative items**).

Learning is done based on the machine learning method, **sequential covering** (see Sect. 3.4.1). The algorithm is given in Fig. 9.10. In each iteration, the algorithm **LearnRule()** (Fig. 9.10) generates a **perfect dis-**

junct that covers as many positive items as possible and does not cover any negative item in E (Examples). Then, all the examples whose positive items are covered by the rule are removed. The next iteration starts. The input to LearnRule() is E. Once all the positive items are covered, the rule is returned (line 6), which consists of an ordered list of learned disjuncts.

The function **LearnDisjunct()** performs the actual generation of perfect disjuncts (Fig. 9.11). It works as follows: It first chooses a **Seed** example (line 1), which is the shortest example. In the case of Fig. 9.9, it is E2. It then generates the initial candidate disjuncts. Let us explain using a generic *Seed*, which can be represented as follows:

$t_1\ t_2\ \ldots\ t_k$ <target item> $t_{k+1}\ t_{k+2}\ \ldots\ t_n$,

where t_i is a token and <target item> is a labeled target item. We call $t_1\ t_2$ $\ldots\ t_k$ the prefix sequence of the target item, and $t_{k+1}t_{k+2}\ldots t_n$ the suffix sequence of the target item. The initial candidate disjuncts for the start rule are t_k, and its matching wildcards. Let us use seven wildcards, *_Numeric_*, *_AlphaNum_*, *_Alphabetic_*, *_Capitalized_*, *_AllCaps_*, *_HtmlTag_*, and *_Punctuation_*. Their meanings are self-explanatory. For the example E2 of Fig. 9.9, the following candidate disjuncts are generated:

D1: *SkipTo(( )*
D2: *SkipTo(_Punctuation_)*

In line 4 of LearnDisjunct(), the function BestDisjunct() selects the best disjunct using a set of heuristics given in Fig. 9.13.

In this case, **D1** is selected as the best disjunct. **D1** is a perfect disjunct, i.e., it only covers positive items in E2 and E4 but not any negative items in E. **D1** is returned from LearnDisjunct(), which also ends the first iteration of LearnRule(). E2 and E4 are removed (line 4) of Fig. 9.10. The next iteration of LearnRule() is left with E1 and E3. LearnDisjunct() will select E1 as *Seed* as it is shorter. Two candidates are then generated:

D3: *SkipTo(<i>)*
D4: *SkipTo(_HtmlTag_)*

Both these two candidates match early in the uncovered examples, E1 and E3. Thus, they cannot uniquely locate the positive items. Even worse, they can match to negative items in the two already covered examples, E2 and E4. Refinement is thus needed, which aims to specialize a disjunct by adding more **terminals** (a token or one of its matching wildcards) to it. We hope the refined version will be able to uniquely identify the positive items in some examples without matching any negative item in any example in E. Two refinement strategies are used:

Algorithm LearnRule(*Examples*)   // *Examples*: training examples
1   *Rule* ← ∅   // *Rule*: the returned rule
2   **while** *Examples* ≠ ∅ **do**
3   *Disjunct* ← LearnDisjunct(*Examples*);
4   remove all examples in *Examples* covered by *Disjunct*;
5   add *Disjunct* to *Rule*
6   **return** *Rule*

Fig. 9.10. The main learning algorithm – based on sequential covering

Function LearnDisjunct(*Examples*)
1   let *Seed* ∈ *Examples* be the shortest example;
2   *Candidates* ← GetInitialCandidates(*Seed*);
3   **while** *Candidates* ≠ ∅ **do**
4   *D* ← BestDisjunct(*Candidates*);
5   **if** *D* is a perfect disjunct **then**
6   **return** *D*
7   *Candidates* ← *Candidates* ∪ Refine(*D*, *Seed*);
8   remove *D* from *Candidates*;
9   **return** *D*

Fig. 9.11. Learning disjuncts

Function Refine(*D*, *Seed*)
1   *D* is a consecutive landmarks $(l_0, l_1, ..., l_n)$; // l_i is in fact *SkipTo*(l_i)
2   *TopologyRefs* ← *LandmarkRefs* ← ∅;
3   **for** $i = 1$ **to** n **do**   // t_0 or t_1 below may be null
4   **for** each sequence $s = t_0 \, l_i \, t_1$ before the target item in *Seed* **do**
5   *LandmarkRefs* ← *LandmarkRefs* ∪ $\{(l_0, ..., l_{i-1}, t_0 \, l_i, ..., l_n)\}$ ∪
                $\{(l_0, ..., l_{i-1}, x \, l_i, ..., l_n) \mid x$ is a wildcard that matches $t_0\}$
                ∪ $\{(l_0, ..., l_i \, t_1, l_{i+1}, ..., l_n)\}$ ∪
                $\{(l_0, ..., l_i \, x, l_{i+1}, ..., l_n) \mid x$ is a wildcard that matches $t_1\}$
6   **for** each token t between l_{i-1} and l_i before the target item in *Seed* **do**
7   *ToplogyRefs* ← *TopologyRefs* ∪ $\{(l_0, ..., l_i, t, l_{i+1}, ..., l_n)\}$ ∪
                $\{(l_0, ..., l_i, x, l_{i+1}, ..., l_n)\} \mid x$ is a wildcard that matches $t\}$
8   **return** *TopologyRefs* ∪ *LandmarkRefs*

Fig. 9.12. Refining a disjunct to generate more specialized candidates

BestDisjunct () prefer candidates that have:
- more correct matches
- accepts fewer false positives
- fewer wildcards
- longer end-landmarks

Fig. 9.13. Choosing the best disjunct

D7:   *SkipTo*(205) SkipTo(<i>)   **D15**: *SkipTo*(_Numeric_) SkipTo(<i>)
D8:   *SkipTo*(Willow) SkipTo(<i>)   **D16**: *SkipTo*(_Alphabetic_) SkipTo(<i>)
D9:   *SkipTo*(,) SkipTo(<i>)   **D17**: *SkipTo*(_Punctuation_) SkipTo(<i>)
D10: *SkipTo*(<i>) SkipTo(<i>)   **D18**: *SkipTo*(_HtmlTag_) SkipTo(<i>)
D11: *SkipTo*(Glen) SkipTo(<i>)   **D19**: *SkipTo*(_Capitalized_) SkipTo(<i>)
D12: *SkipTo*(1) SkipTo(<i>)   **D20**: *SkipTo*(_AlphaNum_) SkipTo(<i>)
D13: *SkipTo*(-) SkipTo(<i>)   **D21**: *SkipTo*(</i>) SkipTo(<i>)
D14: *SkipTo*(Phone) SkipTo(<i>)

Fig. 9.14. All 15 topology refinements of **D3**

1. **Landmark refinement** (lines 4-5 in Fig. 9.12): Increase the size of a landmark l_i by concatenating a terminal (a token t_0 or t_1, and its matching wildcards) at the beginning or at the end of l_i. If t_0 or t_1 does not exist, it will not be considered. We note that each landmark l_i in the algorithm in Fig. 9.12 actually represents the *SkipTo*(l_i).
2. **Topology refinement** (lines 6-7 in Fig. 9.12): Increase the number of landmarks by adding 1-terminal landmarks, i.e., *t* and its matching wildcards. Note that l_0 is not a landmark, which is used to simplify the algorithm presentation. It represents the beginning of the *Seed* example.

Let us go back to our running example. **D3** is selected as the best disjunct (line 4 of Fig. 9.11). Clearly, **D3** is not a perfect disjunct. Then, refinement is carried out. Landmark refinement produces the following candidates:

D5:   *SkipTo*(- <i>)
D6:   *SkipTo*(_Punctuation_ <i>)

Topology refinement produces the 15 candidates in Fig. 9.14. We can already see that **D5, D10, D12, D13, D14, D15, D18** and **D21** match correctly with E1 and E3 and fail to match on E2 and E4. Using the heuristics in Fig. 9.13, **D5** is selected as the final solution as it has longest last landmark (- <i>). **D5** is then returned by **LearnDisjunct()**. It is possible that no perfect disjunct can be found after all possible refinements have been tried. In this case, an imperfect best disjunct will be returned (line 9 in Fig. 9.11).

Since all the examples are covered, LearnRule() ends and returns the disjunctive (start) rule "either **D1** or **D5**", i.e.,

R7:   either *SkipTo*(( )
        or *SkipTo*(- <i>)

In summary, we note the following:

1. The algorithm presented in this section is by no mean the only possible algorithm. Many variations are possible. Of course, there are also many other entirely different algorithms for wrapper induction.

2. In our discussion above, we used only the *SkipTo*() function in extraction rules. However, in some situations it may not be sufficiently expressive. Therefore, other functions may be added. For example, Stalker also has the *SkipUntil*() function. Its argument is a part of the target item to be extracted, and is not consumed when the rule is applied. That is, the rule stops right before its occurrence.

9.2.3   Identifying Informative Examples

One of the important issues in wrapper learning is the manual labeling of training examples. To ensure accurate learning, a large number of training examples are needed. To manually label them is labor intensive and time consuming. The question is: is it possible to automatically select (unlabelled) examples that are informative for the user to label? Clearly, examples of the same format are of limited use. Examples that represent exceptions are informative as they are different from already labeled examples. **Active learning** is an approach that helps identify **informative unlabeled examples** automatically. Given a set of unlabeled examples U, the approach works as follows in the wrapper induction context:

1. Randomly select a small subset L of unlabeled examples from U
2. Manually label the examples in L, and $U = U - L$;
3. Learn a wrapper W based on the labeled set L;
4. Apply W to U to find a set of informative examples L;
5. Stop if $L = \varnothing$, otherwise go to step 2.

The key is to find informative examples in step 4. In [400], Muslea et al. proposed a method, called **co-testing**, to identify informative examples.

The idea of co-testing is simple. It exploits the fact that there are often multiple ways of extracting the same item. Thus, the system can learn different rules, **forward** and **backward rules**, to locate the same item. Let us use learning of start rules as an example. The rules learned in Sect. 9.2.2 are called *forward rules* because they consume tokens from the beginning of the example to the end. In a similar way, we can also learn *backward rules* that consume tokens from the end of the example to the beginning.

Given an unlabeled example, both the forward rule and backward rule are applied. If the two rules disagree on the beginning of a target item in the example, this example is given to the user to label. The intuition behind is simple. When the two rules agree, the extraction is very likely to be correct. When the two rules do not agree on the example, one of them must be wrong. By giving the user the example to label, we obtain a labeled informative training example.

9.2.4   Wrapper Maintenance

Once a wrapper is generated, it is applied to other Web pages that contain similar data and are formatted in the same ways as the training examples. This introduces new problems.

1. If the site changes, does the wrapper know the change? This is called the **wrapper verification problem**.
2. If the change is correctly detected, how to automatically repair the wrapper? This is called the **wrapper repair problem**.

One way to deal with both problems is to learn the characteristic patterns of the target items, which are then used to monitor the extraction to check whether the extracted items are correct. If they are incorrect, the same patterns can be used to locate the correct items assuming that the page changes are minor formatting changes. This is called **re-labeling**. After re-labeling, **re-learning** is performed to produce a new wrapper. These two tasks are very difficult because contextual and/or semantic information is often needed to detect changes and to find the new locations of the target items. Wrapper maintenance is still an active research area.

9.3.   Instance-Based Wrapper Learning

The wrapper induction method discussed in the previous section requires a set of labeled examples to learn extraction rules. Active learning may be applied to identify informative examples for labeling to reduce the manual labeling effort. In this section, we introduce an instance-based learning approach to wrapper building, which does not learn extraction rules. Instead, it extracts target items in a new instance/page by comparing their prefix and suffix token strings with those of the corresponding items in the labeled examples. At the beginning, the user needs to label only a single example, which is then used to identify target items from unlabeled examples. If some item in an unlabeled example cannot be identified, it is sent for labeling, which is *active learning* but with no additional mechanism. Thus, in this approach the user labels only a minimum number of training examples. The method described here is based on the IDE algorithm in [599], which is given in Fig. 9.15. It consists of three steps:

1. A random example p from a set of unlabeled training examples S is selected for labeling (line 1). The examples here can be a set of detail pages or a set of data records identified from list pages. We will see in Sect. 9.8 that data records in list pages can be identified automatically.

Algorithm IDE(*S*)   // *S* is the set of unlabeled examples.
1. $p \leftarrow$ randomSelect(*S*);   // Randomly select a page *p* from *S*
2. $T_p \leftarrow$ labeling(*p*);   // the user labels the page *p*
3.   *Templates* $\leftarrow \langle T_p \rangle$;   // initialization
4.   **for** each remaining unlabeled example *d* in *S* **do**
5.   **if** ¬(extract(*Templates*, *d*)) **then**
6. $T_d \leftarrow$ labeling(*d*);
7.   insert T_d into *Templates*
8.   **end-if**
9.   **end-for**

Fig. 9.15. The IDE algorithm

2. The user labels/marks the target items in the selected example *p* (line 2). The system also stores a sequence of *k* consecutive *tokens* right before each labeled item (called the *prefix string* of the item) and a sequence of *k* consecutive tokens right after the labeled item (called the *suffix string* of the item). The prefix and suffix strings of all target items form a **template**. Storing the prefix and suffix strings is to avoid keeping the whole page in memory. The value of *k* does not affect the extraction result. If it is too small, the algorithm can always get more tokens from the original page. In practice, we can give *k* a large number, say 30, so that the system does not have to refer back to the original page during extraction. The variable *Templates* keeps all templates (line 3).

3. The algorithm then starts to extract items from unlabeled examples (line 4–9) using the function extract() (line 5). For each unlabeled example *d*, it compares the stored prefix and suffix strings of each target item with the token string of *d* to identify its corresponding item. If some item from *d* cannot be identified, *d* is passed to the user for labeling (line 6) (which is *active learning*), i.e., *d* is an **informative example**.

Let us use an example to show what a template looks like. For example, in the page of Fig. 9.1, we are interested in extracting three items from each product, namely, name, image, and price. The template (T_j) for a labeled example *j* is represented with:

$$T_j = \langle pat_{name}, pat_{image}, pat_{price} \rangle$$

Each pat_i in T_j consists of a prefix string and a suffix string of the item *i*. For example, if the product image is embedded in the following source:

… <table><tr><td> **<img>** </td><td></td> …

then we have (we use *k* = 3 and regard each HTML tag as a token)

$$pat_{img} = (img, prefix:\langle <table><tr><td> \rangle, suffix:\langle </td><td></td> \rangle).$$

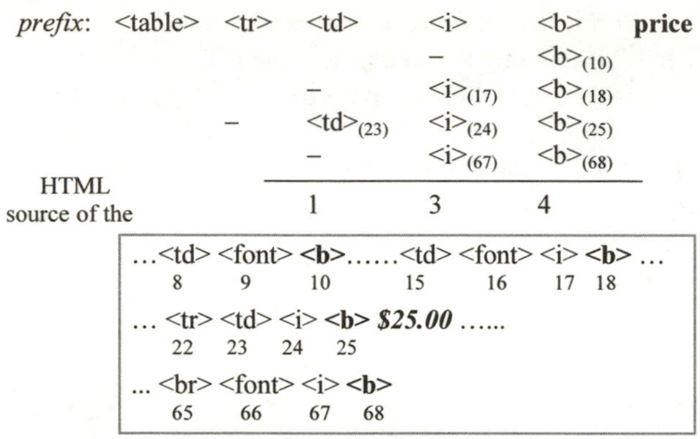

Fig. 9.16. The price is found uniquely.

Extract(*Templates, d*) function: For each unlabeled example *d*, extract() tries to use each saved template *T* (∈ *Templates*) to match with the token string of *d* to identify every target item in *d*. If a sequence of prefix (and respectively suffix) tokens of a target item *g* in *T* matches a sequence of prefix (and suffix) tokens of an item *f* in *d* that **uniquely identifies** *f* in *d*, *f* is regarded as *g*'s corresponding item in *d*. By "uniquely identifies", we mean that only item *f* in *d* matches *g* based on their prefix and suffix strings. An example is given below.

After item *f*, which corresponds to item *g* in *T*, is identified and extracted from *d*, we use the token strings of *d* before *f* and after *f* to find the remaining target items using the same template *T*. This process continues until all the corresponding items of those items in *T* are identified from *d*. If the corresponding item of an item in *T* cannot be uniquely identified from *d*, then the extraction using *T* fails on *d*. The next template in *Templates* is tried. If every template in *Templates* fails on *d*, *d* is sent to the user for labeling (line 6 of Fig. 9.15). The algorithm is fairly straightforward, and thus is omitted. See [599] for more details, which also discusses how to deal with some additional issues, e.g., missing items in a page.

Fig. 9.16 gives an example to show how a target item is *uniquely* identified. Assume that 5 tokens <table><tr><td><i><b> are saved in the prefix string of item price from a labeled example. Given an unlabeled example, after scanning through its token string, we obtain the match situation in Fig. 9.16. That is, we find 4 <b>'s, three <i><b> together, and only one <td><i><b> together, which can match some prefix tokens of price. These are shown in four rows below the saved prefix string. The number within each "( )" is the sequence id of the token (the tag) in the unlabeled example. "−" means no match. The HTML source is given in the box of Fig. 9.16 with sequence id's attached. We observe that the beginning of price is uniquely identified because the sequence of prefix tokens of price, <td><i><b>, has only one match. Note that we do not need to use all the saved tokens in the prefix string of price. This technique is thus called **sufficient match**. We see that <b> is not unique because there are 4 <b>'s. <i><b> is not unique because there are 3 matches.

Once the beginning of item price is found, the algorithm tried to locate the ending of item price in the same way by comparing suffix strings in the opposite direction. After item price is identified and extracted, the algorithm goes to identify other items if they are not extracted.

The final set of templates and the extract() function together form a wrapper, which can be used to extract target items from future examples.

Apart from performing active learning automatically, there are two other interesting features about IDE. Firstly, there is no pre-specified sequence of items to be extracted. For example, the order of items in the HTML source may be: name, price, and image. If at the beginning we can identify item price uniquely in the unlabeled example, we can then start from price and search forward to find item image and search backward to find item name. The final extraction sequence of items may be price, image and name. Secondly, the method exploits local contexts in extraction. It may be the case that from the whole page/data record we are unable to identify a particular item. However, within a local area, it is easy to identify the item. For instance, in the above example, after identifying item price, we only need to search for item image in the rest of the input. Even a similar item appears before price, it will not be considered. Evaluation results in [599] show that this simple technique works very well.

9.4 Automatic Wrapper Generation: Problems

Wrapper generation using supervised learning has two main shortcomings:

1. It is not suitable for a large number of sites due to the manual labeling effort. For example, if a shopping site wants to extract all the products sold on the Web, manual labeling becomes almost an impossible task.
2. Wrapper maintenance is very costly. The Web is a dynamic environment. Sites change constantly. Since wrapper learning systems mainly rely on HTML formatting tags, if a site changes its formatting templates, the existing wrapper for the site will become invalid. As we discussed earlier, automatic verification and repair are still difficult. Doing them manually is very costly if the number of sites involved is large.

Due to these problems, automatic (or unsupervised) extraction has been studied by researchers. Automatic extraction is possible because data records (tuple instances) in a Web site are usually encoded using a very small number of fixed templates. It is possible to find these templates by mining repeated patterns in multiple data records. The rest of the chapter focuses on automatic extraction.

Note that in general we use the term "templates" to refer to hidden templates employed by Web page designers. We use the term "patterns" to refer to regular structures that the system has discovered.

9.4.1   Two Extraction Problems

In Sects. 9.1.2 and 9.1.3, we described an abstract model of structured data on the Web (i.e., nested relations), and a HTML mark-up encoding of the data model respectively. The general goal of data extraction is to recover the hidden schema from the HTML mark-up encoded data. In the rest of the chapter, we focus on two problems, which are really quite similar.

Problem 1: Extraction Based on a Single List Page

Input: A single HTML string S, which contains k non-overlapping substrings $s_1, s_2, \ldots, s_k$ with each s_i encoding an instance of a set type. That is, each s_i contains a collection W_i of m_i (≥ 2) non-overlapping substrings encoding m_i instances of a tuple type.

Output: k tuple types $\sigma_1, \sigma_2, \ldots, \sigma_k$, and k collections $C_1, C_2, \ldots, C_k$ of instances of the tuple types such that for each collection C_i there is a HTML encoding function enc_i such that $enc_i: C_i \rightarrow W_i$ is a bijection.

We use the example in Fig. 9.1 to explain. The input string S is the full Web page (only part of it is shown in Fig. 9.1). In this page, there are two substrings s_1 and s_2 that encode two set instances, i.e., the two sets of data records. s_1 consists of four encodings (displayed horizontally) $enc_1(I_1)$, $enc_1(I_2)$, $enc_1(I_3)$, $enc_1(I_4)$ of four product instances I_1, I_2, I_3, I_4 of a tuple type σ_1, according to some mark-up encoding function enc_1. Similarly, s_2 consists of encodings of some other products (displayed vertically). One important note is that S often contains some other information (not shown in Fig. 9.1) apart from the encoded data. An algorithm needs to work on the string S to find each substring and construct the tuple type by generating a pattern from each substring representing the mark-up encoding function enc_i.

The pattern may be represented as a **regular expression**. Data extraction can be done using the regular expression or the original pattern as we will see in Sect. 9.11.1.

Problem 2: Extraction Based on Multiple Pages

Input: A collection W of k HTML strings, which encodes k instances of the same type.

Output: A type σ, and a collection C of instances of type σ, such that there is a HTML encoding *enc* such that *enc*: $C \to W$ is a bijection.

The input consists of a collection of k encodings $enc(I_1)$, $enc(I_2)$, ..., $enc(I_k)$ of instances I_1, I_2, ..., I_k of a nested type σ, according to some mark-up encoding function *enc*. An algorithm works on the encoded instances and constructs the type by generating a pattern (the encoding function *enc*), which again may be represented as a regular expression and used to extract data from other pages. Note that, for this problem, the input may be a set of detail pages (of a tuple type) or list pages (of a set type).

The next few sections describe several techniques to solve the two problems. As we will see in Sect. 9.10, most techniques for solving problem 1 can also be used for solving problem 2.

9.4.2 Patterns as Regular Expressions

A regular expression can be naturally used to model the HTML encoded version of a nested type. Given an alphabet of symbols Σ and a special token "*#text*" that is not in Σ, a **regular expression** over Σ is a string over $\Sigma \cup \{\textit{#text}, *, ?, |, (,)\}$ defined as follows:

- The empty string ε and all elements of $\Sigma \cup \{\textit{#text}\}$ are regular expressions.
- If A and B are regular expressions, then AB, $(A|B)$ and $(A)^?$ are regular expressions, where $(A|B)$ stands for A or B and $(A)^?$ stands for $(A|\varepsilon)$.
- If A is a regular expression, $(A)^*$ is a regular expression, where $(A)^*$ stands for ε or A or AA or ...

We also use $(A)^+$ as a shortcut for $A(A)^*$, which can be used to model the set type of a list of tuples. $(A)^?$ indicates that A is optional. $(A|B)$ represents a disjunction. If a regular expression does not include $(A|B)$, it is called a **union-free regular expression**. Regular expressions are often employed to represent extraction patterns (or encoding functions). However, extraction patterns do not have to be regular expressions, as we will see later.

Given a regular expression, a nondeterministic finite-state automaton can be constructed and employed to match its occurrences in string sequences representing Web pages. In the process, data items can be extracted, which are text strings represented by *#text*.

9.5   String Matching and Tree Matching

As we can see from both problems in Sect. 9.4.1, the key is to find the encoding template from a collection of encoded instances of the same type. A natural way to do this is to detect repeated patterns from HTML encoding strings. **String matching** and **tree matching** are obvious techniques for the task. Tree matching is useful because HTML encoding strings also form nested structures due to their nested HTML tags. Such nested structures can be modeled as trees, commonly known as **DOM (tag) trees**. DOM stands for **Document Object Model** (http://www.w3.org/DOM/). Below we describe some string matching and tree matching algorithms.

9.5.1   String Edit Distance

String **edit distance** (also known as **Levenshtein distance**) is perhaps the most widely used string matching/comparison technique. The edit distance of two strings, s_1 and s_2, is defined as the minimum number of **point mutations** required to change s_1 into s_2, where a point mutation is one of: (1) change a character, (2) insert a character, and (3) delete a character.

Assume we are given two strings s_1 and s_2. The following recurrence relations define the edit distance, $d(s_1, s_2)$, of two strings s_1 and s_2:

$$d(\varepsilon, \varepsilon) = 0 \qquad\qquad \text{// } \varepsilon \text{ represents an empty string}$$
$$d(s, \varepsilon) = d(\varepsilon, s) = |s| \qquad \text{// } |s| \text{ is the length of string } s$$
$$d(s_{1\_}+c_1, s_{2\_}+c_2) = \min(d(s_{1\_}, s_{2\_}) + p(c_1, c_2), d(s_{1\_}+c_1, s_{2\_}) + 1,$$
$$d(s_{1\_}, s_{2\_}+c_2) + 1),$$

where c_1 and c_2 are the last characters of s_1 ($= s_{1\_}+c_1$) and s_2 ($= s_{2\_}+c_2$) respectively, and $p(c_1, c_2) = 0$ if $c_1 = c_2$; $p(c_1, c_2) = 1$, otherwise.

The first two rules are obvious. Let us examine the last one. Since neither string is empty, each has a last character, c_1 and c_2 respectively. c_1 and c_2 have to be explained in an edit of $s_{1\_}+c_1$ into $s_{2\_}+c_2$. If $c_1 = c_2$, they match with no penalty, i.e., $p(c_1, c_2) = 0$, and the overall edit distance is $d(s_{1\_}, s_{2\_})$. If $c_1 \ne c_2$, then c_1 could be changed into c_2, giving $p(c_1, c_2) = 1$ and an overall cost $d(s_{1\_}, s_{2\_})+1$. Another possibility is to edit $s_{1\_}+c_1$ into $s_{2\_}$ and then insert c_2, giving $d(s_{1\_}+c_1, s_{2\_})+1$. The last possibility is to delete c_1 and edit $s_{1\_}$ into $s_{2\_}+c_2$, giving $d(s_{1\_}, s_{2\_}+c_2)+1$. There are no other alternatives. We take the least expensive, i.e., min. of these alternatives.

From the relations, we can see that $d(s_1, s_2)$ depends only on $d(s_1', s_2')$ where s_1' is a shorter string than s_1, or s_2' is a shorter string than s_2, or both. Thus, the **dynamic programming** technique can be applied to compute the edit distance of two strings.

We can use a two-dimensional matrix, $m[0..|s_1|, 0..|s_2|]$, to hold the edit distances. The low right corner cell $m(|s_1|, |s_2|)$ will furnish the required value of the edit distance $d(s_1, s_2)$. We have

$m[0, 0] = 0$
$m[i, 0] = i, \ \ i =1, 2, ..., |s_1|$
$m[0, j] = j, \ \ j =1, 2, ..., |s_2|$
$m[i, j] = \min(m[i-1, j-1] + p(s_1[i], s_2[j]), m[i-1, j] + 1, m[i, j-1] + 1)$,

> where $i = 1, 2, ..., |s_1|, j = 1, 2, ..., |s_2|$, and $p(s_1[i], s_2[j]) = 0$ if $s_1[i] = s_2[j]$; $p(s_1[i], s_2[j]) = 1$, otherwise.

Once the edit distance computation is completed, we can find the alignment of characters that give the final distance. For this, we need to record which case in the above recursive rule minimizes the distance, and then trace back the path that corresponds to the best alignment. Note that, in many cases, the minimal choice is not unique, and different paths could have been drawn, which indicate alternative optimal alignments.

Example 1: We want to compute the edit distance and find the alignment of the following two strings:

s_1:   X G Y X Y X Y X
s_2:   X Y X Y X Y T X

The edit distance matrix is given in Fig. 9.17. The final edit distance value is 2, which is the value in the bottom right corner cell. Figure 9.17 also shows the trace back path. Notice that a diagonal line means match or change, a vertical line means insertion, and a horizontal line means deletion. Thus, the final alignment of our two strings is:

s_1:   X G Y X Y X Y – X
s_2:   X – Y X Y X Y T X

s_1	X	G	Y	X	Y	X	Y	X
s_2	0	1	2	3	4	5	6	7	8
X	1	0	1	2	3	4	5	6	7
Y	2	1	1	1	2	3	4	5	6
X	3	2	2	2	1	2	3	4	5
Y	4	3	3	2	2	1	2	3	4
X	5	4	4	3	2	2	1	2	3
Y	6	5	5	4	3	2	2	1	2
T	7	6	6	5	4	3	3	2	2
X	8	7	7	6	5	4	3	3	2

Fig. 9.17. The edit distance matrix and back trace path

The time-complexity of the algorithm is $O(|s_1||s_2|)$ (to fill the matrix). The space complexity is also $O(|s_1||s_2|)$. Back trace takes $O(|s_1|+|s_2|)$ time.

The **normalized edit distance** $ND(s_1, s_2)$ is defined as the edit distance divided by the mean length of the two strings:

$$ND(s_1, s_2) = \frac{d(s_1, s_2)}{(|s_1| + |s_2|)/2}.$$

(1)

Another commonly used denominator is $max(|s_1|, |s_2|)$.

Finally, in data extraction, "change a character" may be undesirable (which represents a disjunction in regular expressions). A large distance may be used to disallow it. We will discuss this issue again in Sect. 9.11.2.

9.5.2 Tree Matching

Like string edit distance, tree edit distance between two trees A and B (*labeled ordered rooted trees*) is the cost associated with the minimum set of operations needed to transform A into B. In the classic formulation, the set of operations used to define tree edit distance includes, node removal, node insertion, and node replacement. A cost is assigned to each operation. Solving the tree edit distance problem is to find a minimum-cost *mapping* between two trees. The concept of mapping is formally defined as:

Let X be a tree and let $X[i]$ be the ith node of tree X in a preorder walk of the tree. A *mapping* M between a tree A of size n_1 and a tree B of size n_2 is a set of ordered pairs (i, j), one from each tree, satisfying the following conditions for all $(i_1, j_1), (i_2, j_2) \in M$:

(1) $i_1 = i_2$ *iff* $j_1 = j_2$;
(2) $A[i_1]$ is on the left of $A[i_2]$ *iff* $B[j_1]$ is on the left $B[j_2]$;
(3) $A[i_1]$ is an ancestor of $A[i_2]$ *iff* $B[j_1]$ is an ancestor of $B[j_2]$.

Intuitively, the definition requires that each node appears no more than once in a mapping and the order among siblings and the hierarchical relation among nodes are preserved. Figure 9.18 shows a mapping example.

Several algorithms have been proposed to address the problem of finding the minimum set of operations (i.e., the one with the minimum cost) to transform one tree into another. All the formulations have complexities above quadratic. In [509], a solution based on dynamic programming is presented with the complexity of $O(n_1 n_2 h_1 h_2)$, where n_1 and n_2 are the sizes of the trees and h_1 and h_2 are the heights of the trees. In [95, 532], two improved algorithms are presented, and in [605], it is shown that if the trees are not ordered, the problem is NP-complete.

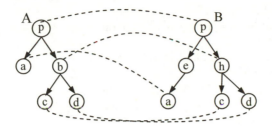

Fig. 9.18. A general tree mapping example

In the above general setting, mapping can cross levels, e.g., node a in tree A and node a in tree B (Fig. 9.18). Replacements are also allowed, e.g., node b in A and node h in B. We now define a restricted tree mapping [572], called **simple tree matching (STM)**, in which no node replacement and no level crossing are allowed. In STM, the aim is to find the maximum matching between two trees (not the edit distance of two trees). This restricted model has been found quite effective for Web data extraction.

Let A and B be two trees, and $i \in A$ and $j \in B$ be two nodes in A and B respectively. A **matching** between two trees is defined to be a mapping M such that, for every pair $(i, j) \in M$ where i and j are non-root nodes, (parent(i), parent(j)) $\in M$. A **maximum matching** is a matching with the maximum number of pairs.

Let $A = R_A:\langle A_1, ..., A_k\rangle$ and $B = R_B:\langle B_1,..., B_n\rangle$ be two trees, where R_A and R_B are the roots of A and B, and A_i and B_j are the ith and jth first-level subtrees of A and B respectively. Let $W(A, B)$ be the number of pairs in the maximum matching of trees A and B. If R_A and R_B contain identical symbols, the maximum matching between A and B (i.e., $W(A, B)$) is $m(\langle A_1, ..., A_k\rangle, \langle B_1, ..., B_n\rangle) + 1$, where $m(\langle A_1, ..., A_k\rangle, \langle B_1, ..., B_n\rangle)$ is the number of pairs in the maximum matching of $\langle A_1, ..., A_k\rangle$ and $\langle B_1, ..., B_n\rangle$. If $R_A \neq R_B$, $W(A, B)) = 0$. Formally, $W(A, B)$ is defined as follows:

$$W(A,B) = \begin{cases} 0 & \text{if } R_A \neq R_B \\ m(\langle A_1,...,A_k\rangle,\langle B_1,...,B_n\rangle)+1 & \text{otherwise} \end{cases};$$

$$m(\langle\rangle, \langle\rangle) = 0 \qquad // \langle\rangle \text{ represents an empty sub-tree list.}$$
$$m(s, \langle\rangle) = m(\langle\rangle, s) = 0 \qquad // s \text{ matches any non-empty sub-tree list}$$
$$m(\langle A_1, ..., A_k\rangle, \langle B_1, ..., B_n\rangle) = \max(m(\langle A_1, ..., A_{k-1}\rangle, \langle B_1, ..., B_{n-1}\rangle) + W(A_k, B_n),$$
$$m(\langle A_1, ..., A_k\rangle, \langle B_1, ..., B_{n-1}\rangle),$$
$$m(\langle A_1, ..., A_{k-1}\rangle, \langle B_1, ..., B_n\rangle)).$$

This definition of m is similar to that of the string edit distance except that here we compute the maximum matching rather than the distance and

Algorithm: STM(A, B)
1.   **if** the roots of the two trees A and B contain distinct symbols **then**
2.   **return** (0)
3.   **else** $k \leftarrow$ the number of first-level sub-trees of A;
4. $n \leftarrow$ the number of first-level sub-trees of B;
5.   Initialization: $m[i, 0] \leftarrow 0$ for $i = 0, \ldots, k$;
                                 $m[0, j] \leftarrow 0$ for $j = 0, \ldots, n$;
6.   **for** $i = 1$ to k **do**
7.   **for** $j = 1$ to n **do**
8. $m[i, j] \leftarrow \max(m[i, j-1], m[i-1, j], m[i-1, j-1]+W[i, j])$,
                     where $W[i, j] \leftarrow$ STM(A_i, B_j)
9.   **end-for**
10.   **end-for**
11.   **return** ($m[k, n]+1$)
12. **end-if**

Fig. 9.19. The simple tree matching (STM) algorithm

that $W(A_k, B_n)$ needs to be computed recursively since A_k and B_n are sub-trees. Clearly, the **dynamic programming** technique is again applicable.

We now give an algorithm for simple tree matching (STM), which computes $W(A, B)$. The algorithm is also called STM (Fig. 9.19). STM is a top-down algorithm. It evaluates the similarity by producing the maximum matching through dynamic programming. The algorithm has the complexity of $O(n_1 n_2)$, where n_1 and n_2 are the sizes of trees A and B respectively.

In line 1, the roots of A and B are compared first. If the roots contain distinct symbols, then the two trees do not match at all. If the roots contain identical symbols, then the algorithm recursively finds the maximum matching between first-level sub-trees of A and B and save it in a W matrix (line 8). Based on the W matrix, a dynamic programming scheme is applied to find the number of pairs in a maximum matching between two trees A and B. We use an example (Fig. 9.20) to explain the algorithm.

To find the maximum matching between trees A and B, their roots (N1 and N15) are compared first. Since N1 and N15 contain identical symbols, $m_{1,15}[4, 2]+1$ is returned as the maximum matching value between trees A and B (line 11). The $m_{1,15}$ matrix is computed based on the $W_{1,15}$ matrix. Each entry in $W_{1,15}$, e.g., $W_{1,15}[i, j]$, is the maximum matching between the ith and jth first-level sub-trees of A and B, which is computed recursively based on its m matrix. For example, $W_{1,15}[4, 2]$ is computed recursively by building the matrices (E)-(H). All the relevant cells are shaded. The zero column and row in m matrices are initializations. Note that we use subscripts for both m and W to indicate the nodes that they are working on.

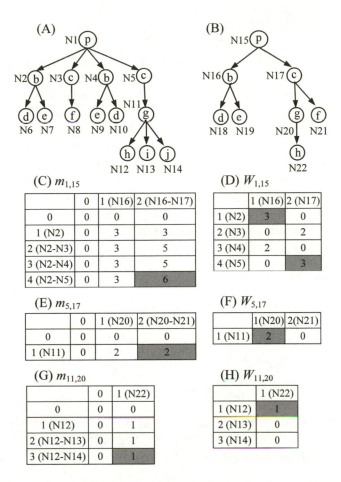

Fig. 9.20. (A) Tree A; (B) Tree B; (C) m matrix for the first level sub-trees of N1 and N15; (D) W matrix for the first level sub-trees of N1 and N15; (E)-(H) m matrixes and W matrixes for the lower level sub-trees.

The **normalized simple tree matching** $NSTM(A, B)$ is obtained by dividing the matching score by the mean number of nodes in the two trees:

$$NSTM(A,B) = \frac{STM(A,B)}{(nodes(A)+nodes(B))/2}. \qquad (2)$$

We may also use $max(nodes(A), nodes(B))$ as the denominator. $nodes(X)$ denotes the number of nodes in tree X.

Similar to string edit distance, after matching computation, we can trace back in the m matrices to find the aligned nodes from the two trees.

9.6 Multiple Alignment

In order to find repeated patterns from HTML strings based on string edit distance or tree matching, we need alignments of strings and trees. We have discussed how to obtain the alignment of two strings or trees. However, a Web page usually contains more than two data records, thus more than two strings or trees need to be aligned. Producing a global alignment of all the strings or trees is crucial. The task is called **multiple alignment**.

In [82], Carrillo and Lipman proposed an optimal multiple alignment based on multidimensional dynamic programming. However, its time complexity is exponential, and is thus not suitable for practical use. Many heuristic methods exist. We describe two of them: the center star method and the partial tree alignment method in [600].

9.6.1 Center Star Method

This is a classic technique [213]. It is commonly used for multiple string alignments, but can be adopted for trees. The method is applied to data extraction based on alignments of HTML strings in [91]. Let the set of strings to be aligned be S. In the method, a string s_c that minimizes

$$\sum_{s_i \in S} d(s_c, s_i) \tag{3}$$

is first selected as the **center string**. $d(s_c, s_i)$ is the distance of two strings. The algorithm then iteratively computes the alignment of rest of the strings with s_c. Spaces are added when needed. The algorithm is in Fig. 9.21.

CenterStar(S)
1.   choose the center star s_c using Equation (3);
2.   initialize the multiple sequence alignment M that contains only s_c;
4.   **for** each s in S-$\{s_c\}$ **do**
5.   let c^* be the aligned version of s_c in M;
6. ·   let s' and $c^{*'}$ be the optimally aligned strings of s and c^*;
7.   add aligned strings s' and $c^{*'}$ into the multiple alignment M;
8.   add spaces to each string in M, except, s' and $c^{*'}$, at locations where new
            spaces are added to c^*
9.   **endfor**
10. **return** multiple string alignment M

Fig. 9.21. The center star algorithm

Example 2: We have three strings, i.e., $S = \{ABC, XBC, XAB\}$. ABC is selected as the center string s_c. Let us align the other strings with ABC.

Iteration 1:   Align $c*$ ($= s_c$) with s =XBC:

$c*'$:   A B C
                       | |
s':   X B C

Update M: A B C →   A B C
                            X B C

Iteration 2:   Align $c*$ with s = XAB:

$c*'$:   – A B C
                      | |
s':   X A B –

Update M: A B C →   – A B C
              X B C   – X B C
                            X A B –

Assume there are k strings in S and all strings have length n, finding the center takes $O(k^2n^2)$ time and the iterative pair-wise alignment takes $O(kn^2)$ time. Thus, the overall time complexity is $O(k^2n^2)$.

For our data extraction task, this method has two shortcomings:

1. the algorithm runs slowly for pages containing many data records and/or data records containing many tags (i.e., long strings) because finding the center string needs $O(k^2n^2)$ time.
2. if the center string (or tree) does not have a particular data item, other data records that contain the same data item may not be aligned properly. For example, the letter X's in the last two strings (in bold) are not aligned in the final result, but they should.

Let us discuss the second point further. As we mentioned in Sect. 9.5.1, giving the cost of 1 for "changing a letter" in edit distance is problematic (e.g., A and X in the first and second strings in the final result) because of optional data items in data records. The problem can be partially dealt with by disallowing "changing a letter" (e.g., giving it a very large cost). However, this introduces another problem. For example, if we align only ABC and XBC, it is not clear which of the following alignment is better.

(1)   A – B C   (2)   – A B C
      – X B C   X – B C

If we also consider the string XAB, then (2) is better. However, the center star method does not have this global view. The partial tree alignment algorithm described below deals with both problems nicely.

9.6.2 Partial Tree Alignment

This method is proposed in [600] for multiple tree alignment in the context of data extraction. It can also be used for aligning multiple strings. For simplicity, we describe the method in the context of trees. The main idea is as follows: The algorithm aligns multiple trees by progressively growing a **seed tree**. The seed tree, denoted by T_s, is initially picked to be the tree with the maximum number of data fields, which is similar to the center string but without the $O(k^2n^2)$ pair-wise tree matching to choose it. The reason for choosing this seed tree is clear as it is more likely for this tree to have a good alignment with data fields in other data records. Then, for each T_i $(i \neq s)$, the algorithm finds for each node in T_i a matching node in T_s. If no match can be found for a node v_i, then the algorithm attempts to expand the seed tree by inserting v_i into T_s. The expanded seed tree T_s is then used in subsequent matching. The insertion is done only if a position for v_i can be uniquely determined in T_s. Otherwise, it is left unmatched. Thus the alignment is **partial**. It represents a **least commitment approach**. Early uncertain commitments can result in undesirable effects for later matches. Note that although the method was designed originally for aligning multiple trees, it can also be adapted for aligning multiple strings.

Partial Alignment of Two Trees

Before presenting the full algorithm for aligning multiple trees, let us first look at the partial alignment of two trees. As indicated above, after T_s and T_i are matched, some nodes in T_i can be aligned with their corresponding nodes of T_s because they match one another. For those nodes in T_i that are not matched, we want to insert them into T_s as they may contain optional data items. There are two possible situations when inserting a new node v_i from T_i into T_s, depending on whether a location in T_s can be uniquely determined to insert v_i. Instead of considering a single node v_i, we can consider each set of unmatched consecutive sibling nodes $v_j...v_m$ from T_i together. Without loss of generality, we assume that the parent of $v_j...v_m$ has a match in T_s and we want to insert $v_j...v_m$ into T_s under the same parent node. We only insert $v_j...v_m$ into T_s if a position for inserting $v_j...v_m$ can be uniquely determined in T_s. Otherwise, they will not be inserted into T_s and left unaligned. The location for inserting $v_j...v_m$ can be uniquely decided:

1. If $v_j...v_m$ have two neighboring siblings in T_i, one on the right and one on the left, that are matched with two consecutive siblings in T_s. Figure 9.22(A) shows such a situation, which gives one part of T_s and one part of T_i. We can see that node c in T_i can be inserted into T_s between node b and node e in T_s because node b and node e in T_s and T_i match. The new (extended) T_s is also shown in Fig. 9.22(A). We note that nodes a, b, c and e may also have their own children. We did not draw them to save space. This applies to all the cases below.

2. If $v_j...v_m$ has only one left neighboring sibling x in T_i and x matches the right most node x in T_s, then $v_j...v_m$ can be inserted after node x in T_s. Figure 9.22(B) illustrates this case.

3. If $v_j...v_m$ has only one right neighboring sibling x in T_i and it matches the left most node x in T_s, then $v_j...v_m$ can be inserted before node x in T_s. This case is similar to the second case above.

Otherwise, we cannot uniquely decide a location for unmatched nodes in T_i to be inserted into T_s. This is shown in Fig. 9.22(C). The unmatched node x in T_i could be inserted into T_s in two positions, between nodes a and b, or between node b and e in T_s. In this situation, we will not insert it into T_s.

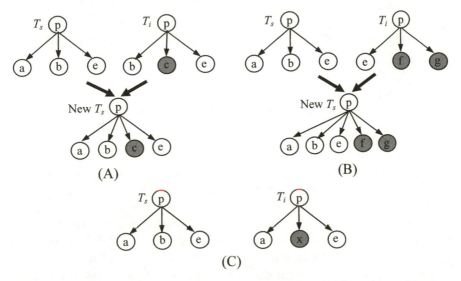

Fig. 9.22. Expand the seed: (A) and (B) unique insertion; (C) insertion ambiguity

Partial Alignment of Multiple Trees

Figure 9.23 gives the full algorithm for multiple tree alignment based on partial alignment of two trees. S is the set of input trees. We use a simple example in Fig. 9.24 to explain the algorithm. S has three example trees.

Lines 1–2 (Fig. 9.23) find the tree with the most data items. It is used as the seed tree T_s. In Fig. 9.24, the seed tree is the first tree (we omitted many nodes on the left of T_1). Line 3 initializes R, which is used to store those trees that are not completely aligned with T_s in each iteration. Line 4 starts the while loop to align every other tree against T_s. Line 5 picks the next unaligned tree, and line 6 does the tree matching. Line 7 finds all the matched pairs by tracing the matrix results of line 6. This function is similar to aligning two strings using edit distance. In Fig. 9.24, T_s and T_2 pro-

Algorithm PartialTreeAlignment(S)
1.   Sort trees in S in descending order of the number of unaligned data items;
2. $T_s \leftarrow$ the first tree (which is the largest) and delete it from S;
3. $R \leftarrow \varnothing$;
4.   **while** ($S \neq \varnothing$) **do**
5. $T_i \leftarrow$ select and delete next tree from S;   // follow the sorted order
6.   STM(T_s, T_i);   // tree matching
7.   AlignTrees(T_s, T_i);   // based on the result from line 6
8.   **if** T_i is not completely aligned with T_s **then**
9.   **if** InsertIntoSeed(T_s, T_i) **then**   // True: some insertions are done
10. $S \leftarrow S \cup R$;
11. $R \leftarrow \varnothing$
12.   endif;
13.   **if** there are still unaligned items in T_i that are not inserted into T_s **then**
14. $R \leftarrow R \cup \{T_i\}$
15.   **endif**;
16.   **endif**;
17.   **endwhile**;
18.   Output data fields from each T_i to a data table based on the alignment results.

Fig. 9.23. The partial tree alignment algorithm

duce one match, node b. Nodes w, c, k and g are not matched to T_s. Lines 8 and 9 attempt to insert the unmatched nodes into T_s. This is the partial tree alignment discussed above. In Fig. 9.24, none of the nodes w, c, k and g in T_2 can be inserted into T_s because no unique location can be found. Thus, it will not pass the if-statement (InsertIntoSeed() returns *false* in line 9 of Fig. 9.23). Lines 13–14 inserts T_2 into R, which is a list of trees that need to be re-matched since some data items are not aligned and not inserted into T_s. In Fig. 9.24, when matching T_3 with T_s in the next iteration, all unmatched nodes c, h and k can be inserted into T_s (line 9). Since there are some insertions, we re-match those trees in R. Line 10 and line 11 put the trees in R into S and reinitializes R. T_3 will not be inserted into R (line 13).

In Fig. 9.24, T_2 is the only tree in R, which will be matched to the new T_s in the next round. Now, every node in T_2 can be matched or inserted, and the process completes. Line 18 of Fig. 9.23 outputs the data items from each tree according to the alignment produced. Note that if there are still un-matched nodes with data after the algorithm completes (e.g., $R \neq \varnothing$), each un-matched data will occupy a single column by itself. Table 1 shows the data table for the trees in Fig. 9.24. We use "1" to indicate a data item.

The complexity of the algorithm is $O(k^2 n^2)$, where k is the number of trees in S and n is the size of each tree (we assume that all the trees are of similar size). However, as reported in [600], in practice, the algorithm almost always goes through S only once (i.e., $R = \varnothing$).

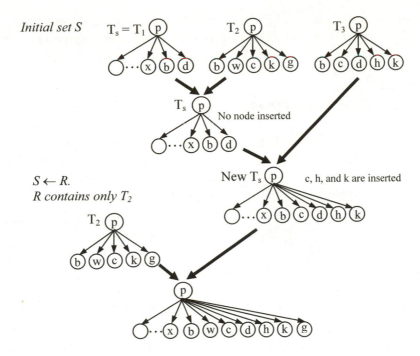

Fig. 9.24. Iterative tree alignment with two iterations

Table 1. Final data table ("1" indicates a data item)

	...	x	b	w	c	d	h	k	g
T_1	...	1	1			1			
T_2			1	1	1			1	1
T_3			1		1	1	1	1	

In fact, to make the algorithm complete, a recursive call should be added after line 17 in Fig. 9.23 to handle the case when $R \neq \varnothing$, i.e., to further align only those trees in R. The following three lines can be added:

18. **if** $R \neq \varnothing$ **then**
19.   PartialTreeAlignment(R)
20. **endif**

This takes care of the situation where some items are not aligned and not inserted. However, it is shown in [600] that this part is usually not needed for data extraction.

We make two remarks about this complete algorithm. First, the recursion will terminate even if no alignment and/or no insertion is made to the seed tree because the seed tree is deleted in each recursion and thus

R becomes smaller and smaller. Second, the algorithm can found multiple templates in the data. The seed tree from each recursion represents a different template.

9.7 Building DOM Trees

DOM (Document Object Model) tree building from input pages is a necessary step for many data extraction algorithms. We describe two methods for building **DOM trees**, which are also commonly called **tag trees** (we will use them interchangeably in this chapter).

Using Tags Alone: Most HTML tags work in pairs. Each pair consists of an open tag and a close tag (indicated by < > and </> respectively). Within each corresponding tag-pair, there can be other pairs of tags, resulting in a nested structure. Building a DOM tree from a page using its HTML code is thus natural. In the tree, each pair of tags is a node, and the nested pairs of tags within it are the children of the node. Two tasks need to be performed:

1. *HTML code cleaning*: Some tags do not require close tags (e.g., <li>, <hr> and <p>) although they have close tags. Hence, additional close tags should be inserted to ensure all tags are balanced. Ill-formatted tags also need to be fixed. Such error tags are usually close tags that cross different nested blocks, e.g., <tr> ... <td> ... </tr> ... </td>, which can be hard to fix if multiple levels of nesting exist. There are open source programs that can be used to clean up HTML pages. One popular program is called **tidy** (available at http://tidy.sourceforge. net/).
2. *Tree building*: We can follow the nested blocks of the HTML tags in the page to build the DOM tree. It is fairly straightforward. We will not discuss it further.

This method works for most pages. However, for some ill-formatted tags, even the tidy program cannot fix. Then, the constructed DOM trees may be wrong, which makes it difficult for subsequent data extraction.

Using Tags and Visual Cues: Instead of analyzing the HTML code to fix errors, rendering or visual information (i.e., the locations on the screen at which tags are rendered) can be used to infer the structural relationship among tags and to construct a DOM tree. This method leads to more robust tree construction due to the high error tolerance of the rendering engines of Web browsers (e.g., Internet Explorer). As long as the browser is able to render a page correctly, its tag tree can be built correctly.

In a Web browser, each HTML element (consisting of an open tag, optional attributes, optional embedded HTML content, and a close tag that may be omitted) is rendered as a rectangle. The visual information can be obtained after the HTML code is rendered by a Web browser. A DOM tree can then be constructed based on the nested rectangles (resulted from nested tags). The details are as follows:

1. Find the four boundaries of the rectangle of each HTML element by calling the rendering engine of a browser, e.g., Internet Explorer.
2. Follow the sequence of open tags and perform containment checks to build the tree. Containment check means checking if one rectangle is contained in another.

Let us use an example to illustrate the process. Assume we have the HTML code on the left of Fig. 9.25. However, there are three errors in the code. The close tag </td> for line 3 is put after the open tag for line 4. Also, the close tags </tr> in line 5 and </td> in line 7 are missing. However, this HTML segment can be rendered correctly in a browser, with the boundary coordinates for each HTML element shown in the middle of Fig. 9.25. Using this visual information, it is easy to build the tree on the right.

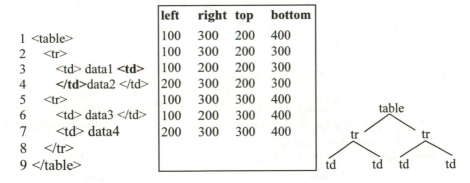

		left	right	top	bottom
1	<table>	100	300	200	400
2	<tr>	100	300	200	300
3	<td> data1 **<td>**	100	200	200	300
4	**</td>**data2 </td>	200	300	200	300
5	<tr>	100	300	300	400
6	<td> data3 </td>	100	200	300	400
7	<td> data4	200	300	300	400
8	</tr>				
9	</table>				

Fig. 9.25. A HTML code segment, boundary coordinates and the resulting tree

9.8   Extraction Based on a Single List Page: Flat Data Records

We are now ready to perform the data extraction task. In this and the next sections, we study the first extraction problem in Sect. 9.4.1, i.e., extraction based on a single list page. This section focuses on a simpler case, i.e., a list (a data region) containing only flat data records (no nesting). We assume that the DOM tree has been built for the page. In Sect. 9.10, we will study the second extraction problem based on multiple input pages. The techniques studied in this section are based on the work in [341, 600].

Given a list page containing multiple lists and each list contains multiple data records (at least two), the following tasks are performed:

1. Identify each list (also called a **data region**), i.e., mine all data regions,
2. Segment data records in each list or data region, and
3. Align data items in the data records to produce a data table for each data region and also a regular expression pattern.

9.8.1   Two Observations about Data Records

Data records in each list (or data region) are encoded using the same HTML mark-up encoding. Finding the data records and its hidden schema means to find repeated patterns and align them. String or tree matching/comparison are natural techniques. The problem, however, is the efficiency because a data record can start from anywhere in a page and has any length. It is prohibitive to try all possibilities. If all data records have exactly the same tag string, then the problem is easier. However, in practice, a set of data records typically does not have exactly the same tag string or data items due to missing or optional items (see Fig. 9.26). The two important observations below help to solve the problem, which are based on the DOM tree structure [341].

Observation 1: A group of data records that contains descriptions of a set of similar objects is typically presented in a contiguous region of a page and is formatted using similar HTML tags. Such a region represents a list or a data region. For example, in Fig. 9.26 two books are presented in one contiguous region.

Observation 2: A list of data records in a region is formed by some child sub-trees of the same parent node. It is unlikely that a data record starts in the middle of a child sub-tree and ends in the middle of another child sub-tree. Instead, it starts from the beginning of a child sub-tree and ends at the end of the same or a later child sub-tree.

For example, Fig. 9.27 shows the DOM tree of the page in Fig. 9.26 (with some parts omitted). In this tree, each data record is wrapped in five TR nodes with their sub-trees under the same parent TBODY. The two data records are in the two dash-lined boxes. It is unlikely that a data record starts from TD* and ends at TD# (Fig. 9.27).

The second observation makes it possible to design an efficient algorithm to identify data records because it limits the tags from which a data record may start and end in a DOM tree.

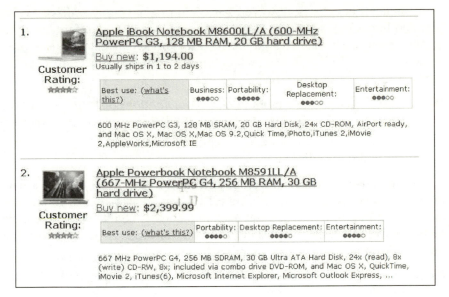

Fig. 9.26. An example of a page segment

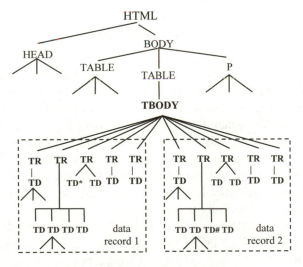

Fig. 9.27. The DOM tree of the page segment in Fig. 9.26

9.8.2   Mining Data Regions

This first step mines every data region in a Web page that contains a list of data records (a set instance). Finding data regions (or individual data records in them) directly is, however, hard. We first mine **generalized nodes** (defined below). A sequence of adjacent generalized nodes forms a **data region**. From each data region, we identify the actual data records (discussed in Sect. 9.8.3). Below, we define generalized nodes and data regions using the DOM (tag) tree:

Definition: A **generalized node** (a *node combination*) of length r consists of r ($r \geq 1$) nodes in the DOM tree with the following two properties:
(1) the nodes all have the same parent;
(2) the nodes are adjacent.

We introduce the generalized node to capture the situation that a data record is contained in several sibling HTML tag nodes rather than one. For example, in Fig. 9.27, we see that each notebook is contained in five table rows (or five TR nodes). We call each node in the HTML tag tree a **tag node** to distinguish it from a generalized node.

Definition: A **data region** is a collection of two or more generalized nodes with the following properties:
(1) the generalized nodes all have the same parent;
(2) the generalized nodes all have the same length;
(3) the generalized nodes are all adjacent;
(4) the similarity between adjacent generalized nodes is greater than a fixed threshold.

For example, in Fig. 9.27, we can form two generalized nodes. The first one consists of the first five children TR nodes of TBODY, and the second one consists of the next five children TR nodes of TBODY. We should note that although the generalized nodes in a data region have the same length (the same number of children nodes of a parent node in the tag tree), their lower level nodes in their sub-trees can be quite different. Thus, they can capture a wide variety of regularly structured objects. We also note that a generalized node may not represent a final data record (see Sect. 9.8.3), but will be used to find the final data records.

To further explain different kinds of generalized nodes and data regions, we make use of an artificial DOM/tag tree in Fig. 9.28. For notational convenience, we do not use actual HTML tag names but ID numbers to denote tag nodes in a tree. The shaded areas are generalized nodes. Nodes 5 and 6 are generalized nodes of length 1 and they together define the data region labeled 1 if the similarity condition (4) is satisfied. Nodes 8, 9 and 10 are also generalized nodes of length 1 and they together define the data region labeled 2 if the similarity condition (4) is satisfied. The pairs of nodes (14, 15), (16, 17) and (18, 19) are generalized nodes of length 2. They together define the data region labeled 3 if the similarity condition (4) is satisfied. It should be emphasized that a data region includes the sub-trees of the component nodes, not just the component nodes alone.

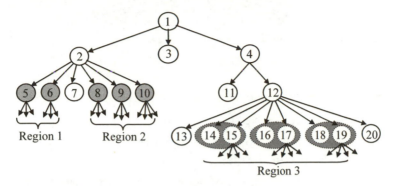

Fig. 9.28. An illustration of generalized nodes and data regions

Comparing Generalized Nodes

In order to find each data region in a Web page, the mining algorithm needs to find the following: (1) Where does the first generalized node of a data region start? For example, in Region 2 of Fig. 9.28, it starts at node 8. (2) How many tag nodes or components does a generalized node in each data region have? For example, in Region 2 of Fig. 9.28, each generalized node has one tag node (or one component).

Let the maximum number of tag nodes that a generalized node can have be K, which is normally a small number (< 10). In order to answer (1), we can try to find a data region starting from each node sequentially. To answer (2), we can try one node, two node combination, …, K node combination. That is, we start from each node and perform all 1-node comparisons, all 2-node comparisons, and so on (see the example below). We then use the comparison results to identify each data region.

The number of comparisons is actually not very large because:

- Due to the two observations in Sect. 9.8.1, we only need to perform comparisons among the children nodes of a parent node. For example, in Fig. 9.28, we do not compare node 8 with node 13.
- Some comparisons done for earlier nodes are the same as for later nodes (see the example below).

We use Fig. 9.29 to illustrate the comparison. There are 10 nodes below the parent node p. We start from each node and perform string (or tree) comparison of all possible combinations of component nodes. Let the maximum number of components that a generalized node can have be 3.

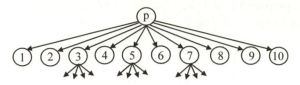

Fig. 9.29. Combination and comparison

Start from node 1: We compute the following string or tree comparisons.

- (1, 2), (2, 3), (3, 4), (4, 5), (5, 6), (6, 7), (7, 8), (8, 9), (9, 10)
- (1-2, 3-4), (3-4, 5-6), (5-6, 7-8), (7-8, 9-10)
- (1-2-3, 4-5-6), (4-5-6, 7-8-9).

(1, 2) means that the tag string of node 1 is compared with the tag string of node 2. The tag string of a node includes all the tags of the sub-tree of the node. (1-2, 3-4) means that the combined tag string of nodes 1 and 2 is compared with the combined tag string of nodes 3 and 4.

Start from node 2: We only compute:

- (2-3, 4-5), (4-5, 6-7), (6-7, 8-9)
- (2-3-4, 5-6-7), (5-6-7, 8-9-10).

We do not need to do 1-node comparisons because they have been done when we started from node 1 above.

Start from node 3: We only need to compute:

- (3-4-5, 6-7-8).

Again, we do not need to do 1-node comparisons. Also, 2-node comparisons are not necessary as they were done when we started at node 1.

We do not need to start from any other node after node 3 because all the computations have been done.

The Overall Algorithm

The overall algorithm (called MDR) is given in Fig. 9.30. It traverses the tag tree from the root downward in a depth-first fashion (lines 5 and 6). *Node* is any tree node. K is the maximum number of tag nodes in a generalized node (10 is sufficient). τ is the similarity threshold. The node comparison can be done either using string edit distance or tree matching (e.g., STM). The similarity threshold can be set empirically.

Line 1 says that the algorithm will not search for data regions if the depth of the sub-tree at *Node* is 2 or 1 as it is unlikely that a data region is formed with only a single level of tag(s).

Algorithm MDR(*Node, K, τ*)
1 **if** TreeDepth(*Node*) >= 3 **then**
2   CombComp(*Node.Children, K*);
3   *DataRegions* ← IdenDRs(*Node, K, τ*);
4   **if** (*UncoveredNodes* ← *Node.Children* − $\bigcup_{DR \in DataRegions} DR$) ≠ ∅ **then**
5   **for** each *ChildNode* ∈ *UncoveredNodes* **do**
6   *DataRegions* ← *DataRegions* ∪ MDR(*ChildNode, K, τ*);
7   **return** *DataRegions*
8 **else return** ∅

Fig. 9.30. The MDR algorithm

At each internal node, the function CombComp() (line 2) performs string (tree) comparisons of various combinations of the children sub-trees, which have been discussed above. The function IdenDRs() (line 3) uses the comparison results to find similar children node combinations (using the similarity threshold *τ*) to obtain generalized nodes and data regions (*DataRegions*) under *Node* (i.e., among the children of *Node*). That is, it decides which combinations represent generalized nodes and where the beginning and end are for each data region. *DataRegions* consists of a set of data regions, and each data region contains a list of tag nodes organized as generalized nodes of the region. IdenDRs() is discussed further below. Line 4 says that if some nodes are not covered by discovered data regions, the algorithm will go down the tree further from these nodes to see whether they contain data regions (lines 5 and 6).

We note that a generalized node may not be a data record, but may contain more than one data record. Figure 9.31 illustrates the point. This data region has eight data records. Each row has two. However, each row will be reported as a generalized node because rows 1–4 are similar. We will explain how to find data records from each generalized node shortly.

Fig. 9.31. A possible configuration of data records

Let us come back to the function IdenDRs(), which is not hard to design and it is omitted. Interested readers can refer to [341]. We only describe two issues that the function needs to consider.

1. It is clear from Fig. 9.28 that there may be several data regions under a single parent node *Node*. Generalized nodes in different data regions may have different number of tag node components.
2. A property about similar strings (or trees) is that if a set of strings (trees), s_1, s_2, s_3, ..., s_n, is similar to one another, then a combination of any number of them is also similar to another combination of them of the same number. IdenDRs should only report generalized nodes of the smallest length that cover a data region. For Fig. 9.31, it only reports each row as a generalized node rather than a combination of two rows (rows 1-2, and rows 3-4).

The computation of the algorithm is dominated by string (or tree) comparison. Assume that the total number of nodes in the tag tree is N, the number of comparisons is in the order of $O(NK^2)$. Since K is normally very small, the computation requirement of the algorithm is low. Visual information (see Sect. 9.8.5) and simple heuristics can be applied to reduce the number of string (tree) comparisons substantially.

9.8.3 Identifying Data Records in Data Regions

As we have discussed above, a generalized node may consist of multiple data records. Figure 9.31 shows an example, where each row is a generalized node that contains two data records. To find data records from each generalized node in a data region, the following observation is useful:

- *If a generalized node contains two or more data records, these data records must be similar in terms of their tag strings.*

This is clear because we assume that a data region contains descriptions of similar data records. Identifying data records from each generalized node in a data region is relatively easy because they are nodes (together with their sub-trees) at the same level as the generalized node, or nodes at a lower level of the DOM/tag tree. This can be done in two steps:

1. Produce one rooted tree for each generalized node: An artificial root node is created first, and all the components (which are sub-trees) of the generalized node are then put as its children.
2. Call the MDR algorithm using the tree built in step 1: Due to the observation above, this step will find the data records if exist. Otherwise, the generalized node is a data record. Two issues needs to be considered.

- The discovered data records should covered all the data items in the original generalized node.
- Each data record should not be too small, e.g., a single number or a piece of text, which is likely to be an entry in a spreadsheet table.

Further details can be found in [341, 601], where handling non-contiguous data records is also discussed. For example, two books are described in two table rows. One row lists the names of the two books in two cells, and the next row lists the other pieces of information about the books also in two cells. This results in the following sequence in the HTML code: name 1, name 2, description 1, description 2.

9.8.4   Data Item Alignment and Extraction

Once data records in each data region are discovered, they are aligned to produce an extraction pattern that can be used to extract data from the current page and also other pages that use the same encoding template. We use the **partial tree alignment algorithm** to perform the task in two steps:

1. Produce a rooted tree for each data record: An artificial root node is created first. The sub-trees of the data record are then put as its children.
2. Align the resulting trees: The trees of all the data records in each data region are aligned using the partial tree alignment method in Sect. 9.6.2. After alignments are done, the final seed tree can be used as the extraction pattern, or be turned into a regular expression.

Conflict Resolution: In tree matching and alignment, it is possible that multiple matches can give the same maximum score, but only one match is correct. Then, we need to decide which one. For example, in Fig. 9.32, node c in tree A can match either the first or the last node c in tree B.

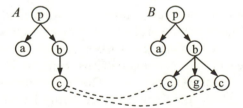

Fig. 9.32. Two trees with more than one possible match: which is correct?

To deal with this problem, we can use data content similarity. Data items that share some common substrings are more likely to match. In many cases, a data item contains both the attribute name and the attribute value. For example, "shopping in 24 hours" and "shopping within a week".

The data content similarity can in fact be considered in the simple tree matching (STM) algorithm in Fig. 9.19 with minor changes to line 1 and line 11. Data contents (data items) should be included as leaves of the DOM trees. When data items are matched, their match score is computed. In [601], the **longest common subsequence** (LCS) is used, but cosine similarity should work too. Let q be the number of words in the LCS of the two data items, and m be the maximal number of words contained in them. Their matching score is computed with q/m.

9.8.5   Making Use of Visual Information

It is quite clear that many visual features that are designed to help people locate and understand information in a Web page can help data extraction. We have already shown that visual cues can be used to construct DOM trees. In fact, they can be exploited everywhere. Here are some examples:

- Reduce the number of string or tree comparisons. If two sub-trees are visual too different, they do not need to be compared.
- Confirm the boundary of data records using the space gap between data records. It is usually the case that gaps between data records are larger than gaps between items within a data record.
- Determine item alignment. Visual alignment (left, right or center) of items can help determine whether two data items should match. Relative positions of the data items in each data record are very helpful too.
- Identify data records based on their contour shapes. This method was exploited to segment data records from search engine results [612].

9.8.6   Some Other Techniques

Both string comparison and tree comparison based methods have been used to solve the data extraction problem. The task described in Sect. 9.8.2 can be done either based on strings or trees. A pure string based method called was proposed in [91]. It finds patterns from the HTML tag string, and then uses the patterns to extract data. The center star method is used to align multiple strings. A more sophisticated string based method, which also deals with nested records, was proposed in [530]. The main problem with string based methods is that they can find many patterns and it is hard to determine which one is correct. Some patterns may cross boundaries of data records. In [91], the user needs to choose the right pattern for extraction. In [530], multiple pages containing similar data records and other methods are used to choose the right pattern. This problem is not a major issue for tree based methods because of tree structures, which eliminate most alternatives. The observations made in Sect. 9.8.1 were also very helpful. In [312], a method based on constraints and the EM algorithm is proposed, which needs to use some information from detail pages to segment data records. Note that a data record usually (not always) has a link to its detail page, which contains the detail description of the object (e.g., a product) represented by the data record.

Visual information is extensively used in [612] to segment snippets (data records) of returned pages from search engines. It uses the contour shape on the left of each snippet, distance from the left boundary of the page, and also the type of each line in a snippet (e.g., text line, link line, empty line, etc.) to determine the similarity of candidate data records.

9.9   Extraction Based on a Single List Page: Nested Data Records

The problem with the method in Sect. 9.8 is that it is not suitable for nested data records, i.e., data records containing nested lists. Since the number of elements in a list of each data record can be different, using a fixed threshold to determine the similarity of data records will not work.

The problem, however, can be dealt with as follows. Instead of traversing the DOM tree top down, we can traverse it post-order. This ensures that nested lists at lower levels are found first based on repeated patterns before going to higher levels. When a nested list is found, its records are **collapsed** to produce a single pattern which replaces the list of data records. When comparisons are made at a higher level, the algorithm only sees the pattern. Thus it is treated as a flat data record. This solves the fixed threshold problem above. We introduce an algorithm below, which is based on the NET system in [351]. A running example is also given to illustrate the process.

The NET algorithm is given in Fig. 9.33. It is basically a post-order tree traversal algorithm. The observations in Sect. 9.8.1 are still applicable here. The function TraverseAndMatch() performs the post-order traversal. During the process, each nested list is collapsed. The function PutDataInTables() (line 3) outputs the extracted data to the user in relational tables (a page may have multiple data regions, and data in each region are put in a separate table). Line 3 can be easily done if the function TraverseAndMatch() saves the nodes whose children form data records.

Line 1 of TraverseAndMatch() says that the algorithm will not search for data records if the depth of the sub-tree from *Node* is 2 or 1 as it is unlikely that a data record is formed with only a single level of tag(s). This

Algorithm NET(*Root*, *τ*)
1   TraverseAndMatch(*Root*, *τ*);
2   **for** each top level node *Node* whose children have aligned data records **do**
3   PutDataInTables(*Node*);
4   **endfor**

Function TraverseAndMatch (*Node*, *τ*)
1   **if** Depth(*Node*) ≥ 3 **then**
2   **for** each *Child* ∈ *Node.Children* **do**
3   TraverseAndMatch(*Child*, *τ*);
4   **endfor**
5   Match(*Node*, *τ*);
6   **endif**

Fig. 9.33. The NET algorithm

Function Match(*Node*, *τ*)
1   *Children* ← *Node.Children*;
2   **while** *Children* ≠ ∅ **do**
3   *ChildFirst* ← select and remove the first child from *Children*;
4   **for** each *ChildR* in *Children* **do**
5   **if** TreeMatch(*ChildFirst*, *ChildR*) > *τ* **then**
6   AlignAndLink();
7   *Children* ← *Children* − {*ChildR*}
8   **endfor**
9   **if** some alignments (or links) have been made with *ChildFirst* **then**
10   GenNodePattern(*ChildFirst*)
11 **endwhile**
12 **If** consecutive child nodes in *Children* are aligned **then**
13   GenRecordPattern(*Node*)

Fig. 9.34. The Match function

parameter can be changed. The Match() function performs tree matching on child sub-trees of *Node* and pattern generation. *τ* is the threshold for a match of two trees that are considered sufficiently similar.

Match(): The Match() function is given in Fig. 9.34. Figure 9.35 shows a running example. In this figure, Ni represents an internal node, and tj represents a terminal (leaf) node with a data item. We use the same shape or shading to indicate matching nodes. We explain the algorithm below.

Given the input node *Node*, line 1 obtains all its children to be matched. In our example, *Children* of *p* are t1, N1, N2, N3, N4, t2, N5, N6, N7, N8, and N9 (with their sub-trees). Lines 2–4 set every pair of child nodes to be matched. The matching is done by TreeMatch(), which uses algorithm

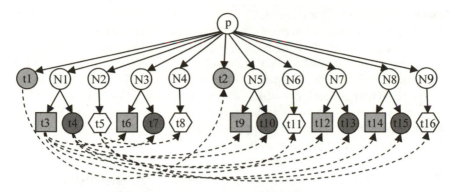

Fig. 9.35. A running example: All matched items are linked

STM() in Fig. 9.19. AlignAndLink() (line 6) aligns and links all matched data items (leaf nodes) in *ChildFirst* and *ChildR*. The links are directional, i.e., from earlier data items to later (matched) data items. If *ChildR* matches *ChildFirst*, *ChildR* is removed from *Chirdren* so that it will not be matched again later (line 7). For our example, after lines 4–11, the resulting matches and links (dashed lines) are given in Fig. 9.35. Assume they all satisfy the match condition in line 5.

In lines 9–10, if some alignments (or links) have been made, the Gen-NodePattern() function generates a **node pattern** for all the nodes (including their sub-trees) that match *ChildFirst*. This function first gets the set of matched nodes *ChildR*'s, and then calls PartialTreeAlignment() in Fig. 9.23 to produce a pattern which is the final seed tree. Note that Partial-TreeAlignment() can be simplified here because most alignments have been done. Only insertions and matching of unaligned items are needed. A node pattern can also be represented as a regular expression.

In lines 12–13, it collapses the sub-trees to produce a global pattern for the data records (which are still unknown). Notice that lines 9–10 already produced the pattern for each child sub-tree. The GenRecordPattern() function simply produces a regular expression pattern for the list of data records. This is essentially a **grammar induction** problem [243].

Grammar induction in our context is to infer a regular expression given a finite set of positive and negative example strings. However, we only have a single positive example (a list of hidden data records). Fortunately, structured data in Web pages are usually highly regular which enables heuristic methods to generate "simple" regular expressions. Here, we introduce such a simple method, which depends on three **assumptions**:

1. The nodes in the first data record at each level must be complete, e.g., in Fig. 9.35, nodes t1, N1 and N2 must all be present.

Function GenRecordPattern(*Node*)

1 *String* ← Assign a distinctive symbol to each set of matched children of *Node*;
2 Initialize a data structure for NFA $N = (Q, \Sigma, \delta, q_0, F)$, where Q is the set of states, Σ is the symbol set containing all symbols appeared in *String*, δ is the transition relation that is a partial function from $Q \times (\Sigma \cup \{\varepsilon\})$ to Q, and F is the set of accept states, $Q \leftarrow \{q_0\}$ (q_0 is the start state), $\delta \leftarrow \varnothing$ and $F \leftarrow \varnothing$;
3 $q_c \leftarrow q_0$;   // q_c is the current state
4 **for** each symbol s in *String* in sequence **do**
5   **if** $\exists$ a transition $\delta(q_c, s) = q_n$ **then**
6          $q_c \leftarrow q_n$ // transit to the next state;
7   **else if** $\exists\ \delta(q_i, s) = q_j$, where $q_i, q_j \in Q$ **then**   // s appeared before
8   **if** $\exists\ \delta(q_f, \varepsilon) = q_i$, where $\delta(q_i, s) = q_j$ and $f \geq c$ **then**
9   TransitTo(q_c, q_f)
10   **else** TransitTo(q_c, q_i)
11             $q_c \leftarrow q_j$
12   **else** create a new state q_{c+1} and a transition $\delta(q_c, s) = q_{c+1}$,
                 i.e., $\delta \leftarrow \delta \cup \{((q_c, \varepsilon), q_{c+1})\}$
13             $Q \leftarrow Q \cup \{q_{c+1}\}$;
14             $q_c \leftarrow q_{c+1}$
15   **if** s is the last symbol in *String* **then**
16   Assign the state with the largest subscript the accept state q_r, $F = \{q_r\}$;
17   TransitTo(q_c, q_r);
18 **endfor**
19 generate a regular expression based on the NFA N;
20 Substitute all the node patterns into the regular expression.

Function TransitTo(q_c, q_s)

1   **while** $q_c \neq q_s$ **do**
2   **if** $\exists\ \delta(q_c, \varepsilon) = q_k$ and $k > c$ **then**
3              $q_c \leftarrow q_k$
4   **else** create a transition $\delta(q_c, \varepsilon) = q_{c+1}$, i.e., $\delta \leftarrow \delta \cup \{((q_c, \varepsilon), q_{c+1})\}$;
5              $q_c \leftarrow q_{c+1}$
6   **endwhile**

Fig. 9.36. Generating regular expressions

2. The first node of every data record at each level must be present, e.g., at the level of t1 and t2, they both must be present, and at the next level, N1, N3, N5, N7 and N8 must be present. Note that the level here is in the hierarchical data organizational sense (not the HTML code sense).
3. Nodes within a single flat data record (no nesting) do not match one another, e.g., N1 and N3 do not appear in the same data record.

The GenRecordPattern() function is given in Fig. 9.36. It generates a regular expression pattern.

Line 1 in Fig. 9.36 simply produces a string for generating a regular expression. For our example, we obtain the following:

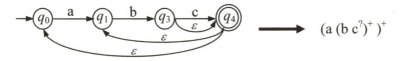

Lines 2–3 initialize a NFA (non-deterministic finite automaton). Lines 4–18 traverses *String* from left to right to construct the NFA. For our example, we obtain the final NFA in Fig. 9.37.

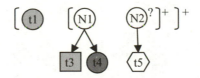

Fig. 9.37. The generated NFA and its regular expression

Line 19 produces a regular expression from the NFA, which is shown in Fig. 9.37 on the right.

Line 20 produces the final pattern (Fig. 9.38) by substituting the node patterns into the regular expression. Here, we use node t1 as the pattern (the seed tree) for nodes t1 and t2, the N1 sub-tree as the pattern for all the linked sub-trees rooted at N1, N3, N5, N7 and N8. The N2 sub-tree is the pattern of the sub-trees rooted at N2, N4, N6 and N9.

$$\left[\boxed{t1} \right] \left[\boxed{N1} \quad N2^{?} \right]^{+} \right]^{+}$$

t3 t4 t5

Fig. 9.38. The regular expression produced from Fig. 9.35

Some additional notes about the algorithm are in order:

- Each child node here represents a sub-tree (e.g., N1, N2, etc). Assumption 1 does not require lower level nodes of each sub-tree in the first data record to be complete (no missing items). We will see an example in Fig. 9.40 and Fig. 9.41.
- Regular expressions produced by the algorithm do not allow disjunctions (i.e., A|B) except (A|ε), which means that A is optional. Such regular expresses are called **union-free regular expressions**. However, disjunctions are possible at lower level matches of the sub-trees. We will discuss the issue of disjunction again in Sect. 9.11.2.
- Function GenRecordPattern() in Fig. 9.37 assumes that under *Node* there is only one data region, which may not be true. The algorithm can

be easily extended to take care of multiple data regions under *Node*. In fact, the NET() algorithm here is a simplified version to present the main ideas. For practical use, it can be significantly enhanced to remove most assumptions if not all.

Finally, the function PutDataInTables() (line 3 of NET() in Fig. 9.33) simply outputs data items in a table, which is straightforward after the data record patterns are found. For the example in Fig. 9.35, the following data table is produced (only terminal nodes contain data):

t1	t3	t4	t5
t1	t6	t7	t8
t2	t9	t10	t11
t2	t12	t13	
t2	t14	t15	t16

Fig. 9.39. The output data table for the example in Fig. 9.35

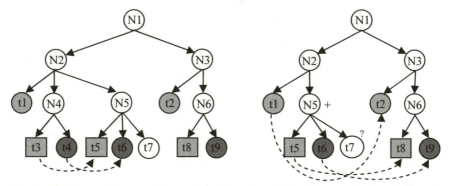

Fig. 9.40. Aligned data nodes are linked   **Fig. 9.41.** Alignment after collapsing

t1	t3	t4	
t1	t5	t6	t7
t2	t8	t9	

Fig. 9.42. The output data table for the example in Fig. 9.40

Let us use a smaller but more complete example (Fig. 9.40) to show that generating a pattern of a lower level list makes it possible for a higher level matching. At the level of N4–N5 (which has the parent N2), t3–t5 and t4–t6 are matched (assume they satisfy the match condition, line 5 of Fig. 9.34). They are aligned and linked (dash lines). N4 and N5 are data records at this level (nested in N2 in this case), in which t7 is optional. N4 and N5 are then collapsed to produce a pattern data record using GenNodePattern() first and then GenRecordPattern(), which does not do anything in this case. t7 is marked with a "?", indicating that it is optional. The pattern data record is N5 (selected based on the PartialTreeAlignment() function). The sub-tree at N4 is then omitted in Fig. 9.41. N5 is marked with a "+" indicating that there is one or more such data records and that the sub-tree of N5 is the pattern. We can see in Fig. 9.41 that the sub-trees rooted at N2 and N3 can now match. The final output data table is given in Fig. 9.42.

9.10 Extraction Based on Multiple Pages

We now discuss the second extraction problem described in Sect. 9.4.1. Given multiple pages with the same encoding template, the system finds patterns from them to be used to extract data from other similar pages. The collection of input pages can be a set of list pages or detail pages. Below, we first see how the techniques described so far can be applied in this setting, and then describe a technique specifically designed for this setting.

9.10.1 Using Techniques in Previous Sections

We discuss extraction of list pages and detail pages separately.

Given a Set of List Pages

Since the techniques described in previous sections are for a single list page, they can obviously be applied to multiple list pages. The pattern discovered from a single page can be used to extract data from the rest of the pages. Multiple list pages may also help improve the extraction. For example, patterns from all input pages may be found separately and merged to produce a single refined pattern. This can deal with the problem that a single page may not contain the complete information.

Given a Set of Detail Pages

In some applications, one needs to extract data from detail pages as they contain more information. For example, in a list page, the information on each product is usually quite brief, e.g., containing only the name, image, and price. However, if an application also needs the product description and customer reviews, one has to extract them from detail pages.

For extraction from detail pages, we can treat each page as a data record and apply the algorithms described in Sect. 9.8 and/or Sect. 9.9. For instance, to apply the NET algorithm, we can simply construct a rooted tree as input to NET as follows: (1) create an artificial root node, and (2) make the DOM tree of each page as a child sub-tree of the artificial root.

9.10.2 RoadRunner Algorithm

We now describe the RoadRunner algorithm [117], which is designed specifically for problem 2. Given a set of pages, each containing one or more data records (i.e., the pages can be list pages or detail pages), the algorithm compares the pages to find similarities and differences, and in the process generating a **union-free regular expression** (i.e., a regular expression without disjunctions) extractor/wrapper. The approach works as follows:

- To start, it takes a random page as the regular expression wrapper W.
- The wrapper W is then refined by matching it sequentially with the HTML code of each remaining page p_i. It generalizes W by solving mismatches between the wrapper W and the page p_i. A mismatch occurs when some token in p_i does not match the grammar of the wrapper.

There are two types of mismatches:

1. Text string mismatches: They indicate data fields or items.
2. Tag mismatches: They indicate

    - optional items, or
    - iterators (a list of repeated patterns):
      In this case, a mismatch occurs at the beginning of a repeated pattern and the end of a list. The system finds the last token of the mismatch position and identifies some candidate repeated patterns from the wrapper and the page p_i by searching forward. It then compares the candidates with the upward portion of the page p_i to confirm.

The algorithm is best explained with an example, which is given in Fig. 9.43. In this figure, page 1 on the left (in HTML code) is the initial wrapper. Page 2 on the right is a new page to be matched with page 1.

Let us look at some matches and mismatches. Lines 1–3 of both pages are the same and thus match. Lines 4 of both pages are text strings and are different. They are thus data items to be extracted. We go down further. Lines 6 of the pages do not match. Line 6 of page 1 matches line 7 of page 2. Thus, <IMG src=.../> is likely to be optional. Line 11 of page 1 and line 12 of page 2 give another mismatch. Since they are text strings, they are thus data items to be extracted. Line 17 of page 1 and line 18 of page 2 are also data items. Another mismatch occurs at line 19 of page 1 and line 20 of page 2. Further analysis will find that we have a list here. The final refined regular expression wrapper is given at the bottom of Fig. 9.43.

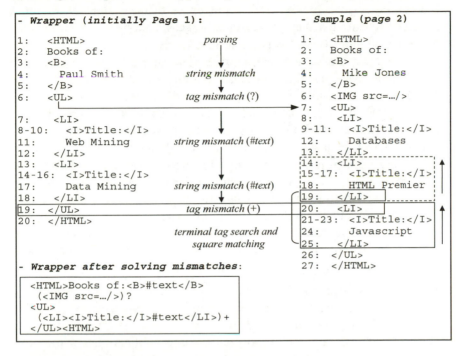

Fig. 9.43. A wrapper generation example

The match algorithm is exponential in the input string length as it has to explore all possibilities. A set of heuristics is introduced to lower the complexity by limiting the space to explore and to backtrack. In [26], a more efficient method is given based on sophisticated tag path analysis.

9.11 Some Other Issues

We now briefly discuss a few other issues that are important to automatic extraction techniques.

9.11.1 Extraction from Other Pages

Once the encoding template pattern is found, it can be used to extract data from other pages that contain data encoded in the same way. There are three ways to perform the extraction:

- Finite-state machines: An encoding template pattern is usually represented as a regular expression. A nondeterministic finite-state automaton can be constructed to match occurrences of the pattern in the input string representing a Web page. In the process, data items are extracted.
- Pattern matching: It is also possible to directly match the string or tree pattern against the input to extract data. This approach is more flexible than finite-state machines because pattern matching allows partial matching. For example, in the page where the pattern is discovered, an optional item does not occur, but it occurs in some other pages. Pattern matching can deal with this easily. In the process, the pattern can be enhanced as well by inserting the new optional item in it.
- Extracting each page independently: The above two approaches can be problematic if the Web site use many different templates to encode its data. If we start to extract after only finding one pattern, then the data encoded using other templates will not be extracted. One solution to this is to find patterns from each page and extract the page using only the discovered patterns from the page. However, handling each page individually is inefficient.

Detecting new templates: To detect new templates without sacrificing efficiency of mining extraction patterns from each page, a pre-screening strategy may be applied. In most applications, the user is interested in only a particular kind of data, e.g., products, research publications, or job postings. It is usually possible to design some simple and efficient heuristics to check whether a page contains such data. If so, a full blown extraction is performed using already generated patterns. If no data is extracted from the page, it is an indication that the page is encoded with a different template. A new mining process can be initiated to discover the new template.

9.11.2   Disjunction or Optional

In automatic extraction, it can be difficult to recognize disjunctions. For example, for the three digital cameras in Fig. 9.44, it is easy to know that "On Sale" is an optional item. However, for the prices (including "Out of stock"), it is hard to decide whether they are optional items or disjuncts of a disjunction. The HTML codes for the three fields are given below,

(1) <b> $250.00 </b>
(2) <b><i> $300.00 </i></b>
(3) <i> Out of stock </i>.

If they are treated as optional items, they are put in three different columns in the output table, but if they are disjuncts, they are put in the same column. In this example, it is easy for a human user to see that they are disjuncts, i.e., (<b> *#text* </b>) | (<b><i> *#text* </i></b>) | (<i> *#text*</i>).

(1)　　　　　　　　　　(2)　　　　　　　　　　(3)

Fig. 9.44. Disjuncts or optional items

There are two main pieces of information that can be used to determine whether they are optional or disjuncts:

1. Visual information: If they are in the same relative location with respect to their objects, then they are more likely to be disjuncts. In the above example, the three items are all at the same relative location.
2. Data type information: If the data items are of the same type, they are more likely to be disjuncts. "$250.00" and "$300.00" are of the same type, but "Out of stock" is not.

In many cases, it can be hard to decide. Fortunately, disjunctive cases are rare on the Web. Even if an extraction system does not deal with disjunction, it does not cause a major problem. For example, if "Out of stock" is identified as optional, it is probably acceptable.

9.11.3 A Set Type or a Tuple Type

Sometimes it can also be difficult to determine whether a list is a tuple type or a set type. For example, if all the lists of a set type have the same number of elements, it is hard to know if they are in fact attributes of a tuple. For instance, the following are three colors of a jacket with different prices. Clearly, they represent a set instance with a list of three tuples:

```
<tr><td><b>Blue:<b></td> <td> $5.00 </td></tr>
<tr><td><b>Yellow:<b></td> <td> $6.50 </td></tr>
<tr><td><b>Pink:<b></td> <td> $10.99 </td></tr>.
```

However, the following specifications of a particular product are obviously the attributes of the product. Without knowing the semantics of the encoded data, it is difficult to know that the above three are a set instance and the following two are attributes of a tuple:

```
<tr><td><b>weight:<b></td> <td> 30 kg </td></tr>
<tr><td><b>height:<b></td> <td> 5 m </td></tr>.
```

If multiple lists of the same type are available, we may have some additional information to make the decision. For instance, one pair of shoes has three colors, and another has four colors. We can be fairly confident that different sets of colors represent set instances (or lists). In the second example, if all products have both height and width, it is more likely that they are attributes. However, these heuristics do not always hold. In some cases, it is hard to decide without understanding of the data semantics.

9.11.4 Labeling and Integration

Once the data is extracted from a page/site and put in tables, it is desirable to label each column (assigning an attribute name to it). Some preliminary studies have been reported in [27, 530]. However, the problem is still very much open. Furthermore, the extracted data from multiple sites may need to be integrated. There are two main integration problems. The first one is **schema matching**, which matches columns of data tables. The second one is **data value/instance match**. For example, in one site, Coca Cola is called "Coke", but in another site it is called "Coca Cola". The problem is: how does the system know that they are the same semantically? In Chap. 10, we will study some data integration techniques.

9.11.5 Domain Specific Extraction

In most applications, the user is only interested in some specific data objects, e.g., products sold online, and for each object, only some specific items are needed, e.g., product name, image, and price. Domain specific information can be exploited to simplify and also to speed up the extraction dramatically. Such information can be utilized in at least two ways.

1. Quickly identify pages that may contain required data. For example, it is fairly easy to design some domain heuristics to determine whether a page contains a list of products (a list page). One heuristic is to detect repeated images and repeated prices in some fixed order and interval. Such heuristics are usually very efficient to execute and can be used to filter out those pages that are unlikely to contain required data. The extraction algorithm, which is slower, will only run on those pages that are very likely to contain target data.
2. Identifying target items in a data record. Based on the characteristics of target items, it may be easy to identify and label the target items. For example, it is often easy to find product names and product images based on simple heuristics. If heuristics are not reliable, machine learn-

ing methods may be applied to learn models to identify target items. For example, in [620], an extended conditional random fields method is used to learn an extraction model, which is then used to extract target items from new data records.

9.12   Discussion

Finally, we discuss the main advantages and disadvantages of wrapper induction and automatic data extraction. The key advantage of wrapper induction is that it extracts only the data that the user is interested in. Due to manual labeling, there is no schema matching problem. However, data value or instance matching is still needed. The main disadvantages are that it is not scalable to a large number of sites due to significant manual efforts, and that maintenance is very costly if sites change frequently.

The main advantages of automatic extraction are that it is scalable to a huge number of sites, and that there is little maintenance cost. The main disadvantage is that it can extract a large amount of unwanted data because the system does not know what is interesting to the user. Also, in some applications, the extracted data from multiple sites need integration, i.e., their schemas as well as values need to be matched, which are difficult tasks. However, if the application domain is narrow, domain heuristics may be sufficient to filter out unwanted data and to perform the integration tasks.

In terms of extraction accuracy, it is reasonable to assume that wrapper induction is more accurate than automatic extraction, although there is no reported large scale study comparing the two approaches.

Bibliographic Notes

Web data extraction techniques can be classified into three main categories: (1) wrapper programming languages and visual platforms, (2) wrapper induction, and (3) automatic data extraction. The first approach provides some specialized pattern specification languages and visual platforms to help the user construct extraction programs. Systems that follow this approach include WICCAP [613], Wargo [457], Lixto [41], etc.

The second approach is wrapper induction, which uses supervised learning to learn data extraction rules from a set of manually labeled positive and negative examples. A theoretical work on wrapper learning based on the PAC learning framework was done by Kushmerick [295]. Example wrapper induction systems include WIEN [296], Softmealy [244], Stalker

[399], WL^2 [108], Thresher [241], IDE [599], [250], etc. Most existing systems are based on inductive learning from a set of labeled examples. IDE [599] employs a simple instance-based learning technique, which performs active learning at the same time so that the user only needs to label a very small number of pages. Related ideas are also used in [90] and [241]. Most existing wrapper induction systems built wrappers based on similar pages from the same site. Zhu et al. [620, 621] reported a system that learns from labeled pages from multiple sites in a specific domain. The resulting wrapper can be used to extract data from other sites. This avoids the labor intensive work of building a wrapper for each site.

The third approach is automatic extraction. In [163], Embley et al. studied the automatic identification of data record boundaries given a list page. The technique uses a set of heuristic rules and domain ontologies. In [75], Buttler et al. proposed additional heuristics to perform the task without using domain ontologies. The MDR algorithm discussed in this chapter was proposed by Liu et al. [341]. It uses string edit distance in pattern finding (incidentally, Lloyd Allison has a great page on string edit distance). An algorithm based on the visual information was given by Zhao et al. [612] for extracting search engine results. These systems, however, do not align or extract data items from data records. Chang et al. [91] reported a semi-automatic system called IEPAD to find extraction patterns from a list page to extract data items. The DeLa system by Wang et al. [530] works in the same framework. The DEPTA system by Zhai and Liu [600] works in a different way. It first segments data records, and then aligns and extracts data items in the data records using the partial tree alignment algorithm. Both DEPTA and IEPAD do not deal with nested data records, which are dealt with in NET [351] and DeLa [530].

The RoadRunner system, which needs multiple pages as input, was proposed by Crescenzi et al. [117]. Its theoretical foundation was given by Grumbach and Mecca [210]. Sects. 9.1 and 9.4 of this chapter are influenced by this paper. The work of RoadRunner was improved by Arasu and Garcia-Molina in their EXALG system [26]. Both systems need multiple input pages with a common schema/template and assume that these pages are given. The pages can be either detail pages or list pages. The method proposed in [312] works in a similar setting.

10 Information Integration

In Chap. 9, we studied data extraction from Web pages. The extracted data is put in tables. For an application, it is, however, often not sufficient to extract data from only a single site. Instead, data from a large number of sites are gathered in order to provide value-added services. In such cases, extraction is only part of the story. The other part is the integration of the extracted data to produce a consistent and coherent database because different sites typically use different data formats. Intuitively, integration means to match columns in different data tables that contain the same type of information (e.g., product names) and to match values that are semantically identical but represented differently in different Web sites (e.g., "Coke" and "Coca Cola"). Unfortunately, limited integration research has been done so far in this specific context. Much of the Web information integration research has been focused on the integration of **Web query interfaces**. This chapter will have several sections on their integration. However, many ideas developed are also applicable to the integration of the extracted data because the problems are similar.

Web query interfaces are used to formulate queries to retrieve needed data from **Web databases** (called the **deep Web**). Figure 10.1 shows two query interfaces from two travel sites, expedia.com and vacation.com. The user who wants to buy an air ticket typically tries many sites to find the cheapest ticket. Given a large number of alternative sites, he/she has to access each individually in order to find the best price, which is tedious. To reduce the manual effort, we can construct a **global query interface** that allows uniform access to disparate relevant sources. The user can then fill in his/her requirements in this single global interface and all the underlying sources (or databases) will be automatically filled and searched. The retrieved results from multiple sources also need to be integrated. Both integration problems, i.e., integration of query interfaces and integration of returned results, are very challenging due to the heterogeneity of Web sites.

Clearly, integration is not peculiar only to the Web. It was, in fact, first studied in the context of relational databases and data warehouse. Hence, this chapter first introduces most integration related concepts using traditional data models (e.g., relational) and then shows how the concepts are tailored to Web applications and how Web specific problems are handled.

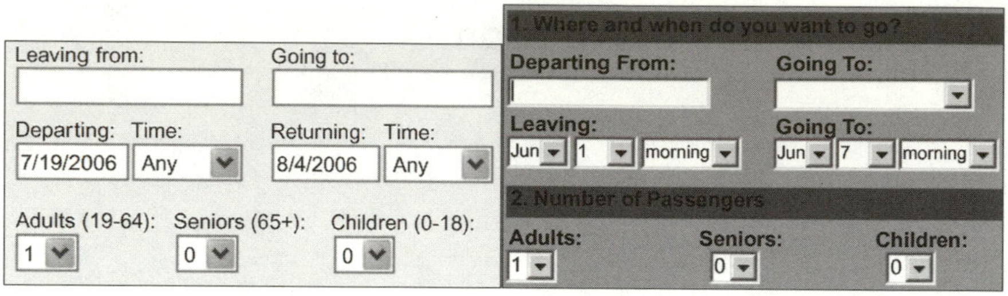

Fig. 10.1. Two examples of Web query interfaces

10.1   Introduction to Schema Matching

Information/data integration has been studied in the database community since the early 1980s [40, 146, 455]. The fundamental problem is **schema matching**, which takes two (or more) database schemas to produce a mapping between **elements** (or **attributes**) of the two (or more) schemas that correspond semantically to each other. The objective is to merge the schemas into a single global schema. This problem arises in building a global database that comprises several distinct but related databases. One application scenario in a company is that each department has its database about customers and products that are related to the operations of the department. Each database is typically designed independently and possibly by different people to optimize database operations required by the functions of the department. This results in different database schemas in different departments. However, to consolidate the data about customers or company operations across the organization in order to have a more complete understanding of its customers and to better serve them, integration of databases is needed. The integration problem is clearly also important on the Web as we discussed above, where the task is to integrate data from multiple sites.

There is a large body of literature on the topic. Most techniques have been proposed to achieve semi-automatic matching in specific domains (see the surveys in [146, 265, 455, 491]). Unfortunately, the criteria and methods used in match operations are almost all based on domain heuristics which are not easily formulated mathematically. Thus, to build a schema matching system, we need to produce mapping heuristics which reflect our understanding of what the user considers to be a good match.

Schema matching is challenging for many reasons. First of all, schemas of identical concepts may have structural and naming differences. Schemas may model similar but not identical contents, and may use different data models. They may also use similar words for different meanings.

Although it may be possible for some specific applications, in general, it is not possible to fully automate all matches between two schemas because some semantic information that determines the matches between two schemas may not be formally specified or even documented. Thus, any automatic algorithm can only generate candidate matches that the user needs to verify, i.e., accept, reject or change. Furthermore, the user should also be allowed to specify matches for elements that the system is not able to find satisfactory match candidates. Let us see a simple example.

Example 1: Consider two schemas, S_1 and S_2, representing two customer relations, Cust and Customer.

S_1	S_2
Cust	Customer
CNo	CustID
CompName	Company
FirstName	Contact
LastName	Phone

We can represent the mapping with a similarity relation, $\cong$, over the power sets of S_1 and S_2, where each pair in $\cong$ represents one element of the mapping. For our example schemas, we may obtain

Cust.CNo $\cong$ Customer.CustID
Cust.CompName $\cong$ Customer.Company
{Cust.FirstName, Cust.LastName} $\cong$ Customer.Contact   ■

There are various types of matching based on the input information [455].

1. **Schema-level only matching:** In this type of matching, only the schema information (e.g. names and data types) is considered. No data instance is available.
2. **Domain and instance-level only matching:** In this type of match, only instance data and possibly the domain of each attribute are provided. No schema is available. Such cases occur quite frequently on the Web, where we need to match corresponding columns of the hidden schemas.
3. **Integrated matching of schema, domain and instance data:** In this type of match, both schemas and instance data (possibly domain information) are available. The match algorithm can exploit clues from all of them to perform matching.

There are existing approaches to all above types of matching. We will focus on the first two types. The third type usually combines the results of techniques from the first two, which we discuss in Sect. 10.5. Before going to the details, we first discuss some pre-processing tasks that usually need to be done before matching.

10.2   Pre-Processing for Schema Matching

For pre-processing, issues such as concatenated words, abbreviations, and acronyms are dealt with. That is, they need to be normalized before being used in matching [227, 358, 559].

Prep 1 (**Tokenization**): This process breaks an item, which can be a schema element (attribute) or attribute value, into atomic words. Such items are usually concatenated words. Delimiters (such as "-", "_", etc.) and case changes of letters are used to suggest the breakdown. For example, we can break "fromCity" into "from City", and "first-name" into "first name". A domain dictionary of words is typically maintained to help the breakdown. Note that if "from", "city", "first" and "name" are not in the dictionary, they will be added to the dictionary. Existing dictionary words are also utilized to suggest the breakdown. For example, "deptcity" will be split into "dept" and "city" if "city" is a word. The dictionary may be constructed automatically, which consists of all the individual words appeared in the given input used in matching, e.g., schemas, instance data and domains. The dictionary is updated as the processing progresses. However, the tokenization step has to be done with care. For example, we have "Baths" and "Bathrooms" if we split "Bath" with "Room" it could be a mistake because "Rooms" could have a very different meaning (the number of rooms in the house). To be sure, we need to ensure that "Bathroom" is not an English word, for which an online English dictionary may be employed.

Prep 2 (**Expansion**): It expands abbreviations and acronyms to their full words, e.g., from "dept" to "departure". The expansion is usually done based on the auxiliary information provided by the user or collected from other sources. Constraints may be imposed to ensure that the expansion is likely to be correct. For example, we may require that the word to be expanded is not in the English dictionary, with at least three letters, and having the same first letter as the expanding word. For example, "CompName" is first converted to (Comp, Name) in tokenization, and then "Comp" is expanded to "Company".

Prep 3 (**Stopword removal and stemming**): These are information retrieval pre-processing methods (see Chap. 6). They can be performed to attribute names and domain values. A domain specific stopword list may also be constructed manually. This step is useful especially in linguistic based matching methods discussed below.

Prep 4 (**Standardization of words**): Irregular words are standardized to a single form (e.g., using WordNet [175]), "colour"→ "color", "Children" → "Child".

10.3 Schema-Level Matching

A schema level matching algorithm relies on information about schema elements, such as name, description, data type and relationship types (such as **part-of**, **is-a**, etc.), constraints and schema structures. Before introducing some matching methods using such information, let us introduce the notion of **match cardinality**, which describes the number of elements in one schema that match the number of elements in the other schema.

In general, given two schemas, S_1 and S_2, within a single match in the match relation one or more elements of S_1 can match one or more elements of S_2. We thus have 1:1, 1:m, m:1 and m:n matches. 1:1 match means that one element of S_1 corresponds to one element of S_2, and 1:m means that one element of S_1 corresponds to a set of m ($m > 1$) elements of S_2.

Example 2: Consider the following schemas:

S_1	S_2
Cust	Customer
CustomID	CustID
Name	FirstName
Phone	LastName

We can find the following 1:1 and 1:m matches:

1:1	CustomID	CustID
1:m	Name	FirstName, LastName

■

m:1 match is similar to 1:m match; m:n match is considerably more complex. An example of an m:n match is to match Cartesian coordinates with polar coordinates. There is little work on such complex matches. Most existing approaches are for 1:1 and 1:m matches.

We now describe some general matching approaches that employ various types of information available in schemas. There are two main types of information in schemas, natural language words and constraints. Thus, there are two main types of approaches to matching.

10.3.1 Linguistic Approaches

They are used to derive match candidates based on the names, comments or descriptions of schema elements [107, 133, 144, 145, 227, 358, 559].

Name Match

N1 – Equality of names: The same name in different schemas often has the same semantics.

N2 – Synonyms: The names of two elements from different schemas are **synonyms**, e.g., Customer ≅ Client. This requires the use of thesaurus and/or dictionaries such as WordNet. In many cases, domain dependent or enterprise specific thesaurus and dictionaries are required.

N3 – Equality of hypernyms: A is a **hypernym** of B if B is **a kind of** A. If X and Y have the same hypernym, they are likely to match. For example, "Car" *is-a* "vehicle" and "automobile" *is-a* "vehicle". Thus, we have Car ≅ vehicle, automobile ≅ vehicle, and Car ≅ automobile.

N4 – Common substrings: Edit distance and similar pronunciation may be used. For example, CustomerID ≅ CustID, and ShipTo ≅ Ship2.

N5 – Cosine similarity: Some names are natural language words or phrases (after pre-processing). Then, text similarity measures are useful. **Cosine similarity** is a popular similarity measure used in information retrieval (see Chap. 6). This method is also very useful for Web query interface integration since the labels of the schema elements are natural language words or phrases (see the query interfaces in Fig. 10.1)

N6 – User provided name matches: The user may provide a domain dependent match dictionary (or table), a thesaurus, and/or an ontology.

Description Match

In many databases, there are comments to schema elements, e.g.,

S_1: CNo   // customer unique number
S_2: CustID   // id number of a customer

These comments can be compared based on the cosine similarity as well.

D1 – Use the cosine similarity to compare comments after stemming and stopword removal.

10.3.2 Constraint Based Approaches

Constraints such as data types, value ranges, uniqueness, relationship types and cardinalities, etc., can be exploited in determining candidate matches [327, 358, 382, 424].

C1: An equivalence or compatibility table for data types and keys that specifies compatibility constraints for two schema elements to match can be provided, e.g., string ≅ varchar, and (primary key) ≅ unique.

Example 3: Consider the following two schemas:

S₁	S₂
Cust	Customer
CNo: int, primary key	CustID: int, unique
CompName: varchar (60)	Company: string
CTname: varchar (15)	Contact: string
StartDate: date	Date: date

Constraints can suggest that "CNo" matches "CustID", and "StartDate" may match "Date". "CompName" in S_1 may match "Company" in S_2 or "Contact" in S_2. Likewise, "CTname" in S_1 may match "Company" or "Contact" in S_2. In both cases, the types match. Although in these two cases, we are unable to find a unique match, the approach helps limit the number of match candidates and may be combined with other matchers (e.g., name and instance matchers). For structured schemas, hierarchical relationships such as **is-a** and **part-of** relationships may be utilized to help match. ∎

In the context of the Web, the constraint information above is often not explicitly available because Web databases are for general public who are unlikely to know what an int, string or varchar is. Thus, these types are never shown in Web pages. However, some information may be inferred from the domain or instance information, which we discuss next.

10.4 Domain and Instance-Level Matching

In this type of matching, value characteristics are exploited to match schema elements [53, 145, 327, 531, 558]. For example, the two attribute names may match according to the linguistic similarity, but they may have different domain value characteristics. Then, they may not be the same but **homonyms**. For example, Location in a real estate sell may mean the address, but could also mean some specific locations, e.g., lakefront property, hillside property, etc.

In many applications, data instances are available, which is often the case in the Web database context. In some applications, although the instance information is not available, the domain information of each attribute may be obtained. This is the case for Web query interfaces. Some attributes in the query interface contain a list of possible values (the domain) for the user to choose from. No type information is explicitly given, but it can often be inferred. We note that the set of value instances of an attribute can be treated in the similar way as a domain. Thus, we will only deal with domains below.

Let us look at two types of domains or types of values: simple domains and composite domains. The domain similarity of two attributes, *A* and *B*, is the similarity of their domains: *dom*(*A*) and *dom*(*B*).

Definition (Simple Domain): A **simple domain** is a domain in which each value has only a single component, i.e., the value cannot be decomposed.

A simple domain can be of any type, e.g., year, time, money, area, month, integer, real, string, etc.

Data Type: If there is no type specification at the schema level, we identify the data type from the domain values. Even if there is a type specification at the schema level for each attribute, we can still refine the type to find more characteristic patterns. For example, the ISBN number of a book may be specified as a string type in a given schema. However, due to its fixed format, it is easy to generate a characteristic pattern from a set of ISBN numbers, e.g., a regular expression. Other examples include phone numbers, post codes, money, etc. Such specialized patterns are more useful in matching compatible attribute types.

We describe two approaches for type identification: semi-automatic [559, 563] and automatic [145, 327] approaches.

Semi-automatic approach: This is done via pattern matching. The pattern for each type may be expressed as a regular expression, which is defined by a human expert. For example, the regular expression for the time type can be defined as "[0–9]{2}:[0–9]{2}" or "dd:dd" (d for digit from 0-9) which recognizes time of the form "03:15". One can use such regular expressions to recognize integer, real, string, month, weekday, date, time, datetime (combination of date and time), etc. To identify the data type, we can simply apply all the regular expression patterns to determine the type.

In some cases, the values themselves may contain some information on the type. For example, values that contain "$" or "US$" indicate the monetary type. For all values that we cannot infer their types, we can assume their domains are of string type with an infinite cardinality.

Automated approach: Machine learning techniques, e.g., grammar induction, may be used to learn the underlying grammar/pattern of the values of an attribute, and then use the grammar to match attribute values of the other schemas. This method is particularly useful for value of fixed format, e.g., zip codes, phone numbers, zip codes, ISBNs, date entries, or money-related entries, if their regular expressions are not specified by the user.

The following methods may be used in matching:

DI 1 – Data types are used as constraints. The method C1 above is applicable here. If the data/domain types of two attributes are not compatible, they should not be matched. We can use a table specifying the degree of compatibility between a set of predefined generic data types, to which data types of schema elements are mapped in order to determine their similarity.

DI 2 – For numerical data, value ranges, averages and variances can be computed to access the level of similarity.

DI 3 – For categorical data, we can extract and compare the set of values in the two domains to check whether the two attributes from different schemas share some common values. For example, if an attribute from S_1 contains many "Microsoft" entries and an attribute in S_2 also contains some "Microsoft"'s, then we can propose them as a match candidate.

DI 4 – For alphanumeric data, string-lengths and alphabetic/non-alphabetic ratios are also helpful.

DI 5 – For textual data, information retrieval methods such as the cosine measure may be used to compare the similarity of all data values in the two attributes.

DI 6 – **Schema element name as value** is another match indicator, which characterizes the cases where matches relate some data instances of a schema with a set of elements (attributes) in another schema. For example, in the airfare domain one schema uses "Economy" and "Business" as instances (values) of the attribute "Ticket Class", while in another interface, "Economy" and "Business" are attributes with the Boolean domain (i.e., "Yes" and "No"). This kind of match can be detected if the words used in one schema as attribute names are among the values of attributes in another schema [133, 563].

Definition (Composite Domain and Attribute): A **composite domain** d of arity k is a set of ordered k-tuples, where the ith component of each tuple is a value from the ith sub-domain of d, denoted as d_i. Each d_i is a simple domain. The arity of domain d is denoted as $arity(d)$ ($= k$). An **attribute** is **composite** if its domain is composite.

A composite domain is usually indicated by its values that contained delimiters of various forms. The delimiters can be punctuation marks (such as ",", "-", "/", "_", etc) and white spaces and some special words such as "to". To detect a composite domain, we can use these delimiters to split a composite domain into simple sub-domains. In order to ensure correctness, we may also want to require that a majority of (composite) values can be consistently split into the same number of components. For example, the date can be expressed as a composite domain with MM/DD/YY.

DI 7 – The similarity of a simple domain and a composite domain is determined by comparing the simple domain with each sub-domain of the composite domain. The similarity of composite domains is established by comparing their component sub-domains.

We note that splitting a composite domain can be quite difficult in the Web context. For example, without sufficient auxiliary information (e.g., information from other sites) it is not easy to split the following: "Dell desktop PC 1.5GHz 1GB RAM 30GB disk space"

10.5   Combining Similarities

Let us call a program that assesses the similarity of a pair of elements from two different schemas based on a particular match criterion a **matcher**. It is typically the case that the more indicators we have the better results we can achieve, because different matchers have their own advantages and also shortcomings. Combining schema-level and instance-level approach will produce better results than each type of approaches alone. This combination can be done in various ways.

Given the set of similarity values, $sim_1(u, v)$, $sim_2(u, v)$, …, $sim_n(u, v)$, of a set of n matchers that compared two schema elements u (from S_1) and v (from S_2), one of the following strategies can be used to combine their similarity values.

1. **Max:** This strategy returns the maximal similarity value of any matcher. It is thus optimistic. Let the combined similarity be $CSim$. Then

$$CSim(u, v) = \max\{sim_1(u, v), sim_2(u, v), …, sim_n(u, v)\} \quad (1)$$

2. **Weighted Sum:** This strategy computes a weighted sum of similarity values of the individual matchers. It needs relative weights which correspond to the expected importance of the matchers:

$$CSim(u, v) = \lambda_1 * sim_1(u, v) + \lambda_2 sim_2(u, v) + … + \lambda_n * sim_n(u, v), \quad (2)$$

where λ_i is a weight coefficient, and usually determined empirically.

3. **Weighted Average:** This strategy computes a weighted average of similarity values of the individual matchers. It also needs relative weights that correspond to the expected importance of the matchers.

$$CSim(u, v) = \frac{\lambda_1 Sim_1(u, v) + \lambda_2 Sim_2(u, v) + … + \lambda_n Sim_n(u, v)}{n} \quad (3)$$

where λ_i is a weight coefficient and is determined experimentally.

4. **Machine Learning:** This approach uses a classification algorithm, e.g., a decision tree, a naïve Bayesian classifier, or SVM, to determine whether two schema elements match each other. In this case, the user needs to label a set of training examples, which is described by a set of attributes and a class. The attributes can be the similarities. Each training example thus represents the similarity values of a pair of schema elements. The class of the example is either Yes or No, which indicates whether the two elements match or not as decided by the user.

There are many other possible approaches. In practice, which method to use involves a significant amount of experimentation and parameter tuning. Note that the combination can also be done in stages for different types of matches. For example, we can combine the instance based similarities first using one method, e.g., **Max**, and then combine schema based similarities using another method, e.g., **Weighted Average**. After that, the final combined similarity computation may use **Weighted Sum**.

10.6 1:*m* Match

The approaches presented above are for 1:1 matches. For 1:*m* match, other techniques are needed [133, 563, 559]. There are mainly two types of 1:*m* matches.

Part-of Type: Each relevant schema element on the many side is a part of the element on the one side. For example, in one schema, we may have an attribute called "Address", while in another schema, we may have three attributes, "Street", "City" and "State". In this case, "Street", "City" or "State" is a part of "Address". That is, the combination of "Street", "City" or "State" forms "Address". Thus, it is a 1:*m* match.

Is-a Type: Each relevant schema element on the many side is a specialization of the schema element on the one side. The content of the attribute on the one side is the union or sum of the contents of the attributes on the many side. For example, "HomePhone" and "CellPhone" in S_2 are specializations of "Phone" in S_1. Another example is the (number of) "Passengers" in Fig. 10.3 (page 397), and the (number of) "Adults", the (number of) "Seniors", and the (number of) "Children" in Fig. 10.1 in the airline ticket domain.

Identifying Part-of 1:*m* Matches: For each attribute A in interface S_1, we first check if it is a composite attribute as described above. If A is a composite attribute, we find a subset of schema elements in S_2 that has a 1:1 correspondence with the sub-attributes of A. For a real application, we may need additional conditions to make the decision (see Sect. 10.8.1).

Identify Is-a 1:*m* Matches: In the case of part-of 1:*m* mappings, the domains of the sub-attributes are typically different. In contrast, the identification of is-a 1:*m* mappings of attributes requires that the domain of each corresponding sub-attribute be similar to that of the general attribute. Name matching of schema elements is useful here. For example, in the case of "Phone" in S_1 and "HomePhone" and "CellPhone" in S_2, the name similarity can help decide 1:*m* mapping. However, this strategy alone is usually not sufficient, e.g., "Passengers" in S_1 and "Adults", "Seniors" and "Children" in S_2 have no name similarity. Additional information is needed. We will show an example in Sect. 10.8.1.

Using the auxiliary information provided by the user is also a possibility. It is not unreasonable to ask the user to provide some information about the domain. For example, a domain ontology that includes a set of concepts and their relationships such as the following (Fig. 10.2) will be of great help:

> Part-of("street", "address")   Is-a("home phone", "phone")
> Part-of("city", "address")   Is-a("cell phone", "phone")
> Part-of("state", "address")   Is-a("office phone", "phone")
> Part-of("country", "address") Is-a("day phone", "phone")

Fig. 10.2. Part-of(X, Y) – X is a part of Y, and Is-a(X, Y) – X is a Y.

10.7   Some Other Issues

10.7.1   Reuse of Previous Match Results

We have mentioned in several places that auxiliary information in addition to the input schemas and data instances, such as dictionaries, thesauri, and user-provided ontology information are very useful in schema matching. The past matching results can also be stored and reused for future matches [356, 455]. Reuse is important because many schemas are very similar to each other and to previously matched schemas. Given a new schema S to be matched with a set of existing schemas E, we may not need to match S with every existing schema in E. There are two slightly different scenarios:

1. Matching of a large number of schemas: If we have a large number of schemas to match, we may not need to perform all pair-wise matches, which have $n(n+1)/2$ of them with n being the number of input schemas. Since most schemas are very similar, the $n(n+1)/2$ number of matches are not necessary.

2. Incremental schema matching: In this scenario, given a set of schemas that has already been matched, when a new schema S needs to be matched with existing matched schemas E, we may not want to use S to match every schema in E using pair-wise matching. This is the same situation as the first case above. If the original match algorithm is not based on pair-wise match, we may not want to run the original algorithm on all the schemas to just match this single schema with them.

For both cases, we want to use old matches to facilitate the discovery of new matches. The key idea is to exploit the **transitive property** of similarity relationship. For example, "Cname" in S_1 matches "CustName" in S_2 as they are both customer names. If "CTname" in the new schema S matches "Cname" in S_1, we may conclude that "CTname" matches "CustName" in S_2. The transitive property has also been used to deal with some difficult matching cases. For example, it may be difficult to map a schema element A directly to a schema element B, but easy to map both A and B to the schema element C in another schema. This helps us decide that A corresponds to B [144, 559, 563].

In the incremental case, we can also use a clustering-based method. For example, if we already have a large number of matches, we can group them into clusters and find a centroid to represent each cluster, in term of schema names and domains. When a new schema needs to be matched, the schema is compared with the centroid rather than with each individual schema in the cluster.

10.7.2 Matching a Large Number of Schemas

The techniques discussed so far are mainly for pair-wise matching of schemas. However, in many cases, we may have a large number of schemas. This is the case for many Web applications because there are many Web databases in any domain or application. With a large number of schemas, new techniques can be applied. We do not need to depend solely on pair-wise matches. Instead, we can use statistical approaches such as data mining to find patterns, correlations and clusters to match the schemas. In the next section, we will see two examples in which clustering and correlation methods are applied.

10.7.3 Schema Match Results

In pair-wise matching, for each element v in S_2, the set of matching elements in S_1 can be decided by one of the following methods [144].

1. **Top N candidates:** The top N elements of S_1 that have the highest similarities are chosen as match candidates. In most cases, $N = 1$ is the natural choice for 1:1 correspondences. Generally, $N > 1$ is useful in interactive mode, i.e., the user can select among several match candidates.
2. **MaxDelta:** The S_1 element with the maximal similarity is determined as
   match candidate plus all S_1 elements with a similarity differing at most
   by a tolerance value t, which can be specified either as an absolute or
   relative value. The idea is to return multiple match candidates when
   there are several S_1 elements with almost the same similarity values.
3. **Threshold:** All S_1 elements with the final combined similarity values
   exceeding a given threshold t are selected.

10.7.4   User Interactions

Due to the difficulty of schema matching, extensive user interaction is often needed in building an accurate matching system for both parameter
tuning and resolving uncertainties

Building the Match System: There are typically many parameters and
thresholds in an integration system, e.g., similarity values, weight coefficients, and decision thresholds, which are usually domain-specific or even
attribute specific. Before the system is used to match other schemas, interactive experiments are needed to tune the parameters by trial-and-errors.

After Matching: Although the parameters are fixed in the system building, their values may not be perfect. Matching mistakes and failures will
still occur: (1) some matched attributes may be wrong (false positive); (2)
some true matches may not be found (false negative). User interactions are
needed to correct the situations and to confirm the correct matches.

10.8   Integration of Web Query Interfaces

The preceding discussions are generic to database integration and Web
data integration. In this and the next sections, we focus on integration in
the Web context. The Web consists of the **surface Web** and the **deep
Web**. The surface Web can be browsed using any Web browser, while the
deep Web consists of databases that can only be accessed through parameterized query interfaces. With the rapid expansion of the Web, there are
now a huge number of deep web data sources. In almost any domain, one
can find a large number of them, which are hosted by e-commerce sites.
Each of such sources usually has a keyword based search engine or a query

interface that allows the user to fill in some information in order to retrieve the needed data. We have seen two query interfaces in Fig. 10.1 for finding airline tickets. We want to integrate multiple interfaces in order to provide the user a **global query interface** [153, 227] so that he/she does not need to manually query each individual source to obtain more complete information. Only the **global interface** needs to be filled with the required information. The individual interfaces are filled and searched automatically.

We focus on query interface integration mainly because there is extensive research in this area, although the returned instance data integration is also of great importance and perhaps even more important due to the fact that the number of sites that provide such structured data is huge and most of them do not have query interfaces but only keyword search or can only be browsed by users (see Chap. 9).

Since query interfaces are different from traditional database schemas, we first define a schema model.

Schema Model of Query Interfaces: In each domain, there is a set of concepts $C = \{c_1, c_2, ..., c_n\}$ that represents the essential information of the domain. These concepts are used in query interfaces to enable the user to restrict the search for some specific instances or objects of the domain. A particular query interface uses a subset of the concepts $S \subseteq C$. A concept i in S may be represented in the interface with a set of attributes (or fields) $f_{i1}, f_{i2}, ..., f_{ik}$. In most cases, each concept is only represented with a single attribute. Each attribute is labeled with a word or phrase, called the **label** of the attribute, which is visible to the user. Each attribute may also have a set of possible values that the user can use in search, which is its **domain**.

All the attributes with their labels in a query interface are called the **schema** of the query interface [227, 608]. Each attribute also has a **name** in the HTML code. The name is attached to a TEXTBOX (which takes the user input). However, this name is not visible to the user. It is attached to the input value of the attribute and returned to the server as the attribute of the input value. The name is often an acronym that is less useful than the label for schema matching. For practical schema integration, we are not concerned with the set of concepts but only the label and name of each attribute and its domain.

Most ideas for schema matching in traditional databases are applicable to Web query interfaces as the schema of a query interface is similar to a schema in databases. However, there are also some important differences [67, 92].

1. **Limited use of acronyms and abbreviations:** Data displayed in Web pages are for the general public to view and must be easy to understand. Hence, the use of acronyms and abbreviations is limited to those very

obvious ones. Enterprise-specific acronyms and abbreviations seldom appear. In the case of a company database, abbreviations are frequently used, which are often hard to understand by human users and difficult to analyze by automated systems. To a certain extent, this feature makes information integration on the Web easier.

2. **Limited vocabulary:** In the same domain, there are usually a limited number of essential attributes that describe each object in the domain. For example, in the book domain, we have the title, author, publisher, ISBN number, etc. For each attribute, there is usually limited ways to express the attribute. The chosen label (describing a data attribute, e.g., "departure city") needs to be short, and easily understood by the general public. Therefore, there are not many ways to express the same attributes. Limited vocabulary also makes statistical approaches possible.

3. **A large number of similar databases:** There are often a large number of sites that offer the same services or sell the same products, which result in a large number of query interfaces and make it possible to use statistical methods. This is not the case in a company because the number of related databases is small. Integration of databases from multiple companies seldom happens.

4. **Additional structure:** The attributes of a Web interface are usually organized in some meaningful ways. For example, related attributes are grouped and put together physically (e.g., "first name" and "last name" are usually next to each other), and there may also be a hierarchical organization of attributes. Such structures also help integration as we will see later. In the case of databases, attributes usually have no structure.

Due to these differences, schema matching of query interfaces can exploit new methods. For example, data mining techniques can be employed as we will see in the next few sub-sections. Traditional schema matching approaches in the database context are usually based on pair-wise matching.

Similar to schema integration, query interface integration also requires mapping of corresponding attributes of all the query interfaces.

Example 4: For the two query interfaces in Fig. 10.3, the attribute correspondences are:

Interface 1 (S_1)	Interface 2 (S_2)
Leaving from	From
Going to	To
Departure date	Departure date
Return date	Return date
Passengers:	Number of tickets
Time	
Preferred cabin	

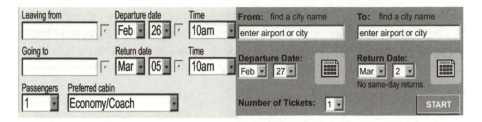

Fig. 10.3. Two query interfaces from the domain of airline ticket reservation

The last two attributes from Interface 1 do not have matching attributes in Interface 2. ∎

The problem of generating the mapping is basically the problem of identifying synonyms in the application domain. However, it is important to note that the synonyms here are domain dependent. A general-purpose semantic lexicon such as WordNet or any thesaurus is not sufficient for the identification of most domain-specific synonyms. For example, it is difficult to infer from WordNet or any thesaurus that "**Passengers**" is synonymous to "**Number of tickets**" in the context of airline ticket reservation. Domain-specific lexicons are not generally available as they are expensive to build. In this section, we discuss three query interface matching techniques. We also describe a method for building a global interface.

10.8.1 A Clustering Based Approach

This technique is a simplified version of the work in [559]. Given a large set of schemas from query interfaces in the same application domain, this technique utilizes a data mining method, **clustering**, to find attribute matches of all interfaces. Three types of information are employed, namely, attribute labels, attribute names and value domains. Let the set of interface schemas be $\{S_1, S_2, \ldots, S_n\}$. The technique works in five steps:

1. Pre-processing the data. It uses the methods given in Sect. 10.2.
2. Computing all pair-wise attribute similarities of u ($\in S_i$) and v ($\in S_j$), $i \neq j$. This produces a similarity matrix.
3. Identify initial 1:m matches.
4. Cluster schema elements based on the similarity matrix. This step discovers 1:1 matches.
5. Generate the final 1:m matches of attributes.

We now discuss each step in turn except the first step.

Computing all Pair-Wise Attribute Similarities: Let u be an attribute of S_i and v be an attribute of S_j ($i \neq j$). This step computes all **linguistic similarities** (denoted by $LingSim(u, v)$) and **domain similarities** (denoted $DomSim(u, v)$). The aggregated similarity (denoted by $AS(u, v)$) is:

$$AS(u,v) = \lambda_{ls} * LingSim(u,v) + \lambda_{ds} * DomSim(u,v), \qquad (4)$$

where λ_{ls} and λ_{ds} are weight coefficients reflecting the relative importance of each component similarity.

The linguistic similarity is based on both attribute labels and attribute names, which give two similarity values, $lSim(u, v)$ and $nSim(u, v)$, representing label and name similarities respectively. Both similarities are computed using the cosine measure as discussed in N5 of Sect. 10.3.1. The two similarities are then combined through a linear combination method similar to Equation (4) above.

Domain similarity of two simple domains d_v and d_u is computed based on the data type similarity (denoted by $typeSim(d_v, d_u)$) and values similarity (denoted by $valueSim(d_v, d_u)$). The final $DomSim$ is again a linear combination of the two values. For the type similarity computation, if the types of domains d_v and d_u are the same, $typeSim(d_v, d_u) = 1$ and 0 otherwise. If $typeSim(d_v, d_u) = 0$, then $valueSim(d_v, d_u) = 0$.

For two domains d_v and d_u of the same type, the algorithm further evaluates their value similarity. Let us consider two character string domains. Let the set of values in d_v be $\{t_1, t_2, \ldots, t_n\}$ and the set of values in d_u be $\{q_1, q_2, \ldots, q_k\}$. $valueSim(d_v, d_u)$ is computed as follows:

1. Calculate all pair-wise value (i.e., (t_i, q_j)) similarities using the cosine measure with one value from each domain.
2. Choose the pair with the maximum similarity among all pairs and delete the corresponding two values from d_v and d_u. For a pair to be considered, its similarity must be greater than a threshold value τ.
3. Repeat step 2 on all remaining values in the domains until no pair of values has a similarity greater than τ.

Let the pairs of values chosen be P. $valueSim(d_v, d_u)$ is then computed using the **Dice function** [136]:

$$valueSim(d_v, d_u) = \frac{2|P|}{|d_v| + |d_u|}. \qquad (5)$$

For two numeric domains, their value similarity is the proportion of the overlapping range of the domains. For an attribute whose domain is unknown, it is assumed that its domain is dissimilar to the domain of any other attribute, be it finite or infinite.

Identify a Preliminary Set of 1:m Mappings: To identify 1:m mappings, the technique exploits the hierarchical organization of the interfaces. The hierarchical organization is determined using the layout and the proximity of attributes as they are likely to be physically close to each other.

Part-of type: To identify the initial set of aggregate 1:m mappings of attributes, it first finds all composite attributes in all interfaces as discussed in Sect. 10.4. For each composite attribute e in S, in every interface other than S, denoted by X, it looks for a set of attributes $f = \{f_1, f_2, \ldots f_r\}$ ($r > 1$) with the same parent p, such that the following conditions hold:

1. f_i's are siblings, i.e., they share the same parent p. The sibling information is derived from the physical proximity in the interface.
2. The label of the parent p of f_i's is highly similar to the label of e.
3. The domains of f_i's have a 1-to-1 mapping with a subset of the subdomains of e based on the high domain similarities.

If there exists such a f in interface X, a 1:m mapping of the part-of type is identified between e and attributes in f.

Is-a type: The identification of is-a 1:m attribute mappings requires that the domain of each corresponding sub-attribute on the m side be similar to that of the general attribute on the one side. More precisely, for each non-composite attribute h in an interface, we look for a set of attributes $f = \{f_1, f_2, \ldots f_r\}$ ($r > 1$) in another interface X, that meets the following conditions:

1. f_i's are siblings of the same parent p, and p does not have any children other than f_i's.
2. The label of the parent p is highly similar to the label of h.
3. The domain of each f_i is highly similar to the domain of h.

If the conditions are met, a 1:m mapping of the is-a type is identified between h and attributes in f.

Cluster the Schema Elements based on the Similarity Matrix: Step 2 produces a similarity matrix M. Let the total number of simple domains in the set of all given query interfaces S be w. We then have a $w \times w$ symmetric similarity matrix. $M[i, j]$ is the aggregated similarity of two attributes i and j. For attributes in the same interface, $M[i, j]$ is infinite, which indicate that they should not be put together into a cluster.

The clustering algorithm used is the hierarchical agglomerative clustering algorithm. The stopping criterion is a similarity threshold. That is, when there is no pair of clusters has the similarity greater than the threshold, the algorithm stops. Each output cluster contains a set of 1:1 attribute mappings from different interfaces.

Obtain Additional 1:m Mapping: The preliminary set of 1:*m* correspondences may not have found all such mappings. The clustering results may suggest additional 1:*m* mappings. The **transitivity** property can be used here. For example, assume that a composite attribute *e* maps to two attributes f_1 and f_2 in another interface in step 3 and the clustering results suggest that f_1 and f_2 map to h_1 and h_2 in yet another interface. Then, *e* also matches h_1 and h_2.

10.8.2 A Correlation Based Approach

This technique also makes use of a large number of interfaces. It is based on the technique in [229]. For pre-processing, the methods discussed in Sect. 10.2 are applied. The approach is based on co-occurrences of schema attributes and the following observations:

1. In an interface, some attributes may be grouped together to form a bigger concept. For example, "first name" and "last name" compose the name of a person. This is called the **grouping relationship**, denoted by a set, e.g., {first name, last name}. Attributes in such a group often co-occur in schemas, i.e., they are **positively correlated**.
2. An attribute group rarely co-occurs in schemas with their synonym attribute groups. For example, "first name" and "last name" rarely co-occur with name in the same query interface. Thus, {first name, last name} and {name} are **negatively correlated**. They represent 2:1 match. Note that a group may contain only one attribute.

Based on the two observations, a correlation-based method to schema matching is in order. Negatively correlated groups represent **synonym groups** or **matching groups**.

Given a set of input schemas $S = \{S_1, S_2, ..., S_n\}$ in the same application domain, where each schema S_i is a transaction of attributes, we want to find all the matches $M = \{m_1, ..., m_v\}$. Each m_j is a complex matching $g_{j1} = g_{j2} = ... = g_{jw}$, where each g_{jk} is an positively correlated attribute group and $g_{jk} \subseteq \bigcup_{i=1}^{n} S_i$. Each m_j represents the synonym relationship of attribute groups $g_{j1}, ..., g_{jw}$. The approach for finding M consists of three steps:

1. **Group discovery:** This step mines co-occurring or positively correlated attribute groups. It is done by first finding the set of all 2-attribute groups (i.e., each group contains only two attributes), denoted by L_2, that are positively correlated according to the input schema set S (one data scan is needed). A 2-attribute group {a, b} is considered positively correlated if $c_p(a, b)$ is greater than a threshold value τ_p, where c_p is a positive correlation measure. The algorithm then extends 2-attribute

groups to 3-attribute groups L_3. A 3-attribute group g is considered posi-tively correlated if every 2-attribute subset of g is in L_2. In general, a k-attribute group g is in L_k if every $(k-1)$-attribute sub-group of g is in L_{k-1}. This is similar to candidate generation in the Apriori algorithm for asso-ciation rule mining (see Chap. 2). However, the method here does not scan the data after all 2-attribute groups have been generated.

Example 5: Let $L_2 = \{\{a, b\}, \{b, c\}, \{a, c\}, \{c, d\}, \{d, f\}\}$, which con-tains all 2-attribute groups that are discovered from the data. $\{a, b, c\}$ is in L_3, but $\{a, c, d\}$ is not because $\{a, d\}$ is not in L_2. ■

2. **Match discovery:** This step mines negatively correlated groups includ-ing those singleton groups. Each discovered positively correlated group is first added into those transactions in S that contain some attributes of the group. That is, for a schema S_i and a group g, if $S_i \cap g \neq \varnothing$, then $S_i = S_i \cup \{g\}$. The final augmented transaction set S is then used to mine negatively correlated groups; which are potential **matching groups**. The procedure for finding all negatively correlated groups is exactly the same as the above procedure for finding positively correlated groups. The only difference is that a different measure is used to determine negative correlations, which will be discussed shortly. A 2-attribute group $\{a, b\}$ is considered negatively correlated if $c_n(a, b)$ is greater than a threshold value τ_n, where c_n is a negative correlation measure.

3. **Matching selection:** The discovered negative correlations may contain conflicts due to the idiosyncrasy of the data. Some correlations may also subsume others. For instance, in the book domain, the mining result may contain both {author} = {first name, last name}, denoted by m_1 and {sub-ject} = {first name, last name}, denoted by m_2. Clearly, m_1 is correct, but m_2 is not. Since {subject} = {author} is not discovered, which should be due to transitivity of synonyms, m_1 and m_2 cannot be both correct. This causes a **conflict**. A match m_j semantically **subsumes** a match m_k, de-noted by $m_j \succeq m_k$, if all the semantic relationships in m_k are contained in m_j. For instance, {arrival city} = {destination} = {to} $\succeq$ {arrival city} = {destination} because the synonym relationship in the second match is subsumed by the first one. Also, {author} = {first name, last name} $\succeq$ {author} = {first name} because the second match is part of the first.

We now present a method to choose the most confident and consistent matches and to remove possibly false ones. Between conflicting matches, we want to select the most negatively correlated one because it is more likely to be a group of genuine synonyms. Thus, a score function is needed, which is defined as the maximum negative correlation values of all 2-attribute groups in the match:

$$score(m_j, c_n) = max \; c_n(g_{j_r}, g_{j_t}), g_{j_r}, g_{j_t} \in m_j, j_r \neq j_t. \tag{6}$$

Combining the score function and semantic subsumption, the matches are ranked based on the following rules:

1. If $score(m_j, c_n) > score(m_k, c_n)$, m_j is ranked higher than m_k.
2. If $score(m_j, c_n) = score(m_k, c_n)$ and $m_j \succeq m_k$, m_j is ranked higher than m_k.
3. Otherwise, m_j and m_k are ranked arbitrarily.

Figure 10.4 gives the MatchingSelection() function. After the highest ranked match m_t in an iteration is selected, the inconsistent parts in the remaining matches are removed (lines 6–10). The final output is the selected n-ary complex matches with no conflict. Note that ranking is redone in each iteration instead of sorting all the matches in the beginning, because after removing some conflicting parts, the ranking may change.

Function MatchingSelection(M, c_n)
1    $R \leftarrow \varnothing$ // R stores the selected n-ary complex matches
2   **while** $M \neq \varnothing$ **do**
4   Let m_t be the highest ranked match in M //select the top ranked match
5        $R \leftarrow R \cup \{m_t\}$
6   **for** each $m_j \in M$ **do**
7            $m_j \leftarrow m_j - m_t$;   // remove the conflicting part
8   **if** $|m_j| < 2$ **then**
9                $M \leftarrow M - \{m_j\}$ // delete m_j if it contains no matching
10   **endfor**
11   **endwhile**
12   **return** R

Fig. 10.4. The MatchingSelection function

Correlation Measures: There are many existing correlation tests in statistic, e.g., χ^2 test and lift, etc. However, it was found that these methods were not suitable for this application. Hence, a new negative correlation measure $corr_n$ for two attributes A_p and A_q was proposed, which is called the **H-measure**. Let us use a contingency table (Fig. 10.5) to define it. f_{ij} in the figure is the co-occurrence frequency count of the corresponding cell:

	A_p	$\neg A_p$	
A_q	f_{11}	f_{10}	f_{1+}
$\neg A_q$	f_{01}	f_{00}	f_{0+}
	f_{+1}	f_{+0}	f_{++}

Fig. 10.5. Contingency table for test of correlation

$$corr_n(A_p, A_q) = H(A_p, A_q) = \frac{f_{01}f_{10}}{f_{+1}f_{1+}}. \tag{7}$$

The positive correlation measure $corr_p$ is defined as (τ_d is a threshold):

$$corr_p(A_p, A_q) = \begin{cases} 1 - H(A_p, A_q) & \frac{f_{11}}{f_{++}} < \tau_d \\ 0 & \text{otherwise.} \end{cases} \tag{8}$$

10.8.3 An Instance Based Approach

This method is based on the technique given in [531]. It matches query interfaces and also the query results. It assumes that:

1. a global schema (GS) for the application domain is given, which represents the key attributes of the domain, and
2. a number of sample data instances under the domain global schema are also available.

This technique only finds 1:1 attribute matches. We use IS to denote the query interface schema and RS the returned result schema. Let us use an example to introduce the key observation exploited in this technique. Figure 10.6 shows an example of an online bookstore. The part labeled **Data Attributes** is the global schema with six attributes {Title, Author, Publisher, ISBN, Publication Date, Format}. The part labeled **Interface** is the query interface with five input elements/attributes. When the keyword query "Harry Potter" is submitted through the Title attribute in the interface, a result page is returned which contains the answer to the query (labeled **Result Page**), which shows three book instances.

Three types of semantic correspondence represented by different lines (dotted, dashed and solid) are also shown in Fig. 10.6. They are respectively, the correspondence between attributes of the global schema and those of the query interface, the correspondence between the attributes of the global schema and those of the instance values in the result page, and the correspondence between attributes in the query interface and those of the instance values in the result pages.

Observation: When a proper query is submitted to the right element of the query interface, the query words are very likely to reappear in the corresponding attribute of the returned results. However, if an improper query is submitted to the Web database there are often few or no returned results.

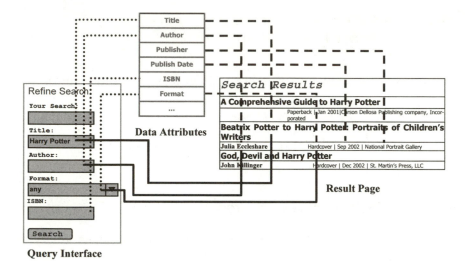

Fig. 10.6. An example of a Web database with its query interface and a result page

In the example shown in Fig. 10.6, the site retrieves only three matches for the query "Harry Potter" when submitted through the "Author" attribute, while it retrieves 228 matches for the same query when submitted to the Title attribute. If "Harry Potter" is submitted to the "ISBN" attribute, there is no returned result. Intuitively, the number of times that query words reappear in the returned results gives us a good indication what attributes match in the interface schema, the global schema, and the result schema.

To obtain the number of reappearing occurrences, each value from the given instances can be submitted to each interface element while keeping default values for the other elements. For each **TEXTBOX** element in the query interface, all attribute values from the given instances are tried exhaustively. For each **SELECT** element, its domain values are limited to a set of fixed options. Then, an option similar to a value in the given instances is found and submitted. Here, "similar" means that the attribute value and the option value have at least one common word. Note that this approach assumes that a data extraction system is available to produce a table from a returned result page (see Chap. 9). Each column has a hidden attribute (i.e., of the result schema).

By counting the number of times that the query words re-occur in each column of the result table, a 3-dimensional **occurrence matrix** (OM) can be constructed. The three dimensions are: global schema (GS) attributes, query interface schema (IS) attributes and result schema (RS) attributes. Each cell $OM[i, j, k]$ contains the sum of the occurrence counts obtained from kth attribute of RS of all the sample query words from the ith attribute of GS when the query words are submitted to the jth attribute of IS.

Intra-Site Schema Matching: We now briefly describe how to match attributes in IS and GS, IS and RS, and GS and RS based on the projected matrices of OM, i.e., $OM_{IG(M×N)}$, $OM_{IR(M×L)}$, and $OM_{GR(N×L)}$, where N is the number of attributes in the global schema, M is the number of elements in the interface schema, and L is the number of columns in the result table. An example $OM_{IG(5×4)}$ matrix is shown in Fig. 10.7 with the correct matching highlighted, GS = {Title$_{GS}$, Author$_{GS}$, Publisher$_{GS}$, ISBN$_{GS}$} and IS = {Author$_{IS}$, Title$_{IS}$, Publisher$_{IS}$, Keyword$_{IS}$, ISBN$_{IS}$}.

We observe from Fig. 10.7 that the highest occurrence count may not represent a correct match. For example, the cell for Author$_{IS}$ and Publisher$_{GS}$ (534) has the highest value in the matrix but Author$_{IS}$ and Publisher$_{GS}$ do not correspond to each other. In general, for a cell m_{ij}, its value in comparison with those of other cells in its row i and its column j is more important than its absolute count.

	Title$_{GS}$	Author$_{GS}$	Publisher$_{GS}$	ISBN$_{GS}$
Author$_{IS}$	93	498	534	0
Title$_{IS}$	451	345	501	0
Publisher$_{IS}$	62	184	468	2
Keyword$_{IS}$	120	248	143	275
ISBN$_{IS}$	0	0	0	258

Fig. 10.7. An example of a $OM_{IG(M×N)}$ matrix with all matches highlighted

The algorithm in [531] uses the **mutual information measure** (MI) to determine correct matches. The mutual information, which measures the mutual dependence of two variables, is defined as follows:

$$MI(x, y) = \Pr(x, y) \log_2 \frac{\Pr(x, y)}{\Pr(x)\Pr(y)}. \tag{9}$$

In our context, x and y are attributes from IS and GS respectively. The probabilities, Pr(x, y), Pr(x) and Pr(y), can be easily computed using the $OM_{IG(M×N)}$ matrix.

The algorithm simply computes the mutual information of every pair of attributes based on the counts in the matrix such as the one in Fig. 10.7. A corresponding mutual information matrix (called **MI matrix**) is then constructed (not shown here). To find 1-1 matches of the two schemas, the algorithm chooses each cell in the MI matrix whose value is the largest among all the values in the same row and the same column. The corresponding attributes of the cell forms a final match.

The paper also has a similar method for finding matches from multiple Web databases, which is called **inter-site schema matching**.

10.9   Constructing a Unified Global Query Interface

Once a set of query interfaces in the same domain is matched, we can automatically construct a *well-designed* **global query interface** that contains all the (or the most significant) distinct attributes of all source interfaces. To build a "good" global interface, three requirements are identified in [154].

1. **Structural appropriateness:** As noted earlier, elements of query interfaces are usually organized in groups (logical units) of related attributes so that semantically related attributes are placed in close vicinity. For example, "Adults", "Seniors", and "Children" of the interfaces shown in Fig. 10.1 are placed together. In addition, multiple related groups of attributes are organized as super-groups (e.g., "Where and when do you want to go?" in Fig. 10.1). This leads to a hierarchical structure for interfaces (see Fig. 10.8), where a leaf in the tree corresponds to an attribute in the interface, an internal node corresponds to a (super)group of attributes and the order among the sibling nodes within the tree resembles the order of attributes in the interface (from left to right). The global query interface should reflect this hierarchical structure of the domain.
2. **Lexical appropriateness:** Labels of elements should be chosen so as to convey the meaning of each individual element and to underline the hierarchical organization of attributes (e.g., the three attributes together with the parent attribute "Number of Passengers" in Fig. 10.1).
3. **Instance appropriateness:** The domain values for each attribute in the global interface must contain the values of the source interfaces.

We will give a high level description of the algorithms in [153, 154] that build the global interface by merging given interfaces based on the above three requirements. The input to the algorithms consists of (1) a set of query interfaces and (2) a global mapping of corresponding attributes in the query interfaces. It is assumed that mapping is organized in clusters as discussed in Sect. 10.8.1. Each cluster contains all the matched attributes from different interfaces. We note that the domain model discovery idea in [227] can be seen as another approach to building global interfaces.

10.9.1   Structural Appropriateness and the Merge Algorithm

Structural appropriateness means to satisfy grouping constraints and ancestor-descendant relationship constraints of the attributes in individual interfaces. These constraints guide the merging algorithm to produce the global interface, which has one attribute for each cluster.

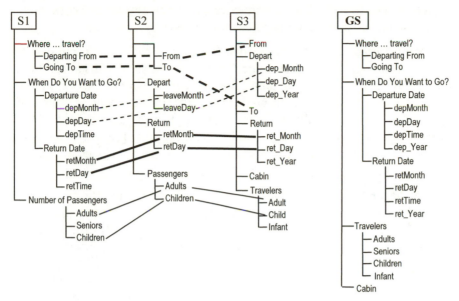

Fig. 10.8. Three input query interfaces (S1, S2, and S3) and the derived global query interface (GS).

Grouping Constraints: Recall that semantically related attributes within an interface are usually grouped together. Grouping constraints require that these attributes should also appear together in the global interface.

As the global interface has an attribute for each cluster, the problem is to partition all the clusters into semantically meaningful subsets (or groups), which are employed to organize attributes in the global interface. For instance, for the example in Fig. 10.8, the following sets of clusters are produced, {c_deptCity, c_destCity}, {c_deptYear, c_deptTime, c_deptDay, c_depMonth}, and {c_Senior, c_Adult, c_Child, c_Infant}, where c_X is a cluster representing X (e.g., c_deptCity and c_destCity are clusters representing departure cities and destination cities, respectively).

The partition is determined by considering each *maximal set* of adjacent sibling leaves in the schema tree of each source interface whose parent is not the root. The leaves whose parent is the root are not considered because no reliable information can be derived. These structural constraints are collected from all source interfaces in order to infer the way that attributes are organized in the global interface. All those sets (or groups) of clusters whose intersection is not empty are merged to form the final groups, which are sequences of attribute clusters that preserve adjacency constraints in all interfaces. For example, {c_Adult, c_Senior, c_Child}, {c_Adult, c_Child}, {c_Adult, c_Child, c_Infant} are merged to produce the final group, [c_Senior, c_Adult, c_Child, c_Infant], which preserves all adjacency constraints. Such a sequence does not always exist. In such a case, a sequence that accommodates most adjacency constraints is sought.

Ancestor-Descendant Relationships: In hierarchical modeling of data the same information can be represented in various ways. For instance, the relationship between "Authors" and "Books" can be captured as either "Authors" having "Books", which makes "Books" a descendant of "Authors", or "Books" having "Authors", which makes "Books" an ancestor of "Authors". This, however, was not found to be a problem [153]. No such conflicting cases were found from a study of 300 query interfaces in eight application domains.

Merge Algorithm: The merge algorithm merges two interfaces at a time to produce the final global interface schema. One of them is the current global interface G. At the beginning, the schema tree with the most levels is chosen as the initial global schema G. Then each other interface is sequentially merged with G. During each merge, G is refined and expanded. The algorithm works in a bottom-up fashion. The merging between leaves is produced based on the clusters. The mapping between internal nodes is based on mappings of their children, which may be either leaf nodes or internal nodes. To meaningfully insert leaves without a match in the correct position, the algorithm relies on groups computed above to infer each leaf position. In our example, we start by merging S_1 and S_3. S_1 is the initial global interface G. Within each group, it is easy to see the position of "Infant", "ret_Year" and "dep_Year" (see Fig. 10.8 on the right). "Cabin" is inserted at the end since leaf children of the root are discarded before merging and then added as children of the root of the integrated schema tree. Additional details can be found in [153].

10.9.2 Lexical Appropriateness

After the interfaces are merged, the attributes in the integrated interface need to be labeled so that (1) the labels of the attributes within a group are consistent and (2) the labels of the internal nodes are consistent with respect to themselves and to the leaf nodes [154].

It can be observed in the query interface of Fig. 10.1 that between the labels of the attributes grouped together there are certain commonalities. For instance, "Adults", "Seniors" and "Children" are all plurals, whereas "Leaving" and "Returning" are gerunds. Ideally, the groups within the global interface should have the same uniformity property. Since the attributes may be from different interfaces, a group of attributes within the unified interface might not correspond to any group in a single interface, which makes it hard to assign consistent labels. To deal with the problem, a strategy called **intersect-and-union** is used, which finds groups with non-empty intersection from different interfaces and then unions them.

Example 6: Consider the example of the three interfaces in Fig. 10.8 with their passenger related groups organized as the table below. It is easy to see a systematic way of building a consistent solution.

Cluster/Interface	c_Adult	c_Senior	c_Child	c_Infant
S_1	Adults	Seniors	Children	
S_2	Adults		Children	
S_3	Adult		Child	Infant

Notice that by combining the labels given by S_1 and S_2 a consistent naming assignment, namely, "Seniors", "Adults" and "Children", can be achieved because the two sets share labels (i.e., "Adults" and "Children") that are consistent with the labels in both sets. This strategy can be iteratively applied until a label is assigned to each attribute in the group.

To deal with minor variations, more relaxed rules for combining attribute labels can be used, e.g., requiring that the set of tokens of the labels to be equal after removal of stopwords (e.g., "Number of Adults" has the same set of tokens as "Adults Number", i.e. {Number, Adults}) and stemming. If a consistent solution for the entire group cannot be found, consistent solutions for subsets of attributes are constructed.

The assignment of consistent labels to the internal nodes uses a set of rules [154] that tries to select a label for each node in such a way that it is generic enough to semantically cover the set of its descendant leaf nodes. For example, the label "Travelers" is obtained in the integrated interface in Fig. 10.8 as follows. First, we know that "Passengers" is more generic than "Number of Passengers" and thus semantically covers both {Seniors, Adults, Children} and {Adults, Children}. Then, "Travelers" is found to be a hypernym of "Passengers" (using WordNet) and thus semantically covers the union of {Seniors, Adults, Children} and {Adults, Children, Infant} which is the desired set {Seniors, Adults, Children, Infant}

10.9.3 Instance Appropriateness

Finally, we discuss how to determine the domain for each attribute in the global schema (interface). A domain has two aspects: the type and the set of values. To determine the domain type of a global attribute, compatibility rules are needed [230]. For instance, if all attributes in a cluster have a finite (infinite) domain then the global attribute will have a finite (infinite) domain. If in the cluster there are both finite and infinite domains, then the domain of the global attribute will be **hybrid** (i.e., users can either select from a list of pre-compiled values or fill in a new value). As a case in point, the "Adults" attribute on the global interface derived from the two interface in Fig. 10.1 will have a finite domain, whereas the attribute "Going to" will have a hybrid domain.

The set of domain values of a global attribute is given by the union of the domains of the attributes in the cluster. Computing the union is not always easy. For example, the values of the domains may have different *scale/unit* (e.g., the price may be in US$ or in Euro). Moreover, the same value may be specified in various ways (e.g., "Chicago O'Hare" vs. "ORD"). Currently, the problem is dealt with using user-provided auxiliary thesauruses [230].

Bibliographic Notes

Schema integration has been studied in the database community since the early 1980s. The main contributions are described in the surveys by Batini et al [40], Doan and Halevy [146], Kalfoglou and Schorlemmer [265], Larson et al. [306], Kashyap and Sheth [269], Rahm and Bernstein [455], Sheth and Larson [488], and Shvaiko and Euzenat [491]. The database integration aspects of this chapter are mainly based on the survey paper by Rahm and Bernstein [455]. Many ideas are also taken from Clifton et al. [105], Cohen [107], Do and Rahm [144], Dhamankar et al. [133], Embley et al. [162], Madhavan et al. [358], Xu and Embley [563], and Yan et al. [566]. Web data integration is considerably more recent. Various ideas on Web information integration in the early part of the chapter are taken from papers by He and Chang [227, 229], and Wu et al. [559].

On Web query interface integration, which perhaps received the most attention in the research community, several methods have been studied in the chapter, which are based on the works of Dragut et al. [153, 154], He and Chang [227, 229], He et al. [230], Wang et al. [531], and Wu et al. [559]. Before matching can be performed, the Web interfaces have to be found and extracted first. This extraction task was investigated by Zhang et al. [609] and He et al. [231].

Another area of research is the ontology, taxonomy or catalog integration. Ontologies (taxonomies or catalogs) are tree structured schemas. They are similar to query interfaces as most interfaces have some hierarchical structures. More focused works on ontology integration include those by Agrawal and Srikant [13], Doan et al. [147], Gal et al. [190], Zhang and Lee [602]. Wache et al. gave a survey of the area in [527].

11 Opinion Mining

In Chap. 9, we studied structured data extraction from Web pages. Such data are usually records retrieved from underlying databases and displayed in Web pages following some fixed templates. The Web also contains a huge amount of information in unstructured texts. Analyzing these texts is of great importance and perhaps even more important than extracting structured data because of the sheer volume of valuable information of almost any imaginable types contained in them. In this chapter, we only focus on mining of opinions on the Web. The task is not only technically challenging because of the need for natural language processing, but also very useful in practice. For example, businesses always want to find public or consumer opinions on their products and services. Potential customers also want to know the opinions of existing users before they use a service or purchase a product. Moreover, opinion mining can also provide valuable information for placing advertisements in Web pages. If in a page people express positive opinions or sentiments on a product, it may be a good idea to place an ad of the product. However, if people express negative opinions about the product, it is probably not wise to place an ad of the product. A better idea may be to place an ad of a competitor's product.

The Web has dramatically changed the way that people express their opinions. They can now post reviews of products at merchant sites and express their views on almost anything in Internet forums, discussion groups, blogs, etc., which are commonly called the **user generated content** or **user generated media**. This online word-of-mouth behavior represents new and measurable sources of information with many practical applications. Techniques are now being developed to exploit these sources to help businesses and individuals gain such information effectively and easily.

The first part of this chapter focuses on three mining tasks of **evaluative texts** (which are documents expressing opinions):

1. **Sentiment classification:** This task treats opinion mining as a text classification problem. It classifies an evaluative text as being positive or negative. For example, given a product review, the system determines whether the review expresses a positive or a negative sentiment of the reviewer. The classification is usually at the document-level. No details are discovered about what people liked or didn't like.

1. **Featured-based opinion mining and summarization:** This task goes to the sentence level to discover details, i.e., what aspects of an object that people liked or disliked. The object could be a product, a service, a topic, an individual, an organization, etc. For example, in a product review, this task identifies product features that have been commented on by reviewers and determines whether the comments are positive or negative. In the sentence, "the battery life of this camera is too short," the comment is on the "battery life" and the opinion is negative. A structured summary will also be produced from the mining results.

2. **Comparative sentence and relation mining:** Comparison is another type of evaluation, which directly compares one object against one or more other similar objects. For example, the following sentence compares two cameras: "the battery life of camera A is much shorter than that of camera B." We want to identify such sentences and extract comparative relations expressed in them.

The second part of the chapter discusses **opinion search** and **opinion spam**. Since our focus is on opinions on the Web, opinion search is naturally relevant, and so is opinion spam. An opinion search system enables users to search for opinions on any object. Opinion spam refers to dishonest or malicious opinions aimed at promoting one's own products and services, and/or at damaging the reputations of those of one's competitors. Detecting opinion spam is a challenging problem because for opinions expressed on the Web, the true identities of their authors are often unknown.

The research in opinion mining only began recently. Hence, this chapter should be treated as statements of problems and descriptions of current research rather than a report of mature techniques for solving the problems. We expect major progresses to be made in the coming years.

11.1 Sentiment Classification

Given a set of evaluative texts D, a sentiment classifier classifies each document $d \in D$ into one of the two classes, **positive** and **negative**. Positive means that d expresses a positive opinion. Negative means that d expresses a negative opinion. For example, given some reviews of a movie, the system classifies them into positive reviews and negative reviews.

The main application of sentiment classification is to give a quick determination of the prevailing opinion on an object. The task is similar but also different from classic topic-based text classification, which classifies documents into predefined topic classes, e.g., politics, science, sports, etc. In topic-based classification, topic related words are important. However, in sentiment classification, topic-related words are unimportant. Instead, sentiment words that indicate positive or negative opinions are important, e.g., great, excellent, amazing, horrible, bad, worst, etc.

The existing research in this area is mainly at the **document-level**, i.e., to classify each whole document as positive or negative (in some cases, the **neutral** class is used as well). One can also extend such classification to the **sentence-level**, i.e., to classify each sentence as expressing a positive, negative or neutral opinion. We discuss several approaches below.

11.1.1 Classification Based on Sentiment Phrases

This method performs classification based on positive and negative sentiment words and phrases contained in each evaluative text. The algorithm described here is based on the work of Turney [521], which is designed to classify customer reviews.

This algorithm makes use of a natural language processing technique called **part-of-speech (POS) tagging**. The part-of-speech of a word is a linguistic category that is defined by its syntactic or morphological behavior. Common POS categories in English grammar are: noun, verb, adjective, adverb, pronoun, preposition, conjunction and interjection. Then, there are many categories which arise from different forms of these categories. For example, a verb can be a verb in its base form, in its past tense, etc. In this book, we use the standard **Penn Treebank POS Tags** as shown in Table 11.1. POS tagging is the task of labeling (or tagging) each word in a sentence with its appropriate part of speech. For details on part-of-speech tagging, please refer to the report by Santorini [472]. The Penn Treebank site is at http://www.cis.upenn.edu/~treebank/home.html.

The algorithm given in [521] consists of three steps:

Step 1: It extracts phrases containing adjectives or adverbs. The reason for doing this is that research has shown that adjectives and adverbs are good indicators of subjectivity and opinions. However, although an isolated adjective may indicate subjectivity, there may be an insufficient context to determine its **semantic (or opinion) orientation**. For example, the adjective "unpredictable" may have a negative orientation in an automotive review, in such a phrase as "unpredictable steering", but it could have a positive orientation in a movie review, in a phrase such as "unpredictable plot". Therefore, the algorithm extracts two consecutive words, where one member of the pair is an adjective/adverb and the other is a context word.

Two consecutive words are extracted if their POS tags conform to any of the patterns in Table 11.2. For example, the pattern in line 2 means

Table 11.1. Penn Treebank part-of-speech tags (excluding punctuation)

Tag	Description	Tag	Description
CC	Coordinating conjunction	PRP$	Possessive pronoun
CD	Cardinal number	RB	Adverb
DT	Determiner	RBR	Adverb, comparative
EX	Existential *there*	RBS	Adverb, superlative
FW	Foreign word	RP	Particle
IN	Preposition or subordinating conjunction	SYM	Symbol
JJ	Adjective	TO	*to*
JJR	Adjective, comparative	UH	Interjection
JJS	Adjective, superlative	VB	Verb, base form
LS	List item marker	VBD	Verb, past tense
MD	Modal	VBG	Verb, gerund or present participle
NN	Noun, singular or mass	VBN	Verb, past participle
NNS	Noun, plural	VBP	Verb, non-3rd person singular present
NNP	Proper noun, singular	VBZ	Verb, 3rd person singular present
NNPS	Proper noun, plural	WDT	Wh-determiner
PDT	Predeterminer	WP	Wh-pronoun
POS	Possessive ending	WP$	Possessive wh-pronoun
PRP	Personal pronoun	WRB	Wh-adverb

that two consecutive words are extracted if the first word is an adverb and the second word is an adjective, but the third word (which is not extracted) cannot be a noun. NNP and NNPS are avoided so that the names of the objects in the review cannot influence the classification.

Table 11.2. Patterns of tags for extracting two-word phrases from reviews

	First word	Second word	Third word (Not Extracted)
1.	JJ	NN or NNS	anything
2.	RB, RBR, or RBS	JJ	not NN nor NNS
3.	JJ	JJ	not NN nor NNS
4.	NN or NNS	JJ	not NN nor NNS
5.	RB, RBR, or RBS	VB, VBD, VBN, or VBG	anything

Example 1: In the sentence "this camera produces beautiful pictures", "beautiful pictures" will be extracted as it satisfies the first pattern. ∎

Step 2: It estimates the semantic orientation of the extracted phrases using the **pointwise mutual information** measure given in Equation 1:

$$PMI(term_1, term_2) = \log_2\left(\frac{Pr(term_1 \wedge term_2)}{Pr(term_1)Pr(term_2)}\right). \tag{1}$$

Here, $Pr(term_1 \wedge term_2)$ is the co-occurrence probability of $term_1$ and $term_2$, and $Pr(term_1)Pr(term_2)$ gives the probability that the two terms co-occur if they are statistically independent. The ratio between $Pr(term_1 \wedge term_2)$ and $Pr(term_1)Pr(term_2)$ is thus a measure of the degree of statistical dependence between them. The log of this ratio is the amount of information that we acquire about the presence of one of the words when we observe the other.

The semantic/opinion orientation (SO) of a phrase is computed based on its association with the positive reference word "excellent" and its association with the negative reference word "poor":

$$SO(phrase) = PMI(phrase, \text{"excellent"}) - PMI(phrase, \text{"poor"}). \tag{2}$$

The probabilities are calculated by issuing queries to a search engine and collecting the number of **hits**. For each search query, a search engine usually gives the number of relevant documents to the query, which is the number of hits. Thus, by searching the two terms together and separately, we can estimate the probabilities in Equation 1. Turney [521] used the AltaVista search engine because it has a NEAR operator, which constrains the search to documents that contain the words within ten words of one another, in either order. Let $hits(query)$ be the number of hits returned. Equation 2 can be rewritten as:

$$SO(phrase) = \log_2\left(\frac{hits(\text{phrase } NEAR \text{ "excellent"})hits(\text{"poor"})}{hits(\text{phrase } NEAR \text{ "poor"})hits(\text{"excellent"})}\right). \tag{3}$$

To avoid division by zero, 0.01 is added to the hits.

Step 3: Given a review, the algorithm computes the average SO of all phrases in the review, and classifies the review as recommended if the average SO is positive, not recommended otherwise.

Final classification accuracies on reviews from various domains range from 84% for automobile reviews to 66% for movie reviews.

11.1.2 Classification Using Text Classification Methods

The simplest approach to sentiment classification is to treat the problem as a topic-based text classification problem. Then, any text classification algorithm can be employed, e.g., naïve Bayesian, SVM, kNN, etc.

The approach was experimented by Pang et al. [428] using movie reviews of two classes, positive and negative. It was shown that using a unigram (a bag of individual words) in classification performed well using either naïve Bayesian or SVM. Test results using 700 positive reviews and 700 negative reviews showed that these two classification algorithms achieved 81% and 82.9% accuracy respectively with 3-fold cross validation. However, neutral reviews were not used in this work, which made the problem easier. No stemming or stopword removal was applied.

11.1.3 Classification Using a Score Function

A custom score function for review sentiment classification was given by Dave et al. [122]. The algorithm consists of two steps:

Step 1: It scores each term in the training set using the following equation,

$$score(t_i) = \frac{\Pr(t_i \mid C) - \Pr(t_i \mid C')}{\Pr(t_i \mid C) + \Pr(t_i \mid C')},\qquad(4)$$

where t_i is a term and C is a class and C' is its complement, i.e., not C, and $\Pr(t_i \mid C)$ is the conditional probability of term t_i in class C. It is computed by taking the number of times that a term t_i occurs in class C reviews and dividing it by the total number of terms in the reviews of class C. A term's score is thus a measure of bias towards either class ranging from −1 and 1.

Step 2: To classify a new document $d_i = t_1...t_n$, the algorithm sums up the scores of all terms and uses the sign of the total to determine the class. That is, it uses the following equation for classification,

$$class(d_i) = \begin{cases} C & eval(d_i) > 0 \\ C' & \text{otherwise,} \end{cases}\qquad(5)$$

where

$$eval(d_i) = \sum_j score(t_j).\qquad(6)$$

Experiments were conducted based on a large number of reviews (more than 13000) of seven types of products. The results showed that the bigrams (consecutive two words) and trigrams (consecutive three words) as terms gave (similar) best accuracies (84.6%–88.3%), on two different review data sets. No stemming or stopword removal was applied.

In this paper, the authors experimented with many alternative classification techniques, e.g., naïve Bayesian, SVM, and several algorithms based on other score functions. They also tried some word substitution strategies to improve generalization, e.g.,

- replace product names with a token ("_productname");
- replace rare words with a token ("_unique");
- replace category-specific words with a token ("_producttypeword");
- replace numeric tokens with NUMBER.

Some linguistic modifications using WordNet, stemming, negation, and collocation were tested too. However, they were not helpful, and usually degraded the classification accuracy.

In summary, the main advantage of document level sentiment classification is that it provides a prevailing opinion on an object, topic or event. The main shortcomings of the document-level classification are:

- It does not give details on what people liked or disliked. In a typical evaluative text such as a review, the author usually writes specific aspects of an object that he/she likes or dislikes. The ability to extract such details is useful in practice.
- It is not easily applicable to non-reviews, e.g., forum and blog postings, because although their main focus may not be evaluation or reviewing of a product, they may still contain a few opinion sentences. In such cases, we need to identify and extract opinion sentences.

There are several variations of the algorithms discussed in this section (see Bibliographic Notes). Apart from these learning based methods, there are also manual approaches for specific applications. For example, Tong [517] reported a system that generates sentiment timelines. The system tracks online discussions about movies and displays a plot of the number of positive sentiment and negative sentiment messages (Y-axis) over time (X-axis). Messages are classified by matching specific phrases that indicate sentiments of the author towards the movie (e.g., "great acting", "wonderful visuals", "uneven editing", "highly recommend it", and "it sucks"). The phrases were manually compiled and tagged as indicating positive or negative sentiments to form a lexicon. The lexicon is specific to the domain (e.g., movies) and must be built anew for each new domain.

11.2 Feature-Based Opinion Mining and Summarization

Although studying evaluative texts at the document level is useful in many cases, it leaves much to be desired. A positive evaluative text on a particular object does not mean that the author has positive opinions on every aspect of the object. Likewise, a negative evaluative text does not mean that the author dislikes everything about the object. For example, in a product review, the reviewer usually writes both positive and negative aspects of the product, although the general sentiment on the product could be positive or negative. To obtain such detailed aspects, we need to go to the sentence level. Two tasks are apparent [245]:

1. Identifying and extracting features of the product that the reviewers have expressed their opinions on, called **product features**. For instance, in the sentence "the picture quality of this camera is amazing," the product feature is "picture quality".
2. Determining whether the opinions on the features are positive, negative or neutral. In the above sentence, the opinion on the feature "picture quality" is positive.

11.2.1 Problem Definition

In general, the opinions can be expressed on anything, e.g., a product, an individual, an organization, an event, a topic, etc. We use the general term **"object"** to denote the entity that has been commented on. The object has a set of **components** (or parts) and also a set of **attributes** (or **properties**). Thus the object can be hierarchically decomposed according to the **part-of** relationship, i.e., each component may also have its sub-components and so on. For example, a product (e.g., a car, a digital camera) can have different components, an event can have sub-events, a topic can have subtopics, etc. Formally, we have the following definition:

Definition (object): An **object** O is an entity which can be a product, person, event, organization, or topic. It is associated with a pair, $O: (T, A)$, where T is a hierarchy or taxonomy of **components** (or **parts**), **sub-components**, and so on, and A is a set of **attributes** of O. Each component has its own set of sub-components and attributes.

Example 2: A particular brand of digital camera is an object. It has a set of components, e.g., lens, battery, view-finder, etc., and also a set of attributes, e.g., picture quality, size, weight, etc. The battery component also has its set of attributes, e.g., battery life, battery size, battery weight, etc. ∎

Essentially, an object is represented as a tree. The root is the object itself. Each non-root node is a component or sub-component of the object. Each link represents a *part-of* relationship. Each node is also associated with a set of attributes. An opinion can be expressed on any node and any attribute of the node.

Example 3: Following Example 2, one can express an opinion on the camera (the root node), e.g., "I do not like this camera", or on one of its attributes, e.g., "the picture quality of this camera is poor". Likewise, one can also express an opinion on one of the camera's components, e.g., "the battery of this camera is bad", or an opinion on the attribute of the component, "the battery life of this camera is too short." ∎

To simplify our discussion, we use the word "**features**" to represent both components and attributes, which allows us to omit the hierarchy. Using features for products is also quite common in practice. For an ordinary user, it is probably too complex to use a hierarchical representation of product features and opinions. We note that in this framework the object itself is also treated as a feature.

Let the evaluative text (e.g., a product review) be r. In the most general case, r consists of a sequence of sentences $r = \langle s_1, s_2, ..., s_m \rangle$.

Definition (explicit and implicit feature): If a feature f appears in evaluative text r, it is called an **explicit feature** in r. If f does not appear in r but is implied, it is called an **implicit feature** in r.

Example 4: "battery life" in the following sentence is an explicit feature:

"The battery life of this camera is too short".

"Size" is an implicit feature in the following sentence as it does not appear in the sentence but it is implied:

"This camera is too large".

Definition (opinion passage on a feature): The **opinion passage** on feature f of an object evaluated in r is a group of consecutive sentences in r that expresses a positive or negative opinion on f.

It is common that a sequence of sentences (at least one) in an evaluative text together expresses an opinion on an object or a feature of the object. Also, it is possible that a single sentence expresses opinions on more than one feature:

"The picture quality is good, but the battery life is short".

Most current research focuses on sentences, i.e., each passage consisting of a single sentence. Thus, in our subsequent discussion, we use **sentences** and **passages** interchangeably.

Definition (explicit and implicit opinion): An **explicit opinion** on feature f is a subjective sentence that directly expresses a positive or negative opinion. An **implicit opinion** on feature f is an objective sentence that implies a positive or negative opinion.

Example 5: The following sentence expresses an explicit positive opinion:

"The picture quality of this camera is amazing."

The following sentence expresses an implicit negative opinion:

"The earphone broke in two days."

Although this sentence states an objective fact (assume it is true), it implicitly expresses a negative opinion on the earphone.   ■

Definition (opinion holder): The **holder** of a particular opinion is a person or an organization that holds the opinion.

In the case of product reviews, forum postings and blogs, opinion holders are usually the authors of the postings, although occasionally some authors cite or repeat the opinions of others. Opinion holders are more important in news articles because they often explicitly state the person or organization that holds a particular view. For example, the opinion holder in the sentence "John expressed his disagreement on the treaty" is "John".

We now put things together to define a **model** of an object and a set of opinions on the object. An object is represented with a finite set of features, $F = \{f_1, f_2, ..., f_n\}$. Each feature f_i in F can be expressed with a finite set of words or phrases W_i, which are **synonyms**. That is, we have a set of corresponding synonym sets $W = \{W_1, W_2, ..., W_n\}$ for the n features. Since each feature f_i in F has a name (denoted by f_i), then $f_i \in W_i$. Each author or **opinion holder** j comments on a subset of the features $S_j \subseteq F$. For each feature $f_k \in S_j$ that opinion holder j comments on, he/she chooses a word or phrase from W_k to describe the feature, and then expresses a positive or negative opinion on it.

This simple model covers most but not all cases. For example, it does not cover the situation described in the following sentence: "the viewfinder and the lens of this camera are too close", which expresses a negative opinion on the distance of the two components. We will follow this simplified model in the rest of this chapter.

This model introduces three main practical problems. Given a collection of evaluative texts D as input, we have:

Problem 1: Both F and W are unknown. Then, in opinion mining, we need to perform three tasks:

Task 1: Identifying and extracting object features that have been commented on in each evaluative text $d \in D$.

Task 2: Determining whether the opinions on the features are positive, negative or neutral.

Task 3: Grouping synonyms of features, as different people may use different words or phrases to express the same feature.

Problem 2: F is known but W is unknown. This is similar to Problem 1, but slightly easier. All the three tasks for Problem 1 still need to be performed, but Task 3 becomes the problem of matching discovered features with the set of given features F.

Problem 3: W is known (then F is also known). We only need to perform Task 2 above, namely, determining whether the opinions on the known features are positive, negative or neutral after all the sentences that contain them are extracted (which is simple).

Clearly, the first problem is the most difficult to solve. Problem 2 is slightly easier. Problem 3 is the easiest, but still realistic.

Example 6: A cellular phone company wants to mine customer reviews on a few models of its phones. It is quite realistic to produce the feature set F that the company is interested in and also the set of synonyms of each feature (although the set might not be complete). Then there is no need to perform Tasks 1 and 3 (which are very challenging problems). ■

Output: The final output for each evaluative text d is a set of pairs. Each pair is denoted by (f, SO), where f is a feature and SO is the semantic or opinion orientation (positive or negative) expressed in d on feature f. We ignore neutral opinions in the output as they are not usually useful.

Note that this model does not consider the strength of each opinion, i.e., whether the opinion is strongly negative (or positive) or weakly negative (or positive), but it can be added easily (see [548] for a related work).

There are many ways to use the results. A simple way is to produce a **feature-based summary** of opinions on the object. We use an example to illustrate what that means.

Example 7: Assume we summarize the reviews of a particular digital camera, *digital_camera_*1. The summary looks like that in Fig. 11.1.

In Fig. 11.1, "picture quality" and (camera) "size" are the product features. There are 123 reviews that express positive opinions about the picture quality, and only 6 that express negative opinions. The <individual review sentences> link points to the specific sentences and/or the whole reviews that give positive or negative comments about the feature.

With such a summary, the user can easily see how the existing customers feel about the digital camera. If he/she is very interested in a particular feature, he/she can drill down by following the <individual review sentences> link to see why existing customers like it and/or what they are not

*Digital_camera*_1:

>   Feature: **picture quality**
>> Positive: 123   <individual review sentences>
>> Negative: 6   <individual review sentences>
>   Feature: **size**
>> Positive: 82   <individual review sentences>
>> Negative: 10   <individual review sentences>
>
> ...

Fig. 11.1. An example of a feature-based summary of opinions

satisfied with. The summary can also be visualized using a bar chart. Figure 11.2(A) shows the feature-based opinion summary of a digital camera.

In the figure, the bars above the *X*-axis in the middle show the percentages of positive opinions on various features (given at the top), and the bars below the *X*-axis show the percentages of negative opinions on the same features.

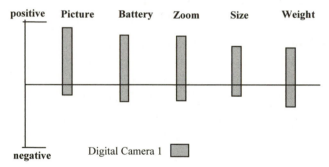

(A) Feature-based summary of opinions on a digital camera

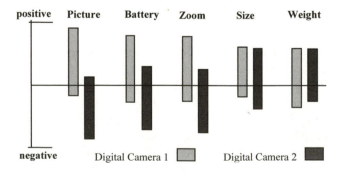

(B) Opinion comparison of two digital cameras

Fig. 11.2. Visualization of feature-based opinion summary and comparison

Comparing the opinion summaries of a few competing products is even more interesting. Figure 11.2(B) shows a visual comparison of consumer opinions on two competing digital cameras. We can clearly see how consumers view different features of each product. Digital camera 1 is clearly superior to digital camera 2. Specifically, most customers have negative opinions about the picture quality, battery and zoom of digital camera 2. However, on the same three features, customers are mostly positive about digital camera 1. Regarding size and weight, customers have similar opinions on both cameras. Thus, the visualization enables users to clearly see how the cameras compare with each other along each feature dimension. ■

Below, we discuss four other important issues.

Separation of Opinions on the Object itself and its Features: It is often useful to separate opinions on the object itself and opinions on the features of the object. The opinion on the object itself reflects the general sentiment of the author (or the opinion holder) on the object, which is what sentiment classification tries to discover at the document level.

Granularity of Analysis: Let us go back to the general representation of an object with a component tree and each component with a set of attributes. We can study opinions at any level.

At level 1: We identify opinions on the object itself and its attributes.
At level 2: We identify opinions on the major components of the object, and also opinions on the attributes of the components.

At other levels, similar tasks can be performed. However, in practice, analysis at level 1 and level 2 are usually sufficient.

Example 8: Given the following review of a camera (the object),

"I like this camera. Its picture quality is amazing. However, the battery life is a little short",

in the first sentence, the positive opinion is at level 1, i.e., a positive opinion on the camera itself. The positive opinion on the picture quality in the second sentence is also at level 1 as "picture quality" is an attribute of the camera. The third sentence expresses a negative opinion on an attribute of the battery (at level 2), which is a component of the camera.   ■

Opinion Holder Identification: In some applications, it is useful to identify and extract opinion holders, i.e., persons or organizations that have expressed certain opinions. As we mentioned earlier, opinion holders are more useful for news articles and other types of formal documents, in which the person or organization that expressed an opinion is usually stated in the text explicitly. However, such holders need to be identified by the system. In the case of the user-generated content on the Web, the opinion holders are often the authors of discussion posts, bloggers, or reviewers, whose login ids are often known although their true identities in the real-world may be unknown. We will not discuss opinion holders in the chapter further due to our focus on the user-generated content on the Web. Interested readers, please refer to [276].

Opinioned Object Identification and Pronoun Resolution: In product reviews, the reviewed objects are usually known. However, this is not the case for opinions expressed in blogs and discussions. For example, in the following post "I have a Canon S50 camera purchased from Amazon. It takes great photos.", two interesting questions can be asked: (1) what object does the post praise? and (2) what "it" means in the second sentence? Clearly, we know that the post praises "Canon S50 camera", which is the problem of **opinioned object identification**, and we also know that "it" here means "Canon S50 camera" too, which is the problem of **pronoun resolution**. However, to automatically discover answers to the questions is a very challenging problem. So far, little work has been done.

11.2.2 Object Feature Extraction

Current research on feature extraction is mainly carried out from online **product reviews**. We focus on such reviews in this subsection as well.

It is a common practice for online merchants (e.g., amazon.com) to ask their customers to review the products that they have purchased. There are also dedicated review sites like epinions.com. There are three main review formats on the Web. Different review formats may need different techniques to perform the feature extraction task.

Format 1 – Pros, cons and the detailed review: The reviewer is asked to describe pros and cons separately and also write a detailed review. An example of such a review is given in Fig. 11.3.

Format 2 – Pros and cons: The reviewer is asked to describe pros and cons separately, but there is not a separate detailed review as in format 1. That is, the details are in pros and cons. An example of such a review is given in Fig. 11.4.

Format 3 – Free format: The reviewer can write freely, i.e., no separation of pros and cons. An example of such a review is given in Fig. 11.5.

For formats 1 and 2, opinion (or semantic) orientations (positive or negative) of the features are known because pros and cons are separated. Only product features need to be identified. For format 3, we need to identify both product features and opinion orientations.

My SLR is on the shelf

by <u>camerafun4</u>. Aug 09 '04

Pros: Great photos, easy to use, very small
Cons: Battery usage; included memory is stingy.

I had never used a digital camera prior to purchasing this Canon A70. I
have always used a SLR ... **Read the full review**

Fig. 11.3. An example of a review of format 1.

"It is a great digital still camera for this century"

September 1 2004.

Pros:
It's small in size, and the rotatable lens is great. It's very easy to use, and
has fast response from the shutter. The LCD ...

Cons:
It almost has no cons. It could be better if the LCD is bigger and it's going
to be best if the model is designed to a smaller size.

Fig. 11.4. An example of a review of format 2.

GREAT Camera., Jun 3, 2004

Reviewer: **jprice174** from Atlanta, Ga.

I did a lot of research last year before I bought this camera... It kinda hurt
to leave behind my beloved nikon 35mm SLR, but I was going to Italy, and
I needed something smaller, and digital.

The pictures coming out of this camera are amazing. The 'auto' feature
takes great pictures most of the time. And with digital, you're not wasting
film if the picture doesn't come out. ...

Fig. 11.5. An example of a review of format 3.

In both formats 2 and 3, reviewers typically use full sentences. However, for format 1, pros and cons tend to be very brief. For example, in Fig.
11.3, under pros, we have "Great photos, easy to use, take videos", which
are elaborated in the detailed review.

Let us deal with pros and cons of format 1 first. The detailed reviews of
format 1 are not used as they are elaborations of pros and cons. Analyzing
short sentence segments in pros and cons produces more accurate results.
Detailed reviews of format 1 are the same as reviews of format 3.

11.2.3   Feature Extraction from Pros and Cons of Format 1

We now describe a supervised pattern learning approach to extract product
features from pros and cons in the reviews of format 1. These patterns are

generated from **label sequential rules** (LSR) (see Sect. 2.9.2). This method is based on the algorithm in [247, 347].

A product feature can be expressed with a noun, adjective, verb or adverb. The labels and their POS tags used in mining LSRs are: {$feature, NN}, {$feature, JJ}, {$feature, VB} and {$feature, RB}, where $feature denotes a feature to be extracted, and NN stands for noun, VB for verb, JJ for adjective, and RB for adverb. They represent both explicit features and implicit feature indicators. We call a word that indicates an implicit feature an **implicit feature indicator**. For example, in the sentence "this camera is too heavy", "heavy" is an adjective and is an implicit feature indicator for feature "weight".

The feature extraction technique is based on the following observation:

- Each sentence segment in pros and cons contains only one feature. Sentence segments are separated by commas, periods, semi-colons, hyphens, '&''s, 'and''s, 'but''s, etc.

Example 9: Pros in Fig. 11.3 can be separated into three segments:

    great photos   ⟨photo⟩
    easy to use   ⟨use⟩
    very small   ⟨small⟩ $\Rightarrow$ ⟨size⟩.

Cons in Fig. 11.3 can be separated into two segments:

    battery usage   ⟨battery⟩
    included memory is stingy   ⟨memory⟩   ■

We can see that each segment describes a product feature, which is listed within ⟨ ⟩. Notice that ⟨small⟩ is an implicit feature indicator and ⟨size⟩ is the implicit feature.

One point to note is that an explicit feature may not be a noun or noun phrase. Verbs can be explicit features as well, e.g., "use" in "easy to use". In general, 60–70% of the features are explicit noun features. A small proportion of explicit features are verbs. 20–30% of the features are implicit features represented by their indicators. Let us now describe the method.

Given a set of reviews, this method consists of the following three steps:

1. **Training data preparation for LSR mining:** It consists of 4 sub-steps:
   - Part-Of-Speech (POS) tagging and sequence generation: For each sentence segment, the algorithm first performs POS tagging, and then produces a sequence. For example, the sentence segment,

    "Included memory is stingy".

   is turned into a sequence with POS tags:

    ⟨{included, VB}{memory, NN}{is, VB}{stingy, JJ}⟩.

- Replace the actual feature words with {$feature, <tag>}, where $feature represents a feature. This replacement is necessary because different products have different features, and the replacement ensures that we can find general language patterns to extract any product feature. After replacement, the above example becomes:

  ⟨{included, VB}{$feature, NN}{is, VB}{stingy, JJ}⟩.

- Use an *n*-gram to produce shorter segments from long ones: For example, the above sequence will generate two trigram sequences:

  ⟨{included, VB}{$feature, NN}{is, VB}⟩
  ⟨{$feature, NN}{is, VB}{stingy, JJ}⟩.

  Trigrams are usually sufficient. The reason for using *n*-grams rather than full sentences is because most product features can be found based on local information and POS tags. Using long sentences tend to generate a large number of spurious rules.

- Perform word stemming: This reduces a word to its stem (see Sect. 6.5.2).

After the four-step pre-processing, the resulting sentence (trigram) segments are saved in a sequence database for label sequential rule mining. In this file, each line contains one processed sequence.

2. **Label sequential rule mining:** A LSR mining system is applied to find all rules that involve a feature, i.e., $feature. An example rule is:

   ⟨{easy, JJ }{to}{*, VB}⟩ → ⟨{easy, JJ}{to}{$feature, VB}⟩.

   Note that both POS tags and words may appear in a rule. A suitable minimum confidence and minimum support should be used, which can be chosen based on experiments. The right-hand-side of the rule is also called a **language pattern**.

3. **Feature extraction:** The resulting language patterns are used to match each sentence segment in a new review to extract product features. That is, the word in the sentence segment that matches $feature in a language pattern is extracted. Three situations are considered in extraction:

   - If a sentence segment satisfies multiple rules, we search for a matching rule in the following order: {$feature, NN}, {$feature, JJ}, {$feature, VB} and {$feature, RB}. The reason for this ordering is that noun features appear more frequently than other types. For rules of the same tag, the rule with the highest confidence is used since higher confidence indicates higher predictive accuracy.

- For sentence segments that no rules apply, nouns or noun phrases produced by a POS tagger are extracted as features if such nouns or noun phrases exist.
- For a sentence segment with only a single word (e.g., "heavy" and "big"), this pattern-based method does not apply. In such cases, the single words are treated as (implicit or explicit) features.

After extraction, we need to deal with several other important problems:

Mapping to Implicit Features: There are many types of implicit feature indicators. Adjectives are perhaps the most common type. Many adjectives modify or describe some specific attributes or properties of objects. For example, the adjective "heavy" usually describes the attribute "weight" of an object. "Beautiful" is normally used to describe (positively) the attribute "look" or "appearance" of an object. By no means, however, does this say that these adjectives only describe such attributes. Their exact meaning can be domain dependent. For example, "heavy" in the sentence "the traffic is heavy" does not describe the "weight" of the traffic.

One way to map indicator words to implicit features is to manually compile a list of such mappings during training data annotation, which can then be used in the same domain in the future. However, it is not clear whether this is an effective approach as little research has been done.

Grouping Synonyms: It is common that people use different words or phrases to describe the same feature. For example, "photo" and "picture" refer to the same feature in digital camera reviews. Identifying and grouping synonyms is essential for practical applications. Although WordNet [175] and other thesaurus dictionaries help to some extent, they are far from sufficient due to the fact that many synonyms are domain dependent. For example, "picture" and "movie" are synonyms in movie reviews. However, they are not synonyms in digital camera reviews as "picture" is more related to "photo" while "movie" refers to "video".

Liu et al. [347] made an attempt using synonyms in WordNet. Carenini et al. [80] proposes a more sophisticated method based on several similarity metrics that require the taxonomy of features to be given. The system merges each discovered feature to a feature node in the taxonomy. The similarity metrics are defined based on string similarity, synonyms and other distances measured using WordNet. Experimental results based on digital camera and DVD reviews show promising results. Clearly, many ideas and techniques described in Chap. 10 for information integration are applicable here.

Granularity of Features: In the sentence segment "great photos", it is easy to decide that "photo" is the feature. However, in "battery usage", we can use either "battery usage" or "battery" as the feature. As we discussed in Sect. 11.2.1, each object has a component/part tree and each component node has a set of attributes. In a practical application, we need to determine the right level of analysis. If it is too general, it may not be useful. If it is too specific, it may result in a large number of features and also make the extraction very difficult and inaccurate.

11.2.4   Feature Extraction from Reviews of Formats 2 and 3

Pros and cons of format 1 mainly consist of short phrases and incomplete sentences. The reviews of formats 2 and 3 usually consist of complete sentences. To extract features from such reviews, the above algorithm can also be applied. However, some preliminary experiments show that it is not effective because complete sentences are more complex and contain a large amount of irrelevant information. Below, we describe an unsupervised method for finding explicit features that are nouns and noun phrases. This method requires a large number of reviews, and consists of two steps:

1. Finding frequent nouns and noun phrases. Nouns and noun phrases (or groups) are identified by using a POS tagger. We then count their frequency and only keep the frequent ones. A frequency threshold can be decided experimentally. The reason for using this approach is that most product features are nouns, and those nouns that are frequently talked about are usually genuine and important features. Irrelevant contents (see Fig. 11.5) in reviews are often diverse, i.e., they are quite different in different reviews. When people comment on product features, the vocabulary that they use converges. Those nouns that are infrequent are likely to be non-features or less important features.
2. Finding infrequent features by making use of sentiment words. **Sentiment words** (also called **opinion words**) are usually adjectives and adverbs that express positive or negative opinions, e.g., great, amazing, bad, and expensive. The idea is as follows: The same opinion word can be used to describe different objects. Opinion words that modify frequent features can be used to find infrequent features. For example, "picture" is found to be a frequent feature, and we have the sentence,

> "The pictures are absolutely amazing."

We also know that "amazing" is a positive opinion word (to be discussed in Sect. 11.2.5). Then "software" may also be extracted as a feature from the following sentence,

> "The software is amazing."

because the two sentences follow the same language pattern and "software" in the sentence is also a noun.

This two-step approach is based on the work of Hu and Liu [245]. At the time this book was written, the shopping site Froogle of the search engine Google implemented a method similar to step 1 of the algorithm. However, it does not restrict frequent terms to be nouns or noun phrases.

The precision of step 1 of the above algorithm was improved by Popescu and Etzioni in [447]. Their algorithm tries to remove those noun phrases that may not be product features. It evaluates each noun phrase by computing a PMI score between the phrase and **meronymy discriminators** associated with the product class, e.g., a scanner class. The meronymy discriminators for the scanner class are, "of scanner", "scanner has", "scanner comes with", etc., which are used to find components or parts of scanners by searching on the Web (see [166] also). The PMI measure is a simplified version of the measure given in Sect. 11.1.1:

$$PMI(f,d) = \frac{hits(f \wedge d)}{hits(f)hits(d)}, \tag{7}$$

where f is a candidate feature identified in step 1 and d is a discriminator. Web search is used to find the number of hits. The idea of this approach is clear. If the PMI value of a candidate feature is too low, it may not be a component of the product because f and d do not co-occur frequently. The algorithm also distinguishes components/parts from attributes/properties using WordNet's *is-a* hierarchy (which enumerates different kinds of properties) and morphological cues (e.g., "-iness", "-ity" suffixes).

Finally, we note that many information extraction techniques are also applicable, e.g., conditional random fields (CRF) [298], hidden Markov models (HMM) [185], and many others. However, no comparative evaluation of these methods on this problem has been reported so far.

11.2.5 Opinion Orientation Classification

For reviews of format 3, we need to classify each sentence that contains a product feature as positive, negative or neutral. This classification may also be needed for reviews of format 2 because although pros and cons are separated in format 2, some sentences containing features are neutral.

We describe two main techniques below. The accuracy is usually reasonable (greater than 80%) if the sentences are either positive or negative, but if neutral sentences are included, the accuracy often drops significantly. Sentences containing negations also pose difficulties.

1. Using **sentiment words** and **phrases**: As explained above, sentiment words and phrases are words and phrases that express positive or negative sentiments (or opinions). They are mostly adjectives and adverbs, but can be verbs and nouns too. Researchers have compiled sets of such words and phrases for adjectives, adverbs, verbs, and nouns respectively. Each set is usually obtained through a bootstrapping process:

   - Manually find a set of seed positive and negative words. Separate seed sets are prepared for adjectives, adverbs, verbs and nouns.
   - Grow each of the seed set by iteratively searching for their synonyms and antonyms in WordNet until convergence, i.e., until no new words can be added to the set. Antonyms of positive (or negative) words will be added to the negative (or positive) set.
   - Manually inspect the results to remove those incorrect words. Although this step is time consuming, it is only an one-time effort.

   Apart from a set of opinion words, there are also **idioms**, which can be classified as positive, negative and neutral as well. Many language patterns also indicate positive or negative sentiments. They can be manually compiled and/or discovered using pattern discovery methods.

   Using the final lists of positive and negative words, phrases, idioms and patterns, each sentence that contains product features can be classified as follows: Sentiment words and phrases in the sentence are identified first. A positive word or phrase is assigned a score of +1 and a negative word or phrase is assigned a score of −1. All the scores are then summed up. If the final total is positive, then the sentence is positive, otherwise it is negative. If a negation word is near a sentiment word, the opinion is reversed. A sentence that contains a "but" clause (sub-sentence that starts with "but", "however", etc.) indicates a sentiment change for the feature in the clause.

   This method is based on the techniques given by Hu and Liu [245], and Kim and Hovy [276]. In [447], Popescu and Etzioni proposed a more complex method, which makes use of syntactical dependencies produced by a parser. Yu and Hatzivassiloglou [584] presented a method similar to that in Sect. 11.1.1 but used a large number of seeds. Recall that Turney [521] used only two seeds (see Sect. 11.1.1), "excellent" for positive and "poor" for negative. The sentence orientation is determined by a threshold of the average score of the words in the sentence. It is not clear which method performs better because there is little comparative evaluation.

   Note that the opinion orientations of many words are **domain** and/or **sentence context dependent**. Such situations are usually hard to deal with. It can be easy in some cases. For example, "small" can be positive or negative. However, if there is a "too" before it, it normally indicates a negative sentiment, e.g., "this camera is too small for me".

2. The methods described in Sect. 11.1 for sentiment classification are applicable here. Using supervised learning, we need to prepare a set of manually labeled positive, negative and neutral sentences as the training data. If sentiment words and phrases, idioms and patterns are used also as attributes, the classification results can be further improved. Sentences containing negations and clauses starting with "but", "however", etc., need special handling since one part of the sentence may be positive and another part may be negative, e.g., "The pictures of this camera are great, but the camera itself is a bit too heavy."

In summary, although many classification techniques have been proposed, little comparative study of these techniques has been reported. A promising approach is to combine these techniques to produce a better classifier.

11.3 Comparative Sentence and Relation Mining

Directly expressing positive or negative opinions on an object is only one form of evaluation. Comparing the object with some other similar objects is another. Comparison is perhaps a more convincing way of evaluation. For example, when a person says that something is good or bad, one often asks "compared to what?" Thus, one of the most important ways of evaluating an object is to directly compare it with some other similar objects.

Comparisons are related to but also different from typical opinions. They have different semantic meanings and different syntactic forms. Comparisons may be subjective or objective. For example, a typical opinion sentence is "the picture quality of camera x is great." A subjective comparison is "the picture quality of camera x is better than that of camera y." An objective comparison is "camera x is 20 grams heavier than camera y", which may be a statement of a fact and may not have an implied opinion on which camera is better.

In this section, we study the problem of identifying comparative sentences and comparative relations (defined shortly) in text documents, e.g., consumer reviews, forum discussions and news articles. This problem is also challenging because although we can see that the above example sentences all contain some indicators, i.e., "better" and "longer", many sentences that contain such words are not comparisons, e.g., "in the context of speed, faster means better". Similarly, many sentences that do not contain such indicators are comparative sentences, e.g., "cellphone X has bluetooth, but cellphone Y does not," and "Intel is way ahead of AMD."

11.3.1 Problem Definition

A **comparative sentence** is a sentence that expresses a relation based on similarities or differences of more than one object. The comparison in a comparative sentence is usually expressed using the **comparative** or the **superlative** form of an adjective or adverb. The comparative is used to state that one thing has more (bigger, smaller) "value" than the other. The superlative is used to say that one thing has the most (the biggest, the smallest) "value". The structure of a comparative consists normally of the stem of an adjective or adverb, plus the suffix *-er*, or the modifier "more" or "less" before the adjective or adverb. For example, in "John is taller than James", "taller" is the comparative form of the adjective "tall". The structure of a superlative consists normally of the stem of an adjective or adverb, plus the suffix *-est*, or the modifier "most" or "least" before the adjective or adverb. In "John is the tallest in the class", "tallest" is the superlative form of the adjective "tall".

A comparison can be between two or more objects, groups of objects, one object and the rest of the objects. It can also be between an object and its previous or future versions.

Types of Important Comparisons: We can classify comparisons into four main types. The first three types are **gradable comparisons** and the last one is the **non-gradable comparison**. The gradable types are defined based on the relationships of *greater* or *less than, equal to,* and *greater* or *less than all others*.

1. *Non-equal gradable comparisons*: Relations of the type *greater* or *less than* that express an ordering of some objects with regard to some of their features, e.g., "the Intel chip is faster than that of AMD". This type also includes user preferences, e.g., "I prefer Intel to AMD".
2. *Equative comparisons*: Relations of the type *equal to* that state two objects are equal with respect to some of their features, e.g., "the picture quality of camera A is as good as that of camera B"
3. *Superlative comparisons*: Relations of the type *greater* or *less than all others* that rank one object over *all* others, e.g., "the Intel chip is the fastest".
4. *Non-gradable comparisons*: Sentences that compare features of two or more objects, but do not grade them. There are three main types:
   - Object *A* is similar to or different from object *B* with regard to some features, e.g., "Coke tastes differently from Pepsi".
   - Object *A* has feature f_1, and object *B* has feature f_2 (f_1 and f_2 are usually substitutable), e.g., "desktop PCs use external speakers but laptops use internal speakers".

- Object *A* has feature *f*, but object *B* does not have, e.g., "cell phone A has an earphone, but cell phone B does not have".

Gradable comparisons can be classified further into two types: **adjectival comparisons** and **adverbial comparisons**. Adjectival comparisons involve comparisons of degrees associated with adjectives, e.g., in "John is taller than Mary," and "John is the tallest in the class". Adverbial comparisons are similar but usually occur after verb phrases, e.g., "John runs faster than James," and "John runs the fastest in the class".

Given an evaluative text *d*, **comparison mining** consists of two tasks:

1. Identify comparative passages or sentences from *d*, and classify the identified comparative sentences into different types or classes.
2. Extract comparative relations from the identified sentences. This involves the extraction of entities and their features that are being compared, and the comparative keywords. Relations in gradable adjectival comparisons can be expressed with

   (*<relationWord>*, *<features>*, *<entityS1>*, *<entityS2>*, *<type>*)

where:
   relationWord: The comparative keyword used to express a comparative relation in a sentence.
   features: a set of features being compared.
   entityS1 and *entityS2*: Sets of entities being compared. Entities in entityS1 appear to the left of the relation word and entities in entityS2 appear to the right of the relation word.
   type: *non-equal gradable*, *equative* or *superlative*.

Example 10: Consider the comparative sentence "Canon's optics is better than those of Sony and Nikon." The extracted relation is:

   (better, {optics}, {Canon}, {Sony, Nikon}, *non-equal gradable*).   ■

We can also design relation representations for adverbial comparisons and non-gradable comparisons. In this section, however, we only focus on adjectival gradable comparisons as there is little study on relation extraction of the other types. For simplicity, we will use *comparative sentences* and *gradable comparative sentences* interchangeably from now on.

Finally, we note that there are other types of comparatives in linguistics that are used to express different types of relations. However, they are relatively rare in evaluative texts and/or less useful in practice. For example, a meta-linguistic comparative compares the extent to which a single object has one property to a greater or lesser extent than another property, e.g., "Ronaldo is angrier than upset" (see [150, 274, 313, 393]).

11.3.2 Identification of Gradable Comparative Sentences

This is a classification problem. A machine learning algorithm is applicable to solve this problem. The main issue is what attributes to use.

An interesting phenomenon about comparative sentences is that such a sentence usually has a comparative keyword. It is shown in [256] that using a set of 83 keywords, 98% of the comparative sentences (recall = 98%) can be identified with a precision of 32% using the authors' data set. Let us see what the keywords are:

1. **Comparative adjectives** (with the POS tag of JJR) and **comparative adverbs** (with the POS tag of RBR), e.g., *more, less, better, longer* and words ending with *-er*.
2. **Superlative adjectives** (with the POS tag of JJS) and **superlative adverbs** (with the POS tag of RBS), e.g., *most, least, best, tallest* and words ending with *-est*.
3. Words like *same, similar, differ* and those used with equative *as*, e.g., *same as, as well as*, etc.
4. Others, such as *favor, beat, win, exceed, outperform, prefer, ahead, than, superior, inferior, number one, up against*, etc.

Note that those words with POS tags of JJR, RBR, JJS and RBS are not used as keywords themselves. Instead, their POS tags, JJR, RBR, JJS and RBS, are treated as four keywords only. There are four exceptions: *more, less, most*, and *least* are treated as individual keywords because their usages are diverse, and using them as individual keywords enables the system to catch their individual usage patterns for classification.

Since keywords alone are able to achieve a very high recall, the following learning approach is used in [255] to improve the precision:

- Use the set of keywords to filter out those sentences that are unlikely to be comparative sentences (do not contain any keywords). The remaining set of sentences R forms the candidate set of comparative sentences.
- Work on R to improve the precision, i.e., to classify the sentences in R into *comparative* and *non-comparative sentences*, and then into different types of comparative sentences.

It is also observed in [255] that comparative sentences have strong patterns involving comparative keywords, which is not surprising. These patterns can be used as attributes in learning. To discover these patterns, **class sequential rule (CSR) mining** (see Sect. 2.9.3) was used. Each training example used for mining CSRs is a pair (s_i, y_i), where s_i is a sequence and y_i is a class, $y_i \in \{comparative, non\text{-}comparative\}$.

Training Data Preparation: The sequences in the training data are generated from sentences. Since we want to find patterns surrounding specific keywords, we use keywords as **pivots** to produce sequences.

Let the set of pivots be P. We generate the training sequence database as follows:

1. For each sentence, only words within the radius of r of the keyword pivot $p_i \in P$ are used to form a sequence. In [256], r is set to 3. Each pivot in a sentence forms a separate sequence.
2. Each word is then replaced with its POS tag. The actual words are not used because the contents of sentences may be very different, but their underlying language patterns can be the same. Using POS tags allow us to capture content independent patterns. There is an exception. For each keyword (except those represented by JJR, RBR, JJS and RBS), the actual word and its POS tag are combined together to form a single item. The reason for this is that some keywords have multiple POS tags depending on their use. Their specific usages can be important in determining whether a sentence is a comparative sentence or not. For example, the keyword "more" can be a comparative adjective (more/JJR) or a comparative adverb (more/RBR) in a sentence.
3. A class is attached to each sequence according to whether the sentence is a comparative or non-comparative sentence.

Example 11: Consider the comparative sentence "this/DT camera/NN has/VBZ significantly/RB more/JJR noise/NN at/IN iso/NN 100/CD than/IN the/DT nikon/NN 4500/CD." It has the keywords "more" and "than". The sequence involving "more" put in the training set is:

(⟨{NN}{VBZ}{RB}{more/JJR}{NN}{IN}{NN}⟩, comparative)

CSR Generation: Using the training data, CSRs can be generated. Recall that a CSR is an implication of the form, $X \rightarrow y$, where X is a sequence and y is a class. Due to the fact that some keywords appear very frequently in the data and some appear rarely, multiple minimum supports are used in mining. The minimum item support for each keyword is computed with *freq** τ, where τ is set to 0.1 (*freq* is the actual frequency of its occurrence). See Sect. 2.7.2 or Sect. 2.8.2 in Chap. 2 for details on mining with multiple minimum supports.

In addition to the automatically generated rules, some manually compiled rules are also used in [255, 256], which are more complex and difficult to generate by current rule mining techniques.

Classifier Building: There are many ways to build classifiers using the discovered CSRs, we describe two methods:

1. Treat all CSRs as a classifier. A CSR simply expresses the conditional probability that a sentence is a comparison if it contains the sequence pattern X. These rules can thus be used for classification. That is, for each test sentence, the algorithm finds all the rules satisfied by the sentence, and then chooses the rule with the highest confidence to classify the sentence. This is basically the "use the strongest rule" method discussed in Sect. 3.5.1 in Chap. 3.

2. Use CSRs as attributes to create a data set and then learn a naïve Bayesian (NB) classifier (or any other types of classifiers) (see Sect. 3.5.2 in Chap. 3). The data set uses the following attribute set:

$Attribute\ Set =$ $\{X \mid X$ is the sequential pattern in CSR $X \rightarrow y\} \cup$
$\{Z \mid Z$ is the pattern in a manual rule $Z \rightarrow y\}$.

The class is not used but only the sequence pattern X (or Z) of each rule. The idea is that these patterns are predictive of the classes. A rule's predictability is indicated by its confidence. The minimum confidence of 60% is used in [255].

Each sentence forms an example in the training data. If the sentence has a particular pattern in the attribute set, the corresponding attribute value is 1, and is 0 otherwise. Using the resulting data, it is straightforward to perform NB learning. Other learning methods can be used as well, but according to [255], NB seems to perform better.

Classify Comparative Sentences into Three Types: This step classifies comparative sentences obtained from the last step into one of the three types or classes, *non-equal gradable*, *equative*, and *superlative*. For this task, the keywords alone are already sufficient. That is, we use the set of keywords as the attribute set for machine learning. If the sentence has a particular keyword in the attribute set, the corresponding attribute value is 1, and otherwise it is 0. SVM gives the best results in this case.

11.3.3 Extraction of Comparative Relations

We now discuss how to extract relation entries/items. Label sequential rules are again used for this task. The algorithm presented below is based on the work in [256], which makes the following assumptions:

1. There is only one relation in a sentence. In practice, this is violated only in a very small number of cases.
2. Entities or features are nouns (includes nouns, plural nouns and proper nouns) and pronouns. These cover most cases. However, a feature can sometimes be a noun used in its verb form or some action described as a

verb (e.g., "Intel costs more"; "costs" is a verb and a feature). Such comparisons are adverbial comparisons and are not considered in [256].

Sequence Data Generation: A sequence database for mining is created as follows: Since we are interested in predicting and extracting items representing *entityS1* (denoted by $entityS1), *entityS2* (denoted by $entityS2), and features (denoted by $feature), which are all called **labels**, we first manually mark/label such words in each sentence in the training data. For example, in the sentence "Intel/NNP is/VBZ better/JJR than/IN amd/NN", the proper noun "Intel" is labeled with $entityS1, and the noun "amd" is labeled with $entityS2. The two labels are then used as pivots to generate sequence data. For every occurrence of a label in a sentence, a separate sequence is created and put in the sequence database. A radius of 4 is used in [256]. The following position words are also added to keep track of the distance between two items in a generated pattern:

1. Distance words = $\{l1, l2, l3, l4, r1, r2, r3, r4\}$, where *li* means distance of *i* to the left of the pivot, and *ri* means the distance of *i* to the right of pivot.
2. Special words *#start* and *#end* are used to mark the start and the end of a sentence.

Example 12: The comparative sentence "Canon/NNP has/VBZ better/JJR optics/NNS than/IN Nikon/NNP" has $entityS1 "Canon", $feature "optics" and $entityS2 "Nikon". The three sequences corresponding to the two entities and one feature put in the database are:

⟨{#start}{/1}{**$entityS1**, NNP}{*r1*}{has, VBZ}{*r2*}{better, JJR}
    {*r3*}{$feature, NNS}{*r4*}{thanIN}⟩

⟨{#start}{/4}{$entityS1, NNP}{/3}{has, VBZ}{/2}{better, JJR} {/1}
    {**$feature**, NNS}{*r1*}{thanIN}{*r2*}{entityS2, NNP}{*r3*} {#end}⟩

⟨{has, VBZ}{/4}{better, JJR}{/3}{$feature, NNS}{/2}{thanIN}
    {/1}{**$entityS2**, NNP}{*r1*}{#end}⟩.

The keyword "than" is merged with its POS tag to form a single item.

LSR Generation: After the sequence database is built, a rule mining system is applied to generate label sequential rules. Note that only those rules that contain one or more labels (i.e., $entityS1, $entityS2, and $feature) will be generated. An example of a LSR rule is as follows

Rule 1: ⟨{*, NN}{VBZ}{JJR}{thanIN}{*, NN}⟩ →
    ⟨{$entityS1, NN}{VBZ}{JJR}{thanIN}{$entityS2, NN}⟩.

Relation Item Extraction: The generated LSRs are used to extract relation items from each input (or test) sentence. One strategy is to use all the rules to match the sentence and to extract the relation items using the rule with the highest confidence. For example, the above rule will label and extract "coke" as entityS1, and "pepsi" as entityS2 from the following sentence:

⟨{coke, NN}{is, VBZ}{definitely, RB}{better, JJR}{thanIN}{pepsi, NN}⟩.

There is no feature in this sentence. The *relationWord* is simply the keyword that identifies the sentence as a comparative sentence. In this case, it is "better." A similar but more complex method is used in [256].

Again, many other methods can also be applied to the extraction, e.g., conditional random fields, hidden Markov models, and others. Results in [256] show that the LSR-based method outperforms conditional random fields. Further research and more comprehensive evaluations are needed to assess the strengths and weaknesses of these methods.

11.4 Opinion Search

Like the general Web search, one can also crawl the user-generated content on the Web and provide an opinion search service. The objective is to enable users to search for opinions on any object. Let us look at some typical opinion search queries:

1. Search for opinions on a particular object or feature of an object, e.g., customer opinions on a digital camera or the picture quality of a digital camera, or public opinions on a political topic. Recall that the object can be a product, organization, topic, etc.
2. Search for opinions of a person or organization (i.e., opinion holder) on a particular object or feature of the object. For example, one may search for Bill Clinton's opinion on abortion or a particular aspect of it. This type of search is particularly relevant to news documents, where individuals or organizations who express opinions are explicitly stated. In the user-generated content on the Web, the opinions are mostly expressed by authors of the postings.

For the first type of queries, the user may simply give the name of the object and/or some features of the object. For the second type of queries, the user may give the name of the opinion holder and also the name of the object. Clearly, it is not appropriate to simply apply keyword matching for either type of queries because a document containing the query words may not have opinions. For example, many discussion and blog posts do not contain opinions, but only questions and answers on some objects. Opinionated documents or sentences need to be identified before search is performed. Thus, the simplest form of opinion search can be keyword-based search applied to the identified opinionated documents/sentences.

As for ranking, traditional Web search engines rank Web pages based on authority and relevance scores. The basic premise is that the top ranked pages (ideally the first page) contain sufficient information to satisfy the user's information need. This may be fine for the second type of queries because the opinion holder usually has only one opinion on the search object, and the opinion is usually contained in a single document or page (in some cases, using a general search engine with an appropriate set of keywords may be sufficient to find answers for such queries). However, for the first type of opinion queries, the top ranked documents only represent the opinions of a few persons. Therefore, they need to reflect the natural distribution of positive and negative sentiments of the whole population. Moreover, in many cases, opinionated documents are very long (e.g., reviews). It is hard for the user to read many of them in order to obtain a complete picture of the prevailing sentiments. Some form of summary of opinions is desirable, which can be either a simple rating average of reviews and proportions of positive and negative opinions, or a sophisticated feature-based summary as we discussed earlier. To make it even easier for the user, two rankings may be produced, one for positive opinions and one for negative opinions.

Providing a feature-based summary for each search query is an ideal solution. An analogy can be drawn from traditional surveys or opinion polls. An opinionated document is analogous to a filled survey form. Once all or a sufficient number of survey forms are collected, some analysts will analyze them to produce a survey summary, which is usually in the form of a bar or pie chart. One seldom shows all the filled survey forms to users (e.g., the management team of an organization or the general public) and asks them to read everything in order to draw their own conclusions. However, automatically generating a feature-based summary for each search object (or query) is a very challenging problem. To build a practical search system, some intermediate solution based on Problem 2 and 3 in Sect. 11.2.1 may be more appropriate.

Opinions also have a temporal dimension. For example, the opinions of people on a particular object, e.g., a product or a topic, may change over time. Displaying the changing trend of sentiments along the time axis can be very useful in many applications.

Finally, like opinion search, comparison search will be useful as well. For example, when you want to register for a free email account, you most probably want to know which email system is best for you, e.g., hotmail, gmail or *Yahoo*! mail. Wouldn't it be nice if you can find comparisons of features of these email systems from existing users by issuing a search query "hotmail vs. gmail vs. yahoo mail."?

11.5 Opinion Spam

In Sect. 6.10, we discussed Web spam, which refers to the use of "illegitimate means" to boost the search rank position of some target Web pages. The reason for spamming is because of the economic and/or publicity value of the rank position of a page returned by a search engine. In the context of opinions on the Web, the problem is similar. It has become a common practice for people to find and to read opinions on the Web for many purposes. For example, if one wants to buy a product, one typically goes to a merchant or review site (e.g., amazon.com) to read some reviews of existing users of the product. If one sees many positive reviews of the product, one is very likely to buy the product. On the contrary, if one sees many negative reviews, he/she will most likely choose another product. Positive opinions can result in significant financial gains and/or fames for organizations and individuals. This, unfortunately, gives good incentives for **opinion spam**, which refers to human activities (e.g., write spam reviews) that try to deliberately mislead readers or automated opinion mining systems by giving undeserving positive opinions to some target objects in order to promote the objects and/or by giving unjust or false negative opinions on some other objects in order to damage their reputation. In this section, we use customer reviews of products as an example to study opinion spam on the Web. Most of the analyses are also applicable to opinions expressed in other forms of user-generated contents, e.g., forum postings, group discussions, and blogs.

11.5.1 Objectives and Actions of Opinion Spamming

As we indicated above, there are two main objectives for writing spam reviews:

1. To promote some target objects, e.g., one's own products.
2. To damage the reputation of some other target objects, e.g., products of one's competitors.

In certain cases, the spammer may want to achieve both objectives, while in others, he/she only aims to achieve one of them because either he/she does not have an object to promote or there is no competition. Another objective is also possible but may be rare. That is, the spammer writes some irrelevant information or false information in order to annoy readers and to fool automated opinion mining systems.

To achieve the above objectives, the spammer usually takes both or one of the actions below:

- Write undeserving positive reviews for the target objects in order to promote them. We call such spam reviews **hype spam**.
- Write unfair or malicious negative reviews for the target objects to damage their reputation. We call such spam review **defaming spam**.

11.5.2 Types of Spam and Spammers

Table 11.3 below gives a simplified view of spam reviews. Spam reviews in regions 1, 3 and 5 are typically written by owners or manufacturers of the product or persons who have direct economic or other interests in the product. Their main goal is to promote the product. Although opinions expressed in reviews of region 1 may be true, reviewers do not announce their conflict of interests.

Spam reviews in regions 2, 4, and 6 are likely to be written by competitors, who give false information in order to damage the reputation of the product. Although opinions in reviews of region 4 may be true, reviewers do not announce their conflict of interests and may have malicious intensions.

Table 11.3. Spam reviews vs. product quality

	Hype spam review	Defaming spam review
Good quality product	1	2
Poor quality product	3	4
In-between good and poor quality product	5	6

Clearly, spam reviews in region 1 and 4 are not so damaging, while spam reviews in regions 2, 3, 5 and 6 are very harmful. Thus, spam detection techniques should focus on identifying reviews in these regions.

Manual and Automated Spam: Spam reviews may be manually written or automatically generated. Writing spam reviews manually is not a simple task if one wants to spam on a product at many review sites and write them differently to avoid being detected by methods that catch near duplicate reviews. Using some language templates, it is also possible to automatically generate many different variations of the same review.

Individual Spammers and Group Spammers: A spammer may act individually (e.g., the author of a book) or as a member of a group (e.g., a group of employees of a company).

Individual spammers: In this case, a spammer, who does not work with anyone else, writes spam reviews. The spammer may register at a review site as a single user, or as "many users" using different user-ids. He/she can also register at multiple review sites and write spam reviews.

Group spammers: A group of spammers works collaboratively to promote a target object and/or to damage the reputation of another object. They may also register at multiple sites and spam at these sites. Group spam can be very damaging because they may take control of the sentiments on the product and completely mislead potential customers.

11.5.3 Hiding Techniques

In order to avoid being detected, spammers may take a variety of precautions. We study individual and group of spammers separately. The lists are by no means exhaustive and should be considered as just examples.

An Individual Spammer

1. The spammer builds up reputation by reviewing other products in the same or different categories/brands that he/she does not care about and give them agreeable ratings and reasonable reviews. Then, he/she becomes a trustworthy reviewer. However, he/she may write spam reviews on the products that he/she really cares about. This hiding method is useful because some sites rank reviewers based on their reviews that are found helpful by readers, e.g., amazon.com. Some sites also have trust systems that allow readers to assign trust scores to reviewers.
2. The spammer registers multiple times at a site using different user-ids and write multiple spam reviews under these user-ids so that their reviews or ratings will not appear as outliers. The spammer may even use different machines to avoid being detected by server log based detection methods that can compare IP addresses of reviewers (discussed below).
3. The spammer gives a reasonably high rating but write a critical (negative) review. This may fool detection methods that find outliers based on ratings alone. Yet, automated review mining systems will pick up all the negative sentiments in the actual review content.
4. Spammers write either only positive reviews on his/her own products or only negative reviews on the products of his/her competitors, but not both. This is to hide from spam detection methods that compare one's reviews on competing products from different brands.

A Group of Spammers

1. Every member of the group reviews the same product to lower the rating deviation.
2. Every member of the group writes a review roughly at the time when the product is launched in order to take control of the product. It is generally not a good idea to write many spam reviews at the same time after many reviews have been written by others because a spike will appear, which can be easily detected.
3. Members of the group write reviews at random or irregular intervals to hide spikes.
4. If the group is sufficiently large, it may be divided into sub-groups so that each sub-group can spam at different web sites (instead of only spam at the same site) to avoid being detected by methods that compare average ratings and content similarities of reviews from different sites.

11.5.4 Spam Detection

So far, little study has been done on opinion spam detection. This subsection outlines some possible approaches. We note that each individual technique below may not be able to reliably detect spam reviews, but it can be treated as a spam indicator. A holistic approach that combines all evidences is likely to be effective. One possible combination method is to treat spam detection as a classification problem. All the individual methods simply compute spam evidences which can be put in a data set from which a spam classifier can be learned. For this approach to work, a set of reviews needs to be manually labeled for learning. The resulting classifier can be used to classify each new review as a spam review or not one.

Review Centric Spam Detection: In this approach, spam detection is based only on reviews. A review has two main parts: rating and content.

Compare content similarity: In order to have the maximum impact, a spammer may write multiple reviews on the same product (using different user-ids) or multiple products of the same brand. He/she may also write reviews at multiple review sites. However, for a single spammer to write multiple reviews that look very different is not an easy task. Thus, some spammers simply use the same review or slight variations of the same review. In a recent study of reviews from amazon.com, it was found that some spammers were so lazy that they simply copied the same review and pasted it for many different products of the same brand. Techniques that can detect near duplicate documents are useful here (see Sect. 6.5.5). For automatically generated spam reviews based on language templates, sophisticated pattern mining methods may be needed to detect them.

Detect rating and content outliers: If we assume that reviews of a product contain only a very small proportion of spam, we can detect possible spam activities based on rating deviations, especially for reviews in region 2 and 3, because they tend to be outliers. For reviews in regions 5 and 6, this method may not be effective.

If a product has a large proportion of spam reviews, it is hard to detect them based on review ratings, even though each spammer may act independently, because they are no longer outliers. In this case, we may need to employ reviewer centric and server centric spam detection methods below. This case is similar to group spam, which is also hard to detect based on content alone because the spam reviews are written by different members of the group and there are a large number of them. Hence, their reviews are not expected to be outliers. However, members of the group may be detected based on reviewer centric detection methods and server centric detection methods. The following methods are also helpful.

Compare average ratings from multiple sites: This method is useful to access the level of spam activities from a site if only a small number of review sites are spammed. For example, if the averages rating at many review sites for a product are quite high but at one site it is quite low, this is an indication that there may be some group spam activities going on.

Detect rating spikes: This method looks at the review ratings (or contents) from the time series point of view. If a number of reviews with similar ratings come roughly at the same time, a spike will appear which indicates a possible group spam.

Reviewer Centric Spam Detection: In this approach, "unusual" behaviors of reviewers are exploited for spam detection. It is assumed that all the reviews of each reviewer at a particular site are known. Most review sites provide such information, e.g., amazon.com, or such information can be found by matching user-ids.

Watch early reviewers: Spammers are often the first few reviewers to review a product because earlier reviews tend to have a bigger impact. Their ratings for each product are in one of the two extremes, either very high or very low. They may do this consistently for a number of products of the same brand.

Detect early remedial actions: For a given product, as soon as someone writes a (or the first) negative review to a product, the spammer gives a positive review just after it, or vice versa.

Compare review ratings of the same reviewer on products from different brands: A spammer often writes very positive reviews for products of one brand (to promote the product) and very negative reviews for similar products of another brand. A rating (or content) comparison will show discrepancies. If some of the ratings also deviate a great deal from the average ratings of the products, this is a good indicator of possible spam.

Compare review times: A spammer may review a few products from different brands at roughly the same time. Such behaviors are unusual for a normal reviewer.

As we mentioned above, detecting a group of spammers is difficult. However, we can reduce their negative impact by detecting each individual member in the group using the above and below methods.

Server centric spam detection: The server log at the review site can be helpful in spam detection as well. If a single person registers multiple times at a Web site having the same IP address, and the person also writes multiple reviews for the same product or even different products using different user-ids, it is fairly certain that the person is a spammer. Using the server log may also detect some group spam activities. For example, if most good reviews of a product are from a particular region where the company that produces the product is located, it is a good indication that these are likely spam.

As more and more people and organizations are using opinions on the Web for decision making, spammers have more and more incentives to express false sentiments in product reviews, discussions and blogs. To ensure the quality of information provided by an opinion mining and/or search system, spam detection is a critical task. Without effective detection, opinions on the Web may become useless. This section analyzed various aspects of opinion spam and outlined some possible detection methods. This may just be the beginning of a long journey of the "arms race" between spam and detection of spam.

Bibliographic Notes

Opinion mining received a great deal of attention recently due to the availability of a huge volume of online documents and user-generated content on the Web, e.g., reviews, forum discussions, and blogs. The problem is intellectually challenging, and also practically useful. The most widely studied sub-problem is sentiment classification, which classifies evaluative texts or sentences as positive, negative, or neutral. Representative works on classification at the document level include those by Turney [521], Pang et al. [428], and Dave et al. [122]. They have been discussed in this chapter. Sentence level subjectivity classification was studied by Hatzivassiloglou and Wiebe [225], which determines whether a sentence is a subjective sentence (but may not express a positive or negative opinion) or a factual one. Sentence level sentiment or opinion classification (positive, negative and neutral) was studied by Hu and Liu [245], Kim and Hovy [276], Wiebe and Riloff [545], among others. Some of these methods were discussed in Sect. 11.2.5. Other related works at both the document and sentence levels include those by Hearst [232], Tong [517], Das and Chen [120], Morinaga et al. [397], Agrawal et al. [10], Nasukawa and Yi [402], Beineke et al. [44], Nigam and Hurst [412], Gamon [191], Gamon et al. [192], Pang and Lee [426, 427], Riloff and Wiebe [462], Wilson et al. [548], etc.

Most sentence level and even document level classification methods are based on word or phrase sentiment identification. Automatic and semi-automatic methods for the purpose have been explored by several researchers. There are basically two types of approaches: (1) corpus-based approaches, and (2) dictionary-based approaches. Corpus-based approaches find co-occurrence patterns of words to determine their sentiments, e.g., the works by Turney [521], Riloff and Wiebe [462], Hatzivassiloglou and McKeown [224], Yu and Hatzivassiloglou [584], and Grefenstette et al. [206]. Dictionary-based approaches use synonyms, antonyms and hierarchies in WordNet to determine word sentiments. Such approaches were studied by Hu and Liu [245], Valitutti et al. [524], Kim and Hovy [276], and Andreevskaia and Bergler [22].

The idea of feature-based opinion summarization was introduced by Hu and Liu [245] to deal with a large number of reviews. Some methods for performing the task were also proposed. Popescu and Etzioni [447], and Carenini et al [80] explored the issues further. Liu et al. [347] studied the problem of product feature extraction from pros and cons, which are short phrases. Yi et al. [577], Ku et al. [291] and Kobayashi et al. [284] investigated similar problems as well. More recent work can be found in [81, 159, 264, 267, 277, 278, 290, 409, 483, 507, 544, 546, 622].

Regarding mining comparative sentences and relations, Jindal and Liu [255, 256] defined the problem and proposed some initial techniques based on data mining and machine learning methods. Research in linguistics on syntax and semantics of comparatives can be found in [150, 274, 393].

The discussions on opinion search and opinion spam are based on a recent study of product reviews from amazon.com by the author's research group.

12 Web Usage Mining

By *Bamshad Mobasher*

With the continued growth and proliferation of e-commerce, Web services, and Web-based information systems, the volumes of **clickstream** and **user data** collected by Web-based organizations in their daily operations has reached astronomical proportions. Analyzing such data can help these organizations determine the life-time value of clients, design cross-marketing strategies across products and services, evaluate the effectiveness of promotional campaigns, optimize the functionality of Web-based applications, provide more personalized content to visitors, and find the most effective logical structure for their Web space. This type of analysis involves the automatic discovery of meaningful patterns and relationships from a large collection of primarily semi-structured data, often stored in Web and applications server access logs, as well as in related operational data sources.

Web usage mining refers to the automatic discovery and analysis of patterns in clickstream and associated data collected or generated as a result of user interactions with Web resources on one or more Web sites [114, 387, 505]. The goal is to capture, model, and analyze the behavioral patterns and profiles of users interacting with a Web site. The discovered patterns are usually represented as collections of pages, objects, or resources that are frequently accessed by groups of users with common needs or interests.

Following the standard data mining process [173], the overall Web usage mining process can be divided into three inter-dependent stages: data collection and pre-processing, pattern discovery, and pattern analysis. In the pre-processing stage, the clickstream data is cleaned and partitioned into a set of user transactions representing the activities of each user during different visits to the site. Other sources of knowledge such as the site content or structure, as well as semantic domain knowledge from site **ontologies** (such as **product catalogs** or **concept hierarchies**), may also be used in pre-processing or to enhance user transaction data. In the pattern discovery stage, statistical, database, and machine learning operations are performed to obtain hidden patterns reflecting the typical behavior of users, as well as summary statistics on Web resources, sessions, and users. In the final stage of the process, the discovered patterns and statistics are further processed, filtered, possibly resulting in aggregate user models that can be

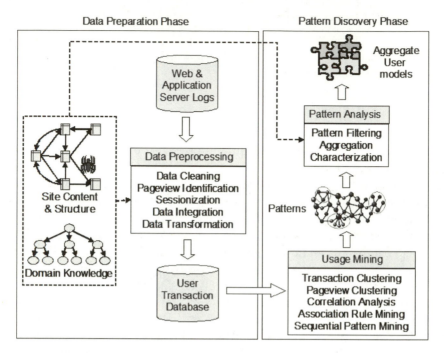

Fig. 12.1. The Web usage mining process

used as input to applications such as **recommendation engines**, visualization tools, and Web analytics and report generation tools. The overall process is depicted in Fig. 12.1.

In the remainder of this chapter, we provide a detailed examination of Web usage mining as a process, and discuss the relevant concepts and techniques commonly used in all the various stages mentioned above.

12.1 Data Collection and Pre-Processing

An important task in any data mining application is the creation of a suitable target data set to which data mining and statistical algorithms can be applied. This is particularly important in Web usage mining due to the characteristics of clickstream data and its relationship to other related data collected from multiple sources and across multiple channels. The data preparation process is often the most time consuming and computationally intensive step in the Web usage mining process, and often requires the use of special algorithms and heuristics not commonly employed in other domains. This process is critical to the successful extraction of useful patterns

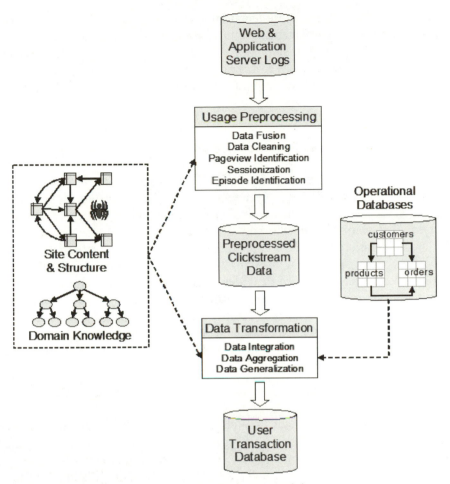

Fig. 12.2. Steps in data preparation for Web usage mining.

from the data. The process may involve pre-processing the original data, integrating data from multiple sources, and transforming the integrated data into a form suitable for input into specific data mining operations. Collectively, we refer to this process as *data preparation*.

Much of the research and practice in usage data preparation has been focused on pre-processing and integrating these data sources for different analysis. Usage data preparation presents a number of unique challenges which have led to a variety of algorithms and heuristic techniques for pre-processing tasks such as data fusion and cleaning, user and session identification, pageview identification [115]. The successful application of data mining techniques to Web usage data is highly dependent on the correct application of the pre-processing tasks. Furthermore, in the context of e-commerce data analysis, these techniques have been extended to allow for the discovery of important and insightful user and site metrics [286].

Figure 12.2 provides a summary of the primary tasks and elements in usage data pre-processing. We begin by providing a summary of data types commonly used in Web usage mining and then provide a brief discussion of some of the primary data preparation tasks.

12.1.1 Sources and Types of Data

The primary data sources used in Web usage mining are the **server log files**, which include **Web server access logs** and **application server logs**. Additional data sources that are also essential for both data preparation and pattern discovery include the site files and meta-data, operational databases, application templates, and domain knowledge. In some cases and for some users, additional data may be available due to client-side or proxy-level (Internet Service Provider) data collection, as well as from external clickstream or **demographic data** sources such as those provided by data aggregation services from ComScore (www.comscore.com), NetRatings (www.nielsen-netratings.com), and Acxiom (www.acxiom.com).

The data obtained through various sources can be categorized into four primary groups [115, 505].

Usage Data: The log data collected automatically by the Web and application servers represents the fine-grained navigational behavior of visitors. It is the primary source of data in Web usage mining. Each hit against the server, corresponding to an HTTP request, generates a single entry in the server access logs. Each log entry (depending on the log format) may contain fields identifying the time and date of the request, the IP address of the client, the resource requested, possible parameters used in invoking a Web application, status of the request, HTTP method used, the user agent (browser and operating system type and version), the referring Web resource, and, if available, client-side **cookies** which uniquely identify a repeat visitor. A typical example of a server access log is depicted in Fig. 12.3, in which six partial log entries are shown. The user IP addresses in the log entries have been changed to protect privacy.

For example, log entry 1 shows a user with IP address "1.2.3.4" accessing a resource: "/classes/cs589/papers.html" on the server (maya.cs. depaul.edu). The browser type and version, as well as operating system information on the client machine are captured in the agent field of the entry. Finally, the referrer field indicates that the user came to this location from an outside source: "http://dataminingresources.blogspot.com/". The next log entry shows that this user has navigated from "papers.html" (as re-

1	2006-02-01 00:08:43 1.2.3.4 - GET /classes/cs589/papers.html - 200 9221 HTTP/1.1 maya.cs.depaul.edu Mozilla/4.0+(compatible;+MSIE+6.0;+Windows+NT+5.1;+SV1;+.NET+CLR+2.0.50727) http://dataminingresources.blogspot.com/
2	2006-02-01 00:08:46 1.2.3.4 - GET /classes/cs589/papers/cms-tai.pdf - 200 4096 HTTP/1.1 maya.cs.depaul.edu Mozilla/4.0+(compatible;+MSIE+6.0;+Windows+NT+5.1;+SV1;+.NET+CLR+2.0.50727) http://maya.cs.depaul.edu/~classes/cs589/papers.html
3	2006-02-01 08:01:28 2.3.4.5 - GET /classes/ds575/papers/hyperlink.pdf - 200 318814 HTTP/1.1 maya.cs.depaul.edu Mozilla/4.0+(compatible;+MSIE+6.0;+Windows+NT+5.1) http://www.google.com/search?hl=en&lr=&q=hyperlink+analysis+for+the+web+survey
4	2006-02-02 19:34:45 3.4.5.6 - GET /classes/cs480/announce.html - 200 3794 HTTP/1.1 maya.cs.depaul.edu Mozilla/4.0+(compatible;+MSIE+6.0;+Windows+NT+5.1;+SV1) http://maya.cs.depaul.edu/~classes/cs480/
5	2006-02-02 19:34:45 3.4.5.6 - GET /classes/cs480/styles2.css - 200 1636 HTTP/1.1 maya.cs.depaul.edu Mozilla/4.0+(compatible;+MSIE+6.0;+Windows+NT+5.1;+SV1) http://maya.cs.depaul.edu/~classes/cs480/announce.html
6	2006-02-02 19:34:45 3.4.5.6 - GET /classes/cs480/header.gif - 200 6027 HTTP/1.1 maya.cs.depaul.edu Mozilla/4.0+(compatible;+MSIE+6.0;+Windows+NT+5.1;+SV1) http://maya.cs.depaul.edu/~classes/cs480/announce.html

Fig. 12.3. Portion of a typical server log

flected in the referrer field of entry 2) to access another resource: "/classes/cs589/papers/cms-tai.pdf". Log entry 3 shows a user who has arrived at the resource "/classes/ds575/papers/hyperlink.pdf" by doing a search on Google using keyword query: "hyperlink analysis for the web survey". Finally, entries 4–6 all correspond to a single click-through by a user who has accessed the resource "/classes/cs480/announce.html". Entries 5 and 6 are images embedded in the file "announce.html" and thus two additional HTTP request are registered as hits in the server log corresponding to these images.

Depending on the goals of the analysis, this data needs to be transformed and aggregated at different levels of abstraction. In Web usage mining, the most basic level of data abstraction is that of a **pageview**. A pageview is an aggregate representation of a collection of Web objects contributing to the display on a user's browser resulting from a single user action (such as a click-through). Conceptually, each pageview can be viewed as a collection of Web objects or resources representing a specific "user event," e.g., reading an article, viewing a product page, or adding a product to the shopping cart. At the user level, the most basic level of behavioral abstraction is that of a **session**. A session is a sequence of pageviews by a single user during a single visit. The notion of a session can be further abstracted by selecting a subset of pageviews in the session that are significant or relevant for the analysis tasks at hand.

Content Data: The content data in a site is the collection of objects and relationships that is conveyed to the user. For the most part, this data is comprised of combinations of textual materials and images. The data sources used to deliver or generate this data include static HTML/XML pages, multimedia files, dynamically generated page segments from scripts, and collections of records from the operational databases. The site content data also includes semantic or structural meta-data embedded within the site or individual pages, such as descriptive keywords, document attributes, semantic tags, or HTTP variables. The underlying domain ontology for the site is also considered part of the content data. Domain ontologies may include conceptual hierarchies over page contents, such as product categories, explicit representations of semantic content and relationships via an ontology language such as RDF, or a database schema over the data contained in the operational databases.

Structure Data: The structure data represents the designer's view of the content organization within the site. This organization is captured via the inter-page linkage structure among pages, as reflected through hyperlinks. The structure data also includes the intra-page structure of the content within a page. For example, both HTML and XML documents can be represented as tree structures over the space of tags in the page. The hyperlink structure for a site is normally captured by an automatically generated "site map." A site mapping tool must have the capability to capture and represent the inter- and intra-pageview relationships. For dynamically generated pages, the site mapping tools must either incorporate intrinsic knowledge of the underlying applications and scripts that generate HTML content, or must have the ability to generate content segments using a sampling of parameters passed to such applications or scripts.

User Data: The operational database(s) for the site may include additional user profile information. Such data may include demographic information about registered users, user ratings on various objects such as products or movies, past purchases or visit histories of users, as well as other explicit or implicit representations of users' interests. Some of this data can be captured anonymously as long as it is possible to distinguish among different users. For example, anonymous information contained in client-side cookies can be considered a part of the users' profile information, and used to identify repeat visitors to a site. Many personalization applications require the storage of prior user profile information.

12.1.2   Key Elements of Web Usage Data Pre-Processing

As noted in Fig. 12.2, the required high-level tasks in usage data pre-processing include the fusion and synchronization of data from multiple log files, data cleaning, pageview identification, user identification, session identification (or sessionization), episode identification, and the integration of clickstream data with other data sources such as content or semantic information, as well as user and product information from operational databases. We now examine some of the essential tasks in pre-processing.

Data Fusion and Cleaning

In large-scale Web sites, it is typical that the content served to users comes from multiple Web or application servers. In some cases, multiple servers with redundant content are used to reduce the load on any particular server. Data fusion refers to the merging of log files from several Web and application servers. This may require global synchronization across these servers. In the absence of shared embedded session ids, heuristic methods based on the "referrer" field in server logs along with various sessionization and user identification methods (see below) can be used to perform the merging. This step is essential in "inter-site" Web usage mining where the analysis of user behavior is performed over the log files of multiple related Web sites [513].

Data cleaning is usually site-specific, and involves tasks such as, removing extraneous references to embedded objects that may not be important for the purpose of analysis, including references to style files, graphics, or sound files. The cleaning process also may involve the removal of at least some of the data fields (e.g. number of bytes transferred or version of HTTP protocol used, etc.) that may not provide useful information in analysis or data mining tasks.

Data cleaning also entails the removal of references due to crawler navigations. It is not uncommon for a typical log file to contain a significant (sometimes as high as 50%) percentage of references resulting from search engine or other crawlers (or spiders). Well-known search engine crawlers can usually be identified and removed by maintaining a list of known crawlers. Other "well-behaved" crawlers which abide by standard robot exclusion protocols, begin their site crawl by first attempting to access to exclusion file "robots.txt" in the server root directory. Such crawlers, can therefore, be identified by locating all sessions that begin with an (attempted) access to this file. However, a significant portion of crawlers references are from those that either do not identify themselves explicitly (e.g., in the "agent" field) or implicitly; or from those crawlers that deliberately masquerade as legitimate users. In this case, identification and removal of crawler references may require the use of heuristic methods that distinguish typical behavior of Web crawlers from those of actual users. Some work has been done on using classification algorithms to build models of crawlers and Web robot navigations [510], but such approaches have so far been met with only limited success and more work in this area is required.

Pageview Identification

Identification of pageviews is heavily dependent on the intra-page structure of the site, as well as on the page contents and the underlying site domain knowledge. Recall that, conceptually, each pageview can be viewed as a collection of Web objects or resources representing a specific "user event," e.g., clicking on a link, viewing a product page, adding a product to the shopping cart. For a static single frame site, each HTML file may have a one-to-one correspondence with a pageview. However, for multi-framed sites, several files make up a given pageview. For dynamic sites, a pageview may represent a combination of static templates and content generated by application servers based on a set of parameters.

In addition, it may be desirable to consider pageviews at a higher level of aggregation, where each pageview represents a collection of pages or objects, for examples, pages related to the same concept category. In e-commerce Web sites, pageviews may correspond to various product-oriented events, such as product views, registration, shopping cart changes, purchases, etc. In this case, identification of pageviews may require *a priori* specification of an "event model" based on which various user actions can be categorized.

In order to provide a flexible framework for a variety of data mining activities a number of attributes must be recorded with each pageview. These attributes include the pageview id (normally a URL uniquely representing the pageview), static pageview type (e.g., information page, product view, category view, or index page), and other metadata, such as content attributes (e.g., keywords or product attributes).

User Identification

The analysis of Web usage does not require knowledge about a user's identity. However, it is necessary to distinguish among different users. Since a user may visit a site more than once, the server logs record multiple sessions for each user. We use the phrase **user activity record** to refer to the sequence of logged activities belonging to the same user.

In the absence of authentication mechanisms, the most widespread approach to distinguishing among unique visitors is the use of client-side cookies. Not all sites, however, employ cookies, and due to privacy concerns, client-side cookies are sometimes disabled by users. IP addresses, alone, are not generally sufficient for mapping log entries onto the set of unique visitors. This is mainly due to the proliferation of ISP proxy servers which assign rotating IP addresses to clients as they browse the Web. It is not uncommon to find many log entries corresponding to a limited number of proxy server IP addresses from large Internet Service Providers such as America Online. Therefore, two occurrences of the same IP address (separated by a sufficient amount of time), in fact, might correspond to two different users. Without user authentication or client-side cookies, it is still possible to accurately identify unique users through a combination of IP addresses and other information such as user agents and referrers [115].

Consider, for instance, the example of Fig. 12.4. On the left, the figure depicts a portion of a partly preprocessed log file (the time stamps are given as hours and minutes only). Using a combination of IP and Agent fields in the log file, we are able to partition the log into activity records for three separate users (depicted on the right).

Time	IP	URL	Ref	Agent
0:01	1.2.3.4	A	-	IE5;Win2k
0:09	1.2.3.4	B	A	IE5;Win2k
0:10	2.3.4.5	C	-	IE6;WinXP;SP1
0:12	2.3.4.5	B	C	IE6;WinXP;SP1
0:15	2.3.4.5	E	C	IE6;WinXP;SP1
0:19	1.2.3.4	C	A	IE5;Win2k
0:22	2.3.4.5	D	B	IE6;WinXP;SP1
0:22	1.2.3.4	A	-	IE6;WinXP;SP2
0:25	1.2.3.4	E	C	IE5;Win2k
0:25	1.2.3.4	C	A	IE6;WinXP;SP2
0:33	1.2.3.4	B	C	IE6;WinXP;SP2
0:58	1.2.3.4	D	B	IE6;WinXP;SP2
1:10	1.2.3.4	E	D	IE6;WinXP;SP2
1:15	1.2.3.4	A	-	IE5;Win2k
1:16	1.2.3.4	C	A	IE5;Win2k
1:17	1.2.3.4	F	C	IE6;WinXP;SP2
1:26	1.2.3.4	F	C	IE5;Win2k
1:30	1.2.3.4	B	A	IE5;Win2k
1:36	1.2.3.4	D	B	IE5;Win2k

User 1

Time	IP	URL	Ref
0:01	1.2.3.4	A	-
0:09	1.2.3.4	B	A
0:19	1.2.3.4	C	A
0:25	1.2.3.4	E	C
1:15	1.2.3.4	A	-
1:26	1.2.3.4	F	C
1:30	1.2.3.4	B	A
1:36	1.2.3.4	D	B

User 2

Time	IP	URL	Ref
0:10	2.3.4.5	C	-
0:12	2.3.4.5	B	C
0:15	2.3.4.5	E	C
0:22	2.3.4.5	D	B

User 3

Time	IP	URL	Ref
0:22	1.2.3.4	A	-
0:25	1.2.3.4	C	A
0:33	1.2.3.4	B	C
0:58	1.2.3.4	D	B
1:10	1.2.3.4	E	D
1:17	1.2.3.4	F	C

Fig. 12.4. Example of user identification using IP + Agent

Sessionization

Sessionization is the process of segmenting the user activity record of each user into sessions, each representing a single visit to the site. Web sites without the benefit of additional authentication information from users and without mechanisms such as embedded session ids must rely on heuristics methods for sessionization. The goal of a sessionization heuristic is to re-construct, from the clickstream data, the actual sequence of actions performed by one user during one visit to the site.

We denote the "conceptual" set of real sessions by R, representing the real activity of the user on the Web site. A sessionization heuristic h attempts to map R into a set of constructed sessions, denoted by C_h. For the ideal heuristic, h^*, we have $C_{h^*} = R$. In other words, the ideal heuristic can re-construct the exact sequence of user navigation during a session. Generally, sessionization heuristics fall into two basic categories: time-oriented or structure-oriented. Time-oriented heuristics apply either global or local time-out estimates to distinguish between consecutive sessions, while structure-oriented heuristics use either the static site structure or the implicit linkage structure captured in the referrer fields of the server logs. Various heuristics for sessionization have been identified and studied [115]. More recently, a formal framework for measuring the effectiveness of such heuristics has been proposed [498], and the impact of different heuristics on various Web usage mining tasks has been analyzed [46].

As an example, two variations of time-oriented heuristics and a basic navigation-oriented heuristic are given below. Each heuristic h scans the user activity logs to which the Web server log is partitioned after user identification, and outputs a set of constructed sessions:

- **h1:** Total session duration may not exceed a threshold θ. Given t_0, the timestamp for the first request in a constructed session S, the request with a timestamp t is assigned to S, iff $t - t_0 \leq \theta$.
- **h2:** Total time spent on a page may not exceed a threshold δ. Given t_1, the timestamp for request assigned to constructed session S, the next request with timestamp t_2 is assigned to S, iff $t_2 - t_1 \leq \delta$.
- **h-ref:** A request q is added to constructed session S if the referrer for q was previously invoked in S. Otherwise, q is used as the start of a new constructed session. Note that with this heuristic it is possible that a request q may potentially belong to more than one "open" constructed session, since q may have been accessed previously in multiple sessions. In this case, additional information can be used for disambiguation. For example, q could be added to the most recently opened session satisfying the above condition.

Time	IP	URL	Ref
0:01	1.2.3.4	A	-
0:09	1.2.3.4	B	A
0:19	1.2.3.4	C	A
0:25	1.2.3.4	E	C
1:15	1.2.3.4	A	-
1:26	1.2.3.4	F	C
1:30	1.2.3.4	B	A
1:36	1.2.3.4	D	B

User 1

Session 1

0:01	1.2.3.4	A	-
0:09	1.2.3.4	B	A
0:19	1.2.3.4	C	A
0:25	1.2.3.4	E	C

Session 2

1:15	1.2.3.4	A	-
1:26	1.2.3.4	F	C
1:30	1.2.3.4	B	A
1:36	1.2.3.4	D	B

Fig. 12.5. Example of sessionization with a time-oriented heuristic

Time	IP	URL	Ref
0:01	1.2.3.4	A	-
0:09	1.2.3.4	B	A
0:19	1.2.3.4	C	A
0:25	1.2.3.4	E	C
1:15	1.2.3.4	A	-
1:26	1.2.3.4	F	C
1:30	1.2.3.4	B	A
1:36	1.2.3.4	D	B

User 1

Session 1

0:01	1.2.3.4	A	-
0:09	1.2.3.4	B	A
0:19	1.2.3.4	C	A
0:25	1.2.3.4	E	C
1:26	1.2.3.4	F	C

Session 2

1:15	1.2.3.4	A	-
1:30	1.2.3.4	B	A
1:36	1.2.3.4	D	B

Fig. 12.6. Example of sessionization with the h-ref heuristic

An example of the application of sessionization heuristics is given in Fig. 12.5 and Fig. 12.6. In Fig. 12.5, the heuristic h_1, described above, with $\theta = 30$ minutes has been used to partition a user activity record (from the example of Fig. 12.4) into two separate sessions.

If we were to apply h_2 with a threshold of 10 minutes, the user record would be seen as three sessions, namely, A→B→C→E, A, and F→B→D. On the other hand, Fig. 12.6 depicts an example of using h-ref heuristic on the same user activity record. In this case, once the request for F (with time stamp 1:26) is reached, there are two open sessions, namely, A→B→C→E and A. But F is added to the first because its referrer, C, was invoked in session 1. The request for B (with time stamp 1:30) may potentially belong to both open sessions, since its referrer, A, is invoked both in session 1 and in session 2. In this case, it is added to the second session, since it is the most recently opened session.

Episode identification can be performed as a final step in pre-processing of the clickstream data in order to focus on the relevant subsets of page-views in each user session. An **episode** is a subset or subsequence of a session comprised of semantically or functionally related pageviews. This task may require the automatic or semi-automatic classification of pageviews into different functional types or into concept classes according to a domain ontology or concept hierarchy. In highly dynamic sites, it may also be necessary to map pageviews within each session into "service-based" classes according to a concept hierarchy over the space of possible parameters passed to script or database queries [47]. For example, the analysis may ignore the quantity and attributes of an item added to the shopping cart, and focus only on the action of adding the item to the cart.

Path Completion

Another potentially important pre-processing task which is usually performed after sessionization is **path completion**. Client- or proxy-side caching can often result in missing access references to those pages or objects that have been cached. For instance, if a user returns to a page A during the same session, the second access to A will likely result in viewing the previously downloaded version of A that was cached on the client-side, and therefore, no request is made to the server. This results in the second reference to A not being recorded on the server logs. **Missing references** due to caching can be heuristically inferred through path completion which relies on the knowledge of site structure and referrer information from server logs [115]. In the case of dynamically generated pages, form-based applications using the HTTP POST method result in all or part of the user input parameter not being appended to the URL accessed by the user (though, in the latter case, it is possible to recapture the user input through **packet sniffers** which listen to all incoming and outgoing TCP/IP network traffic on the server side).

A simple example of missing references is given in Fig. 12.7. On the left, a graph representing the linkage structure of the site is given. The dotted arrows represent the navigational path followed by a hypothetical user. After reaching page E, the user has backtracked (e.g., using the browser's "back" button) to page D and then B from which she has navigated to page C. The back references to D and B do not appear in the log file because these pages where cached on the client-side (thus no explicit server request was made for these pages). The log file shows that after a request for E, the next request by the user is for page C with a referrer B. In other words, there is a gap in the activity record corresponding to user's navigation from page E to page B. Given the site graph, it is possible to infer the two missing references (i.e., E → D and D → B) from the site structure and the referrer information given above. It should be noted that there are, in general, many (possibly infinite), candidate completions (for example, consider the sequence E → D, D → B, B → A, A → B). A simple heuristic

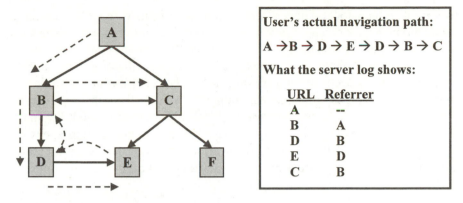

Fig. 12.7. Missing references due to caching.

that can be used for disambiguating among candidate paths is to select the one requiring the fewest number of "back" references.

Data Integration

The above pre-processing tasks ultimately result in a set of user sessions (or episodes), each corresponding to a delimited sequence of pageviews. However, in order to provide the most effective framework for pattern discovery, data from a variety of other sources must be integrated with the preprocessed clickstream data. This is particularly the case in e-commerce applications where the integration of both user data (e.g., demographics, ratings, and purchase histories) and product attributes and categories from operational databases is critical. Such data, used in conjunction with usage data, in the mining process can allow for the discovery of important business intelligence metrics such as **customer conversion ratios** and **lifetime values** [286].

In addition to user and product data, e-commerce data includes various product-oriented events such as shopping cart changes, order and shipping information, **impressions** (when the user visits a page containing an item of interest), **click-throughs** (when the user actually clicks on an item of interest in the current page), and other basic metrics primarily used for data analysis. The successful integration of these types of data requires the creation of a site-specific "event model" based on which subsets of a user's clickstream are aggregated and mapped to specific events such as the addition of a product to the shopping cart. Generally, the integrated e-commerce data is stored in the final transaction database. To enable full-featured Web analytics applications, this data is usually stored in a data warehouse called an **e-commerce data mart**. The e-commerce data mart is a multi-dimensional database integrating data from various sources, and at different levels of aggregation. It can provide pre-computed e-metrics along multiple dimensions, and is used as the primary data source for OLAP (Online Analytical Processing), for data visualization, and in data selection for a variety of data mining tasks [71, 279]. Some examples of such metrics include frequency or monetary value of purchases, average size of market baskets, the number of different items purchased, the number of different item categories purchased, the amount of time spent on pages or sections of the site, day of week and time of day when a certain activity occurred, response to recommendations and online specials, etc.

12.2   Data Modeling for Web Usage Mining

Usage data pre-processing results in a set of n pageviews, $P = \{p_1, p_2, \cdots, p_n\}$, and a set of m **user transactions**, $T = \{t_1, t_2, \cdots, t_m\}$, where each t_i in T is a subset of P. **Pageviews** are semantically meaningful entities to which mining tasks are applied (such as pages or products). Conceptually, we view each transaction t as an l-length sequence of ordered pairs:

$$t = \langle (p_1^t, w(p_1^t)), (p_2^t, w(p_2^t)), \cdots, (p_l^t, w(p_l^t)) \rangle,$$

where each $p_i^t = p_j$ for some j in $\{1, 2, \cdots, n\}$, and $w(p_i^t)$ is the weight associated with pageview p_i^t in transaction t, representing its significance. The weights can be determined in a number of ways, in part based on the type of analysis or the intended personalization framework. For example, in **collaborative filtering** applications which rely on the profiles of similar users to make recommendations to the current user, weights may be based on user ratings of items. In most Web usage mining tasks the weights are either binary, representing the existence or non-existence of a pageview in the transaction; or they can be a function of the duration of the pageview in the user's session. In the case of time durations, it should be noted that usually the time spent by a user on the last pageview in the session is not available. One commonly used option is to set the weight for the last pageview to be the mean time duration for the page taken across all sessions in which the pageview does not occur as the last one. In practice, it is common to use a normalized value of page duration instead of raw time duration in order to account for user variances. In some applications, the log of pageview duration is used as the weight to reduce the noise in the data.

For many data mining tasks, such as clustering and association rule mining, where the ordering of pageviews in a transaction is not relevant, we can represent each user transaction as a vector over the n-dimensional space

Pageviews

	A	B	C	D	E	F
user0	15	5	0	0	0	185
user1	0	0	32	4	0	0
user2	12	0	0	56	236	0
user3	9	47	0	0	0	134
user4	0	0	23	15	0	0
user5	17	0	0	157	69	0
user6	24	89	0	0	0	354
user7	0	0	78	27	0	0
user8	7	0	45	20	127	0
user9	0	38	57	0	0	15

Sessions / users

Fig. 12.8. An example of a user-pageview matrix (or transaction matrix)

of pageviews. Given the transaction t above, the transaction vector **t** (we use a bold face lower case letter to represent a vector) is given by:

$$\mathbf{t} = \left(w_{p_1}^t, w_{p_2}^t, \cdots, w_{p_n}^t \right),$$

where each $w_{p_i}^t = w(p_j^t)$, for some j in $\{1, 2, \cdots, n\}$, if p_j appears in the transaction t, and $w_{p_i}^t = 0$ otherwise. Thus, conceptually, the set of all user transactions can be viewed as an $m \times n$ **user-pageview matrix** (also called the **transaction matrix**), denoted by *UPM*.

An example of a hypothetical user-pageview matrix is depicted in Fig. 12.8. In this example, the weights for each pageview is the amount of time (e.g., in seconds) that a particular user spent on the pageview. In practice, these weights must be normalized to account for variances in viewing times by different users. It should also be noted that the weights may be composite or aggregate values in cases where the pageview represents a collection or sequence of pages and not a single page.

Given a set of transactions in the user-pageview matrix as described above, a variety of unsupervised learning techniques can be applied to obtain patterns. These techniques such as clustering of transactions (or sessions) can lead to the discovery of important user or visitor segments. Other techniques such as item (e.g., pageview) clustering and association or sequential pattern mining can find important relationships among items based on the navigational patterns of users in the site.

As noted earlier, it is also possible to integrate other sources of knowledge, such as semantic information from the content of Web pages with the Web usage mining process. Generally, the textual features from the content of Web pages represent the underlying semantics of the site. Each pageview p can be represented as a r-dimensional feature vector, where r is the total number of extracted features (words or concepts) from the site in a global dictionary. This vector, denoted by $\mathbf{p}$, can be given by:

$$\mathbf{p} = \left(fw^p(f_1), fw^p(f_2), ..., fw^p(f_r) \right)$$

where $fw^p(f_j)$ is the weight of the jth feature (i.e., f_j) in pageview p, for $1 \leq j \leq r$. For the whole collection of pageviews in the site, we then have an $n \times r$ **pageview-feature matrix** $PFM = \{\mathbf{p}_1, \mathbf{p}_2, ..., \mathbf{p}_n\}$. The integration process may, for example, involve the transformation of user transactions (in user-pageview matrix) into "content-enhanced" transactions containing the semantic features of the pageviews. The goal of such a transformation is to represent each user session (or more generally, each user profile) as a vector of semantic features (i.e., textual features or concept labels) rather than as a vector over pageviews. In this way, a user's session reflects not only the pages visited, but also the significance of various concepts or context features that are relevant to the user's interaction.

While, in practice, there are several ways to accomplish this transformation, the most direct approach involves mapping each pageview in a transaction to one or more content features. The range of this mapping can be the full feature space, or feature sets (composite features) which in turn may represent concepts and concept categories. Conceptually, the transformation can be viewed as the multiplication of the user-pageview matrix UPM, defined earlier, with the pageview-feature matrix PFM. The result is a new matrix, $TFM = \{\mathbf{t}_1, \mathbf{t}_2, ..., \mathbf{t}_m\}$, where each $\mathbf{t}_i$ is a r-dimensional vector over the feature space. Thus, a user transaction can be represented as a content feature vector, reflecting that user's interests in particular concepts or topics.

As an example of content-enhanced transactions, consider Fig. 12.9 which shows a hypothetical matrix of user sessions (user-pageview matrix) as well as a document index for the corresponding Web site conceptually represented as a **term-pageview matrix**. Note that the transpose of this term-pageview matrix is the pageview-feature matrix. The user-pageview matrix simply reflects the pages visited by users in various sessions. On the other hand, the term-pageview matrix represents the concepts that appear in each page. For simplicity we have assumed that all the weights are binary (however, note that in practice weights in the user transaction data are usually not binary and represent some measure of significance of the page in that transaction; and the weights in the term-pageview matrix are usually a function of term frequencies).

In this case, the corresponding **content-enhanced transaction matrix** (derived by multiplying the user-pageview matrix and the transpose of the term-pageview matrix) is depicted in Fig. 12.10. The resulting matrix

	A.html	B.html	C.html	D.html	E.html
user1	1	0	1	0	1
user2	1	1	0	0	1
user3	0	1	1	1	0
user4	1	0	1	1	1
user5	1	1	0	0	1
user6	1	0	1	1	1

	A.html	B.html	C.html	D.html	E.html
web	0	0	1	1	1
data	0	1	1	1	0
mining	0	1	1	1	0
business	1	1	0	0	0
intelligence	1	1	0	0	1
marketing	1	1	0	0	1
ecommerce	0	1	1	0	0
search	1	0	1	0	0
information	1	0	1	1	1
retrieval	1	0	1	1	1

Fig. 12.9. Examples of a user-pageview matrix (top) and a term-pageview matrix (bottom)

	web	data	mining	business	intelligence	marketing	ecommerce	search	information	retrieval
user1	2	1	1	1	2	2	1	2	3	3
user2	1	1	1	2	3	3	1	1	2	2
user3	2	3	3	1	1	1	2	1	2	2
user4	3	2	2	1	2	2	1	2	4	4
user5	1	1	1	2	3	3	1	1	2	2
user6	3	2	2	1	2	2	1	2	4	4

Fig. 12.10. The content-enhanced transaction matrix from matrices of Fig. 12.9

shows, for example, that users 4 and 6 are more interested in Web information retrieval, while user 3 is more interested in data mining.

Various data mining tasks can now be performed on the content-enhanced transaction data. For example, clustering the enhanced transaction matrix of Fig. 12.10 may reveal segments of users that have common interests in different concepts as indicated from their navigational behaviors.

If the content features include relational attributes associated with items on the Web site, then the discovered patterns may reveal user interests at the deeper semantic level reflected in the underlying properties of the items that are accessed by the users on the Web site. As an example, consider a site containing information about movies. The site may contain pages related to the movies themselves, as well as attributes describing the properties of each movie, such as actors, directors, and genres. The mining process may, for instance, generate an association rule such as: {"British", "Romance", "Comedy" ⇒ "Hugh Grant"}, suggesting that users who are interested in British romantic comedies may also like the actor Hugh Grant (with a certain degree of confidence). Therefore, the integration of semantic content with Web usage mining can potentially provide a better understanding of the underlying relationships among objects.

12.3   Discovery and Analysis of Web Usage Patterns

The types and levels of analysis, performed on the integrated usage data, depend on the ultimate goals of the analyst and the desired outcomes. In this section we describe some of the most common types of pattern discovery and analysis techniques employed in the Web usage mining domain and discuss some of their applications.

12.3.1   Session and Visitor Analysis

The statistical analysis of pre-processed session data constitutes the most common form of analysis. In this case, data is aggregated by predetermined units such as days, sessions, visitors, or domains. Standard statistical techniques can be used on this data to gain knowledge about visitor behavior. This is the approach taken by most commercial tools available for Web log analysis. Reports based on this type of analysis may include information about most frequently accessed pages, average view time of a page, average length of a path through a site, common entry and exit points, and other aggregate measures. Despite a lack of depth in this type of analysis, the resulting knowledge can be potentially useful for improving the system performance, and providing support for marketing decisions. Furthermore, commercial Web analytics tools are increasingly incorporating a variety of data mining algorithms resulting in more sophisticated site and customer metrics.

Another form of analysis on integrated usage data is Online Analytical Processing (OLAP). OLAP provides a more integrated framework for analysis with a higher degree of flexibility. The data source for OLAP analysis is usually a multidimensional data warehouse which integrates usage, content, and e-commerce data at different levels of aggregation for each dimension. OLAP tools allow changes in aggregation levels along each dimension during the analysis. Analysis dimensions in such a structure can be based on various fields available in the log files, and may include time duration, domain, requested resource, user agent, and referrers.

This allows the analysis to be performed on portions of the log related to a specific time interval, or at a higher level of abstraction with respect to the URL path structure. The integration of e-commerce data in the data warehouse can further enhance the ability of OLAP tools to derive important business intelligence metrics [71]. The output from OLAP queries can also be used as the input for a variety of data mining or data visualization tools.

12.3.2 Cluster Analysis and Visitor Segmentation

Clustering is a data mining technique that groups together a set of items having similar characteristics. In the usage domain, there are two kinds of interesting clusters that can be discovered: user clusters and page clusters.

Clustering of user records (sessions or transactions) is one of the most commonly used analysis tasks in Web usage mining and Web analytics. Clustering of users tends to establish groups of users exhibiting similar browsing patterns. Such knowledge is especially useful for inferring user demographics in order to perform market segmentation in e-commerce applications or provide **personalized Web content** to the users with similar interests. Further analysis of user groups based on their demographic attributes (e.g., age, gender, income level, etc.) may lead to the discovery of valuable business intelligence. **Usage-based clustering** has also been used to create Web-based "user communities" reflecting similar interests of groups of users [423], and to learn user models that can be used to provide dynamic recommendations in Web personalization applications [390].

Given the mapping of user transactions into a multi-dimensional space as vectors of pageviews (see Fig. 12.8), standard clustering algorithms, such as k-means, can partition this space into groups of transactions that are close to each other based on a measure of distance or similarity among the vectors (see Chap. 4). Transaction clusters obtained in this way can represent user or visitor segments based on their navigational behavior or other attributes that have been captured in the transaction file. However, transaction clusters by themselves are not an effective means of capturing the aggregated view of common user patterns. Each transaction cluster may potentially contain thousands of user transactions involving hundreds of pageview references. The ultimate goal in clustering user transactions is to provide the ability to analyze each segment for deriving business intelligence, or to use them for tasks such as personalization.

One straightforward approach in creating an aggregate view of each cluster is to compute the **centroid** (or the mean vector) of each cluster. The dimension value for each pageview in the mean vector is computed by finding the ratio of the sum of the pageview weights across transactions to the total number of transactions in the cluster. If pageview weights in the original transactions are binary, then the dimension value of a pageview p in a cluster centroid represents the percentage of transactions in the cluster in which p occurs. Thus, the centroid dimension value of p provides a measure of its significance in the cluster. Pageviews in the centroid can be sorted according to these weights and lower weight pageviews can be filtered out. The resulting set of pageview-weight pairs can be viewed as an "aggregate usage profile" representing the interests or behavior of a significant group of users.

More formally, given a transaction cluster cl, we can construct the aggregate profile pr_{cl} as a set of **pageview-weight** pairs by computing the centroid of cl:

$$pr_{cl} = \{(p, weight(p, pr_{cl})) \mid weight(p, pr_{cl}) \geq \mu\}, \tag{1}$$

where:
- the significance weight, $weight(p, pr_{cl})$, of the page p within the aggregate profile pr_{cl} is given by

$$weight(p, pr_{cl}) = \frac{1}{|cl|} \sum_{s \in cl} w(p, s); \tag{2}$$

- $|cl|$ is the number of transactions in cluster cl;
- $w(p, s)$ is the weight of page p in transaction vector s of cluster cl; and
- the threshold μ is used to focus only on those pages in the cluster that appear in a sufficient number of vectors in that cluster.

Each such profile, in turn, can be represented as a vector in the original n-dimensional space of pageviews. This aggregate representation can be used directly for predictive modeling and in applications such as recommender systems: given a new user, u, who has accessed a set of pages, P_u, so far, we can measure the similarity of P_u to the discovered profiles, and recommend to the user those pages in matching profiles which have not yet been accessed by the user.

As an example, consider the transaction data depicted in Fig. 12.11 (left). For simplicity we assume that feature (pageview) weights in each transaction vector are binary (in contrast to weights based on a function of pageview duration). We assume that the data has already been clustered using a standard clustering algorithm such as k-means, resulting in three clusters of user transactions. The table on the right of Fig. 12.11 shows the aggregate profile corresponding to cluster 1. As indicated by the pageview weights, pageviews B and F are the most significant pages characterizing the common interests of users in this segment. Pageview C, however, only appears in one transaction and might be removed given a filtering thresh-

		A	B	C	D	E	F
	user 1	0	0	1	1	0	0
Cluster 0	user 4	0	0	1	1	0	0
	user 7	0	0	1	1	0	0
	user 0	1	1	0	0	0	1
	user 3	1	1	0	0	0	1
Cluster 1	user 6	1	1	0	0	0	1
	user 9	0	1	1	0	0	1
	user 2	1	0	0	1	1	0
Cluster 2	user 5	1	0	0	1	1	0
	user 8	1	0	1	1	1	0

Aggregated Profile for Cluster 1	
Weight	Pageview
1.00	B
1.00	F
0.75	A
0.25	C

Fig. 12.11. Derivation of aggregate profiles from Web transaction clusters

old greater than 0.25. Such patterns are useful for characterizing user or customer segments. This example, for instance, indicates that the resulting user segment is clearly interested in items B and F and to a lesser degree in item A. Given a new user who shows interest in items A and B, this pattern may be used to infer that the user might belong to this segment and, therefore, we might recommend item F to that user.

Clustering of pages (or items) can be performed based on the usage data (i.e., starting from the user sessions or transaction data), or based on the content features associated with pages or items (keywords or product attributes). In the case of content-based clustering, the result may be collections of pages or products related to the same topic or category. In usage-based clustering, items that are commonly accessed or purchased together can be automatically organized into groups. It can also be used to provide permanent or dynamic HTML pages that suggest related hyperlinks to the users according to their past history of navigational or purchase activities.

A variety of stochastic methods have also been proposed recently for clustering of user transactions, and more generally for user modeling. For example, recent work in this area has shown that **mixture models** are able to capture more complex, dynamic user behavior. This is, in part, because the observation data (i.e., the user-item space) in some applications (such as large and very dynamic Web sites) may be too complex to be modeled by basic probability distributions such as a normal or a multinomial distribution. In particular, each user may exhibit different "types" of behavior corresponding to different tasks, and common behaviors may each be reflected in a different distribution within the data.

The general idea behind mixture models (such as a **mixture of Markov models**) is as follow. We assume that there exist k types of user behavior (or k user clusters) within the data, and each user session is assumed to be generated via a generative process which models the probability distributions of the observed variables and hidden variables. First, a user cluster is chosen with some probability. Then, the user session is generated from a Markov model with parameters specific to that user cluster. The probabilities of each user cluster is estimated, usually via the **EM** [127] algorithm, as well as the parameters of each mixture component. Mixture-based user models can provide a great deal of flexibility. For example, a mixture of first-order Markov models [76] not only can probabilistically cluster user sessions based on similarities in navigation behavior, but also characterize each type of user behavior using a first-order Markov model, thus capturing popular navigation paths or characteristics of each user cluster. A mixture of hidden Markov models was proposed in [580] for modeling clickstream of Web surfers. In addition to user-based clustering, this approach can also be used for automatic page classification. Incidentally, mixture models have been discussed in Sect. 3.7 in the context of naïve Bayesian classification. The EM algorithm is used in the same context in Sect. 5.1.

Mixture models tend to have their own shortcomings. From the data generation perspective, each individual observation (such as a user session) is generated from one and only one component model. The probability assignment to each component only measures the uncertainty about this assignment. This assumption limits this model's ability of capturing complex user behavior, and more seriously, may result in overfitting.

Probabilistic Latent Semantic Analysis (PLSA) provides a reasonable solution to the above problem [240]. In the context of Web user navigation, each observation (a user visiting a page) is assumed to be generated based on a set of unobserved (hidden) variables which "explain" the user-page observations. The data generation process is as follows: a user is selected with a certain probability, next conditioned on the user, a hidden variable is selected, and then the page to visit is selected conditioned on the chosen hidden variable. Since each user usually visits multiple pages, this data generation process ensures that each user is explicitly associated with multiple hidden variables, thus reducing the overfitting problems associated with the above mixture models. The PLSA model also uses the EM algorithm to estimate the parameters which probabilistically characterize the hidden variables underlying the co-occurrence observation data, and measure the relationship among hidden and observed variables.

This approach provides a great deal of flexibility since it provides a single framework for quantifying the relationships between users, between items, between users and items, and between users or items and hidden variables that "explain" the observed relationships [254]. Given a set of n user profiles (or transaction vectors), $UP = \{u_1, u_2, \ldots, u_n\}$, and a set of m items (e.g., pages or products), $I = \{i_1, i_2, \ldots, i_m\}$, the PLSA model associates a set of unobserved factor variables $Z = \{z_1, z_2, \ldots, z_q\}$ with observations in the data (q is specified by the user). Each observation corresponds to a weight $w_{u_k}(i_j)$ for an item i_j in the user profile for a user u_k. This weight may, for example, correspond to the significance of the page in the user transaction or the user rating associated with the item. For a given user u and a given item i, the following joint probability can be derived (see [254] for details of the derivation):

$$\Pr(u,i) = \sum_{k=1}^{q} \Pr(z_k)\Pr(u\,|\,z_k)\Pr(i\,|\,z_k).\tag{3}$$

In order to explain the observations in (UP, I), we need to estimate the parameters $\Pr(z_k)$, $\Pr(u|z_k)$, and $\Pr(i|z_k)$, while maximizing the following likelihood $L(UP, I)$ of the observation data:

$$L(UP,I) = \sum_{u\in UP}\sum_{i\in I} w_u(i)\log\Pr(u,i).\tag{4}$$

The Expectation–Maximization (EM) algorithm is used to perform maximum likelihood parameter estimation. Based on initial values of $\Pr(z_k)$, $\Pr(u|z_k)$, and $\Pr(i|z_k)$, the algorithm alternates between an expectation step and maximization step. In the expectation step, posterior probabilities are computed for latent variables based on current estimates, and in the maximization step the re-estimated parameters are obtained. Iterating the expectation and maximization steps monotonically increases the total likelihood of the observed data $L(UP, I)$, until a local optimal solution is reached. Details of this approach can be found in [254].

Again, one of the main advantages of PLSA model in Web usage mining is that using probabilistic inference with the above estimated parameters, we can derive relationships among users, among pages, and between users and pages. Thus this framework provides a flexible approach to model a variety of types of usage patterns.

12.3.3 Association and Correlation Analysis

Association rule discovery and statistical correlation analysis can find groups of items or pages that are commonly accessed or purchased together. This, in turn, enables Web sites to organize the site content more efficiently, or to provide effective cross-sale product recommendations.

Most common approaches to association discovery are based on the Apriori algorithm (see Sect. 2.2). This algorithm finds groups of items (pageviews appearing in the preprocessed log) occurring frequently together in many transactions (i.e., satisfying a user specified minimum support threshold). Such groups of items are referred to as **frequent itemsets**. Association rules which satisfy a minimum confidence threshold are then generated from the frequent itemsets.

Recall an association rule is an expression of the form $X \rightarrow Y$ [*sup, conf*], where X and Y are itemsets, *sup* is the support of the itemset $X \cup Y$ representing the probability that X and Y occur together in a transaction, and *conf* is the confidence of the rule, defined by $sup(X \cup Y)/sup(X)$, representing the conditional probability that Y occurs in a transaction given that X has occurred in that transaction. More details on association rule discovery can be found in Chap. 2.

The mining of association rules in Web transaction data has many advantages. For example, a high-confidence rule such as

special-offers/, /products/software/ $\rightarrow$ shopping-cart/

might provide some indication that a promotional campaign on software products is positively affecting online sales. Such rules can also be used to optimize the structure of the site. For example, if a site does not provide direct linkage between two pages A and B, the discovery of a rule, A $\rightarrow$ B, would indicates that providing a direct hyperlink from A to B might aid users in finding the intended information. Both association analysis (among products or pageviews) and statistical correlation analysis (generally among customers or visitors) have been used successfully in Web personalization and recommender systems [236, 389].

Indeed, one of the primary applications of association rule mining in Web usage or e-commerce data is in recommendation. For example, in the collaborative filtering context, Sarwar et al. [474] used association rules in the context of a *top-N* recommender system for e-commerce. The preferences of the target user are matched against the items in the antecedent X of each rule, and the items on the right hand side of the matching rules are sorted according to the confidence values. Then the top N ranked items from this list are recommended to the target user (see Sect. 3.5.3).

One problem for association rule recommendation systems is that a system cannot give any recommendations when the dataset is sparse (which is often the case in Web usage mining and collaborative filtering applications). The reason for this sparsity is that any given user visits (or rates) only a very small fraction of the available items, and thus it is often difficult to find a sufficient number of common items in multiple user profiles. Sarwar et al. [474] relied on some standard dimensionality reduction techniques to alleviate this problem. One deficiency of this and other dimensionality reduction approaches is that some of the useful or interesting items may be removed, and therefore, may not appear in the final patterns. Fu et al. [187] proposed two potential solutions to this problem. The first solution is to rank all the discovered rules based on the degree of intersection between the left-hand side of each rule and the user's active session and then to generate the top k recommendations. This approach will relax the constraint of having to obtain a complete match with the left-hand-side of the rules. The second solution is to utilize **collaborative filtering**: the system finds "close neighbors" who have similar interest to a target user and makes recommendations based on the close neighbors' histories.

Lin et al. [337] proposed a **collaborative recommendation** system using association rules. The proposed mining algorithm finds an appropriate number of rules for each target user by automatically selecting the minimum support. The system generates association rules among users (user associations), as well as among items (item associations). If a user minimum support is greater than a threshold, the system generates recommendations based on user associations, else it uses item associations.

Because it is difficult to find matching rule antecedent with a full user profile (e.g., a full user session or transaction), association-based recommendation algorithms typically use a sliding window w over the target user's active profile or session. The window represents the portion of user's history that will be used to predict future user actions (based on matches with the left-hand sides of the discovered rules). The size of this window is iteratively decreased until an exact match with the antecedent of a rule is found. A problem with the naive approach to this algorithm is that it requires repeated search through the rule-base. However, efficient trie-based data structure can be used to store the discovered itemsets and allow for efficient generation of recommendations without the need to generate all association rules from frequent itemsets [389]. Such data structures are commonly used for string or sequence searching applications. In the context of association rule mining, the frequent itemsets are stored in a directed acyclic graph. This **frequent itemset graph** is an extension of the lexicographic tree used in the tree projection mining algorithm of Agarwal, et al. [2]. The graph is organized into levels from 0 to k, where k is the maximum size among all frequent itemsets. Each node at depth d in the graph corresponds to an itemset, X, of size d and is linked to itemsets of size $d+1$ that contain X at level $d+1$. The single root node at level 0 corresponds to the empty itemset. To be able to search for different orderings of an itemset, all itemsets are sorted in lexicographic order before being inserted into the graph. If the graph is used to recommend items to a new target user, that user's active session is also sorted in the same manner before matching with itemsets.

As an example, suppose that in a hypothetical Web site with user transaction data depicted in the left table of Fig. 12.12. Using a minimum support (minsup) threshold of 4 (i.e., 80%), the Apriori algorithm discovers the frequent itemsets given in the right table. For each itemset, the support is also given. The corresponding frequent itemset graph is depicted in Fig. 12.13.

A recommendation engine based on this framework matches the current user session window with the previously discovered frequent itemsets to find candidate items (pages) for recommendation. Given an active session window w and a group of frequent itemsets, the algorithm considers all the frequent itemsets of size $|w| + 1$ containing the current session window by performing a depth-first search of the Frequent Itemset Graph to level $|w|$. The recommendation value of each candidate is based on the confidence of the corresponding association rule whose consequent is the singleton containing the page to be recommended. If a match is found, then the children of the matching node n containing w are used to generate candidate recommendations. In practice, the window w can be incrementally decreased until a match is found with and itemset. For example, given user active session window $<B, E>$, the recommendation generation algorithm, using the graph of Fig. 12.13, finds items A and C as candidate recommendations. The recommendation scores of item A and C are 1 and 4/5, corresponding to the confidences of the rules, $B, E \rightarrow A$ and $B, E \rightarrow C$, respectively.

A problem with using a single global minimum support threshold in association rule mining is that the discovered patterns will not include "rare" but important items which may not occur frequently in the transaction data. This is particularly important when dealing with Web usage data, it is often the case that references to deeper content or product-oriented pages oc-

Transactions	Size 1		Size 2		Size 3		Size 4	
	Itemset	Supp.	Itemset	Supp.	Itemset	Supp.	Itemset	Supp.
A, B, D, E	A	5	A,B	5	A,B,C	4	A,B,C,E	4
A, B, E, C, D	B	5	A,C	4	A,B,E	5		
A, B, E, C	C	4	A,E	5	A,C,E	4		
B, E, B, A, C	E	5	B,C	4	B,C,E	4		
D, A, B, E, C			B,E	5				
			C,E	4				

Fig. 12.12. Web transactions and resulting frequent itemsets (minsup = 4)

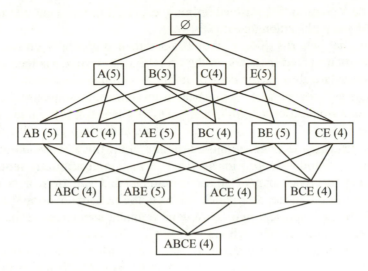

Fig. 12.13. A frequent itemset graph.

cur far less frequently than those of top level navigation-oriented pages. Yet, for effective Web personalization, it is important to capture patterns and generate recommendations that contain these items. A mining method based on **multiple minimum supports** is proposed in [344] that allows users to specify different support values for different items. In this method, the support of an itemset is defined as the minimum support of all items contained in the itemset. For more details on mining using multiple minimum supports, see Sect. 2.4. The specification of multiple minimum supports thus allows frequent itemsets to potentially contain rare items which are deemed important. It has been shown that the use of multiple support association rules in the context of Web personalization can dramatically increase the coverage (or recall) of recommendations while maintaining a reasonable precision [389].

12.3.4 Analysis of Sequential and Navigational Patterns

The technique of sequential pattern mining attempts to find inter-session patterns such that the presence of a set of items is followed by another item in a time-ordered set of sessions or episodes. By using this approach, Web marketers can predict future visit patterns which will be helpful in placing advertisements aimed at certain user groups. Other types of temporal analysis that can be performed on sequential patterns include trend analysis, change point detection, or similarity analysis. In the context of Web usage data, **sequential pattern mining** can be used to capture frequent navigational paths among user trails.

Sequential patterns (SPs) in Web usage data capture the Web page trails that are often visited by users, in the order that they were visited. Sequential patterns are those sequences of items that frequently occur in a sufficiently large proportion of (sequence) transactions. A sequence $\langle s_1 s_2 ... s_n \rangle$ occurs in a transaction $t = \langle p_1, p_2, ..., p_m \rangle$ (where $n \leq m$) if there exist n positive integers $1 \leq a_1 < a_2 < ... < a_n \leq m$, and $s_i = p_{a_i}$ for all i. We say that $\langle cs_1 cs_2 ... cs_n \rangle$ is a **contiguous sequence** in t if there exists an integer $0 \leq b \leq m - n$, and $cs_i = p_{b+i}$ for all $i = 1$ to n. In a **contiguous sequential pattern** (CSP), each pair of adjacent items, s_i and s_{i+1}, must appear consecutively in a transaction t which supports the pattern. A normal sequential pattern can represent non-contiguous frequent sequences in the underlying set of sequence transactions.

Given a sequence transaction set T, the support (denoted by $sup(S)$) of a sequential (respectively, contiguous sequential) pattern S in T is the fraction of transactions in T that contain S. The confidence of the rule $X \to Y$, where X and Y are (contiguous) sequential patterns, is defined as:

$$conf(X \to Y) = sup(X \circ Y) / sup(X),$$

where $\circ$ denotes the concatenation operator.

In the context of Web usage data, CSPs can be used to capture frequent navigational paths among user trails [497]. In contrast, items appearing in SPs, while preserving the underlying ordering, need not be adjacent, and thus they represent more general navigational patterns within the site. Note that sequences and sequential patterns or rules discussed here are special cases of those defined in Sect. 2.9.

The view of Web transactions as sequences of pageviews allows for a number of useful and well-studied models to be used in discovering or analyzing user navigation patterns. One such approach is to model the navigational activities in the Web site as a Markov model: each pageview (or a category) can be represented as a state and the transition probability between two states can represent the likelihood that a user will navigate from one state to the other. This representation allows for the computation of a number of useful user or site metrics. For example, one might compute the probability that a user will make a purchase, given that she has performed a search in an online catalog. **Markov models** have been proposed as the underlying modeling machinery for link prediction as well as for Web prefetching to minimize system latencies [132, 473]. The goal of such approaches is to predict the *next* user action based on a user's previous surfing behavior. They have also been used to discover high probability user navigational trails in a Web site [57]. More sophisticated statistical learning techniques, such as mixtures of Markov models, have also been used to cluster navigational sequences and perform exploratory analysis of users' navigational behavior in a site [76].

More formally, a Markov model is characterized by a set of states $\{s_1, s_2, \ldots, s_n\}$ and a transition probability matrix, $[Pr_{i,j}]_{n \times n}$, where $Pr_{i,j}$ represents the probability of a transition from state s_i to state s_j. Markov models are especially suited for predictive modeling based on contiguous sequences of events. Each state represents a contiguous subsequence of prior events. The order of the Markov model corresponds to the number of prior events used in predicting a future event. So, a kth-order Markov model predicts the probability of next event by looking the past k events. Given a set of all paths R, the probability of reaching a state s_j from a state s_i via a (non-cyclic) path $r \in R$ is the product of all the transition probabilities along the path and is given by $Pr(r) = \prod Pr_{m,m+1}$, where m ranges from i to j − 1. The probability of reaching s_j from s_i is the sum of these path probabilities over all paths: $Pr(j|i) = \sum_{r \in R} Pr(r)$.

As an example of how Web transactions can be modeled as a Markov model, consider the set of Web transaction given in Fig. 12.14 (left). The Web transactions involve pageviews A, B, C, D, and E. For each transaction the frequency of occurrences of that transaction in the data is given in the table's second column (thus there are a total of 50 transactions in the data set). The (absorbing) Markov model for this data is also given in Fig. 12.14 (right). The transitions from the "start" state represent the prior probabilities for transactions starting with pageviews A and B. The transitions into the "final" state represent the probabilities that the paths end with the specified originating pageviews. For example, the transition probability from the state A to B is 16/28 = 0.57 since out of the 28 occurrences of A in transactions, in 16 cases, B occurs immediately after A.

Higher-order Markov models generally provide a higher prediction accuracy. However, this is usually at the cost of lower coverage (or recall) and much higher model complexity due to the larger number of states. In order to remedy the coverage and space complexity problems, Pitkow and Pirolli [446] proposed all-kth-order Markov models (for coverage improvement) and a new state reduction technique, called **longest repeating subsequences** (LRS) (for reducing model size). The use of all-kth-order Markov models generally requires the generation of separate models for each of the k orders: if the model cannot make a prediction using the kth order, it will attempt to make a prediction by incrementally decreasing the model order. This scheme can easily lead to even higher space complexity since it requires the representation of all possible states for each k. Deshpande and Karypis [132] proposed selective Markov models, introducing several schemes in order to tackle the model complexity problems

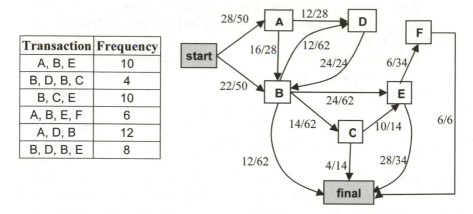

Transaction	Frequency
A, B, E	10
B, D, B, C	4
B, C, E	10
A, B, E, F	6
A, D, B	12
B, D, B, E	8

Fig. 12.14. An example of modeling navigational trails as a Markov

with all-kth-order Markov models. The proposed schemes involve pruning the model based on criteria such as support, confidence, and error rate. In particular, the support-pruned Markov models eliminate all states with low support determined by a minimum frequency threshold.

Another way of efficiently representing contiguous navigational trails is by inserting each trail into a *trie* structure. A good example of this approach is the notion of aggregate tree introduced as part of the WUM (Web Utilization Miner) system [497]. The aggregation service of WUM extracts the transactions from a collection of Web logs, transforms them into sequences, and merges those sequences with the same prefix into the aggregate tree (a trie structure). Each node in the tree represents a navigational subsequence from the root (an empty node) to a page and is annotated by the frequency of occurrences of that subsequence in the transaction data (and possibly other information such as markers to distinguish among repeat occurrences of the corresponding page in the subsequence). WUM uses a mining query language, called MINT, to discover generalized navigational patterns from this trie structure. MINT includes mechanisms to specify sophisticated constraints on pattern templates, such as wildcards with user-specified boundaries, as well as other statistical thresholds such as support and confidence. This approach and its extensions have proved useful in evaluating the navigational design of a Web site [496].

As an example, again consider the set of Web transactions given in the previous example. Figure 12.15 shows a simplified version of WUM's aggregate tree structure derived from these transactions. Each node in the tree represents a navigational subsequence from the root (an empty node) to a page and is annotated by the frequency of occurrences of that subsequence in the session data. The advantage of this approach is that the search for

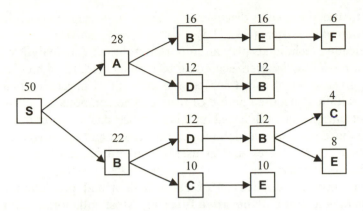

Fig. 12.15. An example of modeling navigational trails in an aggregate tree

navigational patterns can be performed very efficiently and the confidence and support for the navigational patterns can be readily obtained from the node annotations in the tree. For example, consider the contiguous navigational sequence <A, B, E, F>. The support for this sequence can be computed as the support of the last page in the sequence, F, divided by the support of the root node: 6/50 = 0.12, and the confidence of the sequence is the support of F divided by the support of its predecessor, E, or 6/16 = 0.375. If there are multiple branches in the tree containing the same navigational sequence, then the support for the sequence is the sum of the supports for all occurrences of the sequence in the tree and the confidence is updated accordingly. For example, the support of the sequence <D, B> is (12+12)/50 = 0.48, while the confidence is the aggregate support for B divided by the aggregate support for D, i.e., 24/24 = 1.0. The disadvantage of this approach is the possibly high space complexity, especially in a site with many dynamically generated pages.

12.3.5 Classification and Prediction based on Web User Transactions

Classification is the task of mapping a data item into one of several predefined classes. In the Web domain, one is interested in developing a profile of users belonging to a particular class or category. This requires extraction and selection of features that best describe the properties of given the class or category. Classification can be done by using supervised learning algorithms such as decision trees, naive Bayesian classifiers, k-nearest neighbor classifiers, and Support Vector Machines (Chap. 3). It is also possible to use previously discovered clusters and association rules for classification of new users (Sect. 3.5).

Classification techniques play an important role in Web analytics applications for modeling the users according to various predefined metrics. For example, given a set of user transactions, the sum of purchases made by each user within a specified period of time can be computed. A classification model can then be built based on this enriched data in order to classify users into those who have a high propensity to buy and those who do not, taking into account features such as users' demographic attributes, as well their navigational activities.

Another important application of classification and prediction in the Web domain is that of **collaborative filtering**. Most collaborative filtering applications in existing recommender systems use k-nearest neighbor classifiers to predict user ratings or purchase propensity by measuring the correlations between a current (target) user's profile (which may be a set of item ratings or a set of items visited or purchased) and past user profiles in order to find users in the database with similar characteristics or preferences [236]. Many of the Web usage mining approaches discussed earlier can also be used to automatically discover user models and then apply these models to provide personalized content to an active user [386, 445].

Basically, collaborative filtering based on the k-nearest neighbor (kNN) approach involves comparing the activity record for a target user with the historical records T of other users in order to find the top k users who have similar tastes or interests. The mapping of a visitor record to its neighborhood could be based on similarity in ratings of items, access to similar content or pages, or purchase of similar items. In most typical collaborative filtering applications, the user records or profiles are a set of ratings for a subset of items. The identified neighborhood is then used to recommend items not already accessed or purchased by the active user. Thus, there are two primary phases in collaborative filtering: the neighborhood formation phase and the recommendation phase. In the context of Web usage mining, kNN involves measuring the similarity or correlation between the target user's active session **u** (represented as a vector) and each past transaction vector **v** (where $\mathbf{v} \in T$). The top k most similar transactions to **u** are considered to be the neighborhood for the session **u**. More specifically, the similarity between the target user, **u**, and a neighbor, **v**, can be calculated by the **Pearson's correlation coefficient** defined below:

$$sim(\mathbf{u}, \mathbf{v}) = \frac{\sum_{i \in C}(r_{\mathbf{u},i} - \bar{r}_{\mathbf{u}})(r_{\mathbf{v},i} - \bar{r}_{\mathbf{v}})}{\sqrt{\sum_{i \in C}(r_{\mathbf{u},i} - \bar{r}_{\mathbf{u}})^2}\sqrt{\sum_{i \in C}(r_{\mathbf{v},i} - \bar{r}_{\mathbf{v}})^2}}, \tag{5}$$

where C is the set of items that are co-rated by **u** and **v** (i.e., items that have been rated by both of them), $r_{u,i}$ and $r_{v,i}$ are the ratings (or weights) of some item i for the target user $\mathbf{u}$ and a possible neighbor $\mathbf{v}$ respectively, and $\bar{r}_{\mathbf{u}}$ and $\bar{r}_{\mathbf{v}}$ are the average ratings (or weights) of $\mathbf{u}$ and $\mathbf{v}$ respectively. Once similarities are calculated, the most similar users are selected.

It is also common to filter out neighbors with a similarity of less than a specific threshold to prevent predictions being based on very distant or negative correlations. Once the most similar user transactions are identified, the following formula can be used to compute the rating prediction of an item i for target user $\mathbf{u}$.

$$p(\mathbf{u},i) = \bar{r}_{\mathbf{u}} + \frac{\sum_{v \in V} sim(\mathbf{u},\mathbf{v}) \times (r_{v,i} - \bar{r}_{\mathbf{v}})}{\sum_{v \in V} |sim(\mathbf{u},\mathbf{v})|}, \qquad (6)$$

where V is the set of k similar users, $r_{v,i}$ are the ratings of those users on item i, and $sim(\mathbf{u}, \mathbf{v})$ is the Pearson correlation described above. The formula in essence computes the degree of preference of all the neighbors weighted by their similarity and then adds this to the target user's average rating, the idea being that different users may have different "baselines" around which their ratings are distributed.

The problem with the user-based formulation of the collaborative filtering problem is the lack of scalability: it requires the real-time comparison of the target user to all user records in order to generate predictions. A variation of this approach that remedies this problem is called **item-based collaborative filtering** [475]. Item-based collaborative filtering works by comparing items based on their pattern of ratings across users. Again, a nearest-neighbor approach can be used. The kNN algorithm attempts to find k similar items that are co-rated by different users similarly. The similarity measure typically used is the **adjusted cosine similarity** given below:

$$sim(i,j) = \frac{\sum_{u \in U} (r_{u,i} - \bar{r}_{u})(r_{u,j} - \bar{r}_{u})}{\sqrt{\sum_{u \in U} (r_{u,i} - \bar{r}_{u})^2} \sqrt{\sum_{u \in U} (r_{u,j} - \bar{r}_{u})^2}}, \qquad (7)$$

where U is the set of all users, i and j are items, $r_{u,i}$ represents the rating of user $\mathbf{u} \in U$ on item i, and $\bar{r}_{u}$ is the average of the user $\mathbf{u}$'s ratings as before.

Note that in this case, we are computing the pair-wise similarities among items (not users) based on the ratings for these items across all users. After computing the similarity between items we select a set of k most similar items to the target item (i.e., the item for which we are interested in predicting a rating value) and generate a predicted value of user $\mathbf{u}$'s rating by using the following formula

$$p(\mathbf{u}, i) = \frac{\sum_{j \in J} r_{\mathbf{u},j} \times sim(i, j)}{\sum_{j \in J} sim(i, j)}, \tag{8}$$

where J is the set of k similar items, $r_{\mathbf{u},j}$ is the rating of user $\mathbf{u}$ on item j, and $sim(i, j)$ is the similarity between items i and j as defined above. It is also common to ignore items with negative similarity to the target item. The idea here is to use the user's own ratings for the similar items to extrapolate the prediction for the target item.

12.4 Discussion and Outlook

Web usage mining has emerged as the essential tool for realizing more personalized, user-friendly and business-optimal Web services. Advances in data pre-processing, modeling, and mining techniques, applied to the Web data, have already resulted in many successful applications in adaptive information systems, personalization services, Web analytics tools, and content management systems. As the complexity of Web applications and user's interaction with these applications increases, the need for intelligent analysis of the Web usage data will also continue to grow.

Usage patterns discovered through Web usage mining are effective in capturing item-to-item and user-to-user relationships and similarities at the level of user sessions. However, without the benefit of deeper domain knowledge, such patterns provide little insight into the underlying reasons for which such items or users are grouped together. Furthermore, the inherent and increasing heterogeneity of the Web has required Web-based applications to more effectively integrate a variety of types of data across multiple channels and from different sources.

Thus, a focus on techniques and architectures for more effective integration and mining of content, usage, and structure data from different sources is likely to lead to the next generation of more useful and more intelligent applications, and more sophisticated tools for Web usage mining that can derive intelligence from user transactions on the Web.

Bibliographic Notes

Web usage mining as a complete process, integrating various stages of data mining cycle, including data preparation, pattern discovery, and interpretation, was initially introduced by Cooley et al. [114]. This initial work was later extended by Srivastava, et al. [505].

Proper data preparation is an essential activity that enables the discovery of actionable knowledge from usage data. A complete discussion of the stages and tasks in data preparation for Web usage mining can be found in the paper by Cooley et al. [115]. One of these tasks is that of sessionization of the user activity records in the log data which is generally accomplished through the use of various heuristics. Several heuristics were defined by Cooley et al. [115]. Berendt et al. [46] and Spiliopoulou et al. [498] introduced several additional sessionization heuristics, and developed a comprehensive framework for the evaluation of these heuristics in the context of various applications in Web usage mining. Much of the discussion of Sect. 12.1 is based on these sources.

One of the primary applications of Web usage mining has been in Web personalization and predictive user modeling. Initially, Web usage mining as a tool for personalization was introduced by Mobasher et al. [388]. More recent surveys of issues and techniques related to personalization based on Web usage mining can be found in the papers by Pierrakos et al. [445], Mobasher [386], and Anand and Mobasher [20].

Another important application of Web usage mining is the analysis of customer and visitor behavior in e-commerce and for Web marketing. Web usage mining applied to e-commerce data enables the discovery of important business intelligence metrics such as customer conversion ratios and lifetime values. A good discussion of lessons and challenges in e-business data analysis can be found in the paper by Kohavi et al. [286].

References

1. S. Adali, T. Liu, and M. Magdon-Ismail. Optimal Link Bombs are Uncoordinated. In *Proc. of 1st Intl. Workshop on Adversarial Information Retrieval on the Web*, 2005.
2. R. Agarwal, C. Aggarwal, and V. Prasad. A Tree Projection Algorithm for Generation of Frequent Itemsets. In *Proc. of the High Performance Data Mining Workshop*, 1999.
3. C. Aggarwal, F. Al-Garawi, and P. Yu. Intelligent Crawling on the World Wide Web with Arbitrary Predicates. In *Proc. of 10th Intl. World Wide Web Conf. (WWW'01)*, pp. 96–105, 2001.
4. C. Aggarwal, C. Propiuc, J. L. Wolf, P. S. Yu, and J. S. Park. A Framework for Finding Projected Clusters in High Dimensional Spaces. In *Proc. of Intl. Conf. on Management of Data (SIGMOD'99)*, pp. 407–418, 1999.
5. C. Aggarwal, and P. S. Yu. Finding Generalized Projected Clusters in High Dimensional Spaces. In *Proc. of the ACM SIGMOD Intl. Conf. on Management of Data (SIGMOD'00)*, pp. 70–81, 2000.
6. E. Agichtein. Confidence Estimation Methods for Partially Supervised Relation Extraction. In *Proc. of SIAM Intl. Conf. on Data Mining (SDM06)*, 2006
7. R. Agrawal, J. R. Bayardo, and R. Srikant. Athena: Mining-Based Interactive Management of Text Databases. In *Proc. of Extending Database Technology (EDBT'00)*, pp. 365–379, 2000
8. R. Agrawal, J. Gehrke, D. Gunopulos and P. Raghavan. Automatic Subspace Clustering for High Dimensional Data for Data Mining Applications. In *Proc. of the ACM SIGMOD Intl. Conf. on Management of Data (SIGMOD'98)*, pp. 94–105, 1998.
9. R. Agrawal, T. Imielinski, and A. N. Swami. Mining Association Rules between Sets of Items in Large Databases. In *Proc. of the ACM SIGMOD Intl. Conf. on Management of Data (SIGMOD'93)*, pp. 207–216, 1993.
10. R. Agrawal, S. Rajagopalan, R. Srikant, and Y. Xu. Mining Newsgroups Using Networks Arising from Social Behavior. In *Proc. of the 12th Intl. World Wide Web Conf. (WWW'03)*, pp. 529–535, 2003.
11. R. Agrawal and R. Srikant. Fast Algorithms for Mining Association Rules. In *Proc. of the 20th Intl. Conf. on Very Large Data Bases (VLDB'94)*, pp. 487–499, 1994.
12. R. Agrawal and R. Srikant. Mining Sequential Patterns. In *Proc. of the Intl. Conf. on Data Engineering (ICDE'95)*, pp. 3–14, 1995.
13. R. Agrawal and R. Srikant. On Integrating Catalogs. In *Proc. of the Tenth Intl. World Wide Web Conf. (WWW'01)*, pp. 603–612, 2001.
14. L. von Ahn, M. Blum, N. Hopper, and J. Langford. CAPTCHA: Using Hard AI Problems for Security. In *Proc. of Eurocrypt*, pp. 294–311, 2003.
15. R. Akavipat, L.-S. Wu, and F. Menczer. Small World Peer Networks in Distributed Web Search. In *Alt. Track Papers and Posters Proc. 13th Intl. World Wide Web Conf.*, pp. 396–397, 2004.
16. B. Aleman-Meza, M. Nagarajan, C. Ramakrishnan, L. Ding, P. Kolari, A. Sheth, I. Arpinar, A. Joshi, and T. Finin. Semantic Analytics on Social Networks: Experiences

in Addressing the Problem of Conflict of Interest Detection. In *Proc. of the 15th Intl. Conf. on World Wide Web* (*WWW'06*), 2006.

17. C. Alpert, A. Kahng and S. Yao. Spectral Partitioning: The More Eigenvectors, the Better. *Discrete Applied Mathematics*, 90, pp. 3–5, 1999.

18. B. Amento, L. Terveen, and W. Hill. Does Authority Mean Quality? Predicting Expert Quality Ratings of Web Documents. In *Proc. of the 23rd ACM SIGIR Conf. on Research and Development in Information Retrieval*, pp. 296–303, 2000.

19. E. Amitay, D. Carmel, A. Darlow, R. Lempel and A. Soffer. The Connectivity Sonar: Detecting Site Functionality by Structural Patterns. In *Proc. of the 14th ACM Conf. on Hypertext and Hypermedia*, pp. 38–47, 2003.

20. S. S. Anand, and B. Mobasher. Intelligent Techniques for Web Personalization. In *Intelligent Techniques for Web Personalization*, B. Mobasher and S. S. Anand (eds.), *Lecture Notes in AI* (*LNAI 3169*), Springer, 2005.

21. R. Andersen, and K. J. Lang. Communities from Seed Sets. In *Proc. of the 15th Intl. Conf. on World Wide Web* (*WWW'06*), 2006.

22. A. Andreevskaia and S. Bergler. Mining WordNet for Fuzzy Sentiment: Sentiment Tag Extraction from WordNet Glosses. In *Proc. of 11th Conf. of the European Chapter of the Association for Computational Linguistics* (*EACL'06*), pp. 209–216, 2006.

23. M. Antonie, and O. Zaiane. Text Document Categorization by Term Association. In *Proc. of IEEE Intl. Conf. on Data Mining*, 2002.

24. P. Arabie and L. Hubert. An Overview of Combinatorial Data Analysis. In P. Arabie, L. Hubert and G. D. Soets (eds.). *Clustering and Classification*, pp. 5–63, 1996.

25. A. Arasu, J. Cho, H. Garcia-Molina, A. Paepcke, and S. Raghavan. Searching the Web. *ACM Trans. Internet Technology*, 1 (1), pp. 2–43, 2001.

26. A. Arasu and H. Garcia-Molina. Extracting Structured Data from Web Pages. In *Proc. of the ACM SIGMOD Intl. Conf. on Management of Data* (*SIGMOD'03*), pp. 337–348, 2003.

27. L. Arllota, V. Crescenzi, G. Mecca, and P. Merialdo. Automatic Annotation of Data Extraction from Large Web Sites. In *Intl. Workshop on Web and Databases*, 2003.

28. J. A. Aslam and M. Montague. Models for Metasearch. In *Proc. of the 24th Annual Intl. ACM SIGIR Conf. on Research and Development in Information Retrieval* (*SIGIR'01*), pp. 276–284, 2001.

29. J. Ayres, J. Gehrke, T. Yiu, and J. Flannick. Sequential Pattern Mining Using Bitmaps. In *Proc. of the Eighth ACM SIGKDD Intl. Conf. on Knowledge Discovery and Data Mining* (*KDD'02*), pp. 429–435, 2002.

30. R. Baeza-Yates, C. Castillo and V. Lopez. PageRank Increase under Different Collusion Topologies. In *Proc. of the 1st Intl. Workshop on Adversarial Information Retrieval on the Web*, 2005.

31. R. Baeza-Yates and B. Ribeiro-Neto. *Modern Information Retrieval*. Addison–Wesley, 1999.

32. R. Baeza-Yates, F. Saint-Jean, and C. Castillo. Web Dynamics, Age and Page Quality. In *Proc. of String Processing and Information Retrieval*. pp. 117–130, 2002.

33. P. Baldi, P. Frasconi, and P. Smyth. *Modeling the Internet and the Web: Probabilistic Methods and Algorithms*. Wiley, 2003.

34. L. Barabasi and R. Albert. Emergence of Scaling in Random Walk. *Science*, 286, pp. 509–512, 1999.

35. D. Barbará, C. Domeniconi, and N. Kang. Classifying Documents without Labels. In *Proc. of the SIAM Intl. Conf. on Data Mining* (*SDM'04*), 2004.

36. D. Barbará, Y. Li and J. Couto. COOLCAT: an Entropy-Based Algorithm for Categorical Clustering. In *Proc. of the 11th Intl. Conf. on Information and knowledge management* (*CIKM'02*), pp. 582–589, 2002.

37. Z. Bar-Yossef, and M. Gurevich. Random Sampling from a Search Engine's Index. In *Proc. of the 15th Intl. Conf. on World Wide Web (WWW'06)*, 2006.

38. Z. Bar-Yossef, S. Rajagopalan. Template Detection via Data Mining and its Applications. In *Proc. of the 11th Intl World Wide Web Conf. (WWW'02)*, pp. 580–591, 2002.

39. S. Basu, A. Banerjee, and R. J. Mooney: Semi-supervised Clustering by Seeding. In *Proc. of the Nineteenth Intl. Conf. on Machine Learning (ICML'02)*, pp. 27–34, 2002.

40. C. Batini, M. Lenzerini, and S. Navathe. A Comparative Analysis of Methodologies for Database Schema Integration. *ACM Computing Survey* 18(4), pp. 323–364, 1986.

41. R. Baumgartner, S. Flesca, and G. Gottlob. Visual Web Information Extraction with Lixto. In *Proc. of the Intl. Conf. on Very Large Data Bases (VLDB'01)*, pp. 119–128, 2001.

42. R. J. Bayardo. Efficiently Mining Long Patterns from Databases. In *Proc. of the ACM SIGMOD Intl. Conf. on Management of Data (SIGMOD'98)*, pp. 85–93, 1998.

43. R. J. Bayardo, and R. Agrawal. Mining the Most Interesting Rules. In *Proc. of the ACM SIGKDD Intl. Conf. on Knowledge Discovery and Data Mining (KDD'99)*, pp. 145-154, 1999.

44. P. Beineke, T. Hastie, C. Manning, and S. Vaithyanathan. An Exploration of Sentiment Summarization. In: *Proc. of the AAAI Spring Symposium on Exploring Attitude and Affect in Text: Theories and Applications*, 2003.

45. T. C. Bell, A. Moffat, C. G. Nevill-Manning, I. H. Witten, and J. Zobel. Data Compression in Full-Text Retrieval Systems. *Journal of the American Society for Information Science*, 44(9), pp. 508–531, 1993.

46. B. Berendt, B. Mobasher, M. Nakagawa, and M. Spiliopoulou. The Impact of Site Structure and User Environment on Session Reconstruction in Web Usage Analysis. In *Proc. of the KDD'02 WebKDD Workshop*, 2002.

47. B. Berendt and M. Spiliopoulou. Analyzing Navigation Behavior in Web Sites Integrating Multiple Information Systems. *VLDB Journal*, 9(1), pp. 56–75, 2000.

48. M. Berry, S. T. Dumais, and G. W. O'Brien. Using Linear Algebra for Intelligent Information Retrieval. *SIAM: Review*, 37(4), pp. 573–595, 1995.

49. M. J. A. Berry and G. Linoff. *Data Mining Techniques: For Marketing, Sales, and Customer Relationship Management*. Wiley Computer Publishing, 2004.

50. J. C. Bezdek. Cluster Validity with Fuzzy Sets. *J. of Cybernetics*, 3, pp. 58–72. 1974.

51. K. Bharat, and A. Z. Broder: A Technique for Measuring the Relative Size and Overlap of Public Web Search Engines. *Computer Networks*, 30(1–7), pp. 379–388, 1998.

52. K. Bharat and M. Henzinger. Improved Algorithms for Topic Distillation in Hyperlinked Environments. In *Proc. of the 21st ACM SIGIR Conf. on Research and Development in Information Retrieval*, pp. 104–111, 1998

53. A. Bilke, and F. Naumann. Schema Matching Using Duplicates. In *Proc. of Intl. Conf. on Data Engineering (ICDE'05)*, pp. 69–80, 2005.

54. A. Blum, and S. Chawla. Learning from Labeled and Unlabeled Data Using Graph Mincuts. In *Proc. of Intl. Conf. on Machine Learning (ICML'01)*, pp.19–26, 2001.

55. A. Blum and T. Mitchell. Combining Labeled and Unlabeled Data with Co-Training. In *Proc. of Computational Learning Theory*, pp. 92–100, 1998.

56. J. C. De Borda. Mémoire sur les élections au scrutin. *Mémoires de l'Académie Royale des Sciences année*, 1781.

57. J. Borges and M. Levene. Data Mining of User Navigation Patterns. In *Web Usage Analysis and User Profiling*, LNAI 1836, Springer, pp. 92–111, 1999.

58. C. L. Borgman, (ed.) *Scholarly Communication and Bibliometrics*. Sage Publications, Inc., 1990.

59. B. E. Boser, I. Guyon, and V. N. Vapnik. A Training Algorithm for Optimal Margin Classifiers. In *Proc. of the Fifth Annual Workshop on Computational Learning Theory*, 5: pp. 144–152, 1992.

60. P. De Bra and R. Post. Information Retrieval in the World Wide Web: Making Client-Based Searching Feasible, In *Proc. of the 1st Intl. World Wide Web Conf.*, pp. 183–192, 1994.

61. P. S. Bradley, U. Fayyad and C. Reina. Scaling Clustering Algorithms to Large Databases. In *Proc. of Intl. Conf. on Knowledge Discovery and Data Mining (KDD'98)*, pp. 9–15, 1998.

62. L. Breiman, J. H. Friedman, R. Olshen, and C. J. Stone. *Classification and Regression Trees*. Chapman and Hall, New York, 1984.

63. L. Breiman. Bagging Predictors. *Machine Learning*, 24(2), 123–140, 1996.

64. L. Breiman. Prediction Games and Arcing Classifiers. *Technical Report 504*, Statistics Department, University of California at Berkeley, 1997.

65. L. Breiman: Random Forests. *Machine Learning*, 45(1), pp. 5–32, 2001.

66. B. E. Brewington, and G. Cybenko. How Dynamic is the Web? In *Proc. of the 9th Intl. World Wide Web Conf.*, 2000.

67. BrightPlanet.com. *The Deep Web: Surfacing Hidden Value*. Accessible at http://brightplanet.com, July 2000.

68. S. Brin and L. Page. The Anatomy of a Large-Scale Hypertextual Web Search Sngine. *Computer Networks*, 30(1–7), pp. 107–117, 1998.

69. A. Broder, S. Kumar, F. Maghoul, P. Raghavan, S. Rajagopalan, R. Stata, A. Tomkins, and J. Wiener. Graph Structure in the Web. *Computer Networks*, 33(1–6), pp. 309–320, 2000.

70. C. A. Brunk and M. J. Pazzani. An Investigation of Noise-Tolerant Relational Concept Learning Algorithms. In *Proc. of the 8th Intl. Workshop on Machine Learning*, pp. 389–393, 1991.

71. A. Buchner and M. D. Mulvenna. Discovering Internet Marketing Intelligence through Online Analytical Web Usage Mining. In *Proc. of the ACM SIGMOD Intl. Conf. on Management of Data (SIGMOD'99)*, pp. 54–61, 1999.

72. G. Buehrer, S. Parthasarathy, and A. Ghoting. Out-of-core frequent pattern mining on a commodity PC. In *Proc. of the ACM SIGKDD Intl. Conf. on Knowledge Discovery and Data Mining (KDD'06)*, pp. 86 – 95, 2006.

73. D. Burdick, M. Calimlim, and J. Gehrke. MAFIA: A Maximal Frequent Itemset Algorithm for Transactional Databases. In *Proc. of the Intl. Conf. on Data Engineering (ICDE'01)*, pp. 443, 2001.

74. C. J. C. Burges. A Tutorial on Support Vector Machines for Pattern Recognition. *Data Mining and Knowledge Discovery*, 2(2), pp. 955–974, 1998.

75. D. Buttler, L. Liu, and C. Pu. A Fully Automated Object Extraction System for the World Wide Web. In *Proc. of Intl. Conf. on Distributed Computing Systems (ICDCS'01)*, pp. 361–370, 2001.

76. I. V. Cadez, D. Heckerman, C. Meek, P. Smyth, and S. White. Model-Based Clustering and Visualization of Navigation Patterns on a Web Site. *Data Mining Knowledge Discovery* 7(4), pp. 399–424, 2003.

77. D. Cai, S. Yu, J.-R. Wen and W.-Y Ma. Extracting Content Structure for Web Pages based on Visual Representation. In *Proc. of the APWeb'03 Conf.*, Number 2642 in Lecture notes in Computer Science (LNCS), pp. 406–417, 2003

78. D. Cai, S. Yu, J.-R. Wen and W.-Y. Ma. Block-Based Web Search. In *Proc. of the ACM SIGIR Research and Development in Information Retrieval (SIGIR'04)*, pp. 456–463, 2004.

79. Y. Cao, J. Xu, T.-Y. Liu, H. Li, Y. Huang, and H.-W. Hon. Adapting Ranking SVM to Document Retrieval. In *Proc. of the 29th Annual Intl. ACM SIGIR Conf. on Research and Development in Information Retrieval (SIGIR'06)*, pp. 186-193, 2006.

80. G. Carenini, R. Ng, and E. Zwart. Extracting Knowledge from Evaluative Text. In *Proc. of the Third Intl. Conf. on Knowledge Capture (K-CAP'05)*, pp. 11–18. 2005.

81. G. Carenini, R. Ng, and A. Pauls. Interactive Multimedia Summaries of Evaluative Text. In *Proc. of the 10th Intl. Conf. on Intelligent User Interfaces (IUI'06)*, pp. 305–312, 2006.

82. H. Carrillo and D. Lipman. The Multiple Sequence Alignment Problem in Biology. *SIAM Journal Applied Mathematics*, 48(5), pp. 1073–1082, 1988.

83. V. Castelli and T. M. Cover. Classification Rules in the Unknown Mixture Parameter Case: Relative Value of Labeled and Unlabeled Samples. In *Proc. of 1994 IEEE Intern. Symp. Inform. Theory*, 111, 1994.

84. S. Chakrabarti. Integrating the Document Object Model with Hyperlinks for Enhanced Topic Distillation and Information Extraction. In *Proc. of the 13th Intl. World Wide Web Conf. (WWW'01)*, pp. 211–220, 2001.

85. S. Chakrabarti. *Mining the Web. Discovering Knowledge from Hypertext Data*. Morgan Kaufmann, 2003.

86. S. Chakrabarti, B. Dom, S. Kumar, P. Raghavan, S. Rajagopalan, A. Tomkins, D. Gibson, and J. Kleinberg. Mining the Web's Link Structure. *IEEE Computer*, 32(8), pp. 60–67, 1999.

87. S. Chakrabarti, M. van den Berg, and B. Dom. Focused Crawling: A New Approach to Topic-Specific Web Resource Discovery. *Computer Networks*, 31(11–16), pp. 1623–1640, 1999.

88. S. Chakrabarti, B. Dom, P. Raghavan, S. Rajagopalan, D. Gibson, and J. Kleinberg. Automatic Resource Compilation by Analyzing Hyperlink Structure and Associated Text. *Computer Networks* 30(1–7), pp. 65–74, 1998.

89. S. Chakrabarti, K. Puniyani, and S. Das. Optimizing Scoring Functions and Indexes for Proximity Search in Type-annotated Corpora. In *Proc. of the 15th Intl. Conf. on World Wide Web (WWW'06)*, 2006.

90. C-H. Chang, M. Kayed, M. R. Girgis, and K. Shaalan. A Survey of Web Information Extraction Systems. *IEEE Transactions on Knowledge and Data Engineering*, 18(10), pp. 1411–1428, 2006.

91. C-H. Chang and S. Lui. IEPAD. Information Extraction Based on Pattern Discovery. In *Proc. of the Tenth Intl. World Wide Web Conf. (WWW'01)*, pp. 681–688, 2001.

92. K. C.-C. Chang, B. He, C. Li, M. Patel, and Z. Zhang. Structured Databases on the Web: Observations and Implications. *SIGMOD Record*, 33(3), pp. 61-70, 2004.

93. O. Chapelle, B. Schölkopf and A. Zien. (eds.) *Semi-Supervised Learning*. MIT Press, 2006.

94. P. Cheeseman, and J. Stutz. Bayesian Classification (AutoClass): Theory and Results. In *Advances in Knowledge Discovery and Data Mining*, 1996.

95. W. Chen. New Algorithm for Ordered Tree-to-Tree Correction Problem. *J. Algorithms*, 40(2), pp. 135–158, 2001.

96. H. Chen, Y.-M. Chung, M. Ramsey, and C. Yang. A Smart Itsy Bitsy Spider for the Web. *Journal of the American Society for Information Science* 49 (7), 604–618, 1998.

97. S. F. Chen, and J. Goodman. *An Empirical Study of Smoothing Techniques for Language Modeling*. Tech. Rep. TR-10-98, Harvard University, 1998.

98. Y.-Y. Chen, Q. Gan and T. Suel. Local Methods for Estimating PageRank Values. In *Proc. of the Intl. Conf. on Information and Knowledge Management (CIKM'04)*, pp. 381–389, 2004.

99. C. H. Cheng, A. W. Fu and Y Zhang. Entropy-Based Subspace Clustering for Mining Numerical Data. In *Proc. of Knowledge Discovery and Data Mining (KDD'99)*, pp. 84–93, 1999.

100. Y. Cheng and G. Church. Biclustering of Expression Data. In *Proc. ISMB*, pp. 93–103. AAAI Press, 2000.

101. J. Cho and H. Garcia-Molina. The Evolution of the Web and Implications for an Incremental Crawler. In *Proc. of the 26th Intl. Conf. on Very Large Data Bases (VLDB'00)*, 2000.

102. J. Cho, H. Garcia-Molina, and L. Page. Efficient Crawling through URL Ordering. *Computer Networks* 30 (1–7), pp. 161–172, 1998.

103. J. Cho, and S. Roy. Impact of Web Search Engines on Page Popularity. In *Proc. of the 13th Intl. World Wide Web Conf. (WWW'04)*, pp. 20–29, 2004.

104. P. Clark, and T. Niblett. The CN2 Induction Algorithm. *Machine Learning*, 3, pp. 261–283, 1989.

105. C. Clifton, E. Housman, and A. Rosenthal. Experience with a Combined Approach to Attribute-Matching Across Heterogenenous Databases. In: *Proc, IFIP 2.6 Working Conf. Database Semantics*, 1997.

106. W. W. Cohen. Fast Effective Rule Induction. In *Proc. of 12th Intl. Conf. on Machine Learning (ICML'95)*, pp. 115–123, 1995.

107. W. W. Cohen. Integration of Heterogeneous Databases without Common Domains Using Queries Based on Textual Similarity. In *Proc ACM SIGMOD Conf. on Management of Data (SIGMOD'08)*, pp. 201–212, 1998.

108. W. W. Cohen, M. Hurst, and L. S. Jensen. A Flexible Learning System for Wrapping Tables and Lists in Html Documents. In *Proc. of the 11th Intl. World Wide Web Conf. (WWW'02)*, pp. 232–241, 2002.

109. M. Collins and Y. Singer. Unsupervised Models for Named Entity Classification. In *Proc. of Intl. Conf. on Empirical Methods in Natural Language Processing (EMNLP'99)*, pp. 100–110, 1999.

110. M. de Condorcet. *Essai sur l'application de l'analyse a la probabilitie des decisions rendues a la pluralite des voix*, Paris, 1785.

111. G. Cong, W. S. Lee, H. Wu, and B. Liu. Semi-Supervised Text Classification Using Partitioned EM. In *Proc. of Database Systems for Advanced Applications (DASFAA 2004)*: 482–493, 2004.

112. G. Cong, K.-L. Tan, A. K. H. Tung, and X. Xu. Mining Top-k Covering Rule Groups for Gene Expression Data. In *Proc. of ACM SIGMOD Intl. Conf. on Management of Data (SIGMOD'05)*, pp. 670–681, 2005.

113. G. Cong, A. K. H. Tung, X. Xu, F. Pan, and J. Yang. Farmer: Finding Interesting Rule Groups in Microarray Datasets. In *Proc. of the ACM SIGMOD Intl. Conf. on Management of Data (SIGMOD'04)*, pp. 143–154. 2004.

114. R. Cooley, B. Mobasher, and J. Srivastava. Web Mining: Information and Pattern Discovery on the World Wide Web. In *Proc. of the 9th IEEE Intl. Conf. on Tools With Artificial Intelligence (ICTAI'97)*, pp. 558–567, 1997.

115. R. Cooley, B. Mobasher, and J. Srivastava. Data Preparation for Mining World Wide Web Browsing Patterns. *Knowledge and Information Systems*, 1(1), pp. 5–32, 1999.

116. T. Corman, C. Leiserson, R. Rivest, and C. Stein. *Introduction to Algorithms*. MIT Press, 2001.

117. V. Crescenzi, G. Mecca, and P. Merialdo. Roadrunner: Towards Automatic Data Extraction from Large Web Sites. In *Proc. of Very Large Data Bases (VLDB'01)*, pp. 109–118, 2001.

118. N. Cristianini and J. Shawe-Taylor. *An Introduction to Support Vector Machines*. Cambridge University Press, 2000.

119. W. B. Croft. Combining Approaches to Information Retrieval. In W. B. Croft (eds.), *Advances in Information Retrieval: Recent Research from the Center for Intelligent Information Retrieval*, Kluwer Academic Publishers, 2000.

120. S. Das and M. Chen. Yahoo! for Amazon: Extracting Market Sentiment from Stock Message Boards. *APFA'01*, 2001.

121. S. Dasgupta, M.L. Littman, and D. McAllester. PAC Generalization Bounds for Co-Training. *Advances in Neural Information Processing Systems (NIPS)*, 2001.

122. K. Dave, S. Lawrence, and D. Pennock. Mining the Peanut Gallery: Opinion Extraction and Semantic Classification of Product Reviews. In *Proc. of the 12th Intl. World Wide Web Conference (WWW'03)*, pp. 519–528, 2003.

123. B. Davison. Topical Locality in the Web. In *Proc. 23rd Intl. ACM SIGIR Conf. on Research and Development in Information Retrieval*, pp. 272–279, 2000.

124. S. Debnath, P. Mitra, and C. L. Giles. Automatic Extraction of Informative Blocks from Webpages. In *Proc. of the 2005 ACM Symposium on Applied Computing*, pp. 1722–1726, 2005.

125. S. Deerwester, S. T. Dumais, G. W. Furnas, T. K. Landauer, and R. Harshman. Indexing by Latent Semantic Analysis. *Journal of the American Society for Information Science*, 41, pp. 391–407, 1990.

126. M. Degeratu, G. Pant, and F. Menczer. Latency-Dependent Fitness in Evolutionary Multithreaded Web Agents. In *Proc. of GECCO Workshop on Evolutionary Computation and Multi-Agent Systems*, pp. 313–316, 2001.

127. A. P. Dempster, N. M. Laird, and D. B. Rubin Maximum Likelihood from Incomplete Data via the EM algorithm. *Journal of the Royal Statistical Society*, Series B, 39(1), pp. 1–38, 1977.

128. L. Deng, X. Chai, Q. Tan, W. Ng, and D. L. Lee. Spying Out Real User Preferences for Metasearch Engine Personalization. In *Proc. of the Workshop on WebKDD*, 2004.

129. F. Denis. PAC Learning from Positive Statistical Queries. In *Proc. of Intl. Conf. on Algorithmic Learning Theory (ALT'98)*, pp. 112–126, 1998.

130. F. Denis, R. Gilleron and M. Tommasi. Text Classification from Positive and Unlabeled Examples. *IPMU*, 2002.

131. M. Deshpande and G. Karypis. Using conjunction of attribute values for classification. In *Proc. of the ACM Intl. Conf. on Information and Knowledge Management (CIKM'02)*, pp. 356-364, 2002.

132. M. Deshpande and G. Karypis. Selective Markov Models for Predicting Web Page Accesses. *ACM Trans. on Internet Technology*, 4(2), 163–184, 2004.

133. R. Dhamankar, Y. Lee, A. Doan, A. Halevy, and P. Domingos. IMap: Discovering Complex Semantic Matches between Database Schemas. In *Proc. of ACM SIGMOD Intl. Conf. on Management of Data (SIGMOD'04)*, pp. 383–394, 2004.

134. S. Dhillon. Co-Clustering Documents and Words Using Bipartite Spectral Graph Partitioning. In *Proc. of the 7th ACM SIGKDD Intl. Conf. on Knowledge Discovery and Data Mining (KDD'01)*, pp. 269–274, 2001.

135. S. Dhillon, S. Mallela, and D. S. Modha. Information-Theoretic Co-Clustering. In *Proc. of The Ninth ACM SIGKDD Intl. Conf. on Knowledge Discovery and Data Mining (KDD'03)*, pp. 89–98, 2003.

136. L. Dice. Measures of the Amount of Ecologic Association between Species. *Ecology*, 26(3), 1945.

137. J. Diesner and K. M. Carley. Exploration of Communication Networks from the Enron Email Corpus. In *Workshop on Link Analysis, Counterterrorism and Security at SDM'05*, 2005.

138. T. G. Dietterich and G. Bakiri. Solving Multiclass Learning Problems via Error-Correcting Output Codes. *J. of Artificial Intelligence Research*, 2. pp. 263–286, 1995.

139. M. Diligenti, F. Coetzee, S. Lawrence, C. L. Giles, and M. Gori. Focused Crawling Using Context Graphs. In *Proc. of Intl. Conf. on Very Large Databases* (*VLDB'00*), pp. 527–534, 2000.

140. M. Diligenti, M. Gori, and M. Maggini, Web Page Scoring Systems for Horizontal and Vertical Search. In *Proc. of the 11th Intl. World Wide Web Conference* (*WWW'02*), pp. 508–516. 2002.

141. C. Ding, and X. He. Linearized Cluster Assignment via Spectral Ordering. In *Proc. of Int'l Conf. Machine Learning* (*ICML'04*), 2004.

142. C. Ding, X. He, H. Zha, and H. Simon. PageRank, HITS and a Unified Framework for Link Analysis. In *Proc. of SIAM Data Mining Conf.*, 2003.

143. C. Djeraba, O. R. Zaiane, and S. Simoff. (eds.). *Mining Multimedia and Complex Data.* Springer, 2003.

144. H. Do and E. Rahm. Coma: A System for Flexible Combination of Schema Matching Approaches. In *Proc. of the Intl. Conf. on Very Large Data Bases* (*VLDB'02*), pp. 610–621, 2002.

145. A. Doan, P. Domingos, and A. Y. Halevy. Reconciling Schemas of Disparate Data Sources: a Machine-Learning Approach. In *Proc. of the 2001 ACM SIGMOD Intl. Conf. on Management of Data* (*SIGMOD'01*), pp. 509–520, 2001.

146. A. Doan and A. Halevy, Semantic Integration Research in the Database Community: A Brief Survey. *AI Magzine*, 26(1), pp. 83–94, 2005.

147. A. Doan, J. Madhavan, P. Domingos, and A. Y. Halevy: Learning to Map between Ontologies on the Semantic Web. In P*roc. of the 11th Intl. World Wide Web Conference* (*WWW'02*), pp. 662–673, 2002.

148. P. Domingos, and M. J. Pazzani. On the Optimality of the Simple Bayesian Classifier under Zero-One Loss. *Machine Learning* 29(2–3), pp. 103–130, 1997.

149. G. Dong, X. Zhang, L. Wong, and J. Li. CAEP: Classification by Aggregating Emerging Patterns. In *Proc. of Intl. Conf. on Discovery Science*, pp. 30–42, 1999.

150. C. Doran, D. Egedi, B. A. Hockey, B. Srinivas, and M. Zaidel. XTAG System-A Wide Coverage Grammar for English. In *Proc. of Intl. Conf. on Computational Linguistics* (*COLING'94*), pp. 922–928, 1994.

151. J. Dougherty, R. Kohavi, and M. Sahami. Supervised and Unsupervised Discretization of Continuous Features. In *Proc. of the 12th Intl. Conf. on Machine Learning* (*ICML'95*), 1995.

152. A. Douglis. B. Feldmann, Krishnamurthy, and J. C. Mogul. Rate of Change and Other Metrics: a Live Study of the World Wide Web. In *Proc. of USENIX Symp. on Internet Technologies and Systems*, pp. 147–158, 1997.

153. E. Dragut, W. Wu, P. Sistla, C. Yu, and W. Meng. Merging Source Query Interfaces on Web Databases. In *Proc. of the International Conference on Data Engineering* (*ICDE'06*), 2006.

154. E. Dragut, C. Yu, and W. Meng. Meaningful Labeling of Integrated Query Interfaces. In *Proceedings of Very Large Data Bases* (*VLDB'06*), 2006.

155. R. O. Duda, P. E. Hart, and D. G. Stork. *Pattern Classification.* John Wiley & Sons Inc., 2nd edition, 2001

156. M. Dunham. *Data Mining: Introductory and Advanced Topics.* Prentice Hall, 2002.

157. J. C. Dunn. A Fuzzy Relative of the ISODATA Process and Its Use in Detecting Compact Well-Separated Clusters. *Journal of Cybernetics*, 3, pp. 32–57, 1974.

158. J. Eckmann, and E. Moses. Curvature of Co-Links Uncovers Hidden Thematic Layers in the World Wide Web. In *Proc. of the National Academy of Sciences*, pp. 5825–5829, 2002.

159. K. Eguchi, and V. Lavrenko. Sentiment Retrieval Using Generative Models. In *Proc. of the Conf. on Empirical Methods in Natural Language Processing* (*EMNLP'06*),

pp. 345–354, 2006.

160. D. Eichmann. Ethical Web Agents. *Computer Networks*, 28(1–2), pp. 127–136, 1995.

161. P. Elias. Universal Codeword Sets and Representations of the Integers. *IEEE Transactions on Information Theory*, IT–21(2), pp.194–203, 1975.

162. D. W. Embley, D. Jackman, and L. Xu. Multifaceted Exploitation of Metadata for Attribute Match Discovery in Information Integration. In: *Proc Intl. Workshop on Information Integration on the Web*, pp. 110–117, 2001.

163. D. W. Embley, Y. Jiang, and Y. K. Ng. Record-Boundary Discovery in Web Documents. In *Proc. of ACM SIGMOD Intl. Conf. on Management of Data* (*SIGMOD'99*), pp. 467–478, 1999.

164. M. Ester, H.-P. Kriegal, J. Sander and X. Xu. A Density-Based Algorithm for Discovering Clusters in Large Spatial Databases with Noise. In *Proc. of Knowledge Discovery and Data Mining* (*KDD'96*), pp. 226–231, 1996.

165. A. Esuli, and F. Sebastiani. Determining Term Subjectivity and Term Orientation for Opinion Mining. In: *Proc. of Conf. of the European Chapter of the Association for Computational Linguistics* (*EACL'06*), 2006.

166. O. Etzioni, M. Cafarella, D. Downey, S. Kok, A. Popescu, T. Shaked, S. Soderland, D. Weld, and A. Yates. Web-Scale Information Extraction in Knowitall. In *Proc. of the 13th Intl. World Wide Web Conference* (*WWW'04*), pp. 100–110, 2004.

167. B. S. Everitt. *Cluster Analysis*. Heinemann, London, 1974.

168. R. Fagin, R. Kumar, K. S. McCurley, J. Novak, D. Sivakumar, J. A. Tomlin, and D. P. Williamson. Searching the Workplace Web. In *Proc. of the 12th Intl. World Wide Web Conference* (*WWW'03*), pp. 366–375, 2003.

169. W. Fan. On the Optimality of Probability Estimation by Random Decision Trees. In *Proc. of National Conf. on Artificial Intelligence* (*AAAI'04*), pp. 336-341, 2004.

170. W. Fan, S. J. Stolfo, J. Zhang, P. K. Chan: AdaCost: Misclassification Cost-Sensitive Boosting. In *Proc. of the 16th Intl. Conf. on Machine Learning* (*ICML'99*), pp. 97-105, 1999.

171. A. Farahat, T. LoFaro, J. C. Miller, G Rae and L Ward. Authority Rankings from HITS, PageRank, and SALSA: Existence, Uniqueness, and Effect of Initialization. *SIAM Journal on Scientific Computing*, pp. 1181–1201, 2005.

172. U. M. Fayyad, and K. B. Irani. Multi-Interval Discretization of Continuous-Valued Attributes for Classification Learning. In *Proc. of the Intl. Joint Conf. on Artificial Intelligence*, pp. 102–1027, 1993.

173. U. M. Fayyad, G. Piatetsky-Shapiro, and P. Smyth. From Data Mining to Knowledge Discovery: An Overview. In *Advances in Knowledge Discovery and Data Mining*, AAAI/MIT Press, pp. 1–34, 1996.

174. U. M. Fayyad, G. Piatetsky-Shapiro, P. Smyth, and R. Uthurasamy (eds.). *Advances in Knowledge Discovery and Data Mining*. AAAI/MIT Press, 1996.

175. C. Fellbaum. *WordNet: An On-Line Lexical Database*. MIT Press, 1998.

176. D. Fetterly, M. Manasse and M. Najork. Detecting Phrase-Level Duplication on the World Wide Web. In *Proc. of the 28th Annual Intl. ACM SIGIR Conf. on Research and Development in Information Retrieval*, pp. 170–177, 2005.

177. D. Fetterly, M. Manasse, M. Najork, and J. Wiener. A Large-Scale Study of the Evolution of Web Pages. In *Proc. of the 12th Intl. World Wide Web Conf.* (*WWW'03*), pp. 669–678, 2003.

178. D. Fisher. Knowledge Acquisition via Incremental Conceptual Clustering. *Machine Learning*, 2, pp. 139–172, 1987.

179. G. W. Flake, S. Lawrence, and C. L. Giles, Efficient Identification of Web Communities. In *Proc. of the sixth ACM SIGKDD Intl. Conf. on Knowledge discovery and data mining*, pp.150–160, 2000.

180. G. W. Flake, S. Lawrence, C. L. Giles, and F. Coetzee. Self-Organization of the Web and Identification of Communities. *IEEE Computer* 35(3), pp. 66–71, 2002.

181. L. R. Ford Jr. and D. R. Fulkerson. Maximal Flow through a Network. *Canadian Journal Mathematics*, 8: pp. 399–404, 1956.

182. S. Fortunato, A. Flammini, F. Menczer, and A. Vespignani. Topical Interests and the Mitigation of Search Engine Bias. In *Proc. Natl. Acad. Sci.* USA, 103(34), pp. 12684-12689, 2006.

183. S. Fortunato, A. Flammini, and F. Menczer. Scale-Free Network Growth by Ranking. *Phys. Rev. Lett.* 96(21), 2006.

184. E. Fox and J. Shaw. Combination of Multiple Searches. In *Proc. of the Second Text REtrieval Conf.*, pp. 243-252, 1993.

185. D. Freitag and A. McCallum. Information Extraction with HMM Structures Learned by Stochastic Optimization. In *Proc. of National Conf. on Artificial Intelligence (AAAI'00)*, 2000.

186. Y. Freund, and R. E. Schapire. Experiments with a New Boosting Algorithm. In *Proc. of the 13th Intl. Conf. on Machine Learning (ICML'96)*, pp. 148–156, 1996.

187. X. Fu, J. Budzik, and K. J. Hammond. Mining Navigation History for Recommendation. In *Proc. of the Intl. Conf. on Intelligent User Interfaces*, pp. 106–112, 2000.

188. G. P. C. Fung, J. X. Yu, H. Lu, and P. S. Yu. Text Classification without Labeled Negative Documents. In *Proc. 21st Intl. Conf. on Data Engineering (ICDE'05)*, pp. 594–605, 2005.

189. J. Furnkranz and G. Widmer. Incremental Reduced Error Pruning. In *Proc. of the Eleventh Intl. Conf. Machine Learning*, pp. 70–77, 1994.

190. A. Gal, G. Modica, H. Jamil, and A. Eyal. Automatic Ontology Matching Using Application Semantics. *AI Magazine*, 26(1), pp. 21–32, Spring 2005.

191. M. Gamon. Sentiment Classification on Customer Feedback Data: Noisy Data, Large Feature Vectors, and the Role of Linguistic Analysis. In *Proc. of the 20th Intl. Conf. on Computational Linguistics*, pp. 841–847, 2004.

192. M. Gamon, A. Aue, S. Corston-Oliver, and E. K. Ringger. Pulse: Mining Customer Opinions from Free Text. In *Proc. of Intelligent Data Analysis (IDA' 05)*, pp. 121–132, 2005.

193. V. Ganti, J. Gehrke, and R. Ramakrishnan. CACTUS – Clustering Categorical Data Using Summaries. In *Proc. of Knowledge Discovery and Data Mining (KDD'99)*, pp. 73–83, 1999.

194. F. Gasparetti and A. Micarelli. Swarm Intelligence: Agents for Adaptive Web Search. In *Proc. of the 16th European Conf. on Artificial Intelligence (ECAI'04)*, 2004.

195. J. Gehrke, R. Ramakrishnan, and V. Ganti. RainForest - A Framework for Fast Decision Tree Construction of Large Datasets. In *Proc. of Intl. Conf. on Very Large Data Bases (VLDB'98)*, pp. 416-427, 1998.

196. R. Ghani, Combining Labeled and Unlabeled Data for MultiClass Text Categorization. In *Proc. of the Intl. Conf. on Machine Learning (ICML'02)*, pp. 187–194, 2002.

197. D. Gibson, J. M. Kleinberg, and P. Raghavan, Clustering Categorical Data: An Approach Based on Dynamical Systems. In *Proc. of the Intl. Conf. on Very Large Data Bases (VLDB'98)*, pp.311–322, 1998.

198. D. Gibson, J. Kleinberg, and P. Raghavan. Inferring Web Communities from Link Topology. In *Proc. of the 9th ACM Conf. on Hypertext and Hypermedia*, 1998.

199. D. Gibson, K. Punera, and A. Tomkins. The Volume and Evolution of Web Page Templates. In *Special Interest Tracks and Posters of the 14th Intl. Conf. on World Wide Web (WWW '05)*. pp. 830–839, 2005.

200. M. Girvan and M. Newman. Community Structure in Social and Biological Network. In *Proc. of the National Academy of Sciences*, 2001.

201. S. Goldman and Y. Zhou. Enhanced Supervised Learning with Unlabeled Data. In *Proc. of the Intl. Conf. on Machine Learning (ICML'00)*, pp. 327–334, 2000.

202. S. W. Golomb. Run-Length Encodings. *IEEE Transactions on Information Theory*, 12(3), pp. 399–401, July 1966.

203. G. H. Golub, and C. F. Van Loan. *Matrix Computations*. The Johns Hopkins University Press, 1983.

204. I. J. Good. *The Estimation of Probabilities: An Essay on Modern Bayesian Methods*. MIT Press, 1965.

205. J. C. Gower. A General Coefficient of Similarity and Some of its Properties. *Biometrics*, 27, pp. 857–871, 1971.

206. G. Grefenstette, Y. Qu, D. A. Evans, and J. G. Shanahan. Validating the Coverage of Lexical Resources for Affect Analysis and Automatically Classifying New Words along Semantic Axes. In *Proc. of AAAI Spring Symposium on Exploring Attitude and Affect in Text: Theories and Applications*, 2004.

207. G. Grimmett and D. Stirzaker. *Probability and Random Process*. Oxford University Press, 1989.

208. R. L. Grossman, C. Kamath, P. Kegelmeyer, V. Kumar, and R. Namburu, (eds.). *Data Mining for Scientific and Engineering Applications*. Kluwer Academic Publishers, 2001.

209. D. A. Grossman and O. Frieder. *Information Retrieval: Algorithms and Heuristics*, Springer, 2004.

210. S. Grumbach and G. Mecca. In Search of the Lost Schema. In *Proc. of the Intl. Conf. on Database Theory*, pp. 314–331, 1999.

211. S. Guha, R. Rastogi, and K. Shim. CURE: An Efficient Clustering Algorithm for Large Databases. In *Proc. of the ACM SIGMOD Intl. Conf. on Management of Data (SIGMOD'98)*, pp. 73–84, 1998.

212. S. Guha, R. Rastogi, and K. Shim. ROCK: a Robust Clustering Algorithm for Categorical Attributes. In *Proc. of the 15th Intl. Conf. on Data Engineering*, pp. 345–366. 2000.

213. D. Gusfield. *Algorithms on Strings, Trees and Sequences*. Cambridge University Press, 1997.

214. Z. Gyongyi and H. Garcia-Molina. *Web Spam Taxonomy*. Technical Report, Stanford University, 2004.

215. Z. Gyöngyi, and H. Garcia-Molina. Link Spam Alliances. In *Proc. of the 31st Intl Conf. on Very Large Data Bases (VLDB'05)*, pp. 517–528, 2005.

216. Z. Gyongyi, H. Garcia-Molina and J. Pedersen. Combating Web Spam with TrustRank. In *Proc. of 30th Intl. Conf. on Very Large Data Bases (VLDB'04)*, pp. 576–587, 2004.

217. J. Han and Y. Fu. Discovery of Multi-Level Association Rules from Large Databases. In *Proc. of the 21st Intl. Conf. on Very Large Data Bases (VLDB'05)*, pp. 420–431, 1995.

218. J. Han and M. Kamber. *Data Mining: Concepts and Techniques*. Morgan Kaufmann Publishers, San Francisco, 2001.

219. J. Han, J. Pei, B. Mortazavi-Asl, Q. Chen, U. Dayal, and M.-C. Hsu. Freespan: Frequent Pattern-Projected Sequential Pattern Mining. In *Proc. of the 2000 Int. Conf. Knowledge Discovery and Data Mining (KDD'00)*, pp. 355–359, 2000.

220. J. Han, J. Pei, and Y. Yin. Mining Frequent Patterns without Candidate Generation. In *Proc. of the ACM SIGMOD Int. Conf. on Management of Data (SIGMOD'00)*, pp. 1–12, 2000.

221. D. J. Hand, H. Mannila, and P. Smyth. *Principles of Data Mining*. MIT Press, 2001.

222. J. A. Hartigan. *Clustering Algorithms*. John Wiley & Sons Inc. 1975.

223. J. A. Hartigan. Direct Clustering of a Data Matrix. *Journal of the American Statistical Association*, 67(337): pp. 123–129, 1972.

224. V. Hatzivassiloglou and K. McKeown. Predicting the Semantic Orientation of Adjectives. In *Proc. of the 35th Annual Meeting of the Association for Computational Linguistics (ACL-EACL'97)*, pp. 174–181, 1997.

225. V. Hatzivassiloglou, and J. Wiebe. Effects of Adjective Orientation and Gradability on Sentence Subjectivity. In *Proc. of the Intl. Conf. on Computational Linguistics (COLING'00)*, pp. 299–305. 2000.

226. T. Haveliwala. Extrapolation Methods for Accelerating PageRank Computations. In *Proc. of the 12th Intl. World Wide Web Conf. (WWW'03)*, pp. 261–270, 2003.

227. B. He, and K. C.-C. Chang. Statistical Schema Matching across Web Query Interfaces. In *Proc. of the 2003 ACM SIGMOD Intl. Conf. on Management of Data (SIGMOD'03)*, pp. 217–228, 2003.

228. B. He and K. C.-C. Chang. Making Holistic Schema Matching Robust: An Ensemble Approach. In *Proc. of the ACM SIGKDD Intl. Conf. on Knowledge Discovery and Data Mining (KDD'05)*, pp. 429-438, 2005.

229. B. He, K. C.-C. Chang, and J. Han. Discovering Complex Matchings across Web Query Interfaces: A Correlation Mining Approach. In *Proc. of the ACM SIGKDD Conf. on Knowledge Discovery and Data Mining (KDD'04)*, pp. 148–157, 2004.

230. H. He, W. Meng, C. T. Yu, and Z. Wu. WISE-Integrator: An Automatic Integrator of Web Search Interfaces for E-commerce. In *Proc. of Very Large Data Bases (VLDB'03)*, 2003.

231. H. He, W. Meng, C. T. Yu, and Z. Wu. Automatic extraction of web search interfaces for interface schema integration. In *Proc. of WWW Alternate Track Papers and Posters,* pp. 414-415, 2004.

232. M. A. Hearst. Direction-based Text Interpretation as an Information Access Refinement. In P. Jacobs (eds.), *Text-Based Intelligent Systems*. Lawrence Erlbaum Associates, 1992.

233. M. A. Hearst, and J. O. Pedersen. Reexamining the Cluster Hypothesis: Scatter/Gather on Retrieval Results. In *Proc. of the 19th Intl. ACM SIGIR Conf. on Research and Development in Information Retrieval (SIGIR'96)*, 1996.

234. M. Henzinger, A. Heydon, M. Mitzenmacher, and M. Najork. Measuring Search Engine Quality Using Random Walks on the Web. In *Proc. of the 8th Intl. World Wide Web Conf.*, pp. 213–225, 1999.

235. M. Henzinger, A. Heydon, M. Mitzenmacher, and M. Najork. On Near-Uniform URL Sampling. In *Proc. of the 9th Intl. World Wide Web Conf. (WWW'00)*, pp. 295–308, 2000.

236. J. L. Herlocker, J. Konstan, L. Terveen, and J. Riedl. Evaluating Collaborative Filtering Recommender Systems. *ACM Transactions on Information Systems*, 22(1), pp. 5–53, 2004

237. M. Hersovici, M. Jacovi, Y. S. Maarek, D. Pelleg, M. Shtalhaim, and S. Ur. The Shark-Search Algorithm An Application: Tailored Web Site Mapping. In *Proc. of the 7th Intl. World Wide Web Conf. (WWW7)*, pp. 317–326, 1998.

238. A. Heydon and M. Najork. Mercator: A Scalable, Extensible Web Crawler. *World Wide Web* 2(4), pp. 219–229, 1999.

239. A. Hinneburg and D. A. Keim. An Optimal Grid-Clustering: Towards Breaking the Curse of Dimensionality in High-Dimensional Clustering. In *Proc. of Very Large Data Bases (VLDB'99)*, pp. 506–517, 1999.

240. T. Hofmann. Unsupervised Learning by Probabilistic Latent Semantic Analysis. *Machine Learning*, 42(1): pp. 177–196, 2001.

241. A. Hogue and D. Karger. Thresher: Automating the Unwrapping of Semantic Con-

tent from the World Wide Web. In *Proc. of the 14th Intl. World Wide Web Conference* (*WWW'05*), pp. 86–95, 2005.

242. S. C. H. Hoi, R. Jin, M. R. Lyu. Large-Scale Text Categorization by Batch Mode Active Learning. In *Proc. of the 15th Intl. Conf. on World Wide Web* (*WWW'06*), 2006.

243. V. Honavar and G. Slutzki. (eds.). Grammatical Inference. In *Proc. of the Fourth Intl Colloquium on Grammatical Inference*. LNCS 1433. Springer-Verlag, 1998.

244. C.-N. Hsu and M.-T. Dung. Generating Finite-State Transducers for Semi-Structured Data Extraction from the Web. *Inf. System*, 23(9), pp. 521–538, 1998.

245. M. Hu and B. Liu. Mining and Summarizing Customer Reviews. In *Proc. of ACM SIGKDD Intl. Conf. on Knowledge Discovery and Data Mining* (*KDD'04*), pp. 168–177, 2004.

246. M. Hu and B. Liu. Mining Opinion Features in Customer Reviews. In *Proc. of the 19th National Conf. on Artificial Intelligence* (*AAAI'04*), pp. 755–760, 2004.

247. M. Hu and B. Liu. Opinion Feature Extraction Using Class Sequential Rules. In *Proc. of the Spring Symposia on Computational Approaches to Analyzing Weblogs*, 2006.

248. L. Hyafil, and R. L. Rivest. Constructing Optimal Binary Decision Trees is NP-Complete. *Information Processing Letters* 5, pp. 15–17, 1976.

249. H. Ino, M. Kudo, and A. Nakamura. Partitioning of Web Graphs by Community Topology. In *Proc. of the 14th Intl. Conf. on World Wide Web* (*WWW'05*), pp. 661–66, 2005.

250. U. Irmak, and T. Suel. Interactive Wrapper Generation with Minimal User Effort. In *Proc. of the 15th Intl. Conf. on World Wide Web* (*WWW'06*), 2006.

251. T. Jagatic, N. Johnson, M. Jakobsson, and F. Menczer. Social Phishing. *Communications of the ACM*. In press, 2006.

252. A. K. Jain and R. C. Dubes. *Algorithms for Clustering Data*. Prentice Hall, 1988.

253. T. Jiang, L. Wang, and K. Zhang. Alignment of Trees - an Alternative to Tree edit. In *Proc. of Combinatorial Pattern Matching*, pp. 75–86, 1994.

254. X. Jin, Y. Zhou, and B. Mobasher. Web Usage Mining Based on Probabilistic Latent Semantic Analysis. In *Proc. of the ACM SIGKDD Conf. on Knowledge Discovery and Data Mining* (*KDD'04*), pp. 197–205, 2004.

255. N. Jindal, and B. Liu. Identifying Comparative Sentences in Text Documents. In *Proc. of ACM SIGIR Intl. Conf. on Research and Development in Information Retrieval* (*SIGIR'06*), pp. 244–251, 2006.

256. N. Jindal, and B. Liu. Mining Comparative Sentences and Relations. In *Proc. of National Conference on Artificial Intelligence* (*AAAI'06*), 2006.

257. T. Joachims. Text Categorization with Support Vector Machines: Learning with Many Relevant Features. In Machine Learning: In *Proc. of Tenth European Conf. on Machine Learning* (*ECML'98*), pp. 137–142, 1998.

258. T. Joachims. Making Large-Scale SVM Learning Practical. *Advances in Kernel Methods - Support Vector Learning*, B. Schölkopf and C. Burges and A. Smola (eds.), MIT Press, 1999.

259. T. Joachims. Transductive Inference for Text Classification Using Support Vector Machines. In *Proc. of the Intl. Conf. on Machine Learning* (*ICML'99*), pp. 200–209, 1999.

260. T. Joachims. Optimizing Search Engines Using Click-through Data. In *Proc. of the ACM Intl. Conf. on Knowledge Discovery and Data Mining* (*KDD'02*), pp. 133-142, 2002.

261. T. Joachims, Transductive Learning via Spectral Graph Partitioning. In *Proc. of the Intl. Conf. on Machine Learning* (*ICML'03*), pp. 290–297, 2003.

262. R. Jones, B. Rey, O. Madani, and W. Greiner. Generating Query Substitutions. In

Proc. of the 15th Intl. Conf. on World Wide Web (*WWW'06*), 2006.

263. L. P. Kaelbling, M. Littman, and A. Moore. Reinforcement Learning: A survey. *Journal of Artificial Intelligence Research* 4, pp. 237–285, 1996.

264. N. Kaji and M. Kitsuregawa. Automatic Construction of Polarity-Tagged Corpus from HTML Documents. In *Proc. of the COLING/ACL 2006 Main Conference Poster Sessions*, pp. 452–459, 2006.

265. Y. Kalfoglou and M. Schorlemmer. Ontology Mapping: the State of the Art. *The Knowledge Engineering Review Journal,* 18(1), pp. 1–31, 2003.

266. S. D. Kamar, T. Haveliwala, C. D. Manning, and G. H. Golub, Extrapolation Methods for Accelerating PageRank Computations. In *Proc. of the 12th Intl. World Wide Web Conference* (*WWW'03*), pp. 261–270, 2003.

267. H. Kanayama and T. Nasukawa. Fully Automatic Lexicon Expansion for Domain-oriented Sentiment Analysis. In *Proc. of the 2006 Conf. on Empirical Methods in Natural Language Processing* (*EMNLP'06*), pp. 355–363, 2006.

268. J. Kang, and J. F. Naughton, On Schema Matching with Opaque Column Names and Data Values. In *Proc. of the 2003 ACM SIGMOD Intl. Conf. on Management of Data* (*SIGMOD'03*), 2003.

269. V. Kashyap and A. Sheth. Semantic and Schematic Similarities between Database Objects: a Context-Based Approach. In *Proc. of the Intl. Journal on Very Large Data Bases* (*VLDB'96*), 5(4): pp. 276–304, 1996.

270. G. V. Kass. An Exploratory Technique for Investigating Large Quantities of Categorical Data. *Applied Statistics*, 29, pp. 119–127, 1980.

271. L. Kaufman, and P. J. Rousseeuw. *Finding Groups in Data: an Introduction to Cluster Analysis*. John Wiley & Sons, 1990.

272. M. Kearns. Efficient Noise-Tolerant Learning from Statistical Queries. *Journal of the ACM*, 45, pp. 983–1006, 1998.

273. J. S. Kelly. *Social Choice Theory: An Introduction*. Springer-Verlag, 1988.

274. C. Kennedy. Comparatives, Semantics of. In *Encyclopedia of Language and Linguistics*, Second Edition, Elsevier, 2005.

275. M. M. Kessler. Bibliographic Coupling between Scientific Papers. *American Documentation*, 14, 1963.

276. S. Kim and E. Hovy. Determining the Sentiment of Opinions. In *Proc. of the Intl. Conf. on Computational Linguistics* (*COLING'04*), 2004.

277. S.-M. Kim and E. Hovy. Automatic Identification of Pro and Con Reasons in Online Reviews. In *Proc. of the COLING/ACL 2006 Main Conference Poster Sessions,* pp. 483–490, 2006.

278. S.-M. Kim and E. Hovy. Identifying and Analyzing Judgment Opinions. In *Proc. of the Human Language Technology Conference of the North American Chapter of the ACL*, pp. 200–207, 2006.

279. R. Kimball and R. Merz. *The Data Webhouse Toolkit: Building the Web-Enabled Data Warehouse.* John Wiley & Sons, 2000.

280. J. L. Klavans, and S. Muresan. DEFINDER: Rule-Based Methods for the Extraction of Medical Terminology and Their Associated Definitions from On-line Text. In *Proc. of American Medical Informatics Assoc.*, 2000.

281. J. Kleinberg. Authoritative Sources in a Hyperlinked Environment. In *Proc. of the 9th ACM SIAM Symposium on Discrete Algorithms* (*SODA'98*), pp. 668–677, 1998.

282. J. Kleinberg. Authoritative Sources in a Hyperlinked Environment. *Journal of the ACM* 46 (5), pp. 604–632, 1999.

283. M. Klemetinen, H. Mannila, P. Ronkainen, H. Toivonen, and A. I. Verkamo. Finding Interesting Rules from Large Sets of Discovered Association Rules. In *Proc. of the ACM Intl. Conf. on Information and Knowledge Management* (*CIKM'94*), pp. 401-

407, 1994.

284. N. Kobayashi, R. Iida, K. Inui and Y. Matsumoto. Opinion Mining on the Web by Extracting Subject-Attribute-Value Relations. In *Proc. of AAAI-CAAW'06*, 2006.

285. R. Kohavi, B. Becker, and D. Sommerfield, Improving Simple Bayes. In *Proc. of European Conference on Machine Learning (ECML'97)*, 1997.

286. R. Kohavi, L. Mason, R. Parekh, and Z. Zheng. Lessons and Challenges from Mining Retail E-Commerce Data. *Machine Learning*, 57(1–2), pp. 83–113, 2004.

287. T. Kohonen. Self-Organizing Maps. *Series in Information Sciences*, 30, Springer, Heidelberg, Second Edition. 1995.

288. F. Korn, H. V. Jagadish and C. Faloutsos. Efficiently Supporting Ad Hoc Queries in Large Datasets of Time Sequences. In *Proc. ACM SIGMOD International Conference on Management of Data (SIGMOD'97)*, pp 289–300, 1997.

289. R. Kraft, C. C. Chang, F. Maghoul, and Ravi Kumar. Searching with Context. In *Proc. of the 15th Intl. Conf. on World Wide Web (WWW'06)*, 2006.

290. L.-W. Ku, H.-W. Ho, and H.-H. Chen. Novel Relationship Discovery Using Opinions Mined from the Web. In *Proc. of the Twenty-First National Conf. on Artificial Intelligence (AAAI'06)*, 2006.

291. L.-W. Ku, Y.-T. Liang and H.-H. Chen. Opinion Extraction, Summarization and Tracking in News and Blog Corpora. In *Proc. of the AAAI-CAAW'06*, 2006.

292. V. Kumar, A. Grama, A. Gupta, and G. Karypis. *Introduction to Parallel Computing: Design and Analysis of Algorithms*. Benjamin/Cummings, 1994.

293. R. Kumar, P. Raghavan, S. Rajagopalan, and A. Tomkins. Trawling the Web for Emerging Cyber-Communities. In *Proc. of the 8th Intl. World Wide Web Conference (WWW8)*, pp. 1481–1493, 1999.

294. K. Kummamuru, R. Lotlikar, S. Roy, K. Singal, and R. Krishnapuram. 2004. A Hierarchical Monothetic Document Clustering Algorithm for Summarization and Browsing Search Results. In *Proc. of the 13th Intl. Conf. on World Wide Web (WWW'04)*, pp. 658–665, 2004.

295. N. Kushmerick. Wrapper Induction: Efficiency and Expressiveness. *Artificial Intelligence*, 118: pp. 15–68, 2000.

296. N. Kushmerick. *Wrapper Induction for Information Extraction*. Ph.D Thesis. Dept. of Computer Science, University of Washington, TR UW-CSE-97-11-04, 1997.

297. S. H. Kwok and C. C. Yang. Searching the Peer–to–Peer Networks: The Community and their Queries. *Journal of the American Society for Information Science and Technology, Special Topic Issue on Research on Information Seeking*, 55(9), pp.783–793, 2004.

298. J. Lafferty, A. McCallum, and F. Pereira. Conditional Random Fields: Probabilistic Models for Segmenting and Labeling or Sequence Data. In *Proc. of the Intl. Conf. on Machine Learning (ICML'01)*, pp. 282–289, 2001.

299. S. Lam, and J. Reidl. Shilling Recommender Systems for Fun and Profit. In *Proc. of the 13th Int.l World Wide Web Conf. (WWW'04)*, pp. 393-402, 2004.

300. K. Lang. Newsweeder: Learning to Filter Netnews. In *Proc. of the International Conference on Machine Learning (ICML'95)*, pp. 331–339, 1995.

301. P. Langley, W. Iba, and K. Thompson. An Analysis of Bayesian Classifiers. In *Proc. of the 10th National Conf. on Artificial Intelligence (AAAI'92)*, pp. 223–228, 1992.

302. P. Langley. *Elements of Machine Learning*. Morgan Kauffmann, 1996.

303. A. N. Langville and Carl D. Meyer. Deeper Inside PageRank. *Internet Mathematics*, 1(3), pp. 335–380, 2005.

304. A. N. Langville and C. D. Meyer. *Google's PageRank and Beyond: The Science of Search Engine Rankings*. Princeton University Press, 2006.

305. D. T. Larose. *Discovering Knowledge in Data: An Introduction to Data Mining*.

John Wiley, 2004.

306. J. A Larson, S. B Navathe, and R. ElMasri. A Theory of Attribute Equivalence in Databases with Application to Schema Integration. In *Proc. of IEEE Trans Software Engineering* 16(4): pp. 449–463, 1989.

307. S. Lawrence and C. Giles. Accessibility of Information on the Web. *Nature* 400, pp. 107–109, 1999.

308. S. Lawrence, C. L. Giles, and K. Bollaker. Digital Libraries and Autonomous Citation Indexing. *IEEE Computer* 32(6), pp. 67–71, 1999.

309. W. S. Lee, and B. Liu. Learning with Positive and Unlabeled Examples Using Weighted Logistic Regression. In *Proc. of the Twentieth Intl. Conf. on Machine Learning (ICML'03)*, pp. 448–455, 2003.

310. R. Lempel and S. Moran. The Stochastic Approach for Link-Structure Analysis (SALSA) and the TKC Effect. In *Proc. of the Ninth Intl. World Wide Web Conf. (WWW'9)*, pp. 387–401, 2000.

311. A. V. Leouski and W. B. Croft. *An Evaluation of Techniques for Clustering Search Results*. Technical Report IR–76, Department of Computer Science, University of Massachusetts, Amherst, 1996.

312. K. Lerman, L. Getoor, S. Minton, and C. Knoblock. Using the Structure of Web Sites for Automatic Segmentation of Tables. In *Proc. of the ACM SIGMOD Intl. Conf. on Management of Data (SIGMOD'04)*, pp. 119–130, 2004.

313. J. Lerner and M. Pinkal. *Comparatives and Nested Quantification*. CLAUS-Report 21, 1992.

314. N. Lesh, M. J. Zaki, and M. Ogihara. Mining Features for Sequence Classification. In *Proc. of the ACM SIGKDD Intl. Conf. on Knowledge Discovery and Data Mining (KDD'99)*, 1999.

315. F. Letouzey, F. Denis, and R. Gilleron. Learning from Positive and Unlabeled Examples. In *Proc. of the 11th Intl. Conf. on Algorithmic Learning Theory (ALT'00)*, pp. 71–85, 2000.

316. D. Lewis. An Evaluation of Phrasal and Clustered Representations on a Text Categorization Task. In *Proc. of the ACM SIGIR Intl. Conf. on Research and Development in Information Retrieval (SIGIR'92)*, pp. 37–50, 1992.

317. D. Lewis and W. Gale. A Sequential Algorithm for Training Text Classifiers. In *Proc. of the ACM SIGIR Intl. Conf. on Research and Development in Information Retrieval (SIGIR'94)*, pp. 3–12, 1994.

318. H. Li and K. Yamanishi. Document Classification Using a Finite Mixture Model. In *Proc. of the 35th Annual Meeting of the Association for Computational Linguistics*, pp. 39–47, 1997.

319. J. Li, G. Dong, K. Ramamohanarao. Making Use of the Most Expressive Jumping Emerging Patterns for Classification. In *Proc. of Pacific-Asia Conf. on Knowledge Discovery and Data Mining (PAKDD'00)*, pp. 220–232, 2000.

320. J. Li, G. Dong, K. Ramamohanarao, and L. Wong. DeEPs: A New Instance-Based Lazy Discovery and Classification System. *Machine Learning*, 54(2), pp. 99–124 2004.

321. X. L. Li, and B. Liu. Learning to Classify Text Using Positive and Unlabeled Data. In *Proc. of the Eighteenth Intl. Joint Conf. on Artificial Intelligence (IJCAI'03)*, pp. 587–594, 2003.

322. X. L. Li, and B. Liu. Learning from Positive and Unlabeled Examples with Different Data Distributions. In *Proc. of the European Conf. on Machine Learning (ECML'05)*, 2005.

323. X. L. Li, B. Liu and S-K. Ng. Learning to Identify Unexpected Instances in the Test Set. To appear in *Proc. of Intl. Joint Conf. on Artificial Intelligence (IJCAI'06)*, 2006.

324. X. L. Li, T.-H. Phang, M. Hu, and B. Liu. Using Micro Information Units for Internet Search. In *Proc. of the ACM Intl. Conf. on Information and Knowledge Management (CIKM'02)*, pp. 566–573, 2002.

325. X. Li, B. Liu and P. S. Yu. Discovering Overlapping Communities of Named Entities. In *Proc. of Conf. on Practical Knowledge Discovery and Data Mining (PKDD'06)*, 2006.

326. X. Li, B. Liu and P. S. Yu. Time Sensitive Ranking with Application to Publication Search. *Forthcoming paper*, 2006.

327. W. Li, and C. Clifton. SemInt: a Tool for Identifying Attribute Correspondences in Heterogeneous Databases Using Neural Network. *Data Knowledge Engineering* 33(1), pp. 49–84, 2000.

328. W. Li, J. Han, and J. Pei. CMAR: Accurate and Efficient Classification Based on Multiple Class-Association Rules. In *Proc. of the 2001 IEEE Intl. Conf. on Data Mining (ICDM'01)*, pp. 369–376, 2001.

329. C. Li, J.-R. Wen, and H. Li. Text Classification Using Stochastic Keyword Generation. In *Proc. of Intl. Conf. on Machine Learning (ICML'03)*, pp. 464-471, 2003.

330. G. Lidstone. Note on the General Case of the Bayes-Laplace formula for Inductive or a Posteriori Probabilities. *Transactions of the Faculty of Actuaries*, 8, pp. 182–192, 1920.

331. W. Lin, S. A. Alvarez, and C. Ruiz. Efficient adaptive-support association rule mining for recommender systems. *Data Mining and Knowledge Discovery*, 6, pp. 83-105, 2002.

332. D. Lin. PRINCIPAR-An Efficient, Broad-Coverage, Principle-Based Parser. In *Proc. of the 15th Conf. on Computational Linguistics*, pp. 482–488., 1994.

333. C.-R Lin, and M.-S. Chen: A Robust and Efficient Clustering Algorithm Based on Cohesion Self-merging. In *Proc. of the SIGKDD Intl. Conf. on Knowledge Discovery and Data Mining (KDD'02)*, pp. 582–587, 2002.

334. D.-I. Lin and Z. M. Kedem. Pincer-Search: A New Algorithm for Discovering the Max Mum Frequent Set. In *Proc. of the 6th Intl. Conf. Extending Database Technology (EDBT'98)*, 1998.

335. L.-J. Lin. Self-Improving Reactive Agents Based on Reinforcement Learning, Planning, and Teaching. *Machine Learning* 8, pp. 293–321, 1992.

336. S.-H. Lin, and J.-M. Ho. Discovering Informative Content Blocks from Web Documents. In *Proc. of the ACM SIGKDD Intl. Conf. on Knowledge Discovery and Data Mining (KDD'02)*, pp. 588–593, 2002

337. W. Lin, S. A. Alvarez, and C. Ruiz. Efficient Adaptive-Support Association Rule Mining for Recommender Systems. *Data Mining and Knowledge Discovery*, 6, pp. 83–105, 2002.

338. G. S. Linoff, and M. J. Berry. *Mining the Web: Transforming Customer Data into Customer Value*. John Wiley & Sons. 2002.

339. B. Liu, C. W. Chin, and H. T. Ng. Mining Topic-Specific Concepts and Definitions on the Web. In *Proc. of the 12th Intl. World Wide Web Conf. (WWW'03)*, pp. 251–260, 2003.

340. B. Liu, Yang Dai, Xiaoli Li, Wee Sun Lee and Philip Yu. Building Text Classifiers Using Positive and Unlabeled Examples. In *Proc. of the 3rd IEEE Intl. Conf. on Data Mining (ICDM'03)*, pp. 179–188, 2003.

341. B. Liu, R. Grossman, and Y. Zhai. Mining Data Records in Web Pages. In *Proc. of the ACM SIGKDD Intl. Conf. on Knowledge Discovery and Data Mining (KDD'03)*, pp. 601–606. 2003.

342. B. Liu, W. Hsu, and S. Chen. Using General Impressions to Analyze Discovered Classification Rules. In *Proc. of the ACM SIGKDD Intl. Conf. on Knowledge Dis-

covery and Data Mining (*KDD'97*), pp. 31-36, 1997.

343. B. Liu, W. Hsu, and Y. Ma. Integrating Classification and Association Rule Mining. In *Proc. of Knowledge Discovery and Data Mining* (*KDD'98*), pp. 80–86, 1998.

344. B. Liu, W. Hsu, and Y. Ma. Mining Association Rules with Multiple Minimum Supports. In *Proc. of Intl. Conf. Knowledge Discovery and Data Mining* (*KDD'99*), pp. 337–341, 1999.

345. B. Liu, W. Hsu, and Y. Ma. Pruning and Summarizing the Discovered Associations. In *Proc. of the ACM SIGKDD Intl. Conf. on Knowledge Discovery and Data Mining* (*KDD'99*), pp. 125-134, 1999.

346. B. Liu, W. Hsu, L. Mun, and H. Lee. Finding Interesting Patterns Using User Expectations. *IEEE Transactions on Knowledge and Data Engineering*, 11(6), pp.817–832, 1999.

347. B. Liu, M. Hu, and J. Cheng. Opinion Observer: Analyzing and Comparing Opinions on the Web. In *Proc. of the 14th Intl. World Wide Web Conf.* (*WWW'05*), pp. 342–351, 2005.

348. B. Liu, W. S. Lee, Philip S. Yu and Xiaoli Li. Partially Supervised Classification of Text Documents. In *Proc. of the Nineteenth Intl. Conf. on Machine Learning* (*ICML'02*), pp. 8–12, 2002.

349. B. Liu, Y. Ma, and C-K Wong, Classification Using Association Rules: Weaknesses and Enhancements. In Vipin Kumar, et al, (eds), *Data Mining for Scientific Applications*, 2001.

350. B. Liu, Y. Xia, and P. S. Yu. Clustering through Decision Tree Construction. In *Proc. of the ACM Intl. Conf. on Information and Knowledge Management* (*CIKM'00*), pp. 20–29, 2000.

351. B. Liu and Y. Zhai. NET – A System for Extracting Web Data from Flat and Nested Data Records. In *Proc. of 6th Intl. Conf. on Web Information Systems Engineering* (*WISE'05*), pp. 487–495, 2005.

352. B. Liu, K. Zhao, J. Benkler and W. Xiao. Rule Interestingness Analysis Using OLAP Operations. In *Proc. of the Twelfth ACM SIGKDD Intl. Conf. on Knowledge Discovery and Data Mining* (*KDD'06*), pp. 297–306, 2006.

353. H. Liu and H. Motoda. *Feature Selection for Knowledge Discovery and Data Mining*. Kluwer Academic Publishers, 1998.

354. J. Lu and J. Callan. Content-Based Retrieval in Hybrid Peer-to-Peer Networks. In *Proc. 12th ACM Intl. Conf. on Information and Knowledge Management* (*CIKM'03*), pp. 199–206, 2003

355. L. Ma, N. Goharian, and A. Chowdhury. Extracting Unstructured Data from Template Generated Web Document. In *Proc. of the ACM Intl. Conf. on Information and Knowledge Management* (*CIKM'03*), pp. 512–515, 2003

356. J. Madhavan, P. A. Bernstein, A. Doan, and A. Y. Halevy. Corpus-Based Schema Matching. In *Proc. of International Conference on Data Engineering* (*ICDE'05*), pp. 57–68, 2005.

357. J. B. MacQueen. Some Methods for Classification and Analysis of Multivariate Observations. In *Proc. of the 5th Berkeley Symposium on Mathematical Statistics and Probability*, pp. 281–297, 1967.

358. J. Madhavan, P. A. Bernstein, and E. Rahm. Generic Schema Matching with Cupid. In *Proc 27th Int. Conf. on Very Large Data Bases* (*VLDB'01*), pp. 49–58, 2001.

359. A. G. Maguitman, F. Menczer, H. Roinestad, and A. Vespignani. Algorithmic Detection of Semantic Similarity. In *Proc. 14th Intl. World Wide Web Conf.* (*WWW'05*), pp. 107–116, 2005.

360. L. Manevitz and M. Yousef. One-Class SVMs for Document Classification. *Journal of Machine Learning Research*, 2, pp. 139–154, 2001.

361. H. Mannila, H. Toivonen, and I. Verkamo. Efficient Algorithms for Discovering Association Rules. In *Proc. of Knowledge Discovery in Databases (KDD'94)*, pp. 181–19, AAAI Press 1994

362. C. Manning and H. Schutze. *Foundations of Statistical Natural Language Processing.* The MIT Press, Cambridge, Massachusetts, 1999.

363. B. Markines, L. Stoilova, and F. Menczer. Social Bookmarks for Collaborative Search and Recommendation. In *Proc. of the 21st National Conf. on Artificial Intelligence (AAAI'06)*, 2006.

364. O. McBryan. Genvl and WWWW: Tools for Taming the Web. In O. Nierstrasz (Ed.). In *Proc. of the First Intl. World Wide Web Conf.*, Geneva. CERN, 1994.

365. A. McCallum, and K. Nigam. A Comparison of Event Models for Naïve Bayes Text Classification. In *Proc. of the AAAI–98 Workshop on Learning for Text Categorization.* 1998.

366. A. McCallum, K. Nigam, J. Rennie, and K. Seymore. A Machine Learning Approach to Building Domain-Specific Search Engines. In *Proc. 16th Intl. Joint Conf. on Artificial Intelligence (IJCAI'99)*, pp. 662–667, 1999.

367. R. McCann, B. K. AlShebli, Q. Le, H. Nguyen, L. Vu, and A. Doan: Mapping Maintenance for Data Integration Systems. In *Proc. of Intl. Conf. on Very Large Data Bases (VLDB'05)*: pp. 1018–1030, 2005.

368. F. McSherry. A Uniform Approach to Accelerated PageRank Computation. In *Proc. of the 14th Intl. World Wide Web Conference (WWW'05)*, pp. 575–582, 2005.

369. F. Menczer. ARACHNID: Adaptive Retrieval Agents Choosing Heuristic Neighborhoods for Information Discovery. In *Proc. of the 14th Intl. Conf. on Machine Learning*, pp. 227–235, 1997.

370. F. Menczer. Growing and Navigating the Small World Web by Local   Content. In *Proc. Natl. Acad. Sci.* USA, 99(22), pp. 14014-14019, 2002

371. F. Menczer. The Evolution of Document Networks. In *Proc. Natl. Acad. Sci.* USA 101, pp. 5261-5265, 2004

372. F. Menczer. Lexical and Semantic Clustering by Web Links. *Journal of the American Society for Information Science and Technology*, 55(14), pp. 1261–1269, 2004.

373. F. Menczer. Mapping the Semantics of Web Text and Links. *IEEE Internet Computing* 9 (3), pp. 27–36, 2005.

374. F. Menczer and R. Belew. Adaptive Information Agents in Distributed Textual Environments. In *Proc. of the 2nd Intl. Conf. on Autonomous Agents*, pp. 157–164, 1998.

375. F. Menczer and R. Belew. Adaptive Retrieval Agents: Internalizing Local Context and Scaling up to the Web. *Machine Learning*, 39(2–3), 203–242, 2000.

376. F. Menczer, G. Pant, M. Ruiz, and P. Srinivasan. Evaluating Topic Driven Web Crawlers. In *Proc. 24th Annual Intl. ACM SIGIR Conf. on Research and Development in Information Retrieval*, pp. 241–249, 2001.

377. F. Menczer, G. Pant, and P. Srinivasan. Topical Web Crawlers: Evaluating Adaptive Algorithms. *ACM Transactions on Internet Technology* 4(4), pp. 378–419, 2004.

378. W. Meng, C. Yu, and K.-L. Liu. Building Efficient and Effective Metasearch Engines. *ACM Computing Surveys*, 34(1), pp. 48–84, 2002.

379. D. Meretakis, and B. Wüthrich. Extending Naïve Bayes Classifiers Using Long Itemsets. In *Proc. of the Fifth ACM SIGKDD Intl. Conf. on Knowledge Discovery and Data Mining (KDD'99)*, pp. 165–174, 1999.

380. A. Micarelli and F. Gasparetti (in press). Adaptive Focused Crawling. In P. Brusilovsky, W. Nejdl, and A. Kobsa (eds.), *Adaptive Web*. Springer.

381. R. S. Michalski, I. Mozetic, J. Hong and N. Lavrac. The Multi-Purpose Incremental Learning System AQ15 and its Testing Application to Three Medical Domains. In *Proc. of the National Conf. on Artificial Intelligence (AAAI'86)*, pp. 1041–1047,

1986.

382. T. Milo, S. Zohar. Using schema matching to simplify heterogeneous data translation. In: *Proc. of Intl Conf on Very Large Data Bases (VLDB'98)*, pp. 122–133, 1998.

383. B. Mirkin. *Clustering for Data Mining: A Data Recovery Approach*. Chapman & Hall/CRC, April 29, 2005.

384. N. Misha, D. Ron, and R. Swaminathan. A New Conceptual Clustering Framework. *Machine Learning*, 56(1–3): pp. 115–151, 2004.

385. T. Mitchell. *Machine Learning*. McGraw Hill, 1997.

386. B. Mobasher. Web Usage Mining and Personalization. In Munindar P. Singh (ed.), *Practical Handbook of Internet Computing*. CRC Press, 2005.

387. B. Mobasher. Web Usage Mining. In John Wang (eds.), *Encyclopedia of Data Warehousing and Mining*, Idea Group, 2006.

388. B. Mobasher, R. Cooley and J. Srivastava. Automatic Personalization based on Web Usage Mining. *Communications of the ACM*, 43(8), pp. 142–151, 2000.

389. B. Mobasher, H. Dai, T. Luo, and N. Nakagawa. Effective Personalization Based on Association Rule Discovery from Web Usage Data. In *Proc. of the 3rd ACM Workshop on Web Information and Data Management (WIDM01)*, pp. 9–15, 2001.

390. B. Mobasher, H. Dai, T. Luo, and M. Nakagawa. Discovery and Evaluation of Aggregate Usage Profiles for Web Personalization. *Data Mining and Knowledge Discovery*, 6, pp. 61–82, 2002.

391. B. Mobasher and R. Burke and J. J Sandvig. Model-Based Collaborative Filtering as a Defense against Profile Injection Attacks. In *Proc. of the 21st National Conf. on Artificial Intelligence (AAAI'06)*, 2006.

392. A. Moffat, R. Neal, and I. Witten. Arithmetic Coding Revisited. *ACM Transactions on Information Systems*, pp. 256–294, 1998.

393. F. Moltmann, *Coordination and Comparatives*. Ph.D. dissertation. MIT, Cambridge Ma., 1987.

394. M. Montague, and J. Aslam. Condorcet Fusion for Improved Retrieval. *In Proc. of the Intl. Conf. on Information and Knowledge Management (CIKM'02)*, pp. 538–548, 2002.

395. R. J. Mooney and R. Bunescu. Mining Knowledge from Text Using Information Extraction. *SIGKDD Explorations*, pp. 3–10. 2005.

396. A. Moore. Very Fast EM-based Mixture Model Clustering Using Multiresolution Kd-Trees. In *Proc. of the Neural Info. Processing Systems (NIPS'98)*, pp. 543–549, 1998.

397. S. Morinaga, K. Yamanishi, K. Tateishi, and T. Fukushima. Mining Product Reputations on the Web. In *Proc. of the SIGKDD Intl. Conf. on Knowledge Discovery and Data Mining (KDD'02)*, pp. 341–349, 2002.

398. S. Muggleton. Learning from the Positive Data. *Inductive Logic Programming Workshop*, pp. 358–376, 1996.

399. I. Muslea, S. Minton, and C. A. Knoblock. A Hierarchical Approach to Wrapper Induction. In *Proc. of the Intl. Conf. on Autonomous Agents (AGENTS'99)*, pp. 190–197, 1999.

400. I. Muslea, S. Minton, and C. A. Knoblock. Active Learning with Multiple Views. *Journal of Artificial Intelligence Research*, 1, pp. 1–31, 2006.

401. M. Najork and J. L. Wiener. Breadth-First Search Crawling Yields High Quality Pages. In *Proc. of the 10th Intl. World Wide Web Conf. (WWW'01)*, pp. 114–118, 2001.

402. T. Nasukawa and J. Yi. Sentiment Analysis: Capturing Favorability Using Natural Language Processing. In *Proc. of the K-CAP-03, 2nd Intl. Conf. on Knowledge Capture*, pp. 70–77, 2003.

403. T. Nelson. A File Structure for the Complex, the Changing and the Indeterminate. In

Proc. of the ACM National Conf., pp. 84–100, 1965.

404. A. Ng, M. Jordan, and Y. Weiss. On Spectral Clustering: Analysis and an Algorithm. In *Proc. of the 14th Advances in Neural Information Processing Systems*, pp. 849–856, 2001.

405. A. Ng, A. X. Zheng, and M. I. Jordan. Stable Algorithms for Link Analysis. In *Proc. of the 24th Annual ACM SIGIR Intl. Conf on Research and Development on Information Retrieval (SIGIR'01)*, 2001.

406. R. T. Ng and J. Han. Efficient and Effective Clustering Methods for Spatial Data Mining. In *Proc. of Conf. on Very Large Data Bases (VLDB'94)*, pp. 144–155, 1994.

407. R. T. Ng, J. Han. CLARANS: A Method for Clustering Objects for Spatial Data Mining. *IEEE Transactions Knowledge Data Engineering* 14(5), pp. 1003-1016, 2002.

408. R. T. Ng, L. V. S. Lakshmanan, J. Han, and A. Pang. Exploratory Mining and Pruning Optimizations of Constrained Association Rules. In *Proc. of ACM SIGMOD Intl. Conf. on Management of Data (SIGMOD'98)*, pp. 13–24, 1998.

409. V. Ng, S. Dasgupta and S. M. Niaz Arifin. Examining the Role of Linguistic Knowledge Sources in the Automatic Identification and Classification of Reviews. In. *Proc. of the COLING/ACL 2006 Main Conference Poster Sessions*, pp. 611–618, 2006.

410. Z. Nie, Y. Zhang, J-R. Wen, and W-Y Ma. Object Level Ranking: Bringing Order to Web Objects. In *Proc. of the 14th Intl. World Wide Web Conference (WWW'05)*, pp. 567–574, 2005

411. K. Nigam and R. Ghani. Analyzing the Effectiveness and Applicability of Co-training. In *Proc. of the ACM Intl. Conf. on Information and Knowledge Management (CIKM'00)*, pp. 86–93, 2000.

412. K. Nigam and M. Hurst. Towards a Robust Metric of Opinion. *AAAI Spring Symp. on Exploring Attitude and Affect in Text*, 2004.

413. K. Nigam, A. McCallum, S. Thrun, and T. Mitchell. Text Classification from Labeled and Unlabeled Documents Using EM. *Machine Learning*, 39(2/3), pp. 103–134, 2000.

414. Z. Niu, D. Ji, and C. L. Tan. Word Sense Disambiguation Using Label Propagation Based Semi-Supervised Learning. In *Proc. of the Meeting of the Association for Computational Linguistics (ACL'05)*, 2005.

415. C. Notredame. Recent Progresses in Multiple Sequence Alignment: a Survey. Technical report, *Information Génétique et*, 2002.

416. A. Ntoulas, J. Cho, and C. Olston. What's New on the Web? The Evolution of the Web from a Search Engine Perspective. In *Proc. of the 13th Intl. World Wide Web Conference (WWW'04)*, pp. 1–12, 2004.

417. A. Ntoulas, M. Najork, M. Manasse, and D. Fetterly. Detecting Spam Web Pages through Content Analysis. In *Proc. of the 15th Intl. World Wide Web Conference (WWW'06)*, pp. 83–92, 2006.

418. R. Nuray, and F. Can. Automatic Ranking of Information Retrieval Systems Using Data Fusion. *Information Processing and Management*, 4(3), pp. 595–614, 2006.

419. M. O'Mahony, N. Hurley, N. Kushmerick, and G. Silvestre. Collaborative Recommendation: A Robustness Analysis. *ACM Transactions on Internet Technology* 4(4):344–377, 2004.

420. B. Ozden, S. Ramaswamy, and A. Silberschatz. Cyclic Association Rules. In *Proc. 1998 Int. Conf. Data Engineering (ICDE'98)*, pp. 412–421, 1998.

421. B. Padmanabhan, and A. Tuzhilin. Small is Beautiful: Discovering the Minimal Set of Unexpected Patterns. In *Proc. of the ACM SIGKDD Intl. Conf. on Knowledge Discovery and Data Mining (KDD'00)*, pp. 54-63, 2000.

422. L. Page, S. Brin, R. Motwami, and T. Winograd. *The PageRank Citation Ranking: Bringing Order to the Web*. Technical Report 1999–0120, Computer Science De-

partment, Stanford University, 1999.

423. G. Paliouras, C. Papatheodorou, V. Karkaletsis, and C. D. Spyropoulos. Discovering User Communities on the Internet Using Unsupervised Machine Learning Techniques. *Interacting with Computers Journal*, 14(6), pp. 761–791, 2002.

424. L. Palopoli, D. Sacca, and D. Ursino. An Automatic Technique for Detecting Type Conflicts in Database Schemas. In: *Proc of ACM Intl. Conf on Information and Knowledge Management (CIKM'98)*, pp. 306–313, 1998.

425. S. Pandey, S. Roy, C. Olston, J. Cho and S. Chakrabarti, Shuffling a Stacked Deck: The Case for Partially Randomized Ranking of Search Engine Results. In *Proc. of Very Large Data Bases (VLDB'05)*, pp. 781–792, 2005.

426. B. Pang and L. Lee. A Sentimental Education: Sentiment Analysis Using Subjectivity Summarization based on Minimum Cuts. In *Proc. of the 42nd Meeting of the Association for Computational Linguistics (ACL'04)*, pp. 271—278, 2004.

427. B. Pang and L. Lee, Seeing Stars: Exploiting Class Relationships for Sentiment Categorization with Respect to Rating Scales. In *Proc. of the Meeting of the Association for Computational Linguistics (ACL'05)*, pp. 115–124, 2005.

428. B. Pang, L. Lee, and S. Vaithyanathan. Thumbs up? Sentiment Classification Using Machine Learning Techniques. In *Proc. of the EMNLP'02*, 2002.

429. G. Pant. Deriving Link-Context from Html Tag Tree. In *Proc. of the 8th ACM SIGMOD Workshop on Research Issues in Data Mining and Knowledge Discovery (DMKD'03)*, pp. 49–55, 2003.

430. G. Pant, S. Bradshaw, and F. Menczer. Search Engine – Crawler Symbiosis. In *Proc. of the 7th European Conf. on Research and Advanced Technology for Digital Libraries (ECDL'03)*, 2003.

431. G. Pant and F. Menczer. MySpiders: Evolve your Own Intelligent Web Crawlers. *Autonomous Agents and Multi-Agent Systems*, 5(2), pp. 221–229, 2002.

432. G. Pant and F. Menczer. Topical Crawling for Business Intelligence. In *Proc. of the 7th European Conf. on Research and Advanced Technology for Digital Libraries (ECDL'03)*, pp. 233–244, 2003.

433. G. Pant, and P. Srinivasan. Learning to Crawl: Comparing Classification Schemes. ACM Trans. *Information Systems*, 23(4), pp. 430–462, 2005.

434. G. Pant, P. Srinivasan, and F. Menczer. Exploration versus Exploitation in Topic Driven Crawlers. In *Proc. of the WWW-02 Workshop on Web Dynamics*, 2002.

435. J. S. Park, M.-S. Chen, and P. S. Yu: An Effective Hash Based Algorithm for Mining Association Rules. In *Proc. of the 1995 ACM SIGMOD Intl. Conf. on Management of Data (SIGMOD'95)*, pp. 175–186, 1995.

436. N. Pasquier, Y. Bastide, R. Taouil, and L. Lakhal. Discovering Frequent Closed Itemsets for Association Rules. In *Proc. of the 7th Intl. Conf. on Database Theory*, pp. 398–416, 1999.

437. R. Pastor-Satorras and A. Vespignani. *Evolution and Structure of the Internet*. Cambridge University Press, 2004.

438. M. J. Pazzani, C. Brunk, and G. Silverstein. A Knowledge-Intensive Approach to Learning Relational Concepts. In *Proc. of the Eighth Intl. Workshop on Machine Learning (ML'91)*, pp. 432–436, 1991.

439. J. Pei, J. Han, B. Mortazavi-Asl, H. Pinto, Q. Chen, U. Dayal and M-C. Hsu. PrefixSpan: Mining Sequential Patterns Efficiently by Prefix-Projected Pattern Growth. In *Proc. of the 2001 Int. Conf. Data Engineering (ICDE'01)*, pp. 215–224, 2001.

440. M. Pennock, G. W. Flakes, S. Lawrence, C.L Giles, and E. J Gloves. Winners Don't Take All: Characterizing the Competition for Links on the Web. In *Proc. of National Academy of Science*, 99(8), pp. 5207–5211, 2002.

441. P. Perner. *Data Mining on Multimedia Data*. Springer, 2003.

442. T. P. Pham, H. T. Ng, and W. S. Lee. Word Sense Disambiguation with Semi-Supervised Learning. In *Proc. of the National Conference on Artificial Intelligence (AAAI'05)*. pp. 1093–1098, 2005.

443. G. Piatetsky-Shapiro, and B. Masand. Estimating Campaign Benefits and Modeling Lift. In *Proc. of the ACM SIGKDD Intl. Conf. on Knowledge Discovery and Data Mining (KDD'99)*, pp. 185-193, 1999.

444. G. Piatesky-Shapiro, and C. Matheus. The Interestingness of Deviations. In *Proc. of Knowledge Discovery and Data Mining (KDD'94)*, 1994.

445. G. Pierrakos, G. Paliouras, C. Papatheodorou, and C. Spyropoulos. Web Usage Mining as a Tool for Personalization: a Survey. *User Modeling and User-Adapted Interaction*, 13, pp. 311–372, 2003.

446. J. Pitkow and P. Pirolli. Mining Longest Repeating Subsequences to Predict WWW Surfing. In *Proceedings of the 2nd USENIX Symposium on Internet Technologies and Systems*, 1999.

447. A.-M. Popescu, and O. Etzioni. Extracting Product Features and Opinions from Reviews. In *Proc. of Conference on Empirical Methods in Natural Language Processing (EMNLP'05)*, 2005.

448. J. Ponte, and W. B. Croft. A Language Modeling Approach to Information Retrieval. In *Proc. of the Annual Intl. ACM SIGIR Conf. on Research and Development in Information Retrieval (SIGIR'98)*, pp. 275-281, 1998.

449. M. F. Porter. An Algorithm for Suffix Stripping. *Program*, 14(3), pp 130–137, 1980.

450. D. Pyle. *Business Modeling and Data Mining*. Morgan Kaufmann, 2003.

451. F. Qiu and J. Cho. Automatic Identification of User Interest for Personalized Search. In *Proc. of the 15th Intl. Conf. on World Wide Web (WWW'06)*, 2006.

452. J. R. Quinlan. Learning Logical Definitions from Relations. *Machine Learning*, 5, pp. 239–266, 1990.

453. J. R. Quinlan. *C4.5: Program for Machine Learning*. Morgan Kaufmann, 1992.

454. J. R. Quinlan. Bagging, Boosting, and C4.5. In *Proc. of National Conf. on Artificial Intelligence (AAAI-96)*, pp. 725-730, 1996.

455. E. Rahm, and P. A. Bernstein. A Survey of Approaches to Automatic Schema Matching. *VLDB Journal*, 10, pp. 334–35, 2001.

456. L. Ramaswamy, A. Lyengar, L. Liu, and F. Douglis. Automatic Detection of Fragments in Dynamically Generated Web Pages. In *Proc. of the 13th Intl. World Wide Web Conference (WWW'04)*, pp. 443–454, 2004

457. J. Raposo, A. Pan, M. Alvarez, J. Hidalgo, and A. Vina. The Wargo System: Semi-Automatic Wrapper Generation in Presence of Complex Data Access Modes. In *Proc. of the 13th Intl. Work-shop on Database and Expert Systems Applications*, pp. 313–320, 2002.

458. D. de Castro Reis, P. B. Golgher, A. S. da Silva, and A. H. F. Laender. Automatic Web News Extraction Using Tree Edit Distance. In *Proc. of the 13th Intl. World Wide Web Conference (WWW'04)*, pp. 502–511, 2004

459. J. Rennie and A. McCallum. Using Reinforcement Learning to Spider the Web Efficiently. In *Proc. of the 16th Intl. Conf. on Machine Learning (ICML'99)*, pp. 335–343, 1999.

460. M. Richardson, A. Prakash, and E. Brill. Beyond PageRank: Machine Learning for Static Ranking. In *Proc. of the 15th Intl. Conf. on World Wide Web (WWW'06)*, 2006.

461. C. van Rijsbergen. *Information Retrieval*, Chapter 3, London: Butterworths. Second edition, 1979.

462. E. Riloff and J. Wiebe. Learning Extraction Patterns for Subjective Expressions. In *Proc. of Conference on Empirical Methods in Natural Language Processing (EMNLP'03)*, 2003.

463. R. L. Rivest. Learning Decision Lists. *Machine Learning*, 2(3), pp. 229–246, 1987.

464. S. E. Robertson and K. Sparck-Jones. Relevance Weighting of Search Terms. *Journal of the American Society for Information Science*, 27, pp. 129–146, 1976.

465. S. E. Robertson, S. Walker, and M. Beaulieu. Okapi at TREC–7: Automatic Ad Hoc, Filtering, VLC and Filtering Tracks. In *Proc. of the Seventh Text REtrieval Conference (TREC-7)*, pp. 253–264, 1999.

466. J. Rocchio. Relevant Feedback in Information Retrieval. In G. Salton (eds.). *The Smart Retrieval System – Experiments in Automatic Document Processing*, Englewood Cliffs, NJ, 1971

467. R. Roiger and M. Geatz. *Data Mining: A Tutorial Based Primer*. Addison-Wesley, 2002.

468. O. Parr Rud. *Data Mining Cookbook*. John Wiley & Sons, 2003.

469. D. Rumelhart, G. Hinton, and R. Williams. Learning Internal Representations by Error Propagation. In D. Rumelhart and J. McClelland (eds.), *Parallel Distributed Processing: Explorations in the Microstructure of Cognition*, Volume 1, Chapter 8, pp. 318–362, 1996.

470. G. Salton and C. Buckley. Term-Weighting Approaches in Automatic Retrieval. *Information Processing and Management*, 24(5), pp. 513–525, 1988.

471. G. Salton and M. McGill. *An Introduction to Modern Information Retrieval*. New York, NY: McGraw-Hill, 1983.

472. B. Santorini. *Part-of-Speech Tagging Guidelines for the Penn Treebank Project*. Technical Report MS-CIS-90-47, Department of Computer and Information Science, University of Pennsylvania, 1990.

473. R. R. Sarukkai. Link Prediction and Path Analysis Using Markov Chains. In *Proc. of the 9th Intl. World Wide Web Conf. (WWW'00)*, pp. 377–386. 2000.

474. B. Sarwar, G. Karypis, J. Konstan and J. Riedl. Application of Dimensionality Reduction in Recommender Systems – A Case Study. In *Proc. of the KDD Workshop on WebKDD'2000*, 2000.

475. B. Sarwar, G. Karypis, J. Konstan and J. Riedl. Item-Based Collaborative Filtering Recommendation Algorithms. In *Proc. of the 10th Intl. World Wide Web Conference (WWW'01)*, pp. 285–295, 2001.

476. A. Savasere, E. Omiecinski, and S. B. Navathe. Mining for Strong Negative Associations in a Large Database of Customer Transactions. In *Proc. of the Fourteenth Intl. Conf. on Data Engineering (ICDE'98)*, pp. 494–502, 1998.

477. R. E. Schapire. The Strength of Weak Learnability. *Machine Learning*, 5(2), pp. 197–227, 1990.

478. S. Scholkopf, J. Platt, J. Shawe, A. Smola, and R. Williamson. *Estimating the Support of a High-Dimensional Distribution*. Technical Report MSR-TR-99-87, Microsoft Research, pp. 1443–1471, 1999.

479. B. Scholkopf and A. Smola. *Learning with Kernels*. MIT Press, 2002.

480. A. Scime (eds.). *Web Mining: Applications and Techniques*. Idea Group Inc., 2005.

481. G. L. Scott and H. C. Longuet-Higgins. Feature Grouping by Relocalisation of Eigenvectors of the Proximity Matrix. In *Proc. British Machine Vision Conf.*, pp. 103–108, 1990.

482. M. Seno, and G. Karypis. Finding Frequent Patterns Using Length-Decreasing Support Constraints. *Data Mining and Knowledge Discovery*, 10(3), pp 197–228, 2005.

483. J. G. Shanahan, Y. Qu, and J. Wiebe, (eds.). *Computing Attitude and Affect in Text: Theory and Applications*. Springer. 2005.

484. E. Shannon. A Mathematical Theory of Communication. In *Bell System Technical Journal*, 27: pp. 379–423, 1948.

485. G. Sheikholeslami, S. Chatterjee and A. Zhang. WaveCluster: a Multi-resolution

Clustering Approach for Very Large Spatial Databases. In *Proc. of Very Large Data Bases (VLDB'98)*, pp. 428–439, 1998.

486. D. Shen, J.-T. Sun, Q. Yang, and Z. Chen. A Comparison of Implicit and Explicit Links for Web Page Classification. In *Proc. of the 15th Intl. Conf. on World Wide Web (WWW'06)*, 2006.

487. X. Shen, B. Tan, and C. Zhai. Context-Sensitive Information Retrieval with Implicit Feedback. In *Proc. of the Annual Intl. ACM SIGIR Conf. on Research and Development in Information Retrieval (SIGIR'05)*, pp. 43-50, 2005.

488. A. Sheth and J. Larson. Federated Database Systems for Managing Distributed, Heterogeneous, and Autonomous Databases. *ACM Computing Surveys*, 22(3):183–236, 1990.

489. J. Shi and J. Malik. Normalized Cuts and Image Segmentation. In *Proc. of the IEEE Conf. Computer Vision and Pattern Recognition*, pp. 731–737, 1997.

490. X. Shi and C. C. Yang, Mining Related Queries from Search Engines Query Logs. In *Proc. of the Intl. World Wide Web Conf. (WWW'06)*, pp. 943-944, 2006.

491. P. Shvaiko, and J. Euzenat. A Survey of Schema-Based Matching Approaches. *Journal on Data Semantics*, IV, LNCS 3730, pp. 146–171, 2005.

492. A. Silberschatz, and A. Tuzhilin. What Makes Patterns Interesting in Knowledge Discovery Systems. *IEEE Transactions on Knowledge and Data Engineering*, 8(6), pp. 970–974, 1996.

493. A. Singhal. Modern Information Retrieval: A Brief Overview. *IEEE Data Engineering Bulletin* 24(4), pp. 35-43, 2001.

494. H. Small. Co-Citation in the Scientific Literature: a New Measure of the Relationship between Two Documents. *Journal of American Society for Information Science*, 24(4), pp. 265–269, 1973.

495. R. Song, H. Liu, J. R. Wen, and W. Y. Ma. Learning Block Importance Models for Web Pages. In *Proc. of the 13th Conf. on World Wide Web (WWW'04)*, pp. 203–211, 2004.

496. M. Spiliopoulou. Web Usage Mining for Web Site Evaluation. *Communications of ACM*, 43(8), pp. 127–134, 2000.

497. M. Spiliopoulou, and L. Faulstich. WUM: A Tool for Web Utilization Analysis. In *Proc. of EDBT Workshop at WebDB'98*, pp. 184–203, 1999.

498. M. Spiliopoulou, B. Mobasher, B. Berendt, and M. Nakagawa. A Framework for the Evaluation of Session Reconstruction Heuristics in Web Usage Analysis. *INFORMS Journal of Computing*, 15(2), pp. 171–190, 2003.

499. R. Srikant and R. Agrawal. Mining Generalized Association Rules. In *Proc. of the 21st Int'l Conf. on Very Large Data Bases (VLDB'95)*, pp. 407–419, 1995.

500. R. Srikant and R. Agrawal. Mining Sequential Patterns: Generalizations and Performance Improvements. In *Proc. of the 5th Intl. Conf. Extending Database Technology (EDBT'96)*, pp. 3–17, 1996.

501. R. Srikant and R. Agrawal. Mining Quantitative Association Rules in Large Relational Tables. In *Proc. of the ACM SIGMOD Conf. on Management of Data (SIGMOD'96)*, 1996.

502. R. Srikant, Q. Vu, and R. Agrawal. Mining Association Rules with Item Constraints. In *Proc. of the 3rd Intl. Conf. on Knowledge Discovery and Data Mining (KDD'97)*, pp. 67–73, 1997.

503. P. Srinivasan, J. Mitchell, O. Bodenreider, G. Pant, and F. Menczer. Web Crawling Agents for Retrieving Biomedical Information. In *Proc. of the Intl. Workshop on Agents in Bioinformatics (NETTAB'02)*, 2002.

504. P. Srinivasan, G. Pant, and F. Menczer. A General Evaluation Framework for Topical Crawlers. *Information Retrieval* 8 (3), pp. 417–447, 2005.

505. J. Srivastava, R. Cooley, M. Deshpande, and P. Tan. Web Usage Mining: Discovery and Applications of Usage Patterns from Web Data. *SIGKDD Explorations*, 1(2), pp. 12–23, 2000.

506. M. Steinbach, G. Karypis, and V. Kumar. A Comparison of Document Clustering Techniques. In *Proc. of the KDD Workshop on Text Mining*, 2000.

507. V. Stoyanov and C. Cardie. Toward Opinion Summarization: Linking the Sources. In *Proc. of the Workshop on Sentiment and Subjectivity in Text*, pp. 9–14, 2006.

508. J-T. Sun, X. Wang, D. Shen, H-J. Zeng, and Z. Chen. CWS: A Comparative Web Search System. In *Proc. of the 15th Intl. Conf. on World Wide Web (WWW'06)*, 2006.

509. K.-C. Tai. The Tree-to-Tree Correction Problem. *Journal of the ACM*, 26(3), pp. 422–433, 1979.

510. P. -N. Tan and V. Kumar. Discovery of Web Robot Sessions Based on their Navigational Patterns. *Data Mining and Knowledge Discovery*, 6(1), pp. 9–35, 2002.

511. P. -N. Tan, V. Kumar and J. Srivastava. Selecting the Right Interestingness Measure for Association Patterns. In *Proc. of the ACM SIGKDD Intl. Conf. on Knowledge Discovery and Data Mining (KDD'02)*, pp. 32-41, 2002.

512. P.-N. Tan, M. Steinbach, and V. Kumar. *Introduction to Data Mining*. Addison-Wesley, 2006.

513. D. Tanasa and B. Trousse. Advanced Data Preprocessing for Intersite Web Usage Mining. *IEEE Intelligent Systems*, 19(2), pp. 59–65, 2004.

514. Z.-H. Tang, and J. MacLennan. *Data Mining with SQL Server 2005*. Wiley publishing, Inc. 2005.

515. B M. Thuraisingham. *Web Data Mining and Applications in Business Intelligence and Counter-Terrorism*, CRC Press, 2003.

516. J. Tomlin. A New Paradigm for Ranking Pages on the World Wide Web. In *Proc. of the 12th Intl. World Wide Web Conference (WWW'02)*, pp. 350–355, 2003.

517. R. Tong. An Operational System for Detecting and Tracking Opinions in On-Line Discussion. In *Proc. of SIGIR Workshop on Operational Text Classification*, 2001.

518. M. Toyoda and M. Kitsuregawa. Creating a Web Community Chart for Navigating Related Communities. In *Proc. of the Twelfth ACM Conf. on Hypertext and Hypermedia*, pp. 103–112, 2001.

519. M. Toyoda and M. Kitsuregawa. Extracting Evolution of Web Communities from a Series of Web Archives. In *Proc. of the fourteenth ACM Conf. on Hypertext and Hypermedia*, pp. 28–37, 2003.

520. M. Toyoda, and M. Kitsuregawa. What's Really New on the Web?: Identifying New Pages from a Series of Unstable Web Snapshots. In *Proc. of the 15th Intl. Conf. on World Wide Web (WWW'06)*, 2006.

521. P. Turney. Thumbs Up or Thumbs Down? Semantic Orientation Applied to Unsupervised Classification of Reviews. In *Proc. of the Meeting of the Association for Computational Linguistics (ACL'02)*, pp. 417–424, 2002

522. A. Tuzhilin, and G. Adomavicius. Handling very Large Numbers of Association Rules in the Analysis of Microarray Data. In *Proc. of ACM SIGKDD Intl. Conf. on Knowledge Discovery and Data Mining (KDD'02)*, pp. 396–404, 2002.

523. J. R. Tyler, D. M. Wilkinson, and B. A. Huberman. Email as Spectroscopy: Automated Discovery of Community Structure within Organizations. *Communities and Technologies*, pp. 81–96. 2003.

524. A. Valitutti, C. Strapparava, and O. Stock. Developing Affective Lexical Resources. *Psychnology Journal*, 2(1): pp. 61–83, 2004.

525. V. Vapnik. *The Nature of Statistical Learning Theory*. Springer, 1995.

526. V. Vapnik. *Statistical Learning Theory*. John Wiley, 1998.

527. H. Wache, T. Voegele, U. Visser, H. Stuckenschmidt, G. Schuster, H. Neumann, and

S. Huebner. Ontology-Based Integration of Information – a Survey of Existing Approaches. In *Proc. of the IJCAI Workshop on Ontologies and Information Sharing,* pp. 108–117, 2001.

527. K. Wagstaff and C. Cardie. Clustering with Instance-Level Constraints. In *Proc. of the 17th Intl. Conf. on Machine Learning,* pp. 1103–1110, 2000.

528. J. Wang, J. Han, and J. Pei. Closet+: Searching for the Best Strategies for Mining Frequent Closed Itemsets. In *Proc. of the ACM SIGKDD Intl. Conf. on Knowledge Discovery and Data Mining (KDD'03),* pp. 236–245, 2003.

529. J. Wang and F. H. Lochovsky. Data Extraction and Label Assignment for Web Databases. In *Proc. of the 12th Intl. World Wide Web Conference (WWW'03),* pp. 187–196, 2003.

530. J. Wang, J-R. Wen, F. H. Lochovsky, and W-Y. Ma. Instance-Based Schema Matching for Web Databases by Domain-specific Query Probing. In *Proc. of the Intl. Conf. on Very Large Data Bases (VLDB'04),* pp. 408–419, 2004.

531. J. T.-L. Wang, B. A. Shapiro, D. Shasha, K. Zhang, and K. M. Currey. An algorithm for finding the largest approximately common substructures of two trees. *IEEE Transactions on Pattern Analysis and Machine Intelligence,* 20(8), pp. 889.895, 1998.

532. J. Wang, M. Zaki, H. Toivonen, and D. Shasha. (eds.). *Data Mining in Bioinformatics.* Springer, 2004.

533. K. Wang, Yu He, and J. Han. Mining Frequent Itemsets Using Support Constraints. In *Proc. of 26th Intl. Conf. on Very Large Data Bases (VLDB'00),* pp 43–52, 2000.

534. K. Wang, Y. Jiang, and L. V.S. Lakshmanan. Mining Unexpected Rules by Pushing User Dynamics. In *Proc. of the ACM SIGKDD Intl. Conf. on Knowledge Discovery and Data Mining (KDD'03),* pp. 246-255, 2003.

535. K. Wang, S. Zhou, and Y. He. Growing Decision Trees on Support-Less Association Rules. In *Proc. of the ACM SIGKDD Intl. Conf. on Knowledge discovery and data mining (KDD'00),* pp 265–269, 2000.

536. K. Wang, C. Xu, and B. Liu. 1999. Clustering Transactions Using Large Items. In *Proc. of the Eighth Intl. Conf. on information and Knowledge Management (CIKM'99).* 1999, pp. 483–490.

537. W. Wang, J. Yang and R. Muntz. STING: A Statistical Information Grid Approach to Spatial Data Mining. In *Proc. of Intl. Conf. on Very Large Data Bases (VLDB'97),* pp. 186–195, 1997.

538. W. Wang, J. Yang, and P. S. Yu. WAR: Weighted Association Rules for Item Intensities. *Knowledge and Information Systems,* 6(2), pp. 203–229, 2004.

539. S. Wasserman and K. Faust. *Social Network Analysis.* Cambridge University Press, 1994.

540. I. G. Webb. Discovering Associations with Numeric Variables. In *Proc. of the SIGKDD Intl. Conf. on Knowledge Discovery and Data Mining (KDD'01):* pp. 383–388, 2001.

541. Y. Weiss. Segmentation Using Eigenvectors: a Unifying View. In *Proc. IEEE Intl. Conf. on Computer Vision,* pp. 975–982, 1999.

542. J. Wiebe. Learning Subjective Adjectives from Corpora. In *Proc. of 17th National Conf. on Artificial Intelligence,* pp. 735–740, Austin, USA, 2000.

543. J. Wiebe, and R. Mihalcea. Word Sense and Subjectivity. In *Proc. of the 21st Intl. Conf. on Computational Linguistics and 44th Annual Meeting of the ACL,* pp. 1065–1072, 2006.

544. J. Wiebe, and E. Riloff: Creating Subjective and Objective Sentence Classifiers from Unannotated Texts. In *Proc. of CICLing,* pp. 486–497, 2005.

545. T. Wilson, J. Wiebe and R. Hwa. Recognizing Strong and Weak Opinion Clauses.

Computational Intelligence, 22(2), pp. 73-99, 2006.

547. H. Williams and J. Zobel. Compressing Integers for Fast File Access. *Computer Journal,* 42(3), pp. 193—201, 1999.

548. T. Wilson, J. Wiebe, and J. Hwa. Just How Mad Are You? Finding Strong and Weak Opinion Clauses. In *Proc. of the National Conference on Artificial Intelligence (AAAI'04),* 2004.

549. I. H. Witten and E. Frank. *Data Mining*: *Practical Machine Learning Tools and Techniques with Java Implementations.* Academic Press, 2000.

550. I. H. Witten, C. G. Nevill-Manning, and S. J. Cunningham. Building a Digital Library for Computer Science Research: Technical Issues. In *Proc. of the 19th Australasian Computer Science Conf.,* pp. 534–542, 1996.

551. I. H. Witten, A. Moffat, and T. C. Bell. *Managing Gigabytes*: *Compressing and Indexing Documents and Images.* Academic Press, 1999.

552. D. Wolpert. Stacked Generalization. *Neural Networks* 5, pp. 241–259, 1992.

553. L.-S. Wu, R. Akavipat, and F. Menczer. 6S: Distributing Crawling and Searching Across Web Peers. In *Proc. of the IASTED Int. Conf. on Web Technologies, Applications, and Services,* 2005.

554. L.-S. Wu, R. Akavipat, and F. Menczer. Adaptive Query Routing in Peer Web Search. In *Proc. of the 14th Intl. World Wide Web Conf. (WWW'05),* pp. 1074–1075, 2005.

555. B. Wu and B. Davison. Identifying Link Farm Spam Pages. In *Proc. of the 14th Intl. World Wide Web Conf. (WWW'05),* pp. 820–829, May 2005.

556. B. Wu and B. Davison. Cloaking and Redirection: a Preliminary Study. In *Proc. of the 1st Intl. Workshop on Adversarial Information Retrieval on the Web,* 2005.

557. B. Wu, V. Goel, and B. D. Davison. Topical TrustRank: Using Topicality to Combat Web Spam. In *Proc. of the 15th Intl. Conf. on World Wide Web (WWW'06),* 2006.

558. W. Wu, A. Doan, and C. Yu. WebIQ: Learning from the Web to Match Query Interfaces on the Deep Web. In *Proc. of International Conference on Data Engineering (ICDE'06),* 2006.

559. W. Wu, C. Yu, A. Doan, and W. Meng. An Interactive Clustering-Based Approach to Integrating Source Query Interfaces on the Deep Web. In *Proc. of the ACM SIGMOD Intl. Conf. on Management of Data (SIGMOD'04),* pp. 95–106, 2004.

560. X. Wu, C. Zhang and S. Zhang. Mining both Positive and Negative Association Rules. In *Proc. of 19th Intl. Conf. on Machine Learning,* pp. 658–665, 2002.

561. X. Wu, L. Zhang, and Y. Yu. Exploring Social Annotations for the Semantic Web. In *Proc. of the 15th Intl. Conf. on World Wide Web (WWW'06),* 2006.

562. H. Xiong, P.-N. Tan, and V. Kumar. Mining Strong Affinity Association Patterns in Data Sets with Skewed Support Distribution. In *Proc. of the 3rd IEEE Intl. Conf. on Data Mining (ICDM'03),* pp. 387-394, 2003.

563. L. Xu and D. Embley. Discovering Direct and Indirect Matches for Schema Elements. In *Proc. of Intl. Conf. on Database Systems for Advanced Applications (DASFAA'03),* 2003.

564. X. Xu, M. Ester, H-P. Kriegel and J. Sander. A Non-Parametric Clustering Algorithm for Knowledge Discovery in Large Spatial Databases. In *Proc. of the Intl. Conf. on Data Engineering (ICDE'98),* 1998.

565. X. Yan, H. Cheng, J. Han, and D. Xin: Summarizing Itemset Patterns: a Profile-Based Approach. In *Proc. of the ACM SIGKDD Intl. Conf. on Knowledge Discovery and Data Mining (KDD'05),* pp. 314-323, 2005.

566. L. Yan, R. J. Miller, L. M. Haas, and R. Fagin. Data-Driven Understanding and Refinement of Schema Mappings. In *Proc ACM SIGMOD Intl. Conf. on Management of Data,* pp. 485–496, 2001.

567. C. C. Yang and K. Y. Chan. Retrieving Multimedia Web Objects Based on Page Rank Algorithm. *WWW'05 Poster*, 2005.

568. B. Yang and H. Garcia-Molina. Improving Search in Peer-to-Peer Networks. In *Proc. of the 22nd Intl. Conf. on Distributed Computing Systems (ICDCS'02)*, pp. 5–14. IEEE Computer Society, 2002.

569. B. Yang, and G. Jeh. Retroactive Answering of Search Queries. In *Proc. of the 15th Intl. Conf. on World Wide Web (WWW'06)*, 2006.

570. Q. Yang, T. Y. Li, and K. Wang. Building Association-Rule Based Sequential Classifiers for Web-Document Prediction. *Data Mining Knowledge Discovery* 8(3), pp. 253-273, 2004.

571. J. Yang, W. Wang, and P. Yu. Mining Surprising Periodic Patterns. *Data Mining and Knowledge Discovery*, 9(2), pp. 189–216, 2004.

572. W. Yang. Identifying Syntactic Differences between Two Programs. *Software Practice Experiment*, 21(7), pp. 739–755, 1991.

573. Y. Yang. An Evaluation of Statistical Approaches to Text Categorization. *Journal of Information Retrieval*, 1, pp. 67–88, 1999.

574. Y. Yang, and X. Liu. A Re-Examination of Text Categorization Methods. In *Proc. of the ACM SIGIR Intl. Conf. Research and Development in Information Retrieval (SIGIR'99)*, pp. 42–49, 1999.

575. Y. Yang and J. P. Pedersen. A Comparative Study on Feature Selection in Text Categorization. In *Proc. of the Intl. Conf. on Machine Learning (ICML'97)*, pp. 412–420, 1997.

576. L. Yi, B. Liu, and X. L. Li. Eliminating Noisy Information in Web Pages for Data Mining. In *Proc. of the ACM SIGKDD Intl. Conf. on Knowledge Discovery and Data Mining)*, pp. 296–305, 2003.

577. J. Yi, T. Nasukawa, R. C. Bunescu, and W. Niblack. Sentiment Analyzer: Extracting Sentiments about a Given Topic Using Natural Language Processing Techniques. In *Proc. of the IEEE Conf. on Data Mining (ICDM'03)*, pp. 427–434, 2003.

578. X. Yin, and J. Han. CPAR: Classification based on Predictive Association Rules. In *Proc. of the SIAM Intl. Conf. on Data Mining (SDM'03)*, 2003.

579. X. Yin and W. S. Lee. Using Link Analysis to Improve Layout on Mobile Devices. In *Proc. of the 13th Intl. Conf. on World Wide Web (WWW'04)*, pp. 338–344, 2004.

580. A. Ypma and T. Heskes. Categorization of Web Pages and User Clustering with Mixtures of Hidden Markov Models. In *Proc. of the Workshop on WebKDD-2002*, pp. 35–49, 2002.

581. C. Yu and W. Meng. *Principles of Database Query Processing for Advanced Applications*. Morgan Kaufmann, 1998.

582. H. Yu. General MC: Estimating Boundary of Positive Class from Small Positive Data. In *Proc. of the Intl. Conf. on Data Mining (ICDM'03)*, pp. 693–696, 2003.

583. H. Yu, J. Han and K. Chang. PEBL: Positive Example Based Learning for Web Page Classification Using SVM. In *Proc. of the Knowledge Discovery and Data Mining (KDD'02)*, pp. 239–248., 2002.

584. H. Yu, and V. Hatzivassiloglou. Towards Answering Opinion Questions: Separating Facts from Opinions and Identifying the Polarity of Opinion Sentences. In *Proc. of Intl. Conf. on Empirical Methods for Natural Language Processing (EMNLP'03)*, 2003.

585. P. S. Yu, X. Li, and B. Liu. Adding the Temporal Dimension to Search – A Case Study in Publication Search. In *Proc. of Web Intelligence (WI'05)*, pp. 543–549, 2005.

586. M. Zaki. SPADE: An Efficient Algorithm for Mining Frequent Sequences. *Machine Learning*, 40, pp. 31-60, 2001.

587. M. J. Zaki, and C. C. Aggarwal. XRules: an Effective Structural Classifier for XML Data. In *Proc. of the Ninth ACM SIGKDD Intl. Conf. on Knowledge Discovery and Data Mining (KDD'03)*, pp. 316–325, 2003.
588. M. J. Zaki and C. Hsiao. Charm: An Efficient Algorithm for Closed Association Rule Mining. In *Proc. of SIAM Conf. on Data Mining*, 2002.
589. M. J. Zaki, S. Parthasarathy, M. Ogihara, and W. Li. New Algorithms for Fast Discovery of Association Rules. In *Proc. of the 3rd International Conference on Knowledge Discovery and Data Mining (KDD'97)*, pp 283–286, 1997.
590. M. Zaki, M. Peters, I. Assent, and T. Seidl. CLICKS: an Effective Algorithm for Mining Subspace Clusters in Categorical Datasets. In *Proc. of the SIGKDD Intl. Conf. on Knowledge Discovery and Data Mining (KDD'05)*, pp. 736–742, 2005.
591. O. Zamir, and O. Etzioni. Web Document Clustering: A Feasibility Demonstration. In *Proc. of the 19th Intl. ACM SIGIR Conf. on Research and Development of Information Retrieval (SIGIR'98)*, pp. 46–54, 1998.
592. O. Zamir, and O. Etzioni. Grouper: A Dynamic Clustering Interface to Web Search Results. In *Proc. of the 8th Intl. World Wide Web Conf. (WWW8)*, Toronto, Canada, pp. 1361–1374, 1999.
593. H. Zeng, Q. He, Z. Chen, W. Ma, and J. Ma. Learning to Cluster Web Search Results. In *Proc. of the 27th Intl. ACM SIGIR Conf. on Research and Development in information Retrieval (SIGIR'04)*. pp. 210–217, 2004.
594. H. Zha, C. Ding, M. Gu, X. He, and H. Simon. Spectral Relaxation for K-means Clustering. In *Proc. of Neural Information Processing Systems (NIPS'01)*, pp. 1057–1064, 2001.
595. C. Zhai. Statistical Language Model for Information Retrieval. *Tutorial Notes at the Annual Intl. ACM SIGIR Conf. on Research and Development in Information Retrieval (SIGIR'06)*, 2006.
596. C. Zhai and J. Lafferty. A study of smoothing methods for language models applied to ad hoc information retrieval. In *Proc. of the Annual Intl. ACM SIGIR Conf. on Research and Development in Information Retrieval (SIGIR'01)*, pp. 334-342, 2001.
597. C. Zhai and J. Lafferty. Model-based feedback in the language modeling approach to information retrieval. In *Proc. of the ACM Intl. Conf. on Information and Knowledge Management (CIKM'01)*, 2001.
598. C. Zhai and J. Lafferty. Two-stage language models for information retrieval. In *Proc. of the Annual Intl. ACM SIGIR Conf. on Research and Development in Information Retrieval (SIGIR'02)*, pp. 49-56, 2002.
599. Y. Zhai and B. Liu. Extracting Web Data Using Instance-Based Learning. In *Proc. of 6th Intl. Conf. on Web Information Systems Engineering (WISE'05)*, pp. 318–331, 2005.
600. Y. Zhai and B. Liu. Web Data Extraction based on Partial Tree Alignment. In *Proc. of the 14th Intl. World Wide Web Conference (WWW'05)*, pp. 76–85, 2005.
601. Y. Zhai and B. Liu. Structured Data Extraction from the Web Based on Partial Tree Alignment. To appear in *IEEE Transactions on Knowledge and Data Engineering*, 2006.
602. D. Zhang, and W. S. Lee: Web Taxonomy Integration Using Support Vector Machines. In *Proc. of the 13th Intl. World Wide Web Conference (WWW'04)*, pp. 472–481, 2004.
603. D. Zhang, and W. S. Lee. A Simple Probabilistic Approach to Learning from Positive and Unlabeled Examples. In *Proc. of the 5th Annual UK Workshop on Computational Intelligence*, 2005.
604. H. Zhang, A. Goel, R. Govindan, K. Mason and B. Van Roy. Making Eigenvector-Based Systems Robust to Collusion. In *Proc. of the 3rd Intl. Workshop on Algorithms*

and Models for the Web Graph, pp. 92–104. 2004.

605. K. Zhang, R. Statman and D. Shasha. On the Editing Distance between Unordered Labeled Trees. *Information Processing Letters* 42(3), pp. 133–139, 1992.

606. T. Zhang. The Value of Unlabeled Data for Classification Problems. In *Proc. of the Intl. Conf. on Machine Learning* (*ICML'00*), 2000.

607. T. Zhang and F. Oles. A Probability Analysis on the Value of Unlabeled Data for Classification Problems. In *Proc. of the Intl. Conf. on Machine Learning* (*ICML'00*), 2000.

608. Z. Zhang, B. He, and K. C. -C. Chang. Understanding Web Query Interfaces: Best-Effort Parsing with Hidden Syntax. In *Proc. of International Conference on Management of Data* (*SIGMOD'04*), pp. 107–118, 2004.

609. Z. Zhang, B. He, and K. C.-C. Chang. Understanding Web Query Interfaces: Best-Effort Parsing with Hidden Syntax. In *Proc. of the ACM SIGMOD Intl. Conf. on Management of Data* (*SIGMOD'04*), pp. 107-118, 2004.

610. T. Zhang, R. Ramakrishnan and M. Linvy. BIRCH: an Efficient Data Clustering Method for Very Large Data Bases. In *Proc. of the ACM SIGMOD Intl. Conf. on Management of Data* (*SIGMOD'96*), pp. 103–114, 1996.

611. Q. Zhao, S. C. H. Hoi, T-Y. Liu, S. S Bhowmick, M. R. Lyu, and W-Y. Ma. Time-Dependent Semantic Similarity Measure of Queries Using Historical Click-through Data. In *Proc. of the 15th Intl. Conf. on World Wide Web* (*WWW'06*), 2006.

612. H. Zhao, W. Meng, Z. Wu, V. Raghavan, and C. Yu. Fully Automatic Wrapper Generation for Search Engines. In *Proc. of the 14th Intl. World Wide Web Conference* (*WWW'05*), pp. 66–75, 2005.

613. L. Zhao and N. K. Wee. WICCAP: From Semi-structured Data to Structured Data, pp. 86–93. In *Proc. of the 11th IEEE Intl. Conf. and Workshop on the Engineering of Computer-Based Systems* (*ECBS'04*), 1994.

614. Y. Zhao and G. Karypis. Empirical and Theoretical Comparisons of Selected Criterion Functions for Document Clustering. *Machine Learning*, 55, pp. 311–331, 2003.

615. Y. Zhao and G. Karypis. Hierarchical Clustering Algorithms for Document Datasets. *Data Mining and Knowledge Discovery*, 10(2), pp.141–168, 2005.

616. Z. Zheng, R. Kohavi, and L. Mason. Real World Performance of Association Rule Algorithms. In *Proc. of the ACM SIGKDD Intl. Conf. on Knowledge Discovery and Data Mining* (*KDD'01*), pp. 401–406, 2001.

617. N. Zhong, Y. Yao, and J. Liu (eds.) *Web Intelligence*. Springer, 2003.

618. D. Zhou, E. Manavoglu, J. Li, C. L. Giles, and H. Zha. Probabilistic Models for Discovering E-Communities. In *Proc. of the 15th Intl. Conf. on World Wide Web* (*WWW'06*), 2006.

619. X. Zhu, Z. Ghahramani, and J. Lafferty. Semi-Supervised Learning Using Gaussian Fields and Harmonic Functions. In *Proc. of the Intl. Conf. on Machine Learning* (*ICML'03*), pp. 912–919, 2003.

620. J. Zhu, Z. Nie, J-R. Wen, B. Zhang, and W-Y Ma. 2D Conditional Random Fields for Web information extraction. In *Proc. of the Intl. Conf. on Machine Learning* (*ICML'05*), pp. 1044-1051, 2005.

621. J. Zhu, Z. Nie, J-R. Wen, B. Zhang, and W.-Y. Ma. Simultaneous record detection and attribute labeling in web data extraction. In *Proc. of the ACM SIGKDD Intl. Conf. on Knowledge Discovery and Data Mining* (*KDD'06*), pp. 494–503, 2006.

622. L. Zhuang, F. Jing, X.-Yan Zhu, and L. Zhang. Movie Review Mining and Summarization. To appear in *Proc. of the ACM 15th Conf, on Information and Knowledge Management* (*CIKM'06*), 2006.

Index

Y

Z

Printing: Krips bv, Meppel, The Netherlands
Binding: Stürtz, Würzburg, Germany